The Greek Alphabet

Letters		Names	Letters		Names
Α	α	alpha	Ν	ν	nu
Β	β	beta	Ξ	ξ	xi
Γ	γ	gamma	Ο	o	omicron
Δ	δ	delta	Π	π	pi
Ε	ε	epsilon	Ρ	ρ	rho
Ζ	ζ	zeta	Σ	σ	sigma
Η	η	eta	Τ	τ	tau
Θ	θ	theta	Υ	υ	upsilon
Ι	ι	iota	Φ	ϕ	phi
Κ	κ	kappa	Χ	χ	chi
Λ	λ	lambda	Ψ	ψ	psi
Μ	μ	mu	Ω	ω	omega

List of Special Symbols

The following symbols have the fixed meaning given below when they are printed in **boldface type:**

Symbol	Meaning
$\mathbf{A_n}$	The alternating group of all even permutations on $\{1, 2, \ldots, n\}$
$\mathbf{A}(G)$	The group of automorphisms of a group G
$\mathbf{A}(R)$	The group of ring automorphisms of R
$\mathbf{B}(X)$	The group of bijections from a set X to itself
$\mathbf{C}$	The complex numbers
$\mathbf{D_n}$	The dihedral group of symmetries of a regular n-gon
$\mathbf{e}$	The identity of a group
$\boldsymbol{\varepsilon}$	The identity permutation
$\mathbf{i}$	A complex square root of -1
$\mathbf{j}, \mathbf{k}$	The noncomplex quaternion units
$\mathbf{Q}$	The rational numbers
$\mathbf{R}$	The real numbers
$\mathbf{S_n}$	The symmetric group of all permutations on $\{1, 2, \ldots, n\}$
$\mathbf{X_n}$	The set, $\{1, 2, \ldots, n\}$, of the first n positive integers
$\mathbf{V_m}$	The multiplicative group of invertibles in $\mathbf{Z_m}$
$\mathbf{W_n}$	The set of all n-bit strings
$\mathbf{Z}$	The integers
$\mathbf{Z^+}$	The positive integers
$\mathbf{Z_m}$	The set $\{\bar{0}, \bar{1}, \bar{2}, \ldots, \overline{m-1}\}$

ABSTRACT ALGEBRA

A First Undergraduate Course

THE PRINDLE, WEBER & SCHMIDT SERIES IN DEVELOPMENTAL, BUSINESS, AND TECHNICAL MATHEMATICS

Baley/Holstege, *Algebra: A First Course, Third Edition*
Cass/O'Connor, *Beginning Algebra*
Cass/O'Connor, *Beginning Algebra with Fundamentals*
Cass/O'Connor, *Fundamentals with Elements of Algebra, Second Edition*
Ewen/Nelson, *Elementary Technical Mathematics, Fifth Edition*
Geltner/Peterson, *Geometry for College Students, Second Edition*
Johnston/Willis/Hughes, *Developmental Mathematics, Third Edition*
Johnston/Willis/Lazaris, *Essential Algebra, Sixth Edition*
Johnston/Willis/Lazaris, *Essential Arithmetic, Sixth Edition*
Johnston/Willis/Lazaris, *Intermediate Algebra, Fifth Edition*
Jordan/Palow, *Integrated Arithmetic and Algebra*
Kennedy/Green, *Prealgebra for College Students*
Lee, *Self-Paced Business Mathematics, Fourth Edition*
Mangan, *Arithmetic for Self-Study, Second Edition*
McCready, *Business Mathematics, Sixth Edition*
McKeague, *Basic Mathematics, Third Edition*
McKeague, *Introductory Mathematics*
McKeague, *Prealgebra, Second Edition*
Perez/Weltman/Byrum/Polito, *Beginning Algebra: A Worktext*
Perez/Weltman/Byrum/Polito, *Intermediate Algebra: A Worktext*
Pierce/Tebeaux, *Operational Mathematics for Business, Second Edition*
Proga, *Arithmetic and Algebra, Third Edition*
Proga, *Basic Mathematics, Third Edition*
Rogers/Haney/Laird, *Fundamentals of Business Mathematics*
Szymanski, *Algebra Facts*
Szymanski, *Math Facts: Survival Guide to Basic Mathematics*
Wood/Capell/Hall, *Developmental Mathematics, Fourth Edition*

THE PRINDLE, WEBER & SCHMIDT SERIES IN PRECALCULUS, LIBERAL ARTS MATHEMATICS, AND TEACHER TRAINING

Barnett, *Analytic Trigonometry with Applications, Fifth Edition*
Bean, Sharp, and Sharp, *Precalculus*
Boye, Kavanaugh, and Williams, *Elementary Algebra*
Boye, Kavanaugh, and Williams, *Intermediate Algebra*
Davis, Moran, and Murphy, *Precalculus in Context: Functioning in the Real World*
Drooyan and Franklin, *Intermediate Algebra, Seventh Edition*
Franklin and Drooyan, *Modeling, Functions and Graphs*
Fraser, *Elementary Algebra*
Fraser, *Intermediate Algebra: An Early Functions Approach*
Gantner and Gantner, *Trigonometry*
Gobran, *Beginning Algebra, Fifth Edition*

Grady, Drooyan, and Beckenbach, *College Algebra, Eighth Edition*
Hall, *Beginning Algebra*
Hall, *Intermediate Algebra*
Hall, *Algebra for College Students, Second Edition*
Hall, *College Algebra with Applications, Third Edition*
Holder, *A Primer for Calculus, Sixth Edition*
Huff and Peterson, *College Algebra Activities for the TI-81 Graphics Calculator*
Johnson and Mowry, *Mathematics: A Practical Odyssey*
Kaufmann, *Elementary Algebra for College Students, Fourth Edition*
Kaufmann, *Intermediate Algebra for College Students, Fourth Edition*
Kaufmann, *Elementary and Intermediate Algebra: A Combined Approach*
Kaufmann, *Algebra for College Students, Fourth Edition*
Kaufmann, *Algebra with Trigonometry for College Students, Third Edition*
Kaufmann, *College Algebra, Third Edition*
Kaufmann, *Trigonometry, Second Edition*
Kaufmann, *College Algebra and Trigonometry, Third Edition*
Kaufmann, *Precalculus, Second Edition*
Lavoie, *Discovering Mathematics*
McCown and Sequeira, *Patterns in Mathematics: From Counting to Chaos*
Rice and Strange, *Plane Trigonometry, Sixth Edition*
Riddle, *Analytic Geometry, Fifth Edition*
Ruud and Shell, *Prelude to Calculus, Second Edition*
Sgroi and Sgroi, *Mathematics for Elementary School Teachers*
Swokowski and Cole, *Fundamentals of College Algebra, Eighth Edition*
Swokowski and Cole, *Fundamentals of Algebra and Trigonometry, Eighth Edition*
Swokowski and Cole, *Fundamentals of Trigonometry, Eighth Edition*
Swokowski and Cole, *Algebra and Trigonometry with Analytic Geometry, Eighth Edition*
Swokowski and Cole, *Precalculus: Functions and Graphs, Seventh Edition*
Weltman and Perez, *Beginning Algebra, Second Edition*
Weltman and Perez, *Intermediate Algebra, Third Edition*

THE PRINDLE, WEBER & SCHMIDT SERIES IN CALCULUS AND UPPER-DIVISION MATHEMATICS

Althoen and Bumcrot, *Introduction to Discrete Mathematics*
Andrilli and Hecker, *Linear Algebra*
Burden and Faires, *Numerical Analysis, Fifth Edition*
Crooke and Ratcliffe, *A Guidebook to Calculus with Mathematica*
Cullen, *An Introduction to Numerical Linear Algebra*
Cullen, *Linear Algebra and Differential Equations, Second Edition*
Denton and Nasby, *Finite Mathematics, Preliminary Edition*
Dick and Patton, *Calculus*
Dick and Patton, *Single Variable Calculus*
Dick and Patton, *Technology in Calculus: A Sourcebook of Activities*
Edgar, *A First Course in Number Theory*
Eves, *In Mathematical Circles*
Eves, *Mathematical Circles Revisited*
Eves, *Mathematical Circles Squared*
Eves, *Return to Mathematical Circles*
Faires and Burden, *Numerical Methods*
Finizio and Ladas, *Introduction to Differential Equations*
Finizio and Ladas, *Ordinary Differential Equations with Modern Applications, Third Edition*
Fletcher, Hoyle, and Patty, *Foundations of Discrete Mathematics*
Fletcher and Patty, *Foundations of Higher Mathematics, Second Edition*
Gilbert and Gilbert, *Elements of Modern Algebra, Third Edition*
Gordon, *Calculus and the Computer*

Hartfiel and Hobbs, *Elementary Linear Algebra*
Hartig, *Guidebook to Linear Algebra for Theorist*
Hill, Ellis, and Lodi, *Calculus Illustrated*
Hillman and Alexanderson, *Abstract Algebra: A First Undergraduate Course, Fifth Edition*
Humi and Miller, *Boundary-Value Problems and Partial Differential Equations*
Laufer, *Discrete Mathematics and Applied Modern Algebra*
Leinbach, *Calculus Laboratories Using Derive*
Maron and Lopez, *Numerical Analysis, Third Edition*
Miech, *Calculus with Mathcad*
Mizrahi and Sullivan, *Calculus with Analytic Geometry, Third Edition*
Molluzzo and Buckley, *A First Course in Discrete Mathematics*
Nicholson, *Elementary Linear Algebra with Applications, Third Edition*
Nicholson, *Introduction to Abstract Algebra*
O'Neil, *Advanced Engineering Mathematics, Third Edition*
Pence, *Calculus Activities for Graphic Calculators*
Pence, *Calculus Activities for the TI-81 Graphic Calculator*
Plybon, *An Introduction to Applied Numerical Analysis*
Powers, *Elementary Differential Equations with Boundary-Value Problems*
Powers, *Elementary Differential Equations with Linear Algebra*
Prescience Corporation, *The Student Edition of Theorist*
Riddle, *Calculus and Analytic Geometry, Fourth Edition*
Schelin and Bange, *Mathematical Analysis for Business and Economics, Second Edition*
Sentilles, *Applying Calculus in Economics and Life Science*
Swokowski, *Calculus, Fifth Edition (Late Trigonometry Version)*
Swokowski, *Elements of Calculus with Analytic Geometry: High School Edition*
Swokowski, Olinick, and Pence, *Calculus, Sixth Edition*
Swokowski, Olinick, and Pence, *Calculus of a Single Variable, Second Edition*
Tan, *Applied Finite Mathematics, Fourth Edition*
Tan, *Calculus for the Managerial, Life, and Social Sciences, Third Edition*
Tan, *Applied Calculus, Third Edition*
Tan, *College Mathematics, Third Edition*
Trim, *Applied Partial Differential Equations*
Venit and Bishop, *Elementary Linear Algebra, Third Edition*
Venit and Bishop, *Elementary Linear Algebra, Alternate Second Edition*
Wattenberg, *Calculus in a Real and Complex World*
Wiggins, *Problem Solver for Finite Mathematics and Calculus*
Zill, *Calculus, Third Edition*
Zill, *A First Course in Differential Equations, Fifth Edition*
Zill and Cullen, *Differential Equations with Boundary-Value Problems, Third Edition*
Zill and Cullen, *Advanced Engineering Mathematics*

THE PRINDLE, WEBER & SCHMIDT SERIES IN ADVANCED MATHEMATICS

Ehrlich, *Fundamental Concepts of Abstract Algebra*
Eves, *Foundations and Fundamental Concepts of Mathematics, Third Edition*
Judson, *Abstract Algebra: Theory and Applications*
Kirkwood, *An Introduction to Real Analysis*
Patty, *Foundations of Topology*
Ruckle, *Modern Analysis: Measure Theory and Functional Analysis with Applications*
Sieradski, *An Introduction to Topology and Homotopy*
Steinberger, *Algebra*
Strayer, *Elementary Number Theory*
Troutman and Bautista, *Boundary-Value Problems of Applied Mathematics*

FIFTH EDITION

Abstract Algebra

A First Undergraduate Course

Abraham P. Hillman
Professor Emeritus, University of New Mexico

Gerald L. Alexanderson
Santa Clara University

PWS Publishing Company
Boston, MA

PWS
Publishing Company

20 Park Plaza
Boston, Massachusetts 02116

Sponsoring Editor: *Steve Quigley*
Production Editor: *Monique Calello*
Assistant Editor: *Marnie Pommett*
Editorial Assistant: *John Ward*
Manufacturing Coordinator: *Lisa Flanagan*
Production Service: *Cecile Joyner, The Cooper Company*
Interior Designer: *John Edeen*
Interior Artist: *George Nichols*
Cover Designer: *Julia Gecha*
Cover Illustration: © *Victor Stabin/Image Bank*
Cover Printer: *John P. Pow Company*
Text Printer and Binder: *R. R. Donnelley/Crawfordsville*

This book is printed on acid-free recycled paper.

I(T)P ™

International Thomson Publishing
The trademark ITP is used under license.

PWS Publishing Company is a division of Wadsworth, Inc.

Printed in the United States of America

1 2 3 4 5 6 7 8 9 10 — 99 98 97 96 95 94

Library of Congress Cataloging-in-Publication Data

Hillman, Abraham P.
Abstract algebra: a first undergraduate course/Abraham P. Hillman, Gerald L. Alexanderson. — 5th ed.
p. cm.
Rev. ed. of: A first undergraduate course in abstract algebra. 4th ed.
Includes bibliographical references and index.
ISBN 0-534-19128-2
1. Algebra, Abstract. I. Alexanderson, Gerald L. II. Hillman, Abraham P. First undergraduate course in abstract algebra. III. Title.
QA162.H54 1993 93-16675
512′.02 — dc20 CIP

To Josephine Hillman

PREFACE

The treatment of groups, rings, and fields in this text does not require any previous knowledge of abstract or linear algebra. A table later in this preface illustrates that any topic in *Abstract Algebra: A First Undergraduate Course*, Fifth Edition, can be reached and covered effectively in a one-quarter or one-semester course. Throughout the book, sections marked with an asterisk are optional; their omission would not interrupt the flow of the material.

The structure of this book, the text material, and the problem sets have evolved from extensive class testing, accretion, and revision beginning in 1961. The following principles have guided this effort:

1. Abstraction should be built on a solid foundation of concrete examples.
2. The development of important concepts should be gradual, with a painstaking start. As the students' maturity grows, progress can be more rapid. Each concept should be presented when constructive problems on the topic, not mere busy work, can be assigned. Also, foreshadowing the concept in earlier problem sets helps motivate students to learn the new material and helps provide unifying themes.
3. After results and techniques are introduced, they should be applied as frequently as possible to aid in retention.
4. There is no royal road to effective understanding of any branch of mathematics. Effort is required to master, retain, or apply mathematics. The most important tools for this effort are well-constructed problem sets.

The most notable changes in this fifth edition are as follows:

1. Expanded treatment of the classification of groups of small order;
2. A new section on symmetries of polyhedra;
3. Earlier introduction and use of congruence notation and modular arithmetic;
4. Expanded introductions to chapters;
5. Addition of a set of computer programming projects related to topics in the text;
6. Positioning of the supplementary and challenging problems for all chapters together near the close of the book;

7. Streamlining of the chapter on algebraic coding, which continues to illustrate that matrix theory and modular arithmetic have important applications; and
8. Updating of the annotated bibliography.

CONTENTS The relatively short Chapter 1 provides the techniques of mathematical induction, the properties of the integers needed for group theory, and concrete examples for the abstract concepts to follow. Chapter 2 starts with permutation groups. Since every group is isomorphic to a group of permutations, this is an opportunity to study concrete structures without loss of generality. Also, the noncommutative permutation groups show the importance of deducing properties from the axioms and previous theorems rather than from false analogy with elementary algebra. Permutations and other mappings associated with groups furnish the concrete illustrations for the general mappings presented in Chapter 3 and applied there to isomorphisms and homomorphisms.

Thorough treatments are given for groups in Chapters 2 and 3, for rings and fields in Chapter 4, and for polynomial rings in Chapter 5. Many useful results in number theory and theory of equations are obtained with little extra effort as by-products of the work on abstract structures, which greatly facilitates applications. For example, Lagrange's Theorem for finite groups leads easily to the theorems of Euler and Fermat in number theory, and these are seen to be applicable to the trapdoor functions of public key encryption described in Section 8.3.

Chapter 6 applies field extensions to the impossibility proofs for the classical angle-trisecting and cube-duplication problems. Chapter 7 rounds off the number theoretic material already integrated into previous chapters by adding some other standard topics of an introductory course in number theory. Chapter 8 has applications to algebraic coding theory.

The problems have been one of the strongest features of this book from its first appearance. The authors' long experience with and commitment to problems are demonstrated throughout the text. The range extends all the way from simple exercises that strengthen skills at handling algebraic symbols to interesting, challenging problems. At the early stages problems are paired so that in the odd-numbered case an answer is available to the student, but in the paired even-numbered case there is no such support. Each problem set is constructed so that the level of difficulty rises as one proceeds through the set. Earlier problems tend to be more concrete. As usual, problems with an asterisk are quite challenging.

Many problems include hints such as references to previous results. Others are broken down into parts to facilitate solution; these parts are numbered (i), (ii), (iii), and so on. Where parts can be assigned independently, they are labeled (a), (b), (c), and so on. Solutions to even-numbered problems are available in the Instructor's Manual.

Great care has been taken with the problems in this text. They are designed to lead the student through the ideas from specific manipulative problem solving to the construction of proofs. In addition, the number of problems available to an instructor is unparalleled in other texts at this level. The new fifth edition contains over 1900 problems.

The book can be used for a year course; but if only one term or semester is available, one can still get to any of a number of desired goals by choosing sections to be covered, as illustrated in the following table:

Goal	Sections Needed
Automorphisms of extension fields	1.2–1.6, 2.1–2.13, 3.1–3.3, 4.1–4.5, 4.8, 5.1–5.4, 5.7, 5.8, 5.10, 5.12
Euclidean constructions	1.2–1.6, 2.1–2.12, 3.1–3.3, 4.1–4.5, 5.1–5.5, 6.1–6.3
Number theory through quadratic residues	1.1–1.7, 2.1–2.12, 3.1–3.3, 4.1–4.4, 7.1, 7.3, 7.4
Discrete mathematics for computer science	1.1–1.6, 2.1–2.12, 3.1–3.13
Matrix coding	1.1–1.6, 2.1–2.12, 3.1–3.3, 4.1–4.5, 4.7, 8.1–8.3
Trapdoor functions	1.1–1.7, 2.1–2.12, 3.1–3.3, 4.1–4.5, 8.3

Other more modest goals are even more easily and quickly achieved. If short of time, however, instructors should be willing to proceed fairly rapidly through much of Chapter 1, the contents of which will be reasonably familiar to many students. Also, in such a case one must avoid lingering over interesting problems or moving off into starred sections that are not relevant to subsequent topics.

Where time permits, one can profitably take up the supplementary and challenging problems, designed for the able and interested student who wants a taste of the excitement of exploring problems at a more sophisticated level. Students can even gain some insight into what it is like to participate in the development of mathematics.

In order to convey the very important idea that mathematics is created by real people, there are historical and biographical notes throughout the text. We hope that reading about the lives of major contributors to the development of algebra while at the same time delving into the consequences of their creative efforts will be more meaningful than reading their biographies alone.

We also provide an annotated bibliography as a link with other published material on various topics considered. This is especially helpful in showing the many significant modern applications of groups, rings, fields, and other algebraic structures.

Previous editions of this text have been used successfully both for classes for regular undergraduates and for specially designed courses for prospective and in-service secondary school teachers. Students preparing for graduate work benefit from the opportunities for anticipation and generalization, the availability of supplementary material, and the emphasis on homomorphisms and quotient structures. Some of the advantages to teachers are the large number of problems dealing with concrete algebraic structures, the material on constructions with straightedge and compass, the biographical and historical notes, and the references for applications.

ACKNOWLEDGMENTS

The authors are especially grateful to the following people for their many valuable comments and suggestions for this and the previous editions:

David Anderson
University of Tennessee–Knoxville

Phillip Anderson
Montclair State College

Pasquale Arpaia
St. John Fisher College

Thomas Becker
University of New Orleans

Karl Beers
Ripon College

Fredrick Carter
St. Mary's University

Joseph W. Colen
Jackson State University

Paul M. Cook II
Furman University

James W. Cotts
Southern Utah University

Joseph Czerwinski
Milliken University

R. G. Dean
Stephen F. Austin State University

Daniel Fendel
San Francisco State University

Marjorie Fitting
San Jose State University

Sandra Farmer Givens
Lambuth University

Richard Grassl
University of Northern Colorado

J. Myron Hood
California Polytechnic State University

Jurgen Hurrelbrink
Louisiana State University

W. H. Jamison
Rocky Mountain College

Johnny A. Johnson
University of Houston

Kenneth Kalmanson
Montclair State College

David Lahren
St. Cloud University

John Lavelle
Millersville University

Judith Q. Longyear
Wayne State University

Joseph T. Mathis
William Jewell College

David G. Mead
University of California–Davis

George Mead
McNeese University

David Meredith
San Francisco State University

Matthew Miller
University of South Carolina

Lawrence Naylor
Drake University

Pamela Nelson
Panhandle State University

Mervin E. Newton
Thiel College

Soula O'Bannon
Louisiana State University

Matthew O'Malley
University of Houston

Robert Page
University of Maine

Hiram Paley
University of Illinois at Urbana-Champaign

Donald H. Pilgrim
Luther College

Roy Rakestraw
Oral Roberts University

Don Richard
Union University

Alicia Sevilla
Moravian College

Henrene E. Smoot
Alabama A & M University

George Tintera
Texas A & I University

Emilio Toro
University of Tampa

Clifford N. Wagner
Pennsylvania State University at Harrisburg

Clifton Whyburn
University of Houston–University Park

We also express our appreciation to Josephine Hillman and to the editorial and production staffs of PWS Publishing Company and the Cooper Company — especially Steve Quigley, Marnie Pommett, Monique Calello, Cecile Joyner, and Steven Gray — for their invaluable assistance in the preparation of this edition.

Abraham P. Hillman
Gerald L. Alexanderson

CONTENTS

3 Sets and Mappings 133

4 Rings and Fields 214

5 Polynomials 282

INTRODUCTION

Strange as it may sound, the power of mathematics rests on its evasion of all unnecessary thought and on its wonderful saving of mental operations. —**Ernst Mach**

Each of the natural numbers 1, 2, 3, ... is an abstraction. The number 2, for example, is something abstracted from all pairs of objects. An abstraction of higher order is the identity $(x + 1)(x - 1) = x^2 - 1$. This formula summarizes information about the infinite set of special cases in which x is replaced by any number. Abstract algebra deals with abstractions of even higher order.

Oddly enough, this notion of abstraction came fairly late in the development of mathematics in the West. The Babylonians developed some fairly sophisticated mathematics prior to the Greeks, but their work consisted largely of examples — specific problems worked out in detail and often with considerable ingenuity. They seem to have illustrated general truths rather than stating them explicitly. The Greeks developed their axiomatic method, which involves the derivation of general truths from a fixed set of assumed truths (axioms). But most of this work was concerned with geometry, some with properties of numbers. Algebra as we think of it today was not known to the Greeks, and many centuries elapsed before the development of notation that allowed algebra to flourish.

The essential idea of allowing unknown or unspecified numbers to be represented by letters of the alphabet is attributed to the sixteenth-century French mathematician François Viète (often written Vieta). Others — Regiomontanus, Stifel, and Cardano — had anticipated Viète in using letters for such numbers, but theirs was not a systematic treatment. Subsequent contributions to the development of algebraic notation were made by the better known seventeenth-century mathematician–philosophers, René Descartes and Gottfried von Leibniz. But the principal credit for the early development of algebra

François Viète
1540–1603

Viète, born in Poitou, was one of many individuals who have contributed to mathematics without being what we would think of today as professional mathematicians. He was at various times a member of the parliament of Bretagne and the king's privy councillor. He made numerous contributions to trigonometry and the theory of algebraic equations, but it was his development of algebraic notation that assures him an important place in the history of mathematics, as the most eminent French mathematician of the sixteenth century. He was probably the first mathematician to use the terms "coefficient" and "polynomial."

goes to Viète. This simple but gigantic step made possible the flowering of mathematics in the next century, leading to the development of analytic geometry by Pierre de Fermat and Descartes and to the extensive investigations of problems of solving polynomial equations that continued on into the nineteenth century.

Algebra quickly moved beyond the stage of representing unknown or unspecified numbers by letters. It is now concerned with sets of objects where those objects may be numbers but may also be polynomials, matrices, functions, or the like, and with operations on those objects. In an axiomatic, or abstract, treatment of a given type of algebraic structure, one assumes a small number of basic properties as axioms and then deduces many other properties from the axioms. Thus one deals simultaneously with all the structures satisfying a given set of axioms instead of with each structure individually. This is the approach taken in this text.

Since many readers have not previously encountered abstraction at this level, the treatment will gently introduce concrete examples first, to the extent possible, with the abstract principles following. The reader will also find considerable redundancy. For example, each of the first few "if and only if" problems is restated as a proposition and its converse, as a reminder that in these cases the implication goes both ways. There is redundancy, too, in the simultaneous use of phrases and their symbolic equivalents, such as "the quotient group G/N of G by its normal subgroup N." In later courses, this luxury can no longer be afforded and should no longer be expected.

It is assumed that the reader is acquainted with the rudiments of the underlying logic of proofs, for example, with the principle that things equal to the same thing are equal to each other.

For ease of reference, the inside front cover and its facing page contain a description of basic set, mapping, and logic notation, the Greek alphabet, and a list of special symbols, each of which has a fixed meaning when printed in boldface type.

CHAPTER 1

BASIC PROPERTIES OF THE INTEGERS

All results of the profoundest mathematical investigation must ultimately be expressible in the simple form of properties of the integers. —**Leopold Kronecker**

God created the integers, all else is the work of man.
—**Leopold Kronecker**

Introductory abstract algebra consists mainly of a study of mathematical structures — specifically groups, rings, fields, and a few others. In the next chapter we shall concentrate on what is in some ways the simplest of these and in other ways the richest and most complex: the theory of groups. Before moving into the mainstream of the subject, however, we review some basic background and develop some language and techniques that we can use in studying these algebraic structures. We shall assume here knowledge of the most elementary properties of the set of integers, $\mathbf{Z} = \{\ldots, -3, -2, -1, 0, 1, 2, 3, \ldots\}$; in later chapters we shall return to more sophisticated properties of these numbers after we have developed useful techniques through our study of abstract structures.

Some of the examples given in the sections that follow involve well-known sequences of numbers: the Fibonacci sequence and the Lucas sequence. The first of these was investigated by Fibonacci in connection with a study of the growth of a population of rabbits. The technique he outlined did not prove accurate, but it is essentially the same as a method used today in mathematical biology to study the growth and decline of populations of species. Some predator–prey relations that arise in the mathematical study of ecological problems are recursion relations similar to those defining these two sequences.

1.1 Mathematical Induction

A mathematician, like a painter or a poet, is a maker of patterns. If his patterns are more permanent than theirs, it is because they are made with ideas .... The mathematician's patterns, like the painter's or the poet's, must be beautiful; the ideas, like the colours or the words, must fit together in a harmonious way. Beauty is the first test: there is no permanent place in the world for ugly mathematics. —**G. H. Hardy**

First we consider a subset of the integers $\mathbf{Z}$, namely, the set

$$\mathbf{Z}^+ = \{1, 2, 3, 4, \ldots\}$$

of the positive integers. We assume that $\mathbf{Z}^+$ has the following property:

AXIOM 1 Well Ordering Principle

Every nonempty subset S of $\mathbf{Z}^+$ has a least element.

For example, the set $S = \{2, 4, 6, \ldots\}$ of the positive even integers is a nonempty subset of $\mathbf{Z}^+$, and we see that 2 is the smallest integer in S.

This property of $\mathbf{Z}^+$ is not possessed by the set $\mathbf{Z}$ of all the integers; for example, there is no smallest integer in the subset

$$\{\ldots, -4, -2, 0, 2, 4, 6, \ldots\}$$

of even integers in $\mathbf{Z}$.

A major application of the well ordering principle is in the proof of the following theorem.

THEOREM 1 Principle of Mathematical Induction

Let S be a set of positive integers, that is, a subset of $\mathbf{Z}^+$, with the following two properties:
(a) The positive integer 1 is in S.
(b) If a positive integer k is in S, so is $k + 1$.
Then every positive integer is in S; that is, $S = \mathbf{Z}^+$.

Proof Let T be the set of all positive integers that are not in S. If T is empty, then $S = \mathbf{Z}^+$ and our task requires no more effort. Therefore, we assume that T is nonempty. Then it follows from the well ordering principle that there is a least integer t in T.

By hypothesis (a), 1 is in S; hence 1 is not in T. This means that t is greater than 1. It follows that $t - 1$ is a positive integer. Since $t - 1$ is less than the least integer t in T, $t - 1$ must be in S. Then

$$t = (t - 1) + 1$$

is in S by hypothesis (b). Since t is in T, this contradicts the definition of T as the set of positive integers not in S. Thus the assumption that T is nonempty is untenable; that is, T is empty and $S = \mathbf{Z}^+$. ∎

Next we illustrate the application of the principle of mathematical induction to proofs of statements involving the positive integers.

We start by considering the Fibonacci sequence, 1, 1, 2, 3, 5, 8, 13, 21, 34, 55, 89, 144, ..., which is defined by

$$F_1 = F_2 = 1, \qquad F_{n+2} = F_{n+1} + F_n \quad \text{for } n = 1, 2, 3, \ldots.$$

Which of the terms F_m of this sequence are even? Among the first 12 terms, we see that $F_3 = 2$, $F_6 = 8$, $F_9 = 34$, and $F_{12} = 144$ are the only even integers.

It is natural to conjecture that F_m is even if and only if m is an integral multiple of 3. But verifying this conjecture in 12 or even one million cases will not prove it true for the infinite number of terms. However, with the help of mathematical induction, we now prove the conjecture true for all positive integers.

EXAMPLE 1 Here we prove that F_{3n-2} and F_{3n-1} are odd and that F_{3n} is even for all positive integers n.

Proof Let S be the set of all positive integers n for which the statement is true. We wish to prove that every positive integer is in S; that is, $S = \mathbf{Z}^+$.

When $n = 1$, we have $F_{3n-2} = F_1 = 1$, $F_{3n-1} = F_2 = 1$, and $F_{3n} = F_3 = 2$. Since 1 is odd and 2 is even, the statement holds for $n = 1$; that is, the integer 1 is in S, and condition (a) of the principle of mathematical induction is satisfied.

Next we let k be a positive integer in S; in other words, we assume that F_{3k-2} and F_{3k-1} are odd and that F_{3k} is even. Now

$$F_{3(k+1)-2} = F_{3k+1} = F_{3k} + F_{3k-1},$$

using the definition of the Fibonacci sequence. Then F_{3k+1} is odd, since it is the sum of an even integer and an odd integer. For the same reason,

$$F_{3(k+1)-1} = F_{3k+2} = F_{3k+1} + F_{3k}$$

is odd. Also,

$$F_{3(k+1)} = F_{3k+3} = F_{3k+2} + F_{3k+1}$$

is even, since it is the sum of two odd integers.

This shows that the assumption that k is in S implies that $k + 1$ is in S; thus, condition (b) of Theorem 1 is satisfied. Since both conditions (a) and (b) hold, it follows from Theorem 1 that $S = \mathbf{Z}^+$. This completes the proof. ∎

In proofs by mathematical induction, one generally streamlines the procedure of Example 1 by omitting explicit mention of the set S of integers n for which the statement is true and instead using the procedure we describe next.

ALGORITHM 1 Proof by Mathematical Induction

A proof by mathematical induction that a proposition $P(n)$ is true for all n in $\mathbf{Z}^+ = \{1, 2, \ldots\}$ consists of the following two parts:

Basis Show that $P(1)$ is true.

Inductive Part Assume that $P(k)$ is true with k in $\mathbf{Z}^+$, and use this to show that $P(k+1)$ is true; that is, show that $P(k)$ implies $P(k+1)$ for all k in $\mathbf{Z}^+$.

In the inductive part of a proof by mathematical induction, the assumption that $P(k)$ is true is called the ***inductive hypothesis***. With the help of the ***implication arrow*** defined on the inside front cover of this book, the inductive part can be described as showing that $P(k) \Rightarrow P(k+1)$ for all k in $\mathbf{Z}^+$.

We next illustrate the use of Algorithm 1.

EXAMPLE 2 Let F_n be the nth Fibonacci number. We shall prove that

$$(F_{n+1})^2 - F_n F_{n+2} = (-1)^n$$

for all positive integers n.

Proof When $n = 1$, we see that $(F_{n+1})^2 - F_n F_{n+2} = (F_2)^2 - F_1 F_3 = 1^2 - 1 \cdot 2 = -1$. Since $(-1)^n$ is also -1 when $n = 1$, the statement is true for $n = 1$. This establishes the basis.

Next we assume that k is a positive integer for which the statement is true. We introduce the notation $F_k = a$, $F_{k+1} = b$, $F_{k+2} = c$, $F_{k+3} = d$. Our assumption concerning k is that $b^2 - ac = (-1)^k$. We want to use this assumption to show that the statement is true for $k + 1$; that is, $c^2 - bd = (-1)^{k+1}$.

By definition of the Fibonacci sequence, $d = c + b$ and $c = b + a$. Hence,

$$\begin{aligned} c^2 - bd &= c^2 - b(c+b) = c(c-b) - b^2 \\ &= ca - b^2 = -(b^2 - ac) \\ &= -(-1)^k = (-1)^{k+1}. \end{aligned}$$

Since this is the desired result for $n = k + 1$, the inductive part is accomplished. Together, these parts prove the statement true for all positive integers n. ■

EXAMPLE 3 Here we conjecture a simple formula for the sum

$$A_n = \frac{1}{1 \cdot 3} + \frac{1}{3 \cdot 5} + \frac{1}{5 \cdot 7} + \cdots + \frac{1}{(2n-1)(2n+1)}$$

and prove that the formula holds for all positive integers n.

We begin by calculating some values: $A_1 = 1/(1 \cdot 3) = 1/3$; $A_2 = (1/3) + (1/15) = 6/15 = 2/5$; $A_3 = (1/3) + (1/15) + (1/35) = (2/5) + (1/35) = 15/35 = 3/7$; $A_4 = (3/7) + (1/63) = 28/63 = 4/9$; and $A_5 = (4/9) + (1/99) = 45/99 = 5/11$. We assemble these data in the following table:

n	1	2	3	4	5
A_n	$\frac{1}{3}$	$\frac{2}{5}$	$\frac{3}{7}$	$\frac{4}{9}$	$\frac{5}{11}$

This table leads us to conjecture that $A_n = n/(2n + 1)$. Let us see if mathematical induction can be used to make this conjecture into a theorem — that is, to prove the formula correct for all positive integers n.

We have already seen that the formula holds for $n = 1$; hence the basis has been demonstrated. Now we assume the formula to be correct for $n = k$; specifically, we assume that $A_k = k/(2k + 1)$. Then

$$\begin{aligned} A_{k+1} &= A_k + \frac{1}{[2(k+1) - 1][2(k+1) + 1]} = \frac{k}{2k+1} + \frac{1}{(2k+1)(2k+3)} \\ &= \frac{k(2k+3) + 1}{(2k+1)(2k+3)} = \frac{2k^2 + 3k + 1}{(2k+1)(2k+3)} \\ &= \frac{(2k+1)(k+1)}{(2k+1)(2k+3)} = \frac{k+1}{2k+3}. \end{aligned}$$

We see that $A_{k+1} = (k + 1)/[2(k + 1) + 1]$, which is the formula $A_n = n/(2n + 1)$ for $n = k + 1$. This shows that if the formula is true for k, it is also true for $k + 1$. This finishes the inductive part, and the formula is proved.

It is frequently convenient to have available the following generalizations of the well ordering principle and the principle of mathematical induction.

THEOREM 2 Generalized Well Ordering Principle

Let a be any integer (positive, negative, or zero). Let X be the set of all integers x with $x \geq a$. Then every nonempty subset S of X has a least integer.

THEOREM 3 Generalized Mathematical Induction

Let a be an integer, and let X consist of all integers x with $x \geq a$. Let S be a subset of X such that a is in S and whenever an integer k is in S so is $k + 1$. Then $S = X$.

Theorems 2 and 3 can be deduced from Axiom 1 (the original well ordering principle) and Theorem 1, respectively, by a change of variable. The proofs are left to the reader.

One can use Theorem 3 to establish the following modification of Algorithm 1. This new algorithm is useful in Problem 5(iii).

ALGORITHM 2 Generalized Proof by Induction

A proof by mathematical induction that a proposition $P(n)$ is true for all integers n in a set $S = \{a, a + 1, a + 2, \ldots\}$ consists of the following two parts:

Basis Show that $P(a)$ is true.

Inductive Part Assume that $P(k)$ is true with k in S, and use this inductive hypothesis to show that $P(k+1)$ is true. That is, show that $P(k) \Rightarrow P(k+1)$ is a true implication for all n in S.

Here we present a fairly small problem set on mathematical induction. A larger and perhaps more interesting collection appears in Section 7.2.

Problems

1. Let $F_1, F_2, F_3, \ldots$ be the Fibonacci sequence. Prove that
$$F_{n+1}F_{n+2} - F_nF_{n+3} = (-1)^n$$
for all positive integers n.

2. The Lucas sequence 1, 3, 4, 7, 11, 18, 29, ... is defined by
$$L_1 = 1, \qquad L_2 = 3, \qquad L_{n+2} = L_{n+1} + L_n \quad \text{for } n = 1, 2, \ldots.$$
(i) Calculate $(L_{n+1})^2 - L_nL_{n+2}$ for $n = 1, 2, 3, 4$, and 5.
(ii) Make a conjecture based on the data of part (i). Test your conjecture for $n = 6$ and $n = 7$.
(iii) Prove your conjecture true for all positive integers n by mathematical induction.

3. Let the Lucas sequence 1, 3, 4, 7, ... be as in Problem 2. Make a conjecture concerning the subscripts m for which L_m is even or odd, and prove it true for all the terms of this sequence by mathematical induction.

4. Do as in Problem 2, with $(L_{n+1})^2 - L_nL_{n+2}$ replaced by $L_{n+1}L_{n+2} - L_nL_{n+3}$.

5. Let $T_2, T_3, T_4, \ldots$ be defined by
$$T_n = \left(1 - \frac{1}{2^2}\right)\left(1 - \frac{1}{3^2}\right)\left(1 - \frac{1}{4^2}\right)\cdots\left(1 - \frac{1}{n^2}\right).$$
(i) Note that $T_2 = 3/4$ and $T_3 = (3/4)(8/9) = 2/3$. Find T_4, T_5, T_6, T_7, and T_8.
(ii) Conjecture a simple formula for T_n based on the data for T_2, T_4, T_6, and T_8. Is it true for T_{10}? Is it true for T_3, T_5, and T_7?
(iii) Prove the formula true for all integers $n \geq 2$ by mathematical induction.

6. For every positive integer s, let $c_s = \cos(x/2^s)$. Let
$$P_n = c_1c_2c_3 \cdots c_n.$$
Prove that $P_n = (\sin x)/[2^n \sin(x/2^n)]$ for all positive integers n. [*Hint:* The case $n = 1$ follows from the double angle formula $\sin(2\theta) = 2 \sin \theta \cos \theta$.]

7. Let $A_n = 1 + 3 + 5 + \cdots + (2n - 1)$; that is, let A_n be the sum of the first n positive odd integers. Note that $A_1 = 1$ and $A_2 = 1 + 3 = 4$.
(i) Find A_3, A_4, and A_5.
(ii) Conjecture a simple formula for A_n.
(iii) Prove that formula for all positive integers n.

8. Prove that, for all positive integers n,
(a) $1^3 + 2^3 + 3^3 + \cdots + n^3 = [n(n+1)/2]^2$.

Fibonacci (or Leonardo of Pisa)
1175?–1250?

The Fibonacci numbers are named for a thirteenth-century mathematician, Leonardo of Pisa. This is not as unlikely as it may sound, since he was also called Fibonacci, literally "son of Bonaccio." His principal work, *Liber Abaci*, published in 1202, contained a collection of mathematical ideas probably taken from Arabic sources, along with some original work. It marked the renaissance of mathematical studies in Europe following the Dark Ages.

(b) $1 + x + x^2 + \cdots + x^n = \dfrac{1 - x^{n+1}}{1 - x}$ for $x \neq 1$.

(c) $a + [a + d] + [a + 2d] + \cdots + [a + (n - 1)d] = n[2a + (n - 1)d]/2$.

9. Let $u_0, u_1, u_2, \ldots$ be the sequence

$$0, 1, \frac{1}{2}, \frac{3}{4}, \frac{5}{8}, \frac{11}{16}, \ldots$$

in which the average of two consecutive terms is the term that follows them; that is, $u_{s+2} = (u_{s+1} + u_s)/2$. Let $v_s = u_s - u_{s-1}$. Do the following.
(i) Calculate v_1, v_2, v_3, v_4, and v_5, and conjecture a simple formula for v_n.
(ii) Prove the formula for all positive integers n.
(iii) Show that $u_n = v_1 + v_2 + \cdots + v_n$.
(iv) Use (iii) to find a simple formula for u_n.

10. Let $x_n = (n^3 + 5n)/6$ and $y_n = x_n - x_{n-1}$. Prove the following for all positive integers n.
(i) y_n is an integer. [*Hint:* $n(n + 1)$ is an even integer.]
(ii) x_n is an integer. [*Hint:* Use mathematical induction and part (i).]

11. Conjecture a simple formula for the following sum, and prove it for all positive integers n by mathematical induction:

$$B_n = 1(1!) + 2(2!) + 3(3!) + \cdots + n(n!).$$

[*Note:* $n! = 1 \cdot 2 \cdot 3 \cdots n$ for positive integers n. Also, $B_1 = 1$, $B_2 = 5$, and $B_3 = 23$.]

12. Do as in Problem 11 for the sum

$$C_n = \frac{1}{2!} + \frac{2}{3!} + \frac{3}{4!} + \cdots + \frac{n}{(n + 1)!}.$$

13. In each of the following parts, let $P(n)$ denote the stated proposition. In which part is $P(1)$ false? In which part is $[P(k) \Rightarrow P(k + 1)]$ false for some k in $\mathbf{Z}^+$? In which part is $P(n)$ true for all n in $\mathbf{Z}^+$?
(i) $1^2 + 2^2 + 3^2 + \cdots + n^2 = n(n + 1)(2n + 1)/6$.
(ii) $2^n < n^2 + 2$.
(iii) $1 + 2 + 3 + \cdots + n = (n^2 + n + 1)/2$.

14. Let a_n be the number of representations of the positive integer n as a sum of 1's and 2's, taking order into account. For example, $a_4 = 5$, since the representations of 4 are

$$2 + 2, \quad 2 + 1 + 1, \quad 1 + 2 + 1, \quad 1 + 1 + 2, \quad 1 + 1 + 1 + 1.$$

Prove that $a_n = F_{n+1}$.

François Édouard Anatole Lucas
1842–1891

Although the Fibonacci numbers are named after Leonardo of Pisa, the great exponent of this and other sequences with similar recursion relations was the prolific French number theorist, Lucas. He summarized this aspect of his research in two articles in the first volume of the first mathematical journal published in the United States — "Théorie des fonctions numériques simplement périodiques," *American Journal of Mathematics 1* (1878): 184–240, 289–321. One year earlier he wrote his *Recherches sur plusieurs ouvrages de Léonard de Pise*, where he summarized many of the well-known properties of the Fibonacci numbers — their relationship to Pascal's triangle, Binet's formula, and many others.

★ **15.** Prove that every integer in $\{0, 1, 2, 3, \ldots, 2^n - 1\}$ is expressible in one and only one way in the form

$$c_0 + 2c_1 + 2^2c_2 + 2^3c_3 + \cdots + 2^{n-1}c_{n-1}, \qquad \text{(R)}$$

with each c_i in $\{0, 1\}$. [The representation (R) is closely related to the *binary representation* in which the base is 2 instead of being the base 10 of the decimal system.]

★ **16.** Prove that every integer in the set $\{0, 1, 2, \ldots, 3^n - 1\}$ is expressible uniquely in the *base 3* form

$$c_0 + 3c_1 + 3^2c_2 + \cdots + 3^{n-1}c_{n-1} \quad \text{with } c_i \text{ in } \{0, 1, 2\}.$$

1.2 Multiples and Divisors, Primes in Z

Here we introduce some terminology for the set

$$\mathbf{Z} = \{\ldots, -2, -1, 0, 1, 2, 3, \ldots\}$$

of the integers. This terminology will be used in a more general context in Section 5.3.

DEFINITION 1 **$d \mid m$, Integral Divisor, Integral Multiple**

Let d and m be in **Z**. Then $d \mid m$ means that there exists an n in **Z** such that $m = dn$. The fact that $d \mid m$ can also be expressed by any one of the following statements:

1. d is a **divisor** (or **factor**) of m in **Z**.
2. d is an **integral divisor** of m.
3. m is a **multiple** of d in **Z**.
4. m is an **integral multiple** of d.

For example, $14|42$ because $42 = 14 \cdot 3$ with each of 14, 3, and 42 in $\mathbf{Z}$. Also, $(-15)|45$ because $45 = (-15)(-3)$, with each of -15, -3, and 45 in $\mathbf{Z}$. The positive integral divisors of 30 (or of -30) are

$$1, 2, 3, 5, 6, 10, 15, \text{ and } 30, \tag{1}$$

and the other divisors of ± 30 in $\mathbf{Z}$ are the negatives of the integers of (1). However, 7 is not an integral divisor of 30, since 30/7 is not an integer.

NOTATION 1 **Set $a + dS$**

Let a and d be integers, and let S be a set of integers. Then

$$a + dS = \{a + ds : s \in S\};$$

that is, $a + dS$ is the set of all integers of the form $a + ds$ with s in S. (Naturally, $a + S = \{a + s : s \in S\}$ and $dS = \{ds : s \in S\}$.)

For example, when the set S of Notation 1 is taken to be $\mathbf{Z}^+ = \{1, 2, 3, \ldots\}$, we have

$$1 + \mathbf{Z}^+ = \{2, 3, 4, \ldots\}, \qquad 2\mathbf{Z}^+ = \{2, 4, 6, \ldots\}, \qquad -1 + 2\mathbf{Z}^+ = \{1, 3, 5, \ldots\}.$$

Using $\mathbf{Z} = \{\ldots, -2, -1, 0, 1, 2, \ldots\}$ as the S of Notation 1, we see that

$$2\mathbf{Z} = \{\ldots, -4, -2, 0, 2, 4, \ldots\}$$

is the set of even integers and that

$$1 + 2\mathbf{Z} = \{\ldots, -3, -1, 1, 3, 5, \ldots\}$$

is the set of odd integers. Note that

$$(-2)\mathbf{Z} = \{\ldots, 4, 2, 0, -2, -4, \ldots\} = 2\mathbf{Z},$$

and similarly

$$3\mathbf{Z} = \{\ldots, -6, -3, 0, 3, 6, \ldots\} = (-3)\mathbf{Z}.$$

In general,

$$a + d\mathbf{Z} = \{\ldots, a - 2d, a - d, a, a + d, a + 2d, \ldots\}.$$

LEMMA 1 **Divisor of a Linear Combination**

Let a, b, d, h, and k be integers with $d|a$ and $d|b$. Then $d|(ha + kb)$.

Proof By definition, $d|a$ and $d|b$ imply that there are integers u and v such that $a = du$ and $b = dv$. Then

$$ha + kb = hdu + kdv = d(hu + kv),$$

which shows that $d|(ha + kb)$. ∎

The only expressions for 1 as a product bc of integers are

$$1 = 1 \cdot 1 = (-1)(-1);$$

this tells us that each of 1 and -1 is its own reciprocal and that the only integers with reciprocals in **Z** are 1 and -1. We use this fact in the following result.

LEMMA 2 If $s \mid t$ and $t \mid s$, then $s = \pm t$.

Proof The hypothesis tells us that there are integers b and c such that $t = bs$ and $s = ct$. Substituting from one of these equations into the other, we have $t = bct$, and so $(1 - bc)t = 0$. This implies that either $1 = bc$ or $t = 0$.

If $1 = bc$, then $b = c = \pm 1$ and $s = ct = \pm t$. If $t = 0$, then $s = ct = c \cdot 0 = 0$. In either case, the hypothesis implies that $s = \pm t$, as desired. ■

Every integer a has at least the factorizations $a = a \cdot 1$ and $a = (-a)(-1)$; such a factorization is said to be ***trivial***. That is, $a = bc$ is a trivial factorization if either b or c is in the set $\{1, -1\}$.

DEFINITION 2

Prime in Z

An integer p that is not in $\{1, -1\}$ and has only the trivial factorizations $p = p \cdot 1 = (-p)(-1)$ is a **prime** in **Z**.

It follows from this definition that an integer p is a prime if and only if p does not have a reciprocal in **Z** and any equation $p = bc$, with b and c integers, implies that both b and c are in the set $\{-p, -1, 1, p\}$. One easily sees that the first ten positive primes in **Z** are

$$2, 3, 5, 7, 11, 13, 17, 19, 23, \text{ and } 29.$$

DEFINITION 3

Composite in Z

A nonzero integer that has a nontrivial factorization is a **composite integer**.

For example, 15 is composite since $15 \neq 0$ and $15 = 5 \cdot 3$ is a nontrivial factorization. Other composite integers are ± 4, ± 6, ± 8, ± 9, ± 10, ± 12, and ± 14. We note that 0, 1, and -1 are neither prime nor composite.

The following definition helps us to characterize some important subsets of **Z**.

DEFINITION 4

Closure Under Subtraction

A set S is **closed under subtraction** if $a - b$ is in S whenever a and b are in S. (The elements a and b need not be distinct.)

EXAMPLE 1 **Set Closed Under Subtraction**

Let S be a set closed under subtraction and let 10 and 14 be in S. We wish to show that 0, 2, and -2 are in S.

By Definition 4, we know that $a - b$ is in S when a and b are in S. Letting $a = 10 = b$, we find that $10 - 10 = 0$ is in S. The choice of $a = 14$ and $b = 10$ shows that $14 - 10 = 4$ is in S. Similarly, $10 - 4 = 6$ and $6 - 4 = 2$ are in S. Since 0 and 2 are in S, it follows that $0 - 2 = -2$ is in S.

Problems

1. Which of the following integers are in the set $6\mathbf{Z}$ of integral multiples of 6?
(a) 10; (b) -10; (c) 12; (d) -12; (e) 4002;
(f) -4002; (g) 4003; (h) -4003.

2. Let $P = \{2, 3, 5, 7, 11, 13, 17, 19, 23, 29, 31, 37\}$. Which of the integers of P are in the given set?
(a) $1 + 4\mathbf{Z}$; (b) $-1 + 4\mathbf{Z}$; (c) $1 + 6\mathbf{Z}$; (d) $-1 + 6\mathbf{Z}$.

3. (a) Find all integers a such that $a \mid 1$.
(b) Which integers are integral divisors of 0?
(c) How many integral divisors are there of 3?
(d) How many integers a are there with $a \mid 25$?
(e) Of which integers n is 6 an integral multiple?

4. (a) How many integral divisors are there of a prime p?
(b) What is the least number of integral divisors that a composite integer can have? [See Problem 3(d).]

5. Restate the result in Lemma 1 of this section (without the proof), replacing each use of the symbolism "$x \mid y$" by "x is an integral divisor of y."

6. Restate the result in Lemma 1 (without the proof), replacing each use of "$x \mid y$" by "y is an integral multiple of x."

7. (a) Is $2\mathbf{Z}$ the set of all even integers? Is 0 an even integer?
(b) Is $d \mid 0$ for all integers d? Explain.
(c) List all the integers of $0\mathbf{Z}$, that is, of $d\mathbf{Z}$ with $d = 0$.
(d) What is the only integer m with $0 \mid m$?

(e) Is d in $d\mathbf{Z}$ for all integers d? Explain.
(f) Is $n \mid n$ for all integers n? Explain.
(g) Does $d \mid m$ imply that $(-d) \mid m$? Explain.
(h) Is $9\mathbf{Z} \subseteq 3\mathbf{Z}$? Explain.

8. (a) Does $1\mathbf{Z} = \mathbf{Z}$? Explain.
(b) Is $1 \mid m$ for every integer m? Explain.
(c) List the two integers d such that $d \mid 1$.
(d) Does $(-d)\mathbf{Z} = d\mathbf{Z}$ for every integer d? Explain.
(e) If $d \mid m$, is $d \mid (-m)$? Explain.
(f) Is $n \mid (-n)$ for all integers n? Explain.
(g) Is $3\mathbf{Z} \subseteq 9\mathbf{Z}$? Explain.

9. (i) Find the set A of positive integral divisors of 30.
(ii) Find the set B of positive integral divisors of 21.
(iii) Find $A \cap B$, that is, the set of integers common to A and B.

10. Find the set C of all positive integers d such that both $d \mid 22$ and $d \mid 99$.

11. Let a, b, c, d, and q be integers with $a = qb + c$, $d \mid b$, and $d \mid c$. Prove that $d \mid a$.

12. Let a, b, c, d, and q be integers with $a = qb + c$, $d \mid a$, and $d \mid b$. Prove that $d \mid c$.

13. (a) Prove that the set $8\mathbf{Z}$, of all integral multiples of 8, is closed under subtraction.
(b) Is $8\mathbf{Z}$ closed under addition? In other words, is the sum of two integral multiples of 8 also an integral multiple of 8?

14. (a) Prove that $d\mathbf{Z}$ is closed under subtraction for every integer d.
(b) Is $d\mathbf{Z}$ closed under addition?

15. The integers 21 and 27 are in a set T that is closed under subtraction.
(a) Which of the integers in the set $U = \{-20, -19, -18, \ldots, 19, 20\}$ must be in T?
(b) Can all the integers of U be in T? Explain.

16. The integer 4 is in a set R that is closed under subtraction.
(i) Show that 0 must be in R.
(ii) Show that -4 must be in R.
(iii) Show that 8 must be in R.
(iv) Which of the integers between -25 and 25 must be in R?

17. Which of the integers 0, 1, 2, 3, 4, 5, and 6 is expressible in the form $10x + 14y$, with x and y integers? Justify your answer.

18. Do the previous problem with $10x + 14y$ replaced by $15x - 9y$.

19. Given that $a \mid b$ and $b \mid c$, prove that $a \mid c$.

20. Let a, b, and c be integers and let $a \mid b$. Prove that $(ac) \mid (bc)$.

21. (i) Explain why a positive integer c is composite if and only if $c = ab$ with a and b (not necessarily distinct) integers in

$$1 + \mathbf{Z}^+ = \{2, 3, 4, 5, \ldots\}.$$

(ii) Find the twenty smallest positive composite integers.

22. (i) Find the ten smallest positive prime integers. [*Hint:* Use Problem 21(ii).]
(ii) Explain why the negative $-p$ of a prime integer p is also prime.
(iii) Explain why $-c$ is composite whenever c is composite.

23. Let a, b, c, d, h, k, and q be integers with

$$d = bh + ck \quad \text{and} \quad a = qb + c.$$

Find integers x and y (in terms of h, k, and q) such that $d = ax + by$.

24. Given that each of R and S is a set of integers closed under subtraction, prove that their intersection $T = R \cap S$ is also closed under subtraction. (T consists of the integers that are in both R and S.)

25. For each of the following integers d, find the smallest positive integer n such that $d \mid (10^n - 1)$.
(a) $d = 7$; (b) $d = 11$; (c) $d = 13$; (d) $d = 37$.

26. Do as in Problem 25 for the following values of d.
(a) $d = 77$; (b) $d = 91$; (c) $d = 407$.

27. Find all solutions in integers x and y of the equation

$$xy + 5x - 8y = 79.$$

[It may be helpful to show that the given equation implies that
(i) $y = [39/(x - 8)] - 5$;
(ii) $(x - 8) \mid 39$;
(iii) $x \in \{-31, -5, 5, 7, 9, 11, 21, 47\}$.]

28. Tabulate the solutions in integers x and y of the equation

$$xy - 10x + 21y = 228.$$

29. Is every positive composite integer c expressible as

$$xy + xz + yz + 1 \quad \text{with} \quad x, y, \text{ and } z \text{ in } \mathbf{Z}^+ = \{1, 2, 3, \ldots\}?$$

1.3 The Division Algorithm

Multiplication of positive integers is a form of addition; for example, $4 \cdot 5$ may be thought of as

$$5 + 5 + 5 + 5 \quad \text{or} \quad 4 + 4 + 4 + 4 + 4.$$

Similarly, division of positive integers can be accomplished by repeated subtraction. Thus one tests whether or not 8 is an integral divisor of 48 by repeatedly subtracting 8's and seeing if 0 is obtained after some number of steps. In this case, 0 results after 6 subtractions of 8 from 48; hence, $48 = 6 \cdot 8$ and $8 \mid 48$.

If we start with 53 instead of 48, a positive remainder that is less than 8 is reached after 6 subtractions of 8, and we find that

$$53 = 6 \cdot 8 + 5.$$

Further subtractions of 8 would give negative results, not 0. Hence we see that 8 is not a divisor of 53 in **Z**.

This motivates the following result.

THEOREM 1 Division Algorithm

If a and b are integers and b is positive, there exist unique integers q and r such that

$$a = qb + r, \qquad 0 \leq r < b.$$

Also, $b \mid a$ if and only if $r = 0$. (One says that q is the **quotient** and r is the **remainder** in the division of a by b.)

Proof Let **N** be the set $\{0, 1, 2, \ldots\}$ of the nonnegative integers. Let S be the subset of **N** consisting of all the nonnegative integers that are expressible in the form $a - qb$, with q an integer. We next show that the set S is not empty.

Since b is a positive integer, we have $b \geq 1$. Now

$$|a| \cdot b \geq |a| \cdot 1 = |a| \geq -a;$$

that is, $|a| \cdot b \geq -a$. It follows that $a + |a| \cdot b \geq a - a = 0$ and hence that

$$a - (-|a|)b \geq 0.$$

This inequality tells us that one of the values of q, for which $a - qb$ is in S, is $q = -|a|$; hence S is nonempty. Then it follows from the generalized well ordering principle (Theorem 2 of Section 1.1) that the nonempty set S has a least integer r. Associated with this r is an integer q such that $a - qb = r$.

By definition of S, its least integer r satisfies $0 \leq r$. We show that r also satisfies $r < b$ by assuming $r \geq b$ and obtaining a contradiction. Let $r' = r - b$ and $q' = q + 1$. If $r \geq b$, then $r' \geq 0$ and

$$r' = r - b = (a - qb) - b = a - (q + 1)b = a - q'b.$$

Together $r' \geq 0$ and $r' = a - q'b$ imply that r' is in S. Since $r' = r - b < r$, this contradicts the fact that r is the least integer in S. Hence $r < b$; that is, $0 \leq r < b$.

We next prove that r and q are unique. We start by assuming that

$$a = qb + r = q_1 b + r_1, \qquad 0 \leq r < b, \quad 0 \leq r_1 < b.$$

Without loss of generality, one may assume that $r \geq r_1$. Then

$$0 \leq r - r_1 < b, \qquad r - r_1 = (q_1 - q)b. \tag{1}$$

Since there is no multiple of b between 0 and b. it follows from (1) that $r - r_1 = 0$. Then $r = r_1$, $qb = q_1 b$, and finally $q = q_1$, since $b \neq 0$. Thus r and q are uniquely determined by a and b.

If $r = 0$, then $a = qb + r = qb$, and we have $b \mid a$. Conversely, if $b \mid a$, then there is an integer m such that $a = mb$, and hence the unique q and r are m and 0, respectively. We see that $b \mid a$ if and only if $r = 0$; this completes the proof. ■

COROLLARY **Extended Division Algorithm**

If a and c are in **Z** and $c \neq 0$, there exist unique q and r in **Z**, such that

$$a = qc + r \quad \text{and} \quad 0 \leq r < |c|.$$

This corollary is proved by letting $b = |c|$ and then applying the theorem.

Let m be a fixed positive integer. Then Theorem 1 (the division algorithm) tells us that every integer a is of exactly one of the forms

$$qm,\ qm + 1,\ qm + 2,\ \ldots,\ qm + (m - 1),$$

with q an integer; that is, each integer is in exactly one of the sets

$$m\mathbf{Z},\ 1 + m\mathbf{Z},\ 2 + m\mathbf{Z},\ \ldots,\ (m - 1) + m\mathbf{Z}.$$

When $m = 2$, this statement becomes the familiar fact that every integer either is in the set $2\mathbf{Z} = \{\ldots, -4, -2, 0, 2, 4, \ldots\}$ of even integers or is in the set $1 + 2\mathbf{Z} = \{\ldots, -3, -1, 1, 3, 5, \ldots\}$ of odd integers, but is not in both. With $m = 3$, we have the fact that every integer is in exactly one of the sets

$$3\mathbf{Z},\ 1 + 3\mathbf{Z},\ 2 + 3\mathbf{Z}.$$

DEFINITION 1 **Parity of Integers**

Let a and c be integers. Then a and c have the **same parity** if they are both even or are both odd. If one of these integers is even and the other is odd, they have **opposite parity**.

Thus 15 and 77 have the same parity, since they are both odd. Likewise, -4 and 6 have the same parity because both are even. However, 6 and 15 have opposite parity.

The following notation generalizes the concept of "same parity."

NOTATION 1 **Congruence Modulo m**

Let a, c, and m be integers with m positive. Then the notation $a \equiv c \pmod{m}$ means that a and c have the same remainder in division by m. One reads "$a \equiv c \pmod{m}$" as "a is congruent to c modulo m." Such a relation is called a **congruence**, and m is called the **modulus** of the congruence.

For example, the equations $28 = 5 \cdot 5 + 3$ and $73 = 14 \cdot 5 + 3$ show that 28 and 73 have the same remainder in division by 5 and hence that $28 \equiv 73 \pmod{5}$. Each of 28 and 73 is in $3 + 5\mathbf{Z}$.

THEOREM 2 **Two Ways of Showing Equal Remainders**

Let a, c, and m be integers with m positive. Then $a \equiv c \pmod{m}$ if and only if $m \mid (a - c)$.

Proof We have to establish the following two results:

(i) If $a \equiv c \pmod{m}$, then $m \mid (a - c)$.

(ii) If $m \mid (a - c)$, then $a \equiv c \pmod{m}$.

For both parts, we see from the division algorithm that $a = qm + r$ and $c = q'm + r'$ with q, r, q', and r' in $\mathbf{Z}$, $0 \le r < m$, and $0 \le r' < m$.

For (i), we let it be given that $a \equiv c \pmod{m}$. By definition of congruence, this means that $r = r'$, and hence we have

$$a - c = (qm + r) - (q'm + r') = (q - q')m.$$

Therefore, $m \mid (a - c)$, and (i) is proved.

For (ii), we are given that $m \mid (a - c)$; that is, $a - c = tm$ with t in $\mathbf{Z}$. Then

$$tm = a - c = (qm + r) - (q'm + r') = (q - q')m + (r - r').$$

From this it follows that

$$r - r' = (t - q + q')m.$$

This means that $r - r'$ is an integral multiple of m. But r and r' are in $\{0, 1, \ldots, m - 1\}$, and hence $r - r'$ is in

$$\{-(m - 1), -(m - 2), \ldots, 0, \ldots, m - 2, m - 1\}.$$

Since the only integral multiple of m in this set is 0, we see that $r - r' = 0$. Finally, $r = r'$, which implies that $a \equiv c \pmod{m}$, as desired. ∎

The division algorithm has many applications; one of these is the following result.

THEOREM 3 Closure Under Subtraction

Let S be a set of integers closed under subtraction. Then:

(a) If S is nonempty, 0 is in S.

(b) If a is in S, so is $-a$.

(c) If a and b are in S and q is an integer, $a - qb$ is in S.

(d) S consists of all the integral multiples of some t in $\mathbf{Z}$, or S is empty.

(e) If there is at least one nonzero integer in S, then S consists of all the integral multiples of the smallest positive integer t in S; that is, $S = t\mathbf{Z}$.

Proof If S is nonempty, there is an integer a in S, and $a - a = 0$ is in S by closure under subtraction. Then $0 - a = -a$ is in S. Hence we have proved parts (a) and (b).

Now let a and b be in S. Then $a - qb$ is the result of subtracting q b's from a (if q is positive) or of subtracting $-b$ a total of $-q$ times (if q is negative). In either case, $a - qb$ is in S by closure under subtraction. This proves part (c).

If S is the single element set $\{0\}$, S consists of all the integral multiples of 0. Hence we now assume that there is a nonzero element c in S. Then one of c and $-c$ is positive and is in S. Now the subset T of the positive integers in S is a nonempty set, and it follows from the well ordering principle that T has a least integer t.

Let a be any integer in S. The division algorithm gives us integers q and r such that

$$a = qt + r, \qquad 0 \le r < t.$$

Then $r = a - qt$ is in S by part (c).

Since r is smaller than the least positive integer t in S, r cannot be positive. This and the condition $0 \le r$ imply that $r = 0$. Hence, $a = qt$; that is, every a in S is an integral multiple of the least positive integer t in S.

Conversely, if a is an integral multiple qt of the smallest positive integer t in S, then $a = 0 - (-q)t$ is in S by parts (a) and (c). Hence, S is the set $t\mathbf{Z}$ of all integral multiples of t. ∎

We will generalize the concept of sets of integers closed under subtraction when we deal with *subgroups* (see especially Theorem 1 of Section 2.5) and when *ideals in rings* are introduced (in Section 4.2).

As we shall see in the problem set for this section, we now have alternate notations and terminology for expressing certain facts. Each of these ways of making statements is used frequently in mathematics, and each has its advantages in certain contexts.

Problems

1. For each of the following choices of a and b, find the quotient q and the remainder r in the division of a by b.
(a) $a = 203, b = 17$; (b) $a = -203, b = 17$; (c) $a = 0, b = 17$.

2. Do as in Problem 1 for the following.
(a) $a = 1000, b = 13$; (b) $a = 1001, b = 13$; (c) $a = 1002, b = 13$.

3. Let $S = \{-60, -59, -58, \ldots, -1, 0, 1, \ldots, 58, 59\}$.
(i) Which integers are in both S and $6\mathbf{Z}$?
(ii) Which integers in S have 1 as the remainder when divided by 6?
(iii) Which integers in S are also in $-1 + 6\mathbf{Z}$?
(iv) Which integers n in S satisfy $n \equiv 3 \pmod 6$?

4. Let $T = \{-70, -69, -68, \ldots, -1, 0, 1, \ldots, 68, 69\}$.
(i) Which integers are in both T and $7\mathbf{Z}$?
(ii) Which integers in T have 2 as the remainder in division by 7?
(iii) Which integers in T are also in $3 + 7\mathbf{Z}$?
(iv) Which integers n in T satisfy $n \equiv -1 \pmod 7$?

5. Does $n \equiv 0 \pmod 5$ mean the same thing as $5 \mid n$?

6. Does $n \equiv 0 \pmod m$ mean the same thing as $m \mid n$?

7. Do the following five statements all have the same meaning?
(i) $2 \mid (a - c)$;
(ii) $a \equiv c \pmod 2$;
(iii) a and c have the same parity;
(iv) a and c have the same remainder in division by 2;
(v) a and c are both in $2\mathbf{Z}$ or are both in $1 + 2\mathbf{Z}$.

8. Do the following four statements all have the same meaning?
   (i) $4 \mid (a - c)$;
   (ii) $a \equiv c \pmod 4$;
   (iii) a and c have the same remainder in division by 4;
   (iv) a and c are both in $4\mathbf{Z}$ or are both in $1 + 4\mathbf{Z}$ or are both in $2 + 4\mathbf{Z}$ or are both in $3 + 4\mathbf{Z}$.

9. (a) If a and c are both in $3\mathbf{Z}$, must they have the same parity?
   (b) If a and c are both in $1 + 6\mathbf{Z}$, must they have the same parity?

10. (a) If a and c are both in $4\mathbf{Z}$, must they have the same parity?
    (b) If a and c are both in $1 + 5\mathbf{Z}$, must they have the same parity?

11. (i) Is $2\mathbf{Z}$ closed under subtraction? Explain.
    (ii) Is $1 + 2\mathbf{Z}$ closed under subtraction? Explain.

12. Which of the sets $3\mathbf{Z}$, $1 + 3\mathbf{Z}$, $2 + 3\mathbf{Z}$ is closed under subtraction?

13. Let a be in $1 + 4\mathbf{Z}$, and let b be in $2 + 4\mathbf{Z}$.
    (i) Must $a + b$ be in $3 + 4\mathbf{Z}$?
    (ii) Must $a - b$ be in $3 + 4\mathbf{Z}$?

14. Let a be in $1 + 4\mathbf{Z}$, and let b be in $2 + 4\mathbf{Z}$.
    (i) Is $b - a$ always in $1 + 4\mathbf{Z}$?
    (ii) Is $b + a$ always in $1 + 4\mathbf{Z}$?

15. (i) Why do a and $a + 2d$ have the same parity for all integers a and d?
    (ii) Why do $m - n$ and $m + n$ have the same parity for all m and n in $\mathbf{Z}$?

16. Is $a \equiv a + 3d \pmod 3$ for all a and d in $\mathbf{Z}$?

17. (i) Write 88 in all possible ways as a product cd of positive integers with the same parity.
    (ii) Find all solutions of the equation $x^2 - y^2 = 88$ in positive integers x and y. [*Hint:* $x^2 - y^2 = (x - y)(x + y)$.]

18. Find two solutions in positive integers x and y of the equation $x^2 - y^2 = 77$. [*Hint:* $77 = 1 \cdot 77 = 7 \cdot 11$.]

19. Given that a, x, and y are integers with $x^2 - y^2 = a$ and $2 \mid a$, explain why $4 \mid a$.

20. Why are there no integers x and y with $x^2 - y^2 = 34$?

21. Let a be in $1 + 2\mathbf{Z}^+ = \{3, 5, 7, \ldots\}$. Find one solution in positive integers x and y to the equation $x^2 - y^2 = a$.

22. Let c be in $1 + \mathbf{Z}^+ = \{2, 3, 4, \ldots\}$. Find one solution in positive integers x and y to the equation $x^2 - y^2 = 4c$.

1.4 Common Divisors

Let a and b be nonzero integers. A common integral divisor t of a and b, that is, an integer t such that both $t \mid a$ and $t \mid b$, satisfies the inequalities

$$-|a| \le t \le |a|, \qquad -|b| \le t \le |b|.$$

Hence the set T of such common integral divisors is finite. Since 1 is in T, there is at least one positive integer in T. It follows from the last two statements that there is a largest positive integer in T. If c is a nonzero integer, the largest positive integer d such that both $d \mid c$ and $d \mid 0$ is $d = |c|$.

DEFINITION 1

Greatest Common Divisor

Let a and b be integers, not both zero. The largest positive integer d such that both $d \mid a$ and $d \mid b$ is called the **greatest common divisor (gcd)** of a and b and is denoted by (a, b) [or by $\gcd(a, b)$ when one wants to avoid confusion with other uses of (a, b)]. Also, we let $(0, 0) = 0$.

The set of positive integral divisors of 10 is $A = \{1, 2, 5, 10\}$, and the set of positive integral divisors of 12 is $B = \{1, 2, 3, 4, 6, 12\}$. The set of common positive integral divisors of 10 and 12 is

$$C = A \cap B = \{1, 2\},$$

and hence their greatest common divisor (10, 12) is 2. (For a definition of the "intersection" symbol $\cap$, see the inside front cover of the book.) Some other examples of greatest common divisors are

$$(15, 6) = 3, \quad (26, -10) = 2, \quad (0, -7) = 7, \quad (42, 14) = 14.$$

If a and b are integers, one can readily show the following:

1. $(a, b) = (b, a)$.
2. $(a, b) = (a, -b) = (-a, b) = (-a, -b)$.
3. $(1, b) = 1$.
4. If $b \mid a$ and $b \neq 0$, then $(a, b) = |b|$.

Terminology for an important special case, in which only 1 and -1 are common integral divisors of a and b, is given in the following definition.

DEFINITION 2

Relatively Prime Integers

If $\gcd(r, s) = 1$, the integers r and s are **relatively prime** (or **coprime**).

For example, 22 and -15 are relatively prime, since $(22, -15) = 1$. Likewise, 7 and 24 are relatively prime. We note that r and s may be relatively prime even though one (or each) of r and s is composite.

The examples $(17, 51) = 17$ and $(19, -19) = 19$ show that r and s are not necessarily relatively prime even when one (or each) of r and s is a prime.

Next we state the main results of this section.

THEOREM 1 **Linear Combinations**

Let a and b be in $\mathbf{Z}$ with $a \neq 0$ or $b \neq 0$. Let

$$L = \{n : n = ax + by,\ x \in \mathbf{Z},\ y \in \mathbf{Z}\};$$

that is, let L be the set of all $ax + by$ with x and y integers. Then there is a smallest positive integer t in L, and L consists of all the integral multiples of t; that is, $L = t\mathbf{Z}$.

Proof For definiteness, let $a \neq 0$. Let $n_1 = ax_1 + by_1$ and $n_2 = ax_2 + by_2$ be any two integers in L. Then

$$n_1 - n_2 = a(x_1 - x_2) + b(y_1 - y_2)$$

is in L, since the differences $x_1 - x_2$ and $y_1 - y_2$ of integers are also integers. Hence L is closed under subtraction.

Using $x = 1$ and $y = 0$ in $ax + by$, we see that $a \cdot 1 + b \cdot 0 = a$ is in L. Similarly, $-a$ and b are in L.

Since $a \neq 0$, either a or $-a$ is positive; that is, L has at least one positive integer in it. Then among the positive integers in L there is a smallest positive integer t, by the well ordering principle. Also, L consists of all the integral multiples of t, by Theorem 3(e) of Section 1.3. ■

We are now able to give several other characterizations of the gcd that are frequently used as alternative definitions of this concept. In fact, they will furnish a model for our definition of the gcd of two polynomials in Section 5.7.

THEOREM 2 Characterization of the gcd

Let a and b be in $\mathbf{Z}$ with $a \neq 0$ or $b \neq 0$. Then:

(a) $\gcd(a, b)$ is the smallest positive integer in the set

$$L = \{n : n = ax + by,\ x \in \mathbf{Z},\ y \in \mathbf{Z}\}.$$

(b) $\gcd(a, b)$ is the unique positive common divisor d of a and b such that $c \mid d$ for all common divisors c of a and b.

Proof In Theorem 1, we saw that there is a smallest positive integer t in L, that $t \mid n$ for all n in L, and that a and b are in L. Hence $t \mid a$ and $t \mid b$. Since t is a common divisor of a and b while d is the greatest common divisor, we have $t \leq d$.

Since $t \in L$, we have $t = ah + bk$ with h and k in $\mathbf{Z}$. If both $c \mid a$ and $c \mid b$, then $c \mid t$ by Lemma 1 of Section 1.2. One of the things this tells us is that $d \mid t$. Since d and t are positive, it follows that $d \leq t$. This and the previously shown $t \leq d$ imply that $d = t$.

Having proved that $d = t$, we now know that d is a positive common divisor of a and b with the property that $c \mid d$ for every common divisor c of a and b. If d' is another such integer, then both $d \mid d'$ and $d' \mid d$; and it follows that the positive integers d and d' are equal. This establishes the uniqueness asserted in part (b). ■

COROLLARY **Linear Combinations of Relatively Prime Integers**

(a) Let a and b be fixed integers. Then every integer c is a linear combination $c = ax + by$, with x and y integers, if and only if a and b are relatively prime.
(b) Integers a and b are relatively prime if and only if there exist integers h and k such that $1 = ah + bk$.

The proof of the corollary is left to the reader in Problem 18 below.

It follows from Theorem 2(b) that $c \mid a$ and $c \mid b$ together imply that $c \mid \gcd(a, b)$. Symbolically this may be written as

$$c \mid a \quad \text{and} \quad c \mid b \Rightarrow c \mid \gcd(a, b).$$

THEOREM 3 **Multiples of the Greatest Common Divisor**

Let a and b be in $\mathbf{Z}$ with $a \neq 0$ or $b \neq 0$. Let $d = \gcd(a, b)$. Then an integer n is of the form $ax + by$, with x and y in $\mathbf{Z}$, if and only if $d \mid n$.

Proof Theorem 2 tells us that d is the smallest positive integer in the set $L = \{n : n = ax + by,\ x \in \mathbf{Z},\ y \in \mathbf{Z}\}$; that is, d is the t of Theorem 1. Then Theorem 1 states that $d \mid n$ for all n in L. ∎

THEOREM 4 **Equal gcd's**

Let a, b, q, and c be integers with $a = qb + c$. Then $\gcd(a, b) = \gcd(b, c)$.

Proof It follows from the hypothesis $a = qb + c$ and Lemma 1 of Section 1.2 that every common divisor of b and c is a divisor of a and hence is a common divisor of a and b. Similarly, $c = a - qb$ implies that every common divisor of a and b is a common divisor of b and c. These statements imply that $\gcd(a, b) = \gcd(b, c)$. ∎

Let $a_1, a_2, \ldots, a_n$ be integers that are not all zero. It can be shown that there exists a largest positive integer d such that $d \mid a_i$ for each i; this d is called the *greatest common divisor* of the a_i and is denoted by $\gcd(a_1, a_2, \ldots, a_n)$. Also $\gcd(0, 0, \ldots, 0) = 0$.

Problems

1. Find $d = \gcd(11, 99)$, $e = \gcd(-21, 14)$, $f = \gcd(14, 15)$, and $g = \gcd(1, 0)$.

2. Find $d = \gcd(18, 498)$, $e = \gcd(154, 35035)$, $f = \gcd(30, 1001)$, and $g = \gcd(0, -1)$.

3. (i) Find $r = \gcd(14, 42)$.
(ii) Find $s = \gcd(r, 35)$, where r is as in part (i).

(iii) Find $t = \gcd(42, 35)$.
(iv) Find $u = \gcd(14, t)$, where t is as in part (iii).
(v) Does $s = u = \gcd(14, 42, 35)$?

4. (i) Find $r = (26, 78)$.
(ii) Find $s = (r, 65)$, where r is as in (i).
(iii) Find $t = (78, 65)$.
(iv) Find $u = (26, t)$, where t is as in (iii).
(v) Does $s = u = \gcd(26, 78, 65)$?

5. (a) Which integers in the set $\{1, 2, 3, 4, 5\}$ are relatively prime to 5?
(b) Which integers in the set $\{1, 2, 3, 4, 5, 6\}$ are relatively prime to 6?

6. (a) Which integers in the set $\{1, 2, 3, 4, 5, 6, 7\}$ are relatively prime to 7?
(b) Which integers in the set $\{1, 2, 3, 4, 5, 6, 7, 8\}$ are relatively prime to 8?

7. (a) Given that $\gcd(a, 0) = 1$, what are the possibilities for a?
(b) Given that b is a positive integer with $\gcd(6, b) = 2$, what are the possibilities for the remainder when b is divided by 6?

8. Let c be in $\mathbf{Z}^+$, and let $\gcd(12, c) = 3$. What are the possibilities for the remainder when c is divided by 12?

9. (i) What is the least positive integer c of the form $c = 22x + 55y$, with x and y integers? Explain.
(ii) Let c be the answer to part (i). Find integers x and y such that $c = 22x + 55y$ and $0 < x < 5$.

10. Describe the set of all integers c that are expressible in the form $c = 1470x + 119y$, with x and y integers. [*Hint:* See Theorems 1 and 3.]

11. Let $72 = ah + bk$ with a, b, h, and k in $\mathbf{Z}$. What are the possibilities for the value of $\gcd(a, b)$?

12. Let $72 = 99h + bk$ with b, h, and k in $\mathbf{Z}$. What are the possibilities for the value of $\gcd(99, b)$?

13. (i) What are the positive integral divisors of 19?
(ii) Let a be an integer. Explain why either $\gcd(a, 19) = 1$ or $19 \mid a$, but not both.

14. Let p be a positive prime and q a positive integer. Find $\gcd(p, q)$ under each of the following assumptions.
(a) $1 \le q < p$.
(b) $p \mid q$.
(c) $q = mp + r$, with $0 < r < p$ and with m and r in $\mathbf{Z}$.
(d) q is a prime and $q \ne p$.

15. Let a and p be in $\mathbf{Z}$ with p a positive prime and $p \mid a$. Explain why $\gcd(a, p) = p$.

16. Let a and p be in $\mathbf{Z}$ with p a positive prime and a not an integral multiple of p. Explain why $\gcd(a, p) = 1$, that is, why a and p are relatively prime.

17. Let there be at least two integers in a set S of integers that is closed under subtraction. Explain why S consists of the integral multiples of the least positive integer in S.

18. Let a and b be in $\mathbf{Z}$. Show the following.
(a) If $1 = ah + bk$ with h and k in $\mathbf{Z}$, then $(a, b) = 1$.
(b) If $(a, b) = 1$, then for every n in $\mathbf{Z}$ there exist x and y in $\mathbf{Z}$ such that $n = ax + by$.

(The corollary to Theorem 2 follows straightforwardly from the parts of this problem.)

19. Let t, u, and v be nonzero integers. Let $a = tu$, $b = tv$, and $d = (a, b)$. Show that u and v are relatively prime if and only if $t = \pm d$; that is, show both of the following.
(i) If u and v are relatively prime, then $t = \pm d$.
(ii) If $t = \pm d$, then u and v are relatively prime.
[Part (ii) implies that a rational number a/b always has an equivalent form u/v in which the numerator and denominator are relatively prime.]

20. (i) Explain why the division algorithm implies that every integer a can be expressed in one and only one of the forms

$$3q, \qquad 3q + 1, \qquad 3q + 2,$$

with q an integer.
(ii) Let a and q be integers. Show that 3 is not an integral divisor of $(3q + 1)^2$ or of $(3q + 2)^2$ and hence that $3 \mid (a^2)$ if and only if $3 \mid a$.

21. Prove that $\sqrt{3}$ is not rational, by doing the following parts.
(i) Assume that $a/b = \sqrt{3}$, with a and b integers. Let $d = (a, b)$, $a = du$, and $b = dv$. Explain why $u/v = \sqrt{3}$, with u and v relatively prime.
(ii) Show that $u^2 = 3v^2$, $3 \mid u^2$, $3 \mid u$, and hence $u = 3w$, with w an integer.
(iii) Show that $3w^2 = v^2$ and hence $3 \mid v$.
(iv) Note that $3 \mid v$, together with $3 \mid u$, contradicts the fact that u and v are relatively prime. This contradiction completes the proof that $\sqrt{3}$ is not a rational number a/b.

22. (a) Prove that $\sqrt{2}$ is irrational.
(b) Prove that $\sqrt{5}$ is irrational.

23. Let 170 and -102 both be in a set S that is closed under subtraction.
(a) Explain why S must contain all the integral multiples of 34.
(b) Can the integer 2 be in S? Explain.

24. Describe all the possibilities for the set S of Problem 23.

25. Explain why n and $n + 1$ are relatively prime for all integers n.

26. Explain why the following are true.
(i) If n is an even integer, $\gcd(n, n + 2) = 2$.
(ii) If n is an odd integer, $\gcd(n, n + 2) = 1$.

27. What are the possible values of $\gcd(n, n + 10)$, where n is an integer?

28. Let n be an integer. Give the set of integers that might be values of $\gcd(n, n + 30)$.

29. Show that $\gcd(n - 1, n^2 + n + 1)$ is either 1 or 3 for all $n \in \mathbf{Z}$.

30. Show that $\gcd(n + 1, n^2 - n + 1) \mid 3$ for all $n \in \mathbf{Z}$.

31. Given that u and v are integers, explain why $2 \mid uv(u + v)$.

32. Given that u and v are integers, explain why $6 \mid uv(u^2 - v^2)$. [*Hint:* Use Problem 31 and consider three cases: (i) $3 \mid u$; (ii) $3 \mid v$; (iii) $u = 3a \pm 1$ and $v = 3b \pm 1$, with a and b integers.]

1.5 Euclid's Algorithm

If a, b, and c are fixed integers, it follows from Theorem 3 of Section 1.4 that the equation

$$ax + by = c$$

has integer solutions for x and y if and only if c is of the form kd, with $d = \gcd(a, b)$ and k in $\mathbf{Z}$. This section contains a constructive method for solving such an equation.

We start with the special case in which $k = 1$ and $a \geq b > 0$; in this case Euclid's Algorithm (described below) shows how to find integers x and y such that

$$ax + by = \gcd(a, b).$$

Then in Example 2 we illustrate the solution of more general equations $ax + by = kd$, in which k need not be 1, with the help of Euclid's Algorithm.

Let a and b be positive integers with $a \geq b$. It follows from the division algorithm that there exist integers q and r with

$$a = qb + r \tag{D}$$

and $0 \leq r < b$. If $r = 0$, we have $a = qb$ and hence $b = \gcd(a, b)$. If $r > 0$, it follows from Theorem 4 of Section 1.4 that $\gcd(a, b) = \gcd(b, r)$. This equation reduces the problem of finding $\gcd(a, b)$ to the same problem with smaller positive integers. The well ordering principle helps us show that $\gcd(a, b)$ is found after a finite number of such reductions.

We next look at an example and then give the general technique, which is called **Euclid's Algorithm**. (See the biographical note on Euclid near the close of this section.)

EXAMPLE 1 **gcd(803, 154)**

Here we find gcd(803, 154) and integers x and y such that

$$\gcd(803, 154) = 803x + 154y.$$

Since $803 = 5 \cdot 154 + 33$, we have $(803, 154) = (154, 33)$. Similarly, $154 = 4 \cdot 33 + 22$ implies that $(154, 33) = (33, 22)$, and $33 = 1 \cdot 22 + 11$ implies that $(33, 22) = (22, 11)$. Finally, $22 = 2 \cdot 11$ gives us

$$11 = (22, 11) = (33, 22) = (154, 33) = (803, 154);$$

that is, the last nonzero remainder 11 is the desired gcd(803, 154). Next we seek x and y.

We rearrange the preceding data as follows:

$$803 = 5 \cdot 154 + 33, \tag{a}$$

$$154 = 4 \cdot 33 + 22, \tag{b}$$

$$33 = 1 \cdot 22 + 11, \tag{c}$$

$$22 = 2 \cdot 11.$$

Solving the next to the last equation (c) for the gcd, 11, we have

$$11 = 33 - 22. \tag{d}$$

Then solving (b) for the previous remainder 22 and substituting in (d) gives us

$$11 = 33 - (154 - 4 \cdot 33)$$

$$11 = 5 \cdot 33 - 154. \tag{e}$$

Finally, we use (a) to solve for the first remainder 33 and substitute into (e), thus obtaining

$$11 = 5(803 - 5 \cdot 154) - 154$$

$$11 = 5 \cdot 803 - 26 \cdot 154. \tag{f}$$

Equation (f) shows that $x = 5$ and $y = -26$ is one solution for this problem. (A description of all the solutions is given in Problem 19 at the end of Section 1.6.)

Description of Euclid's Algorithm

Let a_0 and a_1 be integers with $0 < a_1 \le a_0$. We outline a procedure for obtaining $d = \gcd(a_0, a_1)$ and integers x and y such that $d = a_0 x + a_1 y$. It follows from the division algorithm that there exist q_1 and a_2 in $\mathbf{Z}$ with

$$a_0 = q_1 a_1 + a_2, \quad 0 \le a_2 < a_1.$$

If $a_2 \ne 0$, there are q_2 and a_3 in $\mathbf{Z}$ with

$$a_1 = q_2 a_2 + a_3, \quad 0 \le a_3 < a_2.$$

If $a_3 \ne 0$, there are q_3 and a_4 in $\mathbf{Z}$ with

$$a_2 = q_3 a_3 + a_4, \quad 0 \le a_4 < a_3.$$

In this process, the remainders $a_2, a_3, a_4, \ldots$ decrease at each step. Hence we must ultimately have a remainder that is 0, since otherwise there would be a subset $\{a_2, a_3, \ldots\}$ of $\mathbf{Z}^+$ without a least element, and this would contradict the well ordering principle. Letting a_{n+1} be the first zero remainder, we have

$$a_i = q_{i+1} a_{i+1} + a_{i+2} \quad \text{for} \quad i = 0, 1, \ldots, n-2, \tag{1}$$

$$a_{n-1} = q_n a_n. \tag{2}$$

Equation (2) shows that $a_n \mid a_{n-1}$. Since a_n is positive, this implies that $a_n = \gcd(a_{n-1}, a_n)$. Then it follows from the equations in display (1) above and Theorem 4 of Section 1.4 that

$$a_n = (a_{n-1}, a_n) = (a_{n-2}, a_{n-1}) = \cdots = (a_0, a_1).$$

Our remaining task is to show how to obtain integers x and y such that $a_n = \gcd(a_0, a_1) = x a_0 + y a_1$. This will be done by producing integers $x_{n-1}, x_{n-2}, \ldots, x_1, x_0$ such that

$$\gcd(a_0, a_1) = a_n = x_{n-1} a_{n-2} + x_{n-2} a_{n-1}$$

$$= x_{n-2} a_{n-3} + x_{n-3} a_{n-2} = \cdots = x_1 a_0 + x_0 a_1. \tag{3}$$

The last two of these integers, x_1 and x_0, will be the coefficients we seek for the desired linear combination

$$a_n = \gcd(a_0, a_1) = x_1 a_0 + x_0 a_1. \tag{4}$$

The equations in display (1), with i replaced by $j - 2$, can be rewritten as

$$a_j = a_{j-2} - q_{j-1} a_{j-1} \quad \text{for} \quad j = 2, 3, \ldots, n. \tag{5}$$

When $j = n$, this becomes $a_n = a_{n-2} - q_{n-1}a_{n-1}$. Therefore, we let $x_{n-1} = 1$ and $x_{n-2} = -q_{n-1}$ and have

$$a_n = x_{n-1}a_{n-2} + x_{n-2}a_{n-1}. \tag{6}$$

Display (5) with $j = n - 1$ gives us $a_{n-1} = a_{n-3} - q_{n-2}a_{n-2}$. Substituting this into (6) leads to

$$\begin{aligned} a_n &= x_{n-1}a_{n-2} + x_{n-2}(a_{n-3} - q_{n-2}a_{n-2}) \\ &= x_{n-2}a_{n-3} + (x_{n-1} - x_{n-2}q_{n-2})a_{n-2}. \end{aligned}$$

Therefore, we let $x_{n-3} = x_{n-1} - x_{n-2}q_{n-2}$ and have

$$a_n = x_{n-2}a_{n-3} + x_{n-3}a_{n-2}.$$

Similarly, we let

$$x_{n-4} = x_{n-2} - x_{n-3}q_{n-3}, \quad x_{n-5} = x_{n-3} - x_{n-4}q_{n-4}, \quad \ldots, \quad x_0 = x_2 - x_1q_1,$$

and these help us produce the remaining equations of display (3). Equating the first and last expressions in (3), we get

$$a_n = \gcd(a_0, a_1) = x_1a_0 + x_0a_1.$$

For later use, we summarize the procedure for obtaining the x_j by noting that

$$x_{n-1} = 1, \quad x_{n-2} = -q_{n-1}, \quad \text{and} \quad x_j = x_{j+2} - x_{j+1}q_{j+1}$$
$$\text{for} \quad j = n-3, n-4, \ldots, 0. \tag{7}$$

In this notation, some data of Example 1 can be arranged as in the accompanying table for gcd(803, 154).

Table for gcd(803, 154)

n	0	1	2	3	4
a_n	803	154	33	22	11
q_n		5	4	1	2
x_n	−26	5	−1	1	

The entries $a_0 = 803$ and $a_1 = 154$ are the integers whose gcd we seek. The remaining a_i and q_i are found from left to right, using the division algorithm. To do this, we divide a_0 by a_1 and let q_1 be the quotient and a_2 be the remainder; then we divide a_1 by a_2 and let q_2 be the quotient and a_3 be the remainder. This continues until we divide a_{n-1} by a_n and get 0 as a remainder (and q_n as the quotient). The x_i are obtained from *right to left*, using the equations of display (7) with $n = 3$, that is, the equations

$$x_3 = 1, \quad x_2 = -q_3 = -1, \quad x_1 = x_3 - q_2x_2 = 1 - 4(-1) = 5,$$

$$\text{and} \qquad x_0 = x_2 - q_1 x_1 = -1 - 5 \cdot 5 = -26.$$

The last two of these x's are the coefficients $x_1 = 5$ and $x_0 = -26$ in

$$11 = \gcd(a_0, a_1) = x_1 a_0 + x_0 a_1 = 5 \cdot 803 - 26 \cdot 154.$$

In the next example we show how to use Euclid's Algorithm to find integers x and y satisfying

$$ax + by = kd,$$

where a, b, and k are integers and $d = \gcd(a, b)$. Note that here we do not require that a and b be positive or that $k = 1$.

EXAMPLE 2 **Integer Solution of $ax + by = k \cdot \gcd(a, b)$**

Now we find integers x and y such that $803x - 154y = 33$. In Example 1 we found that $\gcd(803, 154) = 11$ and that

$$803 \cdot 5 + 154(-26) = 11.$$

Multiplying both sides by 3, we obtain

$$803 \cdot 15 + 154(-78) = 33 \quad \text{or} \quad 803 \cdot 15 - 154 \cdot 78 = 33.$$

Hence one solution is $x = 15$ and $y = 78$.

Clearly, the technique of Example 2 was successful only because the equation is of the form $ax + by = c$, with c an integral multiple of $\gcd(a, b)$; but Theorem 3 of Section 1.4 tells us that c must be of this form if the equation can be solved for integers x and y.

EXAMPLE 3 **$\gcd(1001, 791) = 1001x + 791y$**

Here we use Euclid's Algorithm to find $d = \gcd(1001, 791)$ and integers x and y such that $d = 1001x + 791y$. Repeated use of the division algorithm leads to

$$\begin{aligned} 1001 &= 1 \cdot 791 + 210 \\ 791 &= 3 \cdot 210 + 161 \\ 210 &= 1 \cdot 161 + 49 \\ 161 &= 3 \cdot 49 + 14 \\ 49 &= 3 \cdot 14 + 7 \\ 14 &= 2 \cdot 7 \end{aligned}$$

These equations show that $7 = \gcd(1001, 791)$ and give us the a_i and q_i in the accompanying table for $\gcd(1001, 791)$.

Table for gcd(1001, 791)

n	0	1	2	3	4	5	6
a_n	1001	791	210	161	49	14	7
q_n		1	3	1	3	3	2
x_n	-62	49	-13	10	-3	1	

Euclid of Alexandria
365?–275? B.C.

Although Euclid is known to be the author of a number of mathematical treatises and it is known that the most famous of these, the *Elements*, was written about 300 B.C., little is known of his life. The dates of his birth and death, his birthplace, and even his nationality are not known.

The *Elements* has certainly been one of the most influential books in Western history, a surprising distinction for something that is essentially a textbook of elementary mathematics. Over 1,000 editions have been issued, the first printed version having appeared in Venice in 1482, not many years after the Gutenberg Bible.

The subject matter of the *Elements* is usually thought to be geometry; however, other topics of elementary mathematics are treated as well, albeit from a geometric point of view. Books VII–IX treat the so-called arithmetic, which is, in fact, what the English call "higher arithmetic" and the Americans call "theory of numbers." Euclid's Algorithm appears at the beginning of Book VII. Book IX contains the proof that there exist an infinite number of primes, a formula for the sum of a geometric progression, and the formula for even perfect numbers (numbers, like 6, that equal the sum of their divisors other than the number itself: $6 = 1 + 2 + 3$).

Euclid probably did not discover many of the results in the *Elements*. Mainly he is credited with organizing existing knowledge, although some of the proofs are thought to be his. The style and organization of the *Elements* have been models for much mathematical writing and remain influential today.

We then obtain the x_i from right to left, using display (7) with $n = 6$; that is,

$$x_5 = 1, \quad x_4 = -q_5 = -3,$$
$$x_3 = x_5 - q_4 x_4 = 1 - 3(-3) = 10,$$
$$x_2 = x_4 - q_3 x_3 = -3 - 1 \cdot 10 = -13,$$
$$x_1 = x_3 - q_2 x_2 = 10 - 3(-13) = 49,$$
$$x_0 = x_2 - q_1 x_1 = -13 - 1 \cdot 49 = -62.$$

Using $x_1 = 49$ and $x_0 = -62$, we have $\gcd(a_0, a_1) = x_1 a_0 + x_0 a_1$, or

$$7 = 49 \cdot 1001 - 62 \cdot 791.$$

EXAMPLE 4 **Integer Solution of $ax + by = c$**
Here we find one of the solutions in integers u and v of the equation

$$1001u + 791v = 35. \tag{8}$$

We note from Example 3 that $\gcd(1001, 791) = 7$ and that

$$1001 \cdot 49 + 791(-62) = 7.$$

Multiplying each term of this equation by 5, we get

$$1001 \cdot 245 + 791(-310) = 35.$$

Hence one solution of equation (8) is $u = 245$, $v = -310$.

As in Example 2, this technique is applicable only because equation (8) is of the form $au + bv = c$, with c an integral multiple of gcd(a, b). If c were not of this form, it would follow from Theorem 3 of Section 1.4 that there were no solutions with u and v integers.

Problems related to the material of this section are among those at the close of Section 1.6.

1.6 Common Multiples

Let a and b be nonzero integers and let M be the set of all common integral multiples of a and b; that is, let M consist of the integers m such that both $a \mid m$ and $b \mid m$. Then $|ab|$ is in M; hence there is at least one positive integer in M.

It follows from Lemma 1 of Section 1.2 that M is closed under linear combinations and therefore is closed under subtraction. Then Theorem 3(e) of Section 1.3 tells us that M consists of all the multiples of the smallest positive integer m of M.

DEFINITION 1 **Least Common Multiple**

If a and b are nonzero integers, their least positive common integral multiple is called their **least common multiple (lcm)** and is denoted by $[a, b]$ or lcm$[a, b]$. Also, $[a, 0] = 0 = [0, b]$ for all integers a and b.

Some examples of least common multiples are $[6, 15] = 30$, $[7, 21] = 21$, $[10, 27] = 270$, and $[23, 0] = 0$. For all integers a, b, and c we have

$$[a, b] = [a, -b] = [-a, b] = [-a, -b],$$

$$[a, ac] = |ac|.$$

THEOREM 1 **How the lcm Is Least**

If $a \mid c$ and $b \mid c$, then lcm$[a, b] \mid c$.

Proof Let $a \mid c$, $b \mid c$, and $m =$ lcm$[a, b]$. If a or b is 0, then $0 \mid c$ implies that $c = 0$, and we also have $m = 0$. In this case $m \mid c$ since $0 \mid 0$. Now we can assume that $a \neq 0$ and $b \neq 0$. As stated in the first two paragraphs of this section, the set M of all common integral multiples of a and b is also the set of all integral multiples of $m = \text{lcm}[a, b]$. Since the hypothesis tells us that c is in M, this means that $m \mid c$. ■

Symbolically, Theorem 1 can be rewritten as

$$a \mid c \quad \text{and} \quad b \mid c \Rightarrow \text{lcm}[a, b] \mid c.$$

THEOREM 2 **lcm of Relatively Prime Integers**

Let $u \neq 0$ and $v \neq 0$. Then $\text{lcm}[u, v] = |uv|$ if and only if $\gcd(u, v) = 1$.

Proof Let $\gcd(u, v) = 1$. Then the Corollary to Theorem 2 of Section 1.4 tells us that there exist integers h and k such that

$$1 = hu + kv.$$

Let $m = \text{lcm}[u, v]$. Then $m = su = tv$ with s and t in $\mathbf{Z}$. We see that

$$s = s \cdot 1 = s(hu + kv) = h(su) + skv = htv + skv = (ht + sk)v; \text{ that is, } v \mid s.$$

Then $v \mid s$ implies that $(uv) \mid (su)$; that is, $(uv) \mid m$. Clearly, uv is a common integral multiple of u and v. Also, the least common multiple m is an integral divisor of all common multiples, and hence $m \mid (uv)$. Together, $(uv) \mid m$ and $m \mid (uv)$ imply that $m = \pm uv$. Since $m > 0$, this tells us that $m = |uv|$.

Conversely, let $\text{lcm}[u, v] = |uv|$. Let $\gcd(u, v) = d$. We wish to show that $d = 1$. By definition of d there exist e and f in $\mathbf{Z}$ such that $u = ed$ and $v = fd$. The hypotheses $u \neq 0$ and $v \neq 0$ imply that $e \neq 0$ and $f \neq 0$. Now

$$efd = uf = ve$$

shows that efd is a common integral multiple of u and v. This implies that $|uv| \leq |efd|$, since the hypothesis is that $|uv|$ is the least positive common multiple. Using $uv = edfd = efd^2$, we have

$$|uv| = |efd^2| \leq |efd|.$$

Since $e \neq 0$ and $f \neq 0$, it follows that the positive integer d is 1; that is, $\gcd(u, v) = 1$. ■

Let a and b be in $\mathbf{Z}^+$. Then $\gcd(a, b)$ can be calculated by Euclid's Algorithm, described in Section 1.5. Using Problems 17 and 18 of this section, one can obtain an analogous algorithm for calculating $\text{lcm}[a, b]$. And from the unique factorization results in Section 1.7, it follows that

$$\gcd(a, b) \cdot \text{lcm}[a, b] = ab$$

for all a and b in $\mathbf{Z}^+$.

Let $a_1, a_2, \ldots, a_n$ be nonzero integers. It can be shown that there exists a smallest positive integer m such that $a_i \mid m$ for each i; this m is called the *least common multiple* of the a_i and is denoted by $\text{lcm}[a_1, a_2, \ldots, a_n]$. Also, $\text{lcm}[a_1, a_2, \ldots, a_n]$ is defined to be 0 if at least one of the a_i is 0.

Problems

Problems 15 and 16 below are cited in Section 5.5.

1. For each of the following choices of c and d find (c, d), $[c, d]$, $(c, d)[c, d]$, and cd.
(a) $c = 8$ and $d = 12$;
(b) $c = 7$ and $d = 35$;
(c) $c = 3$ and $d = 14$.

2. Do as in Problem 1 for each of the following choices of c and d.
(a) $c = 12$ and $d = 0$;
(b) $c = 12$ and $d = 1$;
(c) $c = 12$ and $d = -24$.

3. Find $d = \gcd(864371, 735577)$ and integers x and y such that $864371x + 735577y = d$. [*Hint:* Use Euclid's Algorithm, as illustrated in Example 3, with $a_0 = 864371$ and $a_1 = 735577$.]

4. Find $g = \gcd(980051, 926213)$ and integers x and y such that $980051x + 926213y = g$.

5. Let d be as in Problem 3. Find one of the solutions in integers u and v of the equation $864371u + 735577v = 9d$.

6. Let x, y, and g be as in Problem 4. Find u and v, in terms of x and y, such that $980051u + 926213v = 14g$.

7. Find $\gcd(b, c)$ given that a, b, and c are integers with $\gcd(a, b) = 14$ and
(i) $a = 31b + c$; (ii) $a = 33b - c$; (iii) $a \equiv c \pmod{b}$;
(iv) $a \equiv -c \pmod{b}$.

8. Find $\gcd(a, b)$ for integers a, b, and c with $\gcd(b, c) = 15$ and
(i) $a = 41b + c$; (ii) $a = 43b - c$; (iii) $a \equiv c \pmod{b}$;
(iv) $a \equiv -c \pmod{b}$.

9. Let a, b, and c be integers, with $(a, c) = 1 = (b, c)$. Show that c and ab are relatively prime. [*Hint:* Use the existence of integers e, f, g, and h such that $ae + cf = 1$ and $bg + ch = 1$, and show that there exist integers x and y with $abx + cy = 1$.]

10. Let a, b, and c be integers, with $a \mid c$, $b \mid c$, and $(a, b) = 1$. Show that $(ab) \mid c$. [*Hint:* Explain why there exist integers h, k, u, and v such that $ah + bk = 1$, $c = ua$, and $c = vb$. Then find an integer m, in terms of h, k, u, and v, such that $c = mab$.]

11. Let $a = 39$, $b = 42$, and $c = 45$.
(i) Find (a, b) and (a, c).
(ii) Find $[(a, b), (a, c)]$.
(iii) Find $[b, c]$.
(iv) Find $(a, [b, c])$.
(v) Are the answers the same in parts (ii) and (iv)?

12. Do as in Problem 11, but now with $a = 21$, $b = 35$, and $c = 14$.

13. Let $a = 6$, $b = 15$, and $c = 35$.
(i) Find $[a, b]$.
(ii) Find $r = [[a, b], c]$.

(iii) Find $[b, c]$.
(iv) Find $s = [a, [b, c]]$.
(v) Does $r = s = \text{lcm}[a, b, c]$?

14. Repeat Problem 13, but now with $a = 14$, $b = 6$, and $c = 39$.

15. Given that a, b, and c are integers with $(a, b) = 1$ and $a \mid (bc)$, show that $a \mid c$.

16. Let a and b be relatively prime integers and let n be a positive integer. Prove by mathematical induction that a and b^n are relatively prime.

17. Let a and b be in $\mathbf{Z}^+$ and $m = \text{lcm}[a, b]$. Explain why the following are true.
(i) $a \mid m$ and $b \mid m$.
(ii) There exist r and s in $\mathbf{Z}^+$ such that $ra = m = sb$.
(iii) The smallest n in $\mathbf{Z}^+$ such that $b \mid (na)$ is

$$n = r = \frac{m}{a} = \frac{[a, b]}{a}.$$

18. Let a, b, c, q, and n be in $\mathbf{Z}^+$ with $a = qb + c$. Show the following.
(i) If $b \mid (na)$, then $b \mid (nc)$.
(ii) If $b \mid (nc)$, then $b \mid (na)$.
(iii) The smallest n in $\mathbf{Z}^+$ such that $b \mid (na)$ is also the smallest n in $\mathbf{Z}^+$ such that $b \mid (nc)$.
(iv) $\dfrac{\text{lcm}[a, b]}{a} = \dfrac{\text{lcm}[c, b]}{c}$. [*Hint:* Use Problem 17(iii).]

★ **19.** (i) Show that u and v are integers such that $803u + 154v = 0$ if and only if $u = 14t$ and $v = -73t$, with t an integer.
(ii) Show that x and y are integers such that $803x + 154y = 11$ if and only if $x = 5 + 14t$ and $y = -26 - 73t$, with t an integer.

1.7 Unique Factorization in Z

[Arithmetic] is one of the oldest branches, perhaps the very oldest branch, of human knowledge; and yet some of its most abstruse secrets lie close to its tritest truths. —**H. J. S. Smith**

One way in which the number 1 is special is that it is the only positive integer whose reciprocal is an integer. The other positive integers fall into one of two categories: the primes,

$$2, 3, 5, 7, 11, 13, 17, 19, 23, 29, 31, \ldots; \tag{P}$$

and the composites,

$$4 = 2 \cdot 2, \quad 6 = 2 \cdot 3, \quad 8 = 2 \cdot 2 \cdot 2, \quad 9 = 3 \cdot 3,$$

$$10 = 2 \cdot 5, \quad 12 = 2 \cdot 2 \cdot 3, \quad 14 = 2 \cdot 7, \quad \ldots. \tag{C}$$

Below we prove that every integer greater than 1 is either a prime or the product of a finite number of positive primes and that this factorization is unique, except for rearrangement of the factors. The factorizations in (C) are examples of such representations.

We begin the proof with some preliminary results.

LEMMA 1 Euclid's Lemma

Let a, b, and p be integers with p a prime. If $p \mid (ab)$, then either $p \mid a$ or $p \mid b$.

Proof Our approach is to assume that p is an integral divisor of ab but not of a and to show that these assumptions imply $p \mid b$.

The only integral divisors of the prime p are $-p$, -1, 1, and p. With the assumption that p is not an integral divisor of a, it follows that the only common divisors of a and p are 1 and -1.

Hence $\gcd(a, p) = 1$, and it follows from Corollary (b) to Theorem 2 of Section 1.4 that $ah + pk = 1$ with h and k in **Z**. The hypothesis $p \mid (ab)$ tells us that $ab = pt$ with t in **Z**. Then

$$b = b \cdot 1 = b(ah + pk) = (ba)h + bpk = pth + bpk = p(th + bk).$$

This shows that $p \mid b$, as desired. ∎

LEMMA 2 Generalized Euclid's Lemma

Let $p, a_1, a_2, \ldots, a_n$ be integers with p a prime. If $p \mid (a_1 a_2 \cdots a_n)$, then $p \mid a_i$ for at least one i in $\{1, 2, \ldots, n\}$.

The proof follows readily for $n \geq 2$ by mathematical induction and is left to the reader. (Note that Lemma 1 is the basis case $n = 2$ of Lemma 2.)

LEMMA 3 Let $p, u_1, u_2, \ldots, u_t$ be positive primes and let

$$p \mid (u_1 u_2 \cdots u_t). \tag{1}$$

Then p is equal to at least one of the u_i.

Proof It follows from the hypothesis (1) and Lemma 2 that $p \mid u_i$ for some i. But the only positive integral divisors of a positive prime u_i are 1 and u_i. Since $p \neq 1$, we have $p = u_i$. ∎

The results in the two following theorems are frequently called the *Fundamental Theorem of Arithmetic.*

THEOREM 1 Factorization into Primes

Every integer $n > 1$ is expressible as

$$n = p_1 p_2 \cdots p_r \tag{2}$$

with the p_i positive primes in **Z**.

Proof Let S be the set of all integers $n > 1$ such that n is not expressible in the form (2). We show that S is the empty set by assuming that it is not empty and obtaining a contradiction.

A nonempty set S of positive integers has a least integer m, by the well ordering principle. If m is a prime, it is of the form (2) with $r = 1$. Hence m must be composite, and there exist integers u and v such that

$$m = uv, \qquad 1 < u < m, \quad 1 < v < m.$$

Owing to the minimal nature of m, neither u nor v is in S. Hence both u and v are of the form (2); that is,

$$u = q_1 q_2 \cdots q_s, \qquad v = q_1' q_2' \cdots q_t'$$

with the q_i and q_j' positive primes. Then we have

$$m = uv = q_1 q_2 \cdots q_s q_1' q_2' \cdots q_t'.$$

This contradicts the assumption that m is not of the form (2). Hence S is empty and the result is proved. ■

THEOREM 2 Unique Factorization

Let n have the representations

$$p_1 p_2 \cdots p_r = q_1 q_2 \cdots q_s \tag{3}$$

with the p_i and q_j positive primes. Then $r = s$, and the p_i are the same as the q_j except, possibly, for the order of appearance of the primes.

Proof Let us assume that the result is not true and use this assumption to obtain a contradiction. Then there is a prime that appears more times as a p_i than as a q_j, or vice versa. Let p be such a prime and let it appear a times as a p_i and b times as a q_j. Since $a \neq b$, either $a > b$ or $a < b$. For definiteness, let $a > b$. (We may have $b = 0$.)

Now we cancel b of the factors p from each side of (3) and let $u_1 u_2 \cdots u_t$ be the product of the q_j remaining on the right side of (3) after this cancellation. Since $a > b$, there is at least one p on the new left side. This means that $p \mid (u_1 u_2 \cdots u_t)$. It then follows from Lemma 3 that p is one of the u's, and this contradicts the fact that all the q's that are equal to p were removed when we obtained the u's. This completes the proof of the Fundamental Theorem of Arithmetic. ■

Next we present a unique representation for integers $n > 1$ as products of powers of distinct primes. Let

$$n = q_1 q_2 \cdots q_u, \tag{4}$$

where the q's are positive primes. Let $p_1, p_2, \ldots, p_v$ be the distinct primes of (4) in increasing order (that is, with $p_1 < p_2 < \cdots < p_v$). Collecting the multiple appearance of a p among the q's of (4), we have

$$n = p_1^{e1}p_2^{e2} \cdots p_v^{e_v}, \tag{S}$$

with each e_i a positive integer. It follows from Theorem 2 that the primes p and their exponents in (S) are unique for a given integer $n > 1$.

DEFINITION 1

Standard Factorization

For n in $\{2, 3, 4, \ldots\}$, its **standard factorization** is

$$n = (p_1)^{e_1}(p_2)^{e_2} \cdots (p_v)^{e_v},$$

where the p_i are primes with $2 \le p_1 < p_2 < \cdots < p_v$ and the e_i are in $\mathbf{Z}^+$.

For example, the standard factorization of 720 is $2^4 \cdot 3^2 \cdot 5$, and the positive prime 23 is its own standard factorization. Note that this representation only applies to integers $n > 1$.

One could apply the above factorization to a negative integer $m < -1$ by using the fact that its negative $-m$ is a positive integer with $-m > 1$.

Problems

Problem 13(b) below is cited in Section 6.3.

1. Give the standard factorization for each of the following.
(i) 5040; (ii) 5040^2; (iii) 5040^3.

2. Give the standard factorization for each of the following.
(i) 2042040; (ii) 2042040^2; (iii) 2042040^3.

3. List all the positive integral divisors of each of the following.
(i) 8; (ii) 16; (iii) 32; (iv) 64.

4. Give the number of positive integral divisors of each of the following.
(i) 3^3; (ii) 3^4; (iii) 3^5; (iv) 3^6.

5. Let 2^a3^b be the standard factorization for n. Explain why the following are true.
(i) Let d be a positive integer. Then $d \mid n$ if and only if $d = 2^h3^k$ with h and k integers satisfying $0 \le h \le a$ and $0 \le k \le b$.
(ii) There are $(a + 1)(b + 1)$ positive integral divisors of n.

6. How many positive integral divisors are there of $72 = 2^3 \cdot 3^2$?

7. Let a, b, and c be positive integers. How many positive integral divisors are there of $2^a \cdot 5^b \cdot 11^c$?

8. Let $p_1^{e_1}p_2^{e_2} \cdots p_v^{e_v}$ be the standard factorization of n. Give the number of positive integral divisors of each of the following.
(i) n; (ii) n^2; (iii) n^3.

9. (a) Show that $a \mid b$ implies $a^2 \mid b^2$.

(b) Show that $a^2 \mid b^2$ implies $a \mid b$.
(c) Show that $a \mid b$ if and only if $a^3 \mid b^3$.

10. Let $x, y, z \in \mathbf{Z}^+$ and $x^2 + y^2 = z^2$.
(a) Show that $\gcd(x, y) = 1$ if and only if $\gcd(x, z) = 1$.
★ (b) Does $\gcd(x, y) = \gcd(x, z)$? Explain.

11. State a necessary and sufficient condition on the exponents $e_1, \ldots, e_v$ in the standard factorization (S) of n for the integer $n \geq 2$ to be the square of an integer m. [See Problems 1(ii) and 2(ii).]

12. Do the analogue of Problem 11 for cubes. [See Problems 1(iii) and 2(iii).]

13. Use unique factorization into primes to prove each of the following.
(a) There do not exist integers a and b with $a^2 = 30b^2$, and hence $\sqrt{30}$ is irrational.
(b) There are no integers c and d with $c^3 = 2d^3$, and hence $\sqrt[3]{2}$ is irrational.

14. Use unique factorization to prove the following.
(a) $\sqrt{68}$ is irrational; (b) $\sqrt[3]{24}$ is irrational.

15. Let $m \in \mathbf{Z}^+$. Prove that $\sqrt{m}$ is rational if and only if it is an integer.

★ **16.** Let $m, n \in \mathbf{Z}^+$. Prove that $\sqrt[n]{m}$ is rational if and only if it is an integer.

17. Let a and b be positive integers. Explain why $\gcd(a, b) = 1$ if and only if there is no positive prime p that appears in the standard factorizations of both a and b.

18. Let s and t be any integers. Explain why $\gcd(s, t) = 1$ if and only if there is no prime p with both $p \mid s$ and $p \mid t$.

19. Show that $\gcd(a, bc) = 1$ if and only if

$$\gcd(a, b) = 1 = \gcd(a, c).$$

20. Given that $\gcd(a, b) = 1$, use Problem 18 to explain why $\gcd(a^m, b^n) = 1$ for all positive integers m and n.

21. Let $a = 2^{30} \cdot 5^{21} \cdot 19 \cdot 23^3$ and $b = 2^6 \cdot 3 \cdot 7^4 \cdot 11^2 \cdot 19^5 \cdot 23^7$. Give the standard factorization for $\gcd(a, b)$ and $\operatorname{lcm}[a, b]$.

22. (a) Let $e = 2^{10} \cdot 3^7 \cdot 13^6 \cdot 29^8$ and $f = 2^9 \cdot 5^3 \cdot 11^4 \cdot 13^8$. Give the standard factorization for $\gcd(e, f)$ and $\operatorname{lcm}[e, f]$.
(b) Let each of a and b be a positive integer greater than 1. Explain how to obtain the standard factorization for $\gcd(a, b)$ and $\operatorname{lcm}[a, b]$ from those for a and b.

23. Let m and n be positive integers. Use unique factorization to explain why

$$\gcd(m, n) \cdot \operatorname{lcm}[m, n] = mn.$$

24. Use the preceding problem to explain why $\gcd(a, b) \cdot \operatorname{lcm}[a, b] = |ab|$ for all integers a and b.

25. Given that 307 is a prime and that n is an integer such that

$$(1 \cdot 2 \cdot 3 \cdots 99)n = 307 \cdot 306 \cdot 305 \cdots 209,$$

use Lemma 2 to show that $307 \mid n$.

26. The binomial coefficients in the expansion

$$(x + 1)^n = \binom{n}{n}x^n + \binom{n}{n-1}x^{n-1} + \cdots + \binom{n}{1}x + \binom{n}{0}$$

are known to be positive integers and to be given by

$$\binom{n}{k} = \frac{n(n-1)(n-2)\cdots(n-k+1)}{1 \cdot 2 \cdot 3 \cdots k}.$$

Prove that $p \left| \binom{p}{k} \right.$ for $0 < k < p$, when p is a prime.

27. How many pairs $\{s, t\}$ of positive integers satisfy both $st = 55440$ and $\gcd(s, t) = 1$?

28. Let the standard factorization of n be

$$(p_1)^{e_1}(p_2)^{e_2} \cdots (p_r)^{e_r}.$$

How many pairs $\{s, t\}$ of positive integers satisfy both $st = n$ and $\gcd(s, t) = 1$? (The answer depends only on the number r of positive prime divisors of n.)

Review Problems

1. (i) Find $d = \gcd(8051, 8633)$, using Euclid's Algorithm.
(ii) Check that the d of part (i) is a common divisor of 8051 and 8633.
(iii) Find integers x and y such that $-20 < x < 0$ and

$$d = 8051x + 8633y.$$

2. Find $(751903, 800881)$ and check that it is a common divisor.

3. If $n \equiv 1 \pmod 3$, is $n^2 \equiv 1 \pmod 3$? In other words, if n is in $1 + 3\mathbf{Z}$, is n^2 in $1 + 3\mathbf{Z}$?

4. Given that $n \equiv 2 \pmod 3$, explain why $n^2 \equiv 1 \pmod 3$. [*Hint:* Let $n = 3q + 2$, and find n^2.]

5. Explain why $\gcd(n, n + 4)$ is in $\{1, 2, 4\}$ for all n in $\mathbf{Z}$.

6. Let $30 = 84h + bk$ with b, h, and k integers. What are the possible values of $\gcd(84, b)$?

7. Let S be the set of all common integral multiples of 4 and 6.
(i) Show that S is closed under subtraction.
(ii) What is the smallest positive integer t in S?
(iii) Does $S = t\mathbf{Z}$? Explain.

8. Give an example of positive integers a, b, and c such that $a \mid (bc)$ but neither $a \mid b$ nor $a \mid c$.

9. Let a and b be integers such that $23 \mid (ab)$. Use the fact that 23 is a prime to show that either $23^2 \mid a^2$ or $23^2 \mid b^2$.

10. Let $n \in \mathbf{Z}^+$ and $r = \sqrt[3]{n}$. Use unique factorization into primes to show that r is rational if and only if r is an integer.

CHAPTER 2

GROUPS

Wherever groups disclosed themselves, or could be introduced, simplicity crystallized out of comparative chaos.

—Eric Temple Bell

Just as M. Jourdain in Molière's *Bourgeois Gentilhomme* was surprised to discover that he had been speaking prose without knowing it, we find that we have used groups all of our lives without being aware of them. The most basic number systems are examples of groups, and we all learn to deal with these early on. The history of groups indicates that many properties of them were known and used before groups were properly defined and fully understood.

In algebra the problem that eventually suggested a study of groups arose around 1540, when mathematicians tried to find analogues to the quadratic formula (which was known to the Babylonians) for finding roots to polynomial equations of degree higher than two. In 1770, Joseph Lagrange tried to extend the successful work of Tartaglia and Ferrari with third-degree and fourth-degree equations in order to find a formula for solving fifth-degree equations. But he failed. Subsequent efforts of Ruffini, Abel, and Galois eventually succeeded in showing that these analogues cannot exist; and the machinery for this proof involved groups.

But groups were finding their way into mathematics in other contexts as well. Karl Friedrich Gauss developed a number of properties of groups, without using the terminology, in order to treat congruences and several other number theoretic concepts in his most important work, the *Disquisitiones Arithmeticae* of 1801. Later in the nineteenth century, Felix Klein set up his Erlanger Programm to describe various geometries using the properties left unchanged under specific groups of transformations. By the twentieth century, groups had arisen in fields as disparate as physics (quantum mechanics, elementary particle theory), chemistry (crystallography, molecular structure), architecture (the symmetries of the Moorish decoration at the Alhambra in Granada, for example) and art (the graphic work of the Dutch artist, M. C. Escher). Groups have become a classic example of the universality of mathematical ideas — unifying concepts that clarify the relationship between seemingly entirely different things. They reveal the power of abstraction. Using the structure alone, one learns about phenomena in various disciplines, all at once. The theory is a remarkable demonstration of the power (and the beauty) of mathematics.

We start with a study of permutation groups, since these are the groups that were important in the study of algebraic equations — the area that prompted the earliest developments of group theory in the late eighteenth century.

2.1 Permutations on a Finite Set

Here we begin the study of groups of permutations on a set X. We will see immediately that these groups have the advantage of being concrete, and we will find in Theorem 4 of Section 3.14 that they are also very general. In this chapter, concreteness is achieved by letting X be, for some choice of n, the set $\mathbf{X_n}$ of the first n positive integers. Then in Definition 5 of Section 3.1, we will extend the following definition to arbitrary sets X.

DEFINITION 1

Permutation on the Set $X_n = \{1, 2, \ldots, n\}$

A **permutation** on $\mathbf{X_n} = \{1, 2, \ldots, n\}$ is an assignment to each i in $\mathbf{X_n}$ of a unique a_i in $\mathbf{X_n}$ so that each x in $\mathbf{X_n}$ appears once and only once in the listing

$$a_1, a_2, \ldots, a_n.$$

We may indicate that a permutation θ on $\mathbf{X_n}$ assigns a_i to i, by writing $\theta : i \mapsto a_i$ (which is read as "theta sends i to a_i" or as "under theta, i goes to a_i"). This same fact can be indicated by writing $\theta(i) = a_i$ (which is read as "θ of i equals a_i"). These lead to the following two notations for a permutation θ on $\mathbf{X_n}$:

Arrow Form $\quad \theta : 1 \mapsto a_1,\ 2 \mapsto a_2, \ldots, n \mapsto a_n.$

Function Form $\quad \theta(1) = a_1,\ \theta(2) = a_2, \ldots, \theta(n) = a_n.$

The following notation for this permutation will be helpful in the proofs given in the next section:

2-Row Form $$\theta = \begin{pmatrix} 1 & 2 & \ldots & n \\ a_1 & a_2 & \ldots & a_n \end{pmatrix}.$$

The 2-row form for θ shows that θ sends i to a_i by placing a_i on the second row under the i on the first row.

Next we illustrate these representations and a related tabular form, and then we describe a pictorial representation. Each of these forms will be seen to have advantages in some situations.

EXAMPLE 1 Let θ be the permutation on $\mathbf{X_6} = \{1, 2, 3, 4, 5, 6\}$ whose arrow form is

$$\theta: 1 \mapsto 3, \quad 2 \mapsto 2, \quad 3 \mapsto 6, \quad 4 \mapsto 5, \quad 5 \mapsto 4, \quad 6 \mapsto 1. \tag{1}$$

This may be read as "θ sends 1 to 3, 2 to itself, 3 to 6, 4 to 5, 5 to 4, and 6 to 1." In function form, θ is given by

$$\theta(1) = 3, \quad \theta(2) = 2, \quad \theta(3) = 6, \quad \theta(4) = 5, \quad \theta(5) = 4, \quad \theta(6) = 1. \tag{2}$$

This θ may also be given by the table

x	1	2	3	4	5	6
$y = \theta(x)$	3	2	6	5	4	1

(3)

The permutation θ characterized by table (3) is not changed if we replace the letters x and y with some other letters. A form of expressing θ that does not use letters for the variables is the 2-row form

$$\theta = \begin{pmatrix} 1 & 2 & 3 & 4 & 5 & 6 \\ 3 & 2 & 6 & 5 & 4 & 1 \end{pmatrix}. \tag{4}$$

The permutation θ is also unaltered by any rearrangement of the columns in representation (4). For example, this θ may be given by

$$\theta = \begin{pmatrix} 6 & 2 & 1 & 5 & 4 & 3 \\ 1 & 2 & 3 & 4 & 5 & 6 \end{pmatrix},$$

since in this 2-row form we also have a_i under i whenever θ sends i to a_i.

We next describe a pictorial representation for a permutation.

DEFINITION 2

Diagram for a Permutation

A **diagram** for a permutation θ on $\mathbf{X_n} = \{1, 2, \ldots, n\}$ is a figure in which each integer x in $\mathbf{X_n}$ is represented by a dot and there is an arc (sometimes straight) from x to $\theta(x)$ with an arrowhead pointing toward $\theta(x)$.

The diagram for the permutation θ of Example 1 is given in Figure 2.1.

Permutations α and β on $\mathbf{X_n} = \{1, 2, \ldots, n\}$ are *equal* when each x in $\mathbf{X_n}$ is assigned the same integer in $\mathbf{X_n}$ by α and by β; that is, $\alpha = \beta$ if and only if

FIGURE 2.1

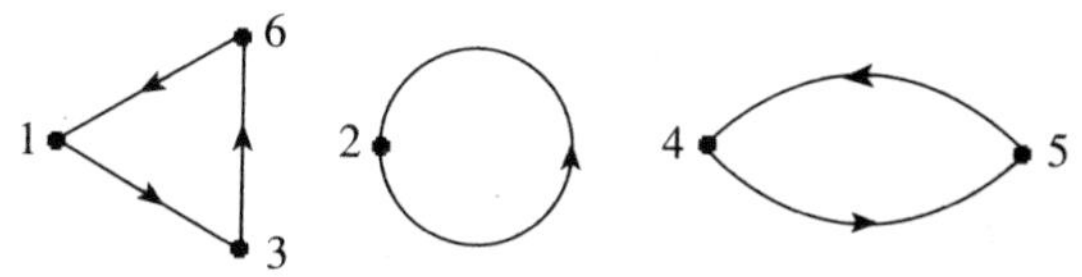

$\alpha(x) = \beta(x)$ for every x in $\mathbf{X_n}$. Hence $\alpha \neq \beta$ if and only if there is at least one c in $\mathbf{X_n}$ with $\alpha(c) \neq \beta(c)$.

Problems

Problem 9 below is cited in Section 2.2.

1. For each of the following parts, tell whether it specifies θ as a permutation on $\mathbf{X_5} = \{1, 2, 3, 4, 5\}$. If not, explain how Definition 1 is not satisfied.
(a) θ: $1 \mapsto 1, 2 \mapsto 3, 3 \mapsto 4, 4 \mapsto 5, 5 \mapsto 1$.
(b) θ: $1 \mapsto 5, 2 \mapsto 1, 3 \mapsto 2, 4 \mapsto 3, 5 \mapsto 4$.
(c) θ: $1 \mapsto 1, 2 \mapsto 2, 3 \mapsto 3, 4 \mapsto 4, 5 \mapsto 5$.

2. Do as in Problem 1 for the following parts.
(a) θ: $1 \mapsto 2, 2 \mapsto 3, 3 \mapsto 4, 4 \mapsto 5, 5 \mapsto 1$.
(b) θ: $1 \mapsto 2, 2 \mapsto 1, 3 \mapsto 4, 4 \mapsto 3, 5 \mapsto 4$.
(c) θ: $1 \mapsto 2, 2 \mapsto 1, 3 \mapsto 3, 4 \mapsto 5, 5 \mapsto 4$.

3. Express each of the two permutations on $\mathbf{X_2} = \{1, 2\}$ in the 2-row form

$$\begin{pmatrix} 1 & 2 \\ a & b \end{pmatrix}.$$

4. Express each of the six permutations on $\mathbf{X_3} = \{1, 2, 3\}$ in the arrow form

$$1 \mapsto a,\ 2 \mapsto b,\ 3 \mapsto c.$$

5. How many permutations are there on $\mathbf{X_n} = \{1, 2, \ldots, n\}$ for the following choices of n?
(a) $n = 4$; (b) $n = 5$.

6. How many permutations are there on $\mathbf{X_6} = \{1, 2, 3, 4, 5, 6\}$?

7. How many permutations θ on $\mathbf{X_4} = \{1, 2, 3, 4\}$ have $\theta(4) = 4$?

8. How many permutations θ on $\mathbf{X_5} = \{1, 2, 3, 4, 5\}$ have $\theta(1) = 1$?

9. Let α and β be the permutations on $\{1, 2, 3\}$ given by

$$\alpha:\ 1 \mapsto 2,\ 2 \mapsto 1,\ 3 \mapsto 3;$$

$$\beta:\ 1 \mapsto 2,\ 2 \mapsto 3,\ 3 \mapsto 1.$$

(i) Find the permutation γ on $\{1, 2, 3\}$ such that $\gamma(x) = z$ whenever $\alpha(x) = y$ and $\beta(y) = z$.
(ii) Find the permutation δ on $\{1, 2, 3\}$ such that $\delta(x) = z$ whenever $\beta(x) = y$ and $\alpha(y) = z$.

10. Let α and β be the permutations on $\{1, 2, 3, 4, 5\}$ given by

$$\alpha:\ 1 \mapsto 2,\ 2 \mapsto 3,\ 3 \mapsto 1,\ 4 \mapsto 5,\ 5 \mapsto 4;$$
$$\beta:\ 1 \mapsto 3,\ 2 \mapsto 1,\ 3 \mapsto 2,\ 4 \mapsto 4,\ 5 \mapsto 5.$$

Find r, s, t, u, and v, so that

$$\gamma:\ 1 \mapsto r,\ 2 \mapsto s,\ 3 \mapsto t,\ 4 \mapsto u,\ 5 \mapsto v.$$

has $\gamma(x) = z$ whenever $\alpha(x) = y$ and $\beta(y) = z$.

Évariste Galois
1811–1832

The story of the life of Galois is certainly one of the most dramatic, tragic, and often told in the history of mathematics. He died at the age of twenty after being wounded in a duel. Fortunately for mathematics, the night before the duel he wrote down his principal mathematical results in a letter to a friend. His work was edited by Liouville and published in 1846. Only then did Galois's ideas become known to the mathematical community at large. Another of his champions was the German mathematician Leopold Kronecker.

Prior to his death, Galois's life was a long series of tragedies, disappointments, and frustrations. He was rejected by the École Polytechnique twice, probably because of unorthodox answers to the entrance examination questions. When he presented a preliminary version of some important results to Cauchy for presentation to the Académie des Sciences, Cauchy lost the paper. Galois's father committed suicide. Shortly after he gave a copy of his work to Fourier in an attempt to win the mathematics prize of the Académie, Fourier died and Galois's paper was lost. Another paper submitted to Poisson was returned because Poisson could not understand it. After gaining admission to the École Normale, Galois was expelled in his second year for writing a letter attacking the king. He finally became an active rebel against the monarchy, was jailed twice, and then was challenged and killed in a duel. His life, though short, was eventful.

Some of the more dramatic facts of Galois's life and the accounts of them have been disputed. (See the article by Rothman in the Annotated Bibliography.) But the fact remains that he did some of the most strikingly original work in mathematics during his brief life, and he deserves a place of high honor in history. He showed that the solvability of an algebraic equation by rational operations and extraction of roots is equivalent to the solvability of an associated group of permutations of its roots. The problem of the insolvability of the fifth-degree polynomial equation had been settled by Abel, but Galois's methods were of great generality and stimulated much of the development of algebra in the nineteenth century. And all of this was accomplished in surprisingly few words. His complete works, published in Paris in 1897, consist of 61 pages. And that includes letters.

11. Let

$$\alpha = \begin{pmatrix} 1 & 2 & 3 & 4 & 5 & 6 \\ 6 & 4 & 1 & 2 & 5 & 3 \end{pmatrix}.$$

Rewrite α with the columns rearranged so that

$$\alpha = \begin{pmatrix} 1 & a & b & 2 & c & d \\ a & b & 1 & c & 2 & d \end{pmatrix}.$$

12. Let

$$\beta = \begin{pmatrix} 1 & 2 & 3 & 4 & 5 & 6 & 7 \\ 5 & 3 & 7 & 6 & 2 & 1 & 4 \end{pmatrix}.$$

Rewrite β in the form

$$\beta = \begin{pmatrix} 1 & a & b & c & d & e & f \\ a & b & c & d & e & f & 1 \end{pmatrix}.$$

13. Give the diagram for the permutation α of Problem 11.

14. Give the diagram for the permutation β of Problem 12.

15. Let α be the permutation of Problem 11, and let the sequence $u_1, u_2, u_3, \ldots$ be determined by

$$u_1 = 1, \quad u_2 = \alpha(u_1), \quad u_3 = \alpha(u_2), \quad u_4 = \alpha(u_3), \quad \ldots .$$

(i) Which integer in $\mathbf{X}_6 = \{1, 2, 3, 4, 5, 6\}$ is the first to be repeated in the sequence $u_1, u_2, \ldots$?

(ii) Which integers in $\mathbf{X}_6$ appear in the sequence $u_1, u_2, \ldots$?

(iii) Does $u_h = u_k$ when $h \equiv k \pmod 3$?

(iv) Which integers appear in the sequence $v_1, v_2, \ldots$ that has $v_1 = 2$ and $v_{j+1} = \alpha(v_j)$ for $j = 1, 2, 3, \ldots$?

(v) Does $v_h = v_k$ when $h \equiv k \pmod 2$?

16. Let β be as in Problem 12. Let a sequence $w_1, w_2, w_3, \ldots$ have $w_1 = 1$ and $w_{j+1} = \beta(w_j)$ for $j = 1, 2, 3, \ldots$.

(i) Which integer is the first to be repeated in the sequence, and which integers appear among the w's?

(ii) What is the smallest positive integer q such that $w_h = w_k$ when $h \equiv k \pmod q$?

(iii) How do the answers to (i) and (ii) change if $w_1 = 3$?

17. Write in the form $1 \mapsto a$, $2 \mapsto b$, $3 \mapsto c$, $4 \mapsto d$ each of the 10 permutations α on $\{1, 2, 3, 4\}$ such that $\alpha(x) = y$ implies that $\alpha(y) = x$, that is, $\alpha[\alpha(x)] = x$ for $x = 1, 2, 3, 4$.

18. How many permutations β are there on $\mathbf{X}_5 = \{1, 2, 3, 4, 5\}$ such that $\beta(x) = y$ implies that $\beta(y) = x$; that is $\beta[\beta(x)] = x$ for all x in $\mathbf{X}_5$?

2.2 Multiplication of Permutations

Next we use composition of functions to define the product of permutations α and β on $\mathbf{X}_n = \{1, 2, \ldots, n\}$, and then we prove the basic properties of this operation. In later sections, we will derive many other properties from those presented here.

Let α and β be permutations on $\mathbf{X}_n = \{1, 2, \ldots, n\}$. We write

$$a \overset{\alpha}{\mapsto} b$$

to indicate that $\alpha(a) = b$ and compress

$$a \overset{\alpha}{\mapsto} b \quad \text{and} \quad b \overset{\beta}{\mapsto} c$$

into

$$a \overset{\alpha}{\mapsto} b \overset{\beta}{\mapsto} c.$$

THEOREM 1 **Closure Under Composition**

Let α and β be permutations on $\mathbf{X_n} = \{1, 2, \ldots, n\}$, and let $\gamma(x) = z$ when $\alpha(x) = y$ and $\beta(y) = z$. Then γ is a permutation on $\mathbf{X_n}$.

Proof Let

$$\alpha = \begin{pmatrix} 1 & 2 & \ldots & n \\ a_1 & a_2 & \ldots & a_n \end{pmatrix}.$$

By definition of a permutation, every one of the numbers $1, 2, \ldots, n$ appears exactly once in the bottom row $a_1, a_2, \ldots, a_n$. We can therefore arrange the columns in the representation for β so that

$$\beta = \begin{pmatrix} a_1 & a_2 & \ldots & a_n \\ b_1 & b_2 & \ldots & b_n \end{pmatrix}.$$

In this form, the top row for β is the bottom row of α. Since β is a permutation, each integer in $\mathbf{X_n}$ occurs exactly once among the b's. Hence

$$\begin{pmatrix} 1 & 2 & \ldots & n \\ b_1 & b_2 & \ldots & b_n \end{pmatrix}$$

is a permutation on $\mathbf{X_n}$. This permutation is γ, since γ is such that

$$k \overset{\alpha}{\mapsto} a_k \overset{\beta}{\mapsto} b_k \text{ implies } k \overset{\gamma}{\mapsto} b_k.$$

It follows that γ is a permutation on $\mathbf{X_n}$. ■

DEFINITION 1

Multiplication of Permutations

If α and β are permutations on $\mathbf{X_n} = \{1, 2, \ldots, n\}$, the **product** (or **composite**) $\alpha\beta$ is the permutation γ such that $\gamma(x) = z$ when $\alpha(x) = y$ and $\beta(y) = z$.

The product $\beta\alpha$ (taken in the other order) need not equal $\alpha\beta$, as we can see from Problem 9 of Section 2.1 or the following example.

EXAMPLE 1 Here we find $\alpha\beta$ and $\beta\alpha$, given that

$$\alpha = \begin{pmatrix} 1 & 2 & 3 \\ 1 & 3 & 2 \end{pmatrix} \quad \text{and} \quad \beta = \begin{pmatrix} 1 & 2 & 3 \\ 3 & 1 & 2 \end{pmatrix}.$$

Rewriting β so that its top row is the same as the bottom row of α and using the technique of the proof of Theorem 1, we have

$$\alpha\beta = \begin{pmatrix} 1 & 2 & 3 \\ 1 & 3 & 2 \end{pmatrix}\begin{pmatrix} 1 & 3 & 2 \\ 3 & 2 & 1 \end{pmatrix} = \begin{pmatrix} 1 & 2 & 3 \\ 3 & 2 & 1 \end{pmatrix}.$$

Similarly,

$$\beta\alpha = \begin{pmatrix} 1 & 2 & 3 \\ 3 & 1 & 2 \end{pmatrix}\begin{pmatrix} 3 & 1 & 2 \\ 2 & 1 & 3 \end{pmatrix} = \begin{pmatrix} 1 & 2 & 3 \\ 2 & 1 & 3 \end{pmatrix}.$$

The next result will be helpful in the work that follows.

LEMMA 1 If

$$\alpha = \begin{pmatrix} u_1 & u_2 & \dots & u_n \\ v_1 & v_2 & \dots & v_n \end{pmatrix} \quad \text{and} \quad \beta = \begin{pmatrix} v_1 & v_2 & \dots & v_n \\ w_1 & w_2 & \dots & w_n \end{pmatrix}$$

are permutations on $\{1, 2, \dots, n\}$, their product, in that order, is

$$\alpha\beta = \begin{pmatrix} u_1 & \dots & u_n \\ v_1 & \dots & v_n \end{pmatrix}\begin{pmatrix} v_1 & \dots & v_n \\ w_1 & \dots & w_n \end{pmatrix} = \begin{pmatrix} u_1 & u_2 & \dots & u_n \\ w_1 & w_2 & \dots & w_n \end{pmatrix}.$$

Proof By definition of the product of permutations,

$$u \overset{\alpha}{\mapsto} v \overset{\beta}{\mapsto} w \text{ implies } u \overset{\alpha\beta}{\mapsto} w.$$

This proves the lemma. ■

DEFINITION 2

Set S_n of Permutations on $\{1, 2, \dots, n\}$

The set of all permutations on $\{1, 2, \dots, n\}$ is denoted by $\mathbf{S_n}$.

The following definition introduces a permutation whose main property is given in Theorem 2 below.

DEFINITION 3

Identity in S_n

The permutation $\varepsilon = \begin{pmatrix} 1 & 2 & \dots & n \\ 1 & 2 & \dots & n \end{pmatrix}$ is called the **identity** permutation on $\mathbf{X_n}$.

Thus the identity permutation ε on $\mathbf{X_n}$ sends i to itself for each i in $\mathbf{X_n}$.

THEOREM 2 Identity Permutation

Let $\varepsilon = \begin{pmatrix} 1 & 2 & \dots & n \\ 1 & 2 & \dots & n \end{pmatrix}$. Then $\alpha\varepsilon = \alpha = \varepsilon\alpha$ for all permutations α on $\mathbf{X_n} = \{1, 2, \dots, n\}$.

Proof Let $\alpha = \begin{pmatrix} 1 & 2 & \dots & n \\ a_1 & a_2 & \dots & a_n \end{pmatrix}$. It follows immediately from Lemma 1 that $\varepsilon\alpha = \alpha$. Rearranging the columns of ε so that its top row is the bottom row of α and using the lemma, we also find that

$$\alpha\varepsilon = \begin{pmatrix} 1 & 2 & \dots & n \\ a_1 & a_2 & \dots & a_n \end{pmatrix}\begin{pmatrix} a_1 & a_2 & \dots & a_n \\ a_1 & a_2 & \dots & a_n \end{pmatrix} = \begin{pmatrix} 1 & 2 & \dots & n \\ a_1 & a_2 & \dots & a_n \end{pmatrix} = \alpha.$$

This completes the proof. ∎

THEOREM 3 Inverse of a Permutation

For any permutation

$$\alpha = \begin{pmatrix} 1 & 2 & \dots & n \\ a_1 & a_2 & \dots & a_n \end{pmatrix},$$

the permutation

$$\beta = \begin{pmatrix} a_1 & a_2 & \dots & a_n \\ 1 & 2 & \dots & n \end{pmatrix}$$

is such that $\alpha\beta = \varepsilon = \beta\alpha$, where ε is the identity permutation on $\{1, 2, \dots, n\}$.

This follows readily from Lemma 1; the proof is left to the reader.

DEFINITION 4

Inverse α^{-1} of a Permutation α

For each α in $\mathbf{S_n}$, the permutation β in $\mathbf{S_n}$ (given in Theorem 3) such that $\alpha\beta = \varepsilon = \beta\alpha$ is called the **inverse** of α and is denoted as α^{-1}.

THEOREM 4 Associativity

For all permutations α, β, and γ on $\{1, 2, \dots, n\}$,

$$(\alpha\beta)\gamma = \alpha(\beta\gamma).$$

Proof Let

$$\alpha = \begin{pmatrix} 1 & 2 & \dots & n \\ a_1 & a_2 & \dots & a_n \end{pmatrix}.$$

Then we can write β and γ as

$$\beta = \begin{pmatrix} a_1 & a_2 & \dots & a_n \\ b_1 & b_2 & \dots & b_n \end{pmatrix}, \quad \gamma = \begin{pmatrix} b_1 & b_2 & \dots & b_n \\ c_1 & c_2 & \dots & c_n \end{pmatrix}.$$

Using Lemma 1, we have

$$\alpha\beta = \begin{pmatrix} 1 & 2 & \dots & n \\ b_1 & b_2 & \dots & b_n \end{pmatrix} \text{ and then } (\alpha\beta)\gamma = \begin{pmatrix} 1 & 2 & \dots & n \\ c_1 & c_2 & \dots & c_n \end{pmatrix}.$$

Similarly,

$$\beta\gamma = \begin{pmatrix} a_1 & a_2 & \dots & a_n \\ c_1 & c_2 & \dots & c_n \end{pmatrix} \text{ and } \alpha(\beta\gamma) = \begin{pmatrix} 1 & 2 & \dots & n \\ c_1 & c_2 & \dots & c_n \end{pmatrix}.$$

Hence $(\alpha\beta)\gamma = \alpha(\beta\gamma)$, as desired. ∎

Problems

Problems 1 and 3 below are cited in Section 2.3.

1. Let ε and θ be the permutations on $\{1, 2\}$ with

$$\varepsilon\colon 1 \mapsto 1,\ 2 \mapsto 2 \quad \text{and} \quad \theta\colon 1 \mapsto 2,\ 2 \mapsto 1.$$

Find each of the products $\varepsilon\varepsilon$, $\varepsilon\theta$, $\theta\varepsilon$, and $\theta\theta$ in the form $1 \mapsto a$, $2 \mapsto b$.

2. Let $\varepsilon = \begin{pmatrix} 1 & 2 & 3 & 4 \\ 1 & 2 & 3 & 4 \end{pmatrix}$, $\alpha = \begin{pmatrix} 1 & 2 & 3 & 4 \\ 2 & 1 & 4 & 3 \end{pmatrix}$, $\beta = \begin{pmatrix} 1 & 2 & 3 & 4 \\ 3 & 4 & 1 & 2 \end{pmatrix}$, and $\gamma = \alpha\beta$.

Complete the following multiplication table of products of permutations in $\{\varepsilon, \alpha, \beta, \gamma\}$; the value of the product xy should be the entry common to the row for x and column for y.

$\cdot$	ε	α	β	γ
ε	ε	α	β	γ
α	α	ε	γ	β
β				
γ				

3. Let $\varepsilon = \begin{pmatrix} 1 & 2 & 3 \\ 1 & 2 & 3 \end{pmatrix}$, $\rho = \begin{pmatrix} 1 & 2 & 3 \\ 2 & 3 & 1 \end{pmatrix}$ and $\phi = \begin{pmatrix} 1 & 2 & 3 \\ 1 & 3 & 2 \end{pmatrix}$.

(i) Find $\rho^2 = \rho\rho$.
(ii) Show that $\phi\rho = \rho^2\phi$.
(iii) Show that $\{\varepsilon, \rho, \rho^2, \phi, \rho\phi, \rho^2\phi\}$ is the set $\mathbf{S}_3$ of all permutations on $\{1, 2, 3\}$.

Baron Augustin-Louis Cauchy
1789–1857

Cauchy was one of the greatest and most prolific of mathematicians, second only to Euler in the volume of his mathematical work. Like Euler, he wrote not only research papers and books but also textbooks. His *Cours d'Analyse de l'École Polytechnique* is a classic text that provided the model for calculus texts for many years. Unlike Euler, however, he was very concerned with rigor. Our present-day treatment of limit and continuity in calculus courses is essentially that of Cauchy. Although Cauchy's work was primarily in analysis, he also made significant contributions to algebra, the theory of numbers (he first proved the conjecture of Fermat, partially proved by Gauss, that every positive integer is the sum of at most three triangular numbers, four squares, five pentagonal numbers, and so on), and even geometry (he gave the first generally accepted proof of Euler's formula for polyhedra, $V - E + F = 2$, a formula actually discovered by Descartes). Cauchy was the first to develop the theory of permutation groups, viewed as something independent of applications to the solution of polynomial equations.

Cauchy shared with Newton a tendency to write on religion as well as mathematics. He was very conservative in matters of religion and politics and was in voluntary exile for a number of years after the exile of King Charles X, to whom he was devoted. In 1830, unwilling to swear an oath to Charles's successor, Louis-Philippe, he was forced to leave France, since he could not hold an academic post there. Not until 1848, under Louis Napoleon, was he awarded a professorship at the École Polytechnique in Paris.

(iv) Complete the following multiplication table for $\mathbf{S}_3$.

$\cdot$	ε	ρ	ρ^2	ϕ	$\rho\phi$	$\rho^2\phi$
ε	ε	ρ	ρ^2	ϕ	$\rho\phi$	$\rho^2\phi$
ρ	ρ	ρ^2	ε	$\rho\phi$	$\rho^2\phi$	ϕ
ρ^2	ρ^2			$\rho^2\phi$		
ϕ						
$\rho\phi$						
$\rho^2\phi$						

4. (i) Does $\alpha\beta = \beta\alpha$ for all permutations α and β on $\{1, 2\}$? (See Problem 1.)
(ii) Does $\alpha\beta = \beta\alpha$ for all permutations α and β on $\{1, 2, 3\}$? (See Problem 3.)

5. Let α be the permutation on $\{1, 2, 3, 4, 5, 6\}$ with

$$\alpha\colon\ 1 \mapsto 6,\ 2 \mapsto 4,\ 3 \mapsto 1,\ 4 \mapsto 2,\ 5 \mapsto 5,\ 6 \mapsto 3.$$

Express each of the following in this arrow form.
(i) $\alpha^2 = \alpha\alpha$.
(ii) $\alpha^2\alpha$.
(iii) $\alpha\alpha^2$.
(iv) The inverse α^{-1} of the permutation α.

6. Let β be the permutation on $\{1, 2, 3, 4, 5, 6, 7\}$ with

$$\beta\colon\ 1 \mapsto 5,\ 2 \mapsto 3,\ 3 \mapsto 7,\ 4 \mapsto 6,\ 5 \mapsto 2,\ 6 \mapsto 1,\ 7 \mapsto 4.$$

Find each of the following in the arrow form.
(a) $\beta^2 = \beta\beta$.
(b) The inverse β^{-1} of the permutation β.

7. Let α be as in Problem 5. Give the diagram (see Definition 2 of Section 2.1) for each of the following.
(a) α; (b) α^2; (c) α^3; (d) α^4; (e) α^5;
(f) α^6; (g) the inverse α^{-1} of α.

8. Let β be as in Problem 6. Give the diagram for each of the following.
(a) β; (b) β^2; (c) β^3; (d) β^4; (e) β^5;
(f) β^6; (g) β^7; (h) the inverse β^{-1} of β.

9. Let α be as in Problems 5 and 7. Does $\alpha^n = \alpha^k$ when $n \equiv k \pmod 6$?

10. Let β be as in Problems 6 and 8. Does $\beta^n = \beta^k$ when $n \equiv k \pmod 7$?

★ 11. Let $y = \theta(x)$ determine a permutation θ on $\{1, 2, \ldots, n\}$. In the sequence $u_1, u_2, \ldots,$ let $u_2 = \theta(u_1)$, $u_3 = \theta(u_2)$, ... Let r be the least positive integer such that u_{r+1} is one of the previous terms $u_1, \ldots, u_r$. Explain why u_{r+1} must equal u_1.

★ 12. Let θ and $u_1, u_2, \ldots, u_r$ be as in the previous problem. Let v_1 be an element of $\{1, 2, \ldots, n\}$ that is not in $A = \{u_1, \ldots, u_r\}$. Let $v_2 = \theta(v_1)$, $v_3 = \theta(v_2)$, ... Explain why none of the v's is in A.

2.3 Abstract Groups

The four Theorems of Section 2.2 state the basic properties of the operation of multiplication (i.e., composition) of permutations on $\{1, 2, \ldots, n\}$. The following definition helps us deduce many other properties in such a way that we can apply the results to any algebraic structure that also has these four basic properties. This illustrates the power of abstract methods.

DEFINITION 1

Group

A **group** G is a set $\hat{G}$ with an operation having the following four properties:

G_1 ***Closure*** If a and b are elements (not necessarily distinct) in $\hat{G}$, there is a unique product ab in $\hat{G}$.

G_2 ***Identity*** There is an element $\mathbf{e}$ in $\hat{G}$ such that

$$a\mathbf{e} = a = \mathbf{e}a$$

for all a in $\hat{G}$.

G_3 ***Inverses*** For each a in $\hat{G}$, there is an h in $\hat{G}$ such that

$$ah = \mathbf{e} = ha.$$

G_4 ***Associativity*** If a, b, and c are elements (not necessarily distinct) in $\hat{G}$, then

$$(ab)c = a(bc).$$

G_1, G_2, G_3, and G_4 are called the *group axioms*; hence a group is a set with an operation satisfying all four group axioms. In other texts the group axioms may be stated in a somewhat different, but equivalent, form.

DEFINITION 2

Operation Table for a Finite Group G

If $\hat{G}$ is finite, the **operation table** for G is a square array with the elements of $\hat{G}$ in some order as row labels and column labels and with ab as the entry in the position common to the row labeled a and the column labeled b.

EXAMPLE 1

Group of Fourth Roots of Unity

Here we show that the operation of complex number multiplication makes the set $\hat{G} = \{1, -1, \mathbf{i}, -\mathbf{i}\}$ of four complex numbers into a group. The multiplication table for $\hat{G}$ is

·	1	−1	**i**	−**i**
1	1	−1	**i**	−**i**
−1	−1	1	−**i**	**i**
i	**i**	−**i**	−1	1
−**i**	−**i**	**i**	1	−1

$\hat{G}$ is closed under the operation since each entry in the table is in $\hat{G}$. One sees that 1 is an identity. Also, each of 1 and -1 is its own inverse while $\mathbf{i}$ and $-\mathbf{i}$ are inverses of each other. The operation for $\hat{G}$ is associative, since multiplication for any complex numbers is associative. Hence $\hat{G}$ and its operation form a group G.

The two following definitions are means of classifying groups.

DEFINITION 3

Order of a Group

If the set $\hat{G}$ of a group G is finite, the number of elements in $\hat{G}$ is called the **order** of G and is sometimes denoted as **ord** G. If $\hat{G}$ is infinite, G is a group of **infinite order**.

DEFINITION 4

Abelian Group

If $xy = yx$ for all x and y in $\hat{G}$, the group G is said to be **abelian**, or **commutative**.

Abelian groups are named after Niels Henrik Abel. (See the biographical note on Abel after this section.)

Clearly, the group of Example 1 above is an abelian group of order 4. The set of positive real numbers, with the operation of multiplication of real numbers, is an infinite abelian group.

A group is more than a set in that a group is a set with an operation satisfying the group axioms. Although it is important that one appreciate the distinction between a group G and its set of elements $\hat{G}$, consistently stressing the difference results in notational awkwardness. Hence we shall frequently write G for $\hat{G}$. For example, we shall refer to "an element of the set $\hat{G}$ of the group G" as "an element of the group G." When this is done, the context will indicate whether G represents a group or its set.

Theorems 1 through 4 of Section 2.2 show that the $n!$ permutations on $\{1, 2, \ldots, n\}$ form a group under the operation of composition of functions.

A function $f(x_1, x_2, \ldots, x_n)$ is said to be symmetric in $x_1, x_2, \ldots, x_n$ if its value is unchanged by all permutations of its n variables. This concept, which will be discussed further in Section 2.8, is responsible for the use of the word *symmetric* in the definition given next.

DEFINITION 5

The Symmetric Group $\mathbf{S_n}$

For every positive integer n, the group $\mathbf{S_n}$ of all the permutations on $\mathbf{X_n} = \{1, 2, \ldots, n\}$ is called the **symmetric group** on $\mathbf{X_n}$.

We note that ord $\mathbf{S_1} = 1$, ord $\mathbf{S_2} = 2$, ord $\mathbf{S_3} = 6$, ord $\mathbf{S_4} = 24$, ord $\mathbf{S_5} = 120$, and in general

$$\text{ord } \mathbf{S_n} = n! = 1 \cdot 2 \cdot 3 \cdots n.$$

The symmetric group $\mathbf{S_2}$ is abelian. (See Problem 1 of Section 2.2.) The symmetric group $\mathbf{S_3}$ is not abelian, since there is at least one pair of permutations α and β on $\{1, 2, 3\}$ such that $\alpha\beta \neq \beta\alpha$. (See Problem 3 of Section 2.2.)

Even in a nonabelian group G, some pairs of elements commute; for example, the definition of the identity $\mathbf{e}$ of G implies that $g\mathbf{e} = \mathbf{e}g$ for all g in G.

Our treatment of group theory includes results concerning specific groups, such as a symmetric group $\mathbf{S_n}$ or the group of Example 1. We shall also deal with theorems about groups G in general — that is, with theorems that are consequences of the group axioms and whose proofs depend in no other way on the nature of the elements of the set G or its operation. Such theorems are part of *abstract group theory*.

As an aid in distinguishing permutation groups from other groups, we generally will use lowercase Greek letters for permutations and lowercase Latin letters for elements of an abstract group.

Some abstract results on groups follow.

LEMMA 1 There is only one element e in a group G such that

$$ge = g = eg$$

for all g in G.

Proof Axiom G_2 tells us that there is at least one element with this property. Let each of e_1 and e_2 designate such an element; that is, let

$$ge_1 = g = e_1 g, \qquad ge_2 = g = e_2 g$$

for all g in G. Then we can replace g with e_2 in $g = e_1 g$ and replace g with e_1 in $ge_2 = g$, thus obtaining

$$e_2 = e_1 e_2 \quad \text{and} \quad e_1 e_2 = e_1.$$

Since each of e_2 and e_1 is equal to $e_1 e_2$, they are equal to each other. This proves the stated uniqueness. ■

Now we can say "the identity $\mathbf{e}$" instead of "an identity $\mathbf{e}$."

THEOREM 1 **Cancellation**

If either $ab = ac$ or $ba = ca$ in a group G, then $b = c$.

Proof We show left cancellation, namely that $ab = ac$ implies $b = c$, and leave right cancellation to the reader as Problem 14 of this section.

Let $\mathbf{e}$ be the identity of G. Axiom G_3 tells us that there is an element h of G such that $ah = \mathbf{e} = ha$. Substituting the equal ac for ab in $h(ab)$, we obtain

$$h(ab) = h(ac).$$

Then it follows from associativity and properties of the identity that

$$(ha)b = (ha)c, \qquad \mathbf{e}b = \mathbf{e}c, \qquad b = c.$$ ■

Note that Theorem 1 deals with either left or right cancellation but not with "mixed" cancellation; that is, an equation $ab = ca$ need not imply $b = c$. (See Problem 13 of this section.)

We are now able to strengthen the result in Lemma 1 above.

THEOREM 2 **Uniqueness of the Identity**

Let u be an element of a group G. If there exists at least one element v of G such that either

$$uv = v \quad \text{or} \quad vu = v,$$

then u is the identity $\mathbf{e}$ of G.

Proof Let $uv = v$. Since $\mathbf{e}v = v$, we then have $uv = \mathbf{e}v$. This and right cancellation give us $u = \mathbf{e}$. Similarly, $vu = v$ implies that $u = \mathbf{e}$. ■

THEOREM 3 Uniqueness of the Inverse

Let a be an element of a group G. Then there is only one element h in G such that $ah = \mathbf{e}$ or $ha = \mathbf{e}$.

Proof Axiom G_3 tells us that there is an h_1 in G such that

$$ah_1 = \mathbf{e} = h_1 a.$$

If there also is an element h_2 in G with $ah_2 = \mathbf{e}$, then $ah_2 = ah_1$, and left cancellation gives us $h_2 = h_1$. Similarly, $h_3 a = \mathbf{e}$ implies that $h_3 = h_1$. ∎

NOTATION 1 Inverse of an Element of a Group

The unique inverse of an element a of a group is designated as a^{-1}.

If g is an element of a group G, closure of G under its operation implies that the product of g by itself is also in G; we use the notation g^2 of ordinary algebra for this product gg. Similarly, we let $g^2 g = g^3$, $g^3 g = g^4$, and so on. To complete the process of defining g^n for all integers n, we let

$$g^0 = \mathbf{e}, \quad g^1 = g, \quad \text{and} \quad g^{-m} = (g^m)^{-1} \quad \text{for } m = 1, 2, 3, \ldots .$$

Using associativity, one sees that

$$gg^2 = g(gg) = (gg)g = g^2 g = g^3.$$

Associativity can be used together with mathematical induction to prove the general rules

$$g^m g^n = g^{m+n}, \qquad (g^m)^n = g^{mn} \tag{E}$$

for any element g of a group G and any integers m and n. It is also clear that $\mathbf{e}^n = \mathbf{e}$ for all integers n. We assume these rules without proof.

The exponent rules (E) apply to commutative and to noncommutative groups, since one can use associativity as a substitute for commutativity in dealing with powers having a fixed base. However, $(ab)^n$ need not equal $a^n b^n$ in a noncommutative group. For example, $(ab)^2 = abab$; and this can be proved to be different from $a^2 b^2 = aabb$ when a and b do not commute. (See Problem 3 of this section.)

In Problems 4 and 5 of this section, we will see that $(ab)^{-1}$ may or may not equal $a^{-1}b^{-1}$ in a noncommutative group. What is a correct formula for $(ab)^{-1}$ that holds in all groups? We can motivate a formula by thinking of the inverse as an "undoing." We undo the combined operation of first putting on socks and then putting on shoes by undoing these in reverse order — that is, first removing the shoes and then removing the socks. This leads us to conjecture that "the inverse of a product is the product of the inverses in reverse order." Next we prove this conjecture.

THEOREM 4 Inverse of a Product

In a group $(a_1 a_2 \cdots a_{n-1} a_n)^{-1} = a_n^{-1} a_{n-1}^{-1} \cdots a_2^{-1} a_1^{-1}$.

Proof Let $g = a_1a_2 \cdots a_{n-1}a_n$ and $h = a_n^{-1}a_{n-1}^{-1} \cdots a_2^{-1}a_1^{-1}$. We prove by induction on n that $gh = \mathbf{e} = hg$ for $n = 1, 2, \ldots$ and hence that $g^{-1} = h$.

Basis When $n = 1$, we have $gh = a_1a_1^{-1} = \mathbf{e} = a_1^{-1}a_1 = hg$. Hence the formula holds for $n = 1$.

Inductive Step Assume that the formula holds for $n = k$; that is, assume that $(a_1 \cdots a_k)(a_k^{-1} \cdots a_1^{-1}) = \mathbf{e}$. Then for $n = k + 1$ we have

$$gh = (a_1 \cdots a_k a_{k+1})(a_{k+1}^{-1}a_k^{-1} \cdots a_1^{-1}) = a_1 \cdots a_k(a_{k+1}a_{k+1}^{-1})a_k^{-1} \cdots a_1^{-1}$$
$$= (a_1 \cdots a_k)\mathbf{e}(a_k^{-1} \cdots a_1^{-1}) = (a_1 \cdots a_k)(a_k^{-1} \cdots a_1^{-1}).$$

It now follows from the hypothesis of the induction that $gh = \mathbf{e}$. Similarly $hg = \mathbf{e}$. This completes the induction and shows that $g^{-1} = h$. ■

The following result facilitates construction of tables for operations of finite groups.

THEOREM 5 **Operation Table for a Group**

Each element occurs exactly once as an entry on each row and on each column of the table for the operation of a finite group $G = \{g_1, \ldots, g_m\}$.

Proof The entries on the row for an element x are $xg_1, xg_2, \ldots, xg_m$. Any element y of G appears as the entry xg with $g = x^{-1}y$. No element can appear more than once, since that would mean that $xg_i = xg_j$ with $g_i \neq g_j$; but this is impossible because of left cancellation. Similarly, one shows that each element of G appears exactly once on any given column. ■

We next show that the multiplication table for a group of order 3 is completely determined once one knows which element is the identity.

EXAMPLE 2 **Unique Table for a Group of Order 3**

Here we assume that there exists a group $G = \{\mathbf{e}, b, c\}$ of order 3 with $\mathbf{e}$ as the identity. We show that

$$bc = \mathbf{e} = cb, \quad b^2 = c, \quad \text{and} \quad b^3 = \mathbf{e} = c^3$$

in such a group, and we tabulate the operation of the group.

Since $b \neq \mathbf{e}$ and $c \neq \mathbf{e}$, it follows from Theorem 2 (Uniqueness of the Identity) that $bc \neq c$ and $bc \neq b$. But bc is some element of G by Axiom G_1 (Closure), and hence $bc = \mathbf{e}$.

We note that $b\mathbf{e} = b$ and $bc = \mathbf{e}$. Since $b^2 = bb$ is different from $b\mathbf{e}$ and from bc by Theorem 5, we must have $b^2 = c$. Then

$$b^3 = bb^2 = bc = \mathbf{e}.$$

Similarly, $cb = \mathbf{e}$, $c^2 = b$, and $c^3 = \mathbf{e}$. Now we can use these facts to give the table for the operation in G.

	$\mathbf{e}$	b	c
$\mathbf{e}$	$\mathbf{e}$	b	c
b	b	c	$\mathbf{e}$
c	c	$\mathbf{e}$	b

We see that no element is repeated among the entries in a given row or column; this is as promised by Theorem 5.

Example 2 does not prove that groups of order 3 exist. It only shows that there is just one possibility for the operation of a group with three elements. One way to show the existence of such a group is to verify that all the axioms of Definition 1 are satisfied by the operation tabulated in Example 2. This would be somewhat tedious, since 27 cases are involved in checking associativity. We prefer to produce a group $\{1, \beta, \gamma\}$ of complex numbers whose multiplication table is that of Example 2 but with $\mathbf{e}$, b, and c replaced by 1, β, and γ, respectively. (See Problem 8 of this section.) Then it follows from associativity of multiplication of complex numbers that the operation given by the table in Example 2 is associative. Alternatively, one can show that $\{\varepsilon, \rho, \rho^2\}$, with ε and ρ given by

$$\varepsilon\colon\ 1 \mapsto 1,\ 2 \mapsto 2,\ 3 \mapsto 3;$$

$$\rho\colon\ 1 \mapsto 2,\ 2 \mapsto 3,\ 3 \mapsto 1,$$

is a group of permutations whose multiplication table is that of Example 2 but with $\mathbf{e}$, b, and c replaced by ε, ρ, and ρ^2, respectively; and then one can use the fact that multiplication of permutations is associative.

DEFINITION 6

Main Diagonal

In the operation table for a finite group G, the **main diagonal** is the diagonal from the upper left corner to the lower right corner; that is, its entries are the squares of the elements of G.

In the table of Example 2, the entries on the main diagonal are $\mathbf{e} = \mathbf{e}^2$, $c = b^2$, and $b = c^2$.

Problems

Problem 9 below is cited in Section 4.1.

1. For each given θ, find θ^{-1}, θ^2, θ^3, θ^4, θ^{100}, and θ^{101}.
   (a) $\theta\colon\ 1 \mapsto 2,\ 2 \mapsto 1$.
   (b) $\theta\colon\ 1 \mapsto 2,\ 2 \mapsto 3,\ 3 \mapsto 1$.
   (c) $\theta\colon\ 1 \mapsto 3,\ 2 \mapsto 4,\ 3 \mapsto 1,\ 4 \mapsto 2$.
2. Do as in Problem 1 for each of the following permutations.
   (a) $\theta\colon\ 1 \mapsto 1,\ 2 \mapsto 3,\ 3 \mapsto 2$.

(b) θ: $1 \mapsto 3, 2 \mapsto 1, 3 \mapsto 2, 4 \mapsto 4$.
(c) θ: $1 \mapsto 2, 2 \mapsto 3, 3 \mapsto 4, 4 \mapsto 1$.

3. Let $\alpha = \begin{pmatrix} 1 & 2 & 3 \\ 2 & 3 & 1 \end{pmatrix}$, $\beta = \begin{pmatrix} 1 & 2 & 3 \\ 2 & 1 & 3 \end{pmatrix}$, and $\gamma = \begin{pmatrix} 1 & 2 & 3 \\ 3 & 1 & 2 \end{pmatrix}$.
(i) Show that $(\alpha\beta)^2 \neq \alpha^2\beta^2$. [*Hint:* Find $\alpha\beta$, $(\alpha\beta)^2$, α^2, β^2, and $\alpha^2\beta^2$.]
(ii) Show that $(\alpha\gamma)^2 = \alpha^2\gamma^2$.

4. For the α, β, and γ of Problem 3, show that $(\alpha\beta)^{-1} \neq \alpha^{-1}\beta^{-1}$ and $(\alpha\gamma)^{-1} = \alpha^{-1}\gamma^{-1}$.

5. Let a and b be elements of a group. Show that $(ab)^{-1} = a^{-1}b^{-1}$ if and only if $ab = ba$; that is, show both of the following.
(i) If $ab = ba$, then $(ab)^{-1} = a^{-1}b^{-1}$.
(ii) If $(ab)^{-1} = a^{-1}b^{-1}$, then $ab = ba$.

6. Prove that $(ab)^2 = a^2b^2$ in a group G if and only if $ab = ba$; that is, prove both of the following.
(i) If $ab = ba$, then $(ab)^2 = a^2b^2$.
(ii) If $(ab)^2 = a^2b^2$, then $ab = ba$.

7. (i) Show that the subset $H = \{1, -1\}$ of $\mathbf{Z}$ is a group under the operation of multiplication of integers.
(ii) Show that the operation table for a group $G = \{\mathbf{e}, a\}$ of order 2 is completely determined when one knows that $\mathbf{e}$ is the identity.
(iii) How can one tell, by looking at the multiplication table of a group G, whether or not G is commutative?
(iv) Is a group of order 2 necessarily commutative? Explain.

8. Let $\beta = (-1 + \mathbf{i}\sqrt{3})/2$ and $\gamma = (-1 - \mathbf{i}\sqrt{3})/2$.
(i) Show that $\{1, \beta, \gamma\}$ is a group (of order 3) under multiplication of complex numbers.
(ii) Explain why the multiplication table for this group must be the table of Example 2 but with $\mathbf{e}$, b, and c replaced by 1, β, and γ, respectively.
(iii) Is a group of order 3 necessarily abelian? Explain.

9. (i) Given that $G = \{\mathbf{e}, b, c, d\}$ is a group of order 4 with

$$\mathbf{e}^2 = b^2 = c^2 = d^2 = \mathbf{e},$$

construct its multiplication table. (In Problem 12 of Section 2.5, we shall see that such a group exists.)
(ii) For each g in G, does $g^m = g^n$ when m and n are integers with the same parity?

10. (i) Assume that a group $G = \{\mathbf{e}, u, v, w\}$ of order 4 exists with $\mathbf{e}$ the identity, $u^2 = v$, and $v^2 = \mathbf{e}$. Construct the table for the operation of such a group.
(ii) Does there exist a group G with the properties given in (i)? Explain.

11. Let $g^2 = \mathbf{e}$ for all g in a group G. Show that G is abelian.

12. (i) Given that $G = \{\mathbf{e}, a, a^2, a^3, a^4\}$ is a group of order 5 with $a^5 = \mathbf{e}$, construct its multiplication table.
(ii) For each g in G, does $g^m = g^n$ when $m \equiv n \pmod 5$?
(iii) Is G abelian?

13. Find permutations α, β, and γ in the symmetric group $\mathbf{S_3}$ such that $\alpha\beta = \gamma\alpha$ and $\beta \neq \gamma$ (and thus show that "mixed cancellation" is not always valid).

14. Prove the "right cancellation" part of Theorem 1 that was left to the reader by showing that $ba = ca$ implies $b = c$ in a group.

Niels Henrik Abel
1802–1829

Abel was born in Findoé, a small town in Norway, one of seven children of the local pastor. The responsibility for supporting the family fell largely on his shoulders at the age of eighteen, when his father died. One year later, he proved the insolvability of the fifth-degree polynomial equation by rational operations and extraction of roots. Thus he succeeded where the greatest mathematicians of the previous 300 years had all failed.

As a student at the University of Christiania (now Oslo), he was recognized for his extraordinary talent and given a stipend by the Norwegian government to continue his studies in Germany and France. He left Norway in 1825 and went first to Berlin, where he met A. L. Crelle. Abel's famous proof on the quintic equation had been published in brief form in Norway in 1824, but the first detailed explanation of his method appeared in the first volume (1826) of Crelle's famous journal, now the *Journal für die Reine und Angewandte Mathematik*. He had sent a copy of the proof in 1824 to Gauss, who ignored it. He therefore decided not to visit Gauss in Göttingen, but went on to Paris from Berlin, where he did meet Cauchy and Legendre, among others. There he submitted a memoir to Cauchy and had the same misfortune that Galois experienced: Cauchy mislaid it! It was not found until after Abel's death and not published until 1841.

Abel returned to Oslo where he continued to do important work in a variety of fields: elliptic functions, abelian functions, and integration in finite terms. Abel sought an academic post without success, but subsisted mainly by giving private lessons. He died of tuberculosis in Norway in 1829. Two days after his death he was appointed to a professorship at the University of Berlin, where they had not yet heard the news.

A heroic statue of Abel by Gustav Vigeland stands in the Royal Park in Oslo. He is shown dramatically standing on two prone and seemingly vanquished figures. It is understood among mathematicians that these represent the general quintic equation and elliptic functions! His accomplishments in these fields were his greatest achievements.

15. Let a, b, and c be elements of a group.
(a) Show that $(a^{-1})^{-1} = a$.
(b) Express $(ab)^{-1}$ in terms of a^{-1} and b^{-1}.
(c) Express $(abc)^{-1}$ in terms of a^{-1}, b^{-1}, and c^{-1}.

16. In a group G, express $(abcd)^{-1}$ in terms of a^{-1}, b^{-1}, c^{-1}, and d^{-1}.

17. In a group G, let $a^3 = \mathbf{e}$ but $a \neq \mathbf{e}$ and $a^2 \neq \mathbf{e}$. Let $H = \{a^n : n \in \mathbf{Z}\}$ consist of all the integral powers $\ldots, a^{-2}, a^{-1}, \mathbf{e}, a, a^2, \ldots$ of a.
(i) Use the division algorithm to show that $H = \{\mathbf{e}, a, a^2\}$.

Paolo Ruffini

1765–1822

Ruffini was an Italian physician who, like Cardano a couple of centuries earlier, did mathematics on the side. He anticipated the idea of a group (which he called a permutation) in his little-known book *Teoria Generale delle Equazioni* (full title, translated: *General Theory of Equations in Which the Algebraic Solution of General Equations of Degree Higher than Four Is Demonstrated to Be Impossible!*), published in Bologna in 1799. In this he claimed to have proved the insolvability of the quintic by radicals, although his proof was not universally accepted. The most efficient modern proofs of insolvability of the general quintic are still quite difficult. Moreover, the mathematics of any period is written with assumptions that certain things are known to prospective readers. Hence there would be a large subjective element in any present-day evaluation of Ruffini's work. However, in spite of his work, there is little doubt that Abel's paper settled the matter for the mathematical community.

(ii) Show that $a^m = \mathbf{e}$ if and only if $3 \mid m$, that is, if and only if $m \equiv 0 \pmod 3$.
(iii) Show that $a^m = a^n$ if and only if $m \equiv n \pmod 3$.

18. Let a be in a group G, let s be the smallest positive integer such that $a^s = \mathbf{e}$, and let $H = \{a^n : n \in \mathbf{Z}\}$.
(i) If n is an integer, does $a^n = a^r$ where r is the remainder in the division of n by s?
(ii) Is $a^h = a^k$ whenever $h \equiv k \pmod s$?
(iii) Does $H = \{\mathbf{e}, a, a^2, \ldots, a^{s-1}\}$?
(iv) If a^h is in H, is its inverse in H?

19. Let a be in a group G, and let $a^n \neq \mathbf{e}$ for all n in $\mathbf{Z}^+$. Show that $a^h \neq a^k$ when h and k are distinct integers.

20. Let a be in a group G, and let $H = \{a^n : n \in \mathbf{Z}\}$. Show the following.
(i) If h and h' are in H, so is hh'.
(ii) The identity $\mathbf{e}$ of G is in H.
(iii) If h is in H, so is h^{-1}.

21. Let G be a finite group. Explain why there are an even number of elements x of G such that $x^2 \neq \mathbf{e}$.

22. Show that a finite group G of even order has at least one element y with $y^2 = \mathbf{e}$ and $y \neq \mathbf{e}$ by showing that G has an odd number of such elements.

23. Given that $abc = \mathbf{e}$ in a group G, show that $bca = \mathbf{e} = cab$.

24. (i) Show that $abcd = \mathbf{e}$ implies $bcda = \mathbf{e}$ in a group G.
(ii) Generalize on Problems 23 and 24(i).

Leopold Kronecker
1823–1891

Kronecker was born near Breslau and studied there at the local gymnasium under the eminent algebraist E. E. Kummer, after which he studied in Berlin under C. G. J. Jacobi, Jakob Steiner, and P. G. L. Dirichlet, among the greatest mathematicians of the nineteenth century. Kronecker is a counterexample to the often stated view that good mathematicians, though very creative in mathematics, are not very adept at other human pursuits. While still a young man, Kronecker fell heir to a sizable fortune; and although he had already pursued mathematical studies, he went on to manage his business interests with great success. For eleven years he was a businessman and did almost no mathematics. Then he entered academic life and eventually succeeded Kummer as professor at the University of Berlin. His scholarly interests also ranged beyond mathematics to include Greek, Latin, Hebrew, and philosophy.

He was a good friend of Karl Weierstrass but carried on a longstanding dispute with him concerning the proper direction of mathematics. Weierstrass was one of the foremost proponents of analysis during that period, while Kronecker advocated the "arithmetization" of mathematics, that is, the treatment of mathematics in terms of the natural numbers. This view led him to make his famous comment: "Die ganze Zahl schuf die liebe Gott, alles Übrige ist Menschenwerk" ("God created the integers, all else is the work of man").

Kronecker was one of the first mathematicians to investigate thoroughly the work of Galois, and he wrote on Galois Theory in a way that made the subject accessible to many. He gave the first axiomatic formulation of abstract groups.

25. Tell why $S = \{a, b, c, d\}$ is not a group under the operation of the following table.

	a	b	c	d
a	c	a	d	b
b	a	b	c	d
c	d	c	b	c
d	b	d	c	a

2.4 Cycle Notation

This section presents the last of the notations commonly used for permutations. It is a convenient notation and one that brings out many important facts about permutations.

DEFINITION 1

s-Cycle, Transposition

Let $A = \{a_1, a_2, \ldots, a_s\}$ be a subset of s elements chosen from $\mathbf{X_n} = \{1, 2, \ldots, n\}$, and let γ be the permutation on $\mathbf{X_n}$ given by the two following properties:

(a) $\gamma(x) = x$ if x is not in A.

(b) $\gamma(a_i) = a_{i+1}$ for $i = 1, 2, \ldots, s-1$, and $\gamma(a_s) = a_1$. This γ is represented by $(a_1 a_2 \cdots a_s)$ and is called an ***s*-cycle** or a **cycle** of **length** s. A 2-cycle is also called a **transposition**.

Every cycle of length 1 represents the identity ε.

Clearly the cycle notation saves space; we will see other advantages later. The cycle notation by itself for a given γ does not tell us which $\mathbf{S_n}$ the cycle is in. For example, the 3-cycle (123) represents any of the permutations

$$\begin{pmatrix} 1 & 2 & 3 \\ 2 & 3 & 1 \end{pmatrix}, \begin{pmatrix} 1 & 2 & 3 & 4 \\ 2 & 3 & 1 & 4 \end{pmatrix}, \ldots, \begin{pmatrix} 1 & 2 & 3 & 4 & \ldots & n \\ 2 & 3 & 1 & 4 & \ldots & n \end{pmatrix}, \ldots$$

of $\mathbf{S_3}, \mathbf{S_4}, \ldots$, respectively. This ambiguity has advantages. For example, one can use any of the techniques of Examples 1 through 4 below to show that, in all the $\mathbf{S_n}$ with $n \geq 3$, one has

$$(12)(13) = (123) \quad \text{and} \quad (13)(12) = (132)$$

and thus show that each of these $\mathbf{S_n}$ is nonabelian.

EXAMPLE 1 In $\mathbf{S_3}$ let $\rho = (123)$ and $\phi = (23)$. Then we represent $\rho\phi$, ρ^2, and $\rho^2\phi$ in cycle notation as follows. We see that

$$1 \overset{\rho}{\mapsto} 2 \overset{\phi}{\mapsto} 3 \quad \text{or} \quad 1 \overset{\rho\phi}{\mapsto} 3,$$
$$3 \overset{\rho}{\mapsto} 1 \overset{\phi}{\mapsto} 1 \quad \text{or} \quad 3 \overset{\rho\phi}{\mapsto} 1,$$
$$2 \overset{\rho}{\mapsto} 3 \overset{\phi}{\mapsto} 2 \quad \text{or} \quad 2 \overset{\rho\phi}{\mapsto} 2.$$

This shows that $\rho\phi = (13)$. Similarly, one finds that $\rho^2 = (132)$ and $\rho^2\phi = (12)$. Thus the six permutations of the symmetric group $\mathbf{S_3}$ are

$$\varepsilon = (1), \qquad \rho = (123), \qquad \rho^2 = (132),$$
$$\phi = (23), \qquad \rho\phi = (13), \qquad \rho^2\phi = (12).$$

Juxtaposition of cycles indicates multiplication. We next illustrate some techniques for finding the product of cycles.

EXAMPLE 2 Here we show that $(12)(13) = (123)$ in $\mathbf{S_n}$ for $n \geq 3$, as follows.

$$(12)(13) = \begin{pmatrix} 1 & 2 & 3 & \ldots & n \\ 2 & 1 & 3 & \ldots & n \end{pmatrix} \begin{pmatrix} 2 & 1 & 3 & 4 & \ldots & n \\ 2 & 3 & 1 & 4 & \ldots & n \end{pmatrix}$$
$$= \begin{pmatrix} 1 & 2 & 3 & 4 & \ldots & n \\ 2 & 3 & 1 & 4 & \ldots & n \end{pmatrix} = (123).$$

EXAMPLE 3 In any $\mathbf{S_n}$ with $n \geq 5$, let $\alpha = (12345)$ and $\beta = \alpha^2$. We show that β is a 5-cycle as follows. We see that

$$1 \overset{\alpha}{\mapsto} 2 \overset{\alpha}{\mapsto} 3 \quad \text{or} \quad 1 \overset{\beta}{\mapsto} 3,$$
$$3 \overset{\alpha}{\mapsto} 4 \overset{\alpha}{\mapsto} 5 \quad \text{or} \quad 3 \overset{\beta}{\mapsto} 5,$$
$$5 \overset{\alpha}{\mapsto} 1 \overset{\alpha}{\mapsto} 2 \quad \text{or} \quad 5 \overset{\beta}{\mapsto} 2,$$
$$2 \overset{\alpha}{\mapsto} 3 \overset{\alpha}{\mapsto} 4 \quad \text{or} \quad 2 \overset{\beta}{\mapsto} 4,$$
$$4 \overset{\alpha}{\mapsto} 5 \overset{\alpha}{\mapsto} 1 \quad \text{or} \quad 4 \overset{\beta}{\mapsto} 1.$$

Also, if x is in $\{1, 2, \ldots, n\}$ but not in $\{1, 2, 3, 4, 5\}$, then

$$x \overset{\alpha}{\mapsto} x \overset{\alpha}{\mapsto} x \quad \text{or} \quad x \overset{\beta}{\mapsto} x.$$

Hence β is the 5-cycle (13524).

EXAMPLE 4 Now we show that $(123)(234) = (13)(24)$ in $\mathbf{S_n}$ for $n \geq 4$. Let $\alpha = (123)$, $\beta = (234)$, $\gamma = (13)$, $\delta = (24)$. Then

$$1 \overset{\alpha}{\mapsto} 2 \overset{\beta}{\mapsto} 3 \quad \text{or} \quad 1 \overset{\alpha\beta}{\mapsto} 3; \qquad 1 \overset{\gamma}{\mapsto} 3 \overset{\delta}{\mapsto} 3 \quad \text{or} \quad 1 \overset{\gamma\delta}{\mapsto} 3.$$

Hence $\alpha\beta$ and $\gamma\delta$ both send 1 to 3. Similarly, we see that $\alpha\beta$ and $\gamma\delta$ have the same value at every x in $\{1, 2, \ldots, n\}$, and hence $\alpha\beta = \gamma\delta$.

We next introduce an important concept.

DEFINITION 2

Disjoint Cycles

Cycles $(a_1a_2 \ldots a_r)$, $(b_1b_2 \ldots b_s)$, ..., $(m_1m_2 \ldots m_v)$ in $\mathbf{S_n}$ are **disjoint** if no integer in $\{1, 2, \ldots, n\}$ appears more than once in the listing

$$a_1, a_2, \ldots, a_r, b_1, b_2, \ldots, b_s, \ldots, m_1, m_2, \ldots, m_v.$$

For example, (13), (24), and (567) are three disjoint cycles in any $\mathbf{S_n}$ with $n \geq 7$. However, (143) and (2356) are not disjoint, since 3 appears in both of these cycles.

Let θ be the product (23)(456) of disjoint cycles in $\mathbf{S_7}$. Then the diagram for θ is shown in Figure 2.2.

The connected pieces of this diagram correspond to the disjoint cycles in the representation (1)(23)(456)(7) for θ as a product of disjoint cycles. The proof of the following result contains an algorithm for expressing any permutation in $\mathbf{S_n}$ as such a product.

FIGURE 2.2

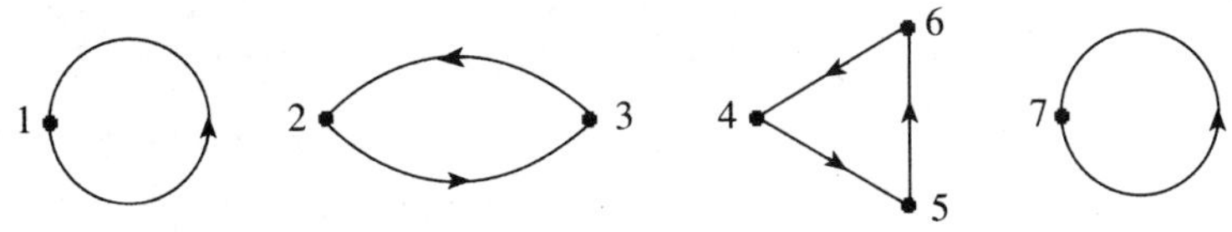

THEOREM 1 Product of Disjoint Cycles

Every θ in $\mathbf{S_n}$ is either a cycle or a product

$$(a_1a_2 \ldots a_r)(b_1b_2 \ldots b_s) \cdots (m_1m_2 \ldots m_v)$$

of disjoint cycles.

Proof Let $a_1 = 1$, $a_2 = \theta(a_1)$, $a_3 = \theta(a_2)$, and so on. Since $\mathbf{X_n} = \{1, 2, \ldots, n\}$ is finite, there must be repetitions among these a's. Let r be the smallest positive integer such that a_{r+1} is one of the previous terms $a_1, \ldots, a_r$.

Then $a_1, \ldots, a_r$ are distinct terms, and it follows from the definition of a permutation θ that $\theta(a_1), \ldots, \theta(a_r)$ are distinct. Since $\theta(a_i) = a_{i+1}$, it follows that $a_2, \ldots, a_{r+1}$ are distinct. Therefore $a_{r+1} = a_1$, since this is the only remaining way in which a_{r+1} can equal an earlier term.

Let A be the subset $\{a_1, \ldots, a_r\}$ of $\mathbf{X_n}$. If $r < n$, let b_1 be the least number in $\mathbf{X_n}$ that is not in A. Then let $\theta(b_1) = b_2$, $\theta(b_2) = b_3$, and so on. As was true with the a's, the first repetition among the b's is a reappearance of b_1 as some b_{s+1}. Let $B = \{b_1, \ldots, b_s\}$.

We next show that no integer in B is in A by assuming that $b_k = a_j$ for some j and k and obtaining a contradiction. First we note that k cannot be 1 since b_1 is not in A.

If $j \geq 2$, it follows from $b_k = \theta(b_{k-1})$, $a_j = \theta(a_{j-1})$, $b_k = a_j$, and the fact that θ is a permutation that $b_{k-1} = a_{j-1}$. If $j = 1$, it follows similarly that $b_{k-1} = a_r$. In either case, the assumption that b_k is in A implies that b_{k-1} is in A. Then this implies that b_{k-2} is in A. Continuing in this manner, we find that the assumption that b_k is in A leads to the contradiction that b_1 is in A. Hence A and B have no integer in common.

If there are integers in $\mathbf{X_n}$ that are not in A or B, we let c_1 be the smallest such integer and continue the process; that is, we let $\theta(c_1) = c_2$, $\theta(c_2) = c_3$, and so on.

We stop this process when $\mathbf{X_n}$ is exhausted. Then θ can be written in the form

$$\begin{pmatrix} a_1 & a_2 & \ldots & a_{r-1} & a_r & b_1 & b_2 & \ldots & b_{s-1} & b_s & \ldots & m_1 & \ldots & m_{v-1} & m_v \\ a_2 & a_3 & \ldots & a_r & a_1 & b_2 & b_3 & \ldots & b_s & b_1 & \ldots & m_2 & \ldots & m_v & m_1 \end{pmatrix}$$

and hence θ is the product of disjoint cycles

$$\theta = (a_1a_2 \ldots a_r)(b_1b_2 \ldots b_s)(c_1c_2 \ldots c_t) \cdots (m_1m_2 \ldots m_v). \quad \text{(P)}$$

■

A given permutation may have several expressions as a product of disjoint cycles. For example, in $\mathbf{S_4}$ we have $(1) = (1)(2)(3)(4)$ and

$$(12)(34) = (12)(43) = (21)(34) = (21)(43) = (34)(12)$$
$$= (34)(21) = (43)(12) = (43)(21).$$

However, we can choose one representation as the preferred form and thus avoid confusion and be able to list permutations without the danger of duplication. This is done by removing unnecessary 1-cycles from the representation (P) of the proof of Theorem 1 and imposing conditions fulfilled by the algorithm of that proof, as we do next.

DEFINITION 3 Standard Form

The **standard form** for the identity is (1). A cycle $(x_1x_2 \dots x_s)$ is in **standard form** if x_1 is the smallest integer in the set $\{x_1, x_2, \dots, x_s\}$. The **standard form** for a permutation $\theta \neq (1)$ is a cycle in standard form or a product

$$(a_1a_2 \dots a_r)(b_1b_2 \dots b_s)(c_1c_2 \dots c_t) \cdots (m_1m_2 \dots m_v)$$

of disjoint cycles each in standard form such that there are no 1-cycles and

$$a_1 < b_1 < c_1 < \cdots < m_1.$$

The product (123)(234) is not in standard form since (123) and (234) are not disjoint. Example 4 above shows that (123)(234) = (13)(24). Since (13)(24) meets the conditions for a standard form, it is the standard form for (123)(234).

The product (543)(17) of disjoint cycles is not in standard form, since 5 is not the least integer in $\{5, 4, 3\}$. We note that

$$(543)(17) = (354)(17) = (17)(354).$$

The product (354)(17) is not in standard form, since the lead number 3 of the first factor is greater than the lead number 1 of the second factor. The product (17)(354) meets all the conditions of Definition 3 to be the standard form for (543)(17).

If θ is in $\mathbf{S}_n$, the sum of the lengths of the cycles in its standard form cannot exceed n. For example, in $\mathbf{S}_4$ the sum of the lengths is at most 4. Therefore, each of the 24 permutations in $\mathbf{S}_4$ has a standard form of one of the following types:

$$(1),\ (ab),\ (abc),\ (abcd),\ (ab)(cd).$$

In our discussion above, we have noted the equality (354)(17) = (17)(354), which is a special case of the following result.

THEOREM 2 Disjoint Cycles Commute

Let $\alpha = (a_1a_2 \dots a_r)$ and $\beta = (b_1b_2 \dots b_s)$ be disjoint cycles. Then $\alpha\beta = \beta\alpha$.

Proof Since no a_j is a b_k, $\alpha(b_k) = b_k$ and $\beta(a_j) = a_j$ for all j and k. Hence

$$a_1 \overset{\alpha}{\mapsto} a_2 \overset{\beta}{\mapsto} a_2 \quad \text{or} \quad a_1 \overset{\alpha\beta}{\mapsto} a_2,$$
$$a_1 \overset{\beta}{\mapsto} a_1 \overset{\alpha}{\mapsto} a_2 \quad \text{or} \quad a_1 \overset{\beta\alpha}{\mapsto} a_2.$$

This shows that a_1 is sent to a_2 by both $\alpha\beta$ and $\beta\alpha$. Similarly, one shows that each of $\alpha\beta$ and $\beta\alpha$ sends a_2 to $a_3, \ldots, a_{r-1}$ to a_r, a_r to a_1, b_1 to $b_2, \ldots, b_{r-1}$ to b_r, and b_r to b_1. Also, each of $\alpha\beta$ and $\beta\alpha$ sends x to x if x is not an a_i or a b_j. Thus $\alpha\beta = \beta\alpha$. ■

The following result is very important in the study of permutation groups.

THEOREM 3 **Product of 2-Cycles**

For $n > 1$, every permutation in $\mathbf{S}_n$ is a transposition or a product of transpositions (which need not be disjoint).

Proof Let θ be in $\mathbf{S}_n$. If $\theta = (1)$, an expression for θ as a product of transpositions is $(1) = (12)(12)$. If $\theta \neq (1)$, one obtains a representation of the desired form by replacing each cycle with length greater than 2 in the standard form of θ with a product of transpositions, using the easily verified formula

$$(d_1 d_2 d_3 \ldots d_s) = (d_1 d_s)(d_2 d_s) \cdots (d_{s-1} d_s).$$

(See Problem 5 below for cases of this formula.) ■

Problems

1. Give the standard forms for the following.
   (a) (54321); (b) (234)(156); (c) (125)(345);
   (d) the eight permutations in $\mathbf{S}_4$ with standard forms of the type (abc);
   (e) the three permutations in $\mathbf{S}_4$ with standard forms of the type $(ab)(cd)$.
2. Give the standard forms for the following.
   (a) (2143); (b) (23)(14); (c) (126)(3456);
   (d) the six permutations in $\mathbf{S}_4$ with standard forms of the type (ab);
   (e) the six permutations in $\mathbf{S}_4$ with standard forms of the type $(abcd)$.
3. For each of the following cycles γ, express each of $\gamma^{-1}, \gamma^2, \gamma^3, \gamma^4, \gamma^5$, and γ^6 in the standard form.
   (a) (123); (b) (1234); (c) (12345); (d) (123456).
4. Let $\gamma = (123 \ldots s)$.

(i) What is the smallest positive integer m such that $\gamma^m = (1)$?
(ii) Does $\gamma^{-1} = \gamma^{s-1}$?
(iii) Express γ^{-1} in standard form.

5. Express each of the following products in standard form.
(i) $(13)(23)$; (ii) $(14)(24)(34)$; (iii) $(15)(25)(35)(45)$;
(iv) $(16)(26)(36)(46)(56)$.

6. Express each of the following products in standard form.
(i) $(12)(13)$; (ii) $(12)(13)(14)$; (iii) $(12)(13)(14)(15)$; (iv) $(12345)(16)$.

7. Express $(123 \ldots s)$ as a product of transpositions.

8. Express $(a_1a_2 \ldots a_s)$ as a product of transpositions.

9. (i) Express $(abc)^2$ as a 3-cycle.
(ii) Show that (xyz) is the square of a 3-cycle.
(iii) Express $(abc)^5$ as a 3-cycle.
(iv) Show that (xyz) is the fifth power of a 3-cycle.

10. (i) Express $(abcde)^2$ as a 5-cycle.
(ii) Show that $(vwxyz)$ is the square of a 5-cycle.
(iii) Express $(abcde)^3$ as a 5-cycle.
(iv) Show that $(vwxyz)$ is the cube of a 5-cycle.

11. Let $\alpha = (a_1a_2a_3 \ldots a_{2r})$. Write α^2 as the product of two r-cycles.

12. Let $\beta = (b_1b_2 \ldots b_{3s})$. Write β^3 as the product of three s-cycles.

13. For which positive integers n is $\mathbf{S}_n$ abelian?

14. (a) Which elements α of $\mathbf{S}_3$ are of the form $\alpha = \beta^2$, with β in $\mathbf{S}_3$?
(b) Which elements α of $\mathbf{S}_3$ are of the form $\alpha = \gamma^3$, with γ in $\mathbf{S}_3$?

15. Let $\mathbf{A}_3$ be the subset $\{(1), (123), (132)\}$ of $\mathbf{S}_3$. Show the following.
(i) Each α of $\mathbf{A}_3$ is expressible as $\alpha = \beta\gamma$, with β and γ transpositions. (β and γ cannot be disjoint.)
(ii) $\mathbf{A}_3$ is closed under composition of permutations.
(iii) α^{-1} is in $\mathbf{A}_3$ for every α in $\mathbf{A}_3$.

16. Express each of 12 of the 24 permutations of $\mathbf{S}_4$ as a product $\beta\gamma$ of two (not necessarily disjoint) transpositions.

17. Let $\alpha = \gamma_1\gamma_2 \cdots \gamma_{2h}$ and $\beta = \delta_1\delta_2 \cdots \delta_{2k}$, where the γ_i and δ_j are (not necessarily disjoint) transpositions. Is $\alpha\beta$ expressible as the product of an even number of transpositions? Explain. (Here h and k are positive integers.)

18. (i) Is every transposition $\gamma = (ab)$ its own inverse? Explain.
(ii) Let $\alpha = \gamma_1\gamma_2 \cdots \gamma_{2h}$, where each γ_i is a transposition. Is α^{-1} the product of an even number of transpositions? Explain.

19. Let $\alpha = (1234)$.
(i) Does $\alpha^4 = (1)$?
(ii) Does $\alpha^h = \alpha^k$ when $h \equiv k \pmod 4$?
(iii) How many distinct permutations are there among the powers

$$\ldots, \alpha^{-3}, \alpha^{-2}, \alpha^{-1}, \alpha^0, \alpha, \alpha^2, \alpha^3, \ldots ?$$

20. Let θ be the s-cycle $(123 \ldots s)$.
(i) What is the smallest positive integer m such that

$$\theta^h = \theta^k \quad \text{when} \quad h \equiv k \pmod m?$$

(ii) How many distinct elements are there in $\{\theta^n : n \in \mathbf{Z}\}$?

21. Let $\theta = (12345)$. Show that the subset
$$H = \{(1), \theta, \theta^2, \theta^3, \theta^4\}$$
of $\mathbf{S_5}$ is closed under multiplication, by making a multiplication table for H.

22. Let θ be the s-cycle $(12 \ldots s)$. What is the smallest integer r such that the set $\{(1), \theta, \theta^2, \ldots, \theta^{r-1}\}$ is closed under multiplication?

23. Let $\theta = (12)(345)$. Find the smallest positive integer r such that $\theta^r = (1)$ and the smallest positive integer m such that $\theta^h = \theta^k$ whenever $h \equiv k \pmod{m}$.

24. Let $\theta = \alpha\beta$, where α and β are disjoint cycles of lengths 9 and 6, respectively. Find the smallest positive integer s such that $\theta^s = (1)$. Is $\theta^h = \theta^k$ whenever $h \equiv k \pmod{s}$?

25. Let $\theta = \alpha\beta$, where α and β are disjoint cycles of lengths r and s, respectively. Let t be an integer. Prove that $\theta^t = (1)$ if and only if $(\operatorname{lcm}[r, s]) \mid t$.

26. Let $\theta = (12345)(567)$. Find the smallest positive integer q such that $\theta^q = (1)$. [*Hint:* First find the standard form for θ.]

27. Find the smallest positive integer m such that $\theta^m = (1)$ for all θ in $\mathbf{S_4}$.

28. Find the smallest positive integer n such that $\theta^n = (1)$ for all θ in $\mathbf{S_5}$.

29. Complete the following table.

θ	(1)	(123)	(132)	(23)	(13)	(12)
$(12)\theta(12)$		(132)			(23)	

It may be helpful to use the multiplication table for $\mathbf{S_3}$ on the inside back cover.

30. Complete the following table.

θ	(1)	(123)	(132)	(23)	(13)	(12)
$(123)\theta(132)$						

It may be helpful to use the multiplication table for $\mathbf{S_3}$ on the inside back cover of the book.

31. A permutation θ with $\theta^2 = (1)$ is called an *involution*.

(i) Explain why the standard form for an involution θ involves no cycle with length greater than 2.

★ (ii) Let $T(n)$ be the number of involutions in $\mathbf{S_n}$. Prove that
$$T(n+1) = T(n) + nT(n-1) \quad \text{for } n > 1.$$

2.5 Subgroups in a Group

Expressed in cycle notation, the multiplication table for the symmetric group $\mathbf{S_3}$ is shown in Table 2.1.

TABLE 2.1 The Symmetric Group S_3

	(1)	(123)	(132)	(23)	(13)	(12)
(1)	(1)	(123)	(132)	(23)	(13)	(12)
(123)	(123)	(132)	(1)	(13)	(12)	(23)
(132)	(132)	(1)	(123)	(12)	(23)	(13)
(23)	(23)	(12)	(13)	(1)	(132)	(123)
(13)	(13)	(23)	(12)	(123)	(1)	(132)
(12)	(12)	(13)	(23)	(132)	(123)	(1)

Let us consider the subset $T = \{(1), (23)\}$ of $\mathbf{S}_3$. If we strike out of the table all the rows and columns except those for (1) and (23), what remains is the following multiplication table for T:

	(1)	(23)
(1)	(1)	(23)
(23)	(23)	(1)

Does the operation of this table make T into a group? Let us check the axioms. Since each entry in this smaller table is (1) or (23), the set T is closed under multiplication. The permutation (1) is an identity of T since it is an identity in the larger set $\mathbf{S}_3$. The small table also shows that each element of T is its own inverse. Finally, the elements of the subset T obey the associative law, since they are elements of the group $\mathbf{S}_3$. These facts show that T is also a group under the operation of composition of permutations.

We generalize on this example as follows.

DEFINITION 1

Subgroup in a Group

Let H consist of some or all of the elements of a group G. Let the product ab of elements of H be the same as when a and b are thought of as elements of G. If H is a group under this operation, H is called a **subgroup** in G.

LEMMA 1 **Identity and Inverses in a Subgroup**

Let H be a subgroup in G. Then the identity of H is the identity of G, and the inverse of an h of H is the inverse of h as an element of G.

Proof Let $\mathbf{e}$ be the identity of H. Then $\mathbf{ee} = \mathbf{e}$ in H and also in G, since multiplication in H is the same as in G. It then follows from Uniqueness of the Identity (Theorem 2 of Section 2.3) that $\mathbf{e}$ is the identity of G.

Similarly, Uniqueness of the Inverse (Theorem 3 of Section 2.3) implies that h^{-1} in H is the same as in G. ■

LEMMA 2 Sufficient Conditions for a Subgroup

Let H be a nonempty subset closed under the operation of a group G. For every h in H, let its inverse h^{-1} in G be in H. Then H is a subgroup in G.

Proof Associativity for G automatically implies associativity, under the same operation, for any subset of G. Since H is nonempty, there is some h in H, and the hypothesis tells us that h^{-1} is in H. Then $\mathbf{e} = hh^{-1}$ is in H by the closure assumption. Hence H is a group under the operation of G; that is, H is a subgroup in G. ■

The single element subset $\{\mathbf{e}\}$ is easily seen to be a subgroup in every group G. It is also clear that every group G is a subgroup in itself. Therefore, a group with more than one element has at least two subgroups.

DEFINITION 2 Trivial Subgroup, Proper Subgroup

The singleton subgroup $\{\mathbf{e}\}$ in a group G is called the **trivial subgroup**. A **proper subgroup** in G is a subgroup H with H a proper subset of G. (The only **improper subgroup** in G is G itself.)

Frequently, the following result provides the easiest means of verifying that a given subset is a subgroup.

THEOREM 1 Subgroup Conditions

In a group G, let H be a nonempty subset closed under division; that is, let hk^{-1} be in H whenever h and k are in H. Then H is a subgroup in G.

Proof Since H is nonempty, it has at least one element a. It is given that hk^{-1} is in H when h and k are in H. Letting each of h and k equal a, we see that the identity $\mathbf{e} = aa^{-1}$ of G is in H.

For every k in H the hypothesis now tells us that $\mathbf{e}k^{-1} = k^{-1}$ is in H. It then implies that, for all h and k, we have $h(k^{-1})^{-1} = hk$ in H; that is, H is closed under multiplication. Now it follows from Lemma 2 that H is a subgroup in G. ■

The definitions and theorems that follow provide means for obtaining some subgroups in noncommutative groups. They also furnish information concerning the structure of such groups.

DEFINITION 3

Center in a Group

The **center** C in a group G is the subset consisting of the c in G such that $cx = xc$ for all x in G; that is, $C = \{c \in G : cx = xc \text{ for all } x \text{ in } G\}$.

THEOREM 2 **Center as a Subgroup**

The center C in a group G is a subgroup in G.

The proof of this theorem is left to the reader as Problem 19 of this section.

DEFINITION 4

Centralizer of a in G

Let a be a fixed element of a group G. The **centralizer** of a in G is the subset C_a of G such that g is in C_a if and only if $ag = ga$.

THEOREM 3 **Centralizers and the Center**

In a group G, let C be the center and C_a be the centralizer of a. Then:
(a) C is a subgroup in C_a for every a in G.
(b) C_a is a subgroup in G for every a in G.
(c) An element g is in C if and only if g is in C_a for every a in G.
(d) An element g is in C if and only if $C_g = G$.

The proof of this result is left to the reader as Problem 21 of this section.

EXAMPLE 1 **The Center and a Centralizer in $\mathbf{S}_3$**

Here we find the center C and the centralizer C_α of $\alpha = (132)$ in the symmetric group $\mathbf{S}_3$.

With the help of Table 2.1, one sees that the identity (1) is the only element of $\mathbf{S}_3$ which commutes with every element of $\mathbf{S}_3$; that is, $C = \{(1)\}$. Also, one sees that in $\mathbf{S}_3$ only (1), (123), and (132) commute with (132); that is, $C_\alpha = \{(1), (123), (132)\}$. We note that C is a subgroup in C_α and C_α is a subgroup in G. Thus this example illustrates some relationships stated in Theorem 3.

Problems

1. Make a multiplication table for the subset
$$A = \{(1), (123), (132)\}$$
of $\mathbf{S}_3$, and show that A is a subgroup in $\mathbf{S}_3$.
2. Show that the subset $B = \{(1), (12)\}$ is a subgroup in $\mathbf{S}_3$.
3. Let $D = \{(1), (123), (132), (23)\}$ and $F = \{(1), (123), (23), (13), (12)\}$.
   (a) Why is D not a subgroup in $\mathbf{S}_3$?
   (b) Why is F not a subgroup in $\mathbf{S}_3$?
4. Let $E = \{(1), (123), (132), (13)\}$. Show that E is not a subgroup in $\mathbf{S}_3$.
5. In a group G, let $a^2 = \mathbf{e} = b^2$ and $ab = ba$ with $a \neq \mathbf{e}$, $b \neq \mathbf{e}$, and $a \neq b$. Show that $H = \{\mathbf{e}, a, b, ab\}$ is a subgroup in G.
6. Is there a subgroup H in $\mathbf{S}_3$ such that (132) is in H but (123) is not? Explain.
7. Let $\beta = (1234)$ and $H = \{(1), \beta, \beta^2, \beta^3\}$.
   (i) Make the multiplication table for H.
   (ii) Use the table to show that H is a subgroup in $\mathbf{S}_4$.
   (iii) Is β an element of a subgroup in $\mathbf{S}_4$ that has fewer than four elements? Explain.
   (iv) Show that $\{(1), \beta^2\}$ is a subgroup in H. Is it also a subgroup in $\mathbf{S}_4$?
8. Let $\alpha = (123)$. Find the subgroup H of lowest order in $\mathbf{S}_3$ such that α is in H.
9. Let H be a subgroup having order 2 in $\mathbf{S}_3$.
   (i) Explain why H must be of the form $\{(1), \beta\}$ with $\beta^2 = (1)$.
   (ii) List all the subgroups with order 2 in $\mathbf{S}_3$.
10. Let H be a subgroup having order 3 in $\mathbf{S}_3$.
    (i) Explain why H must be of the form $\{(1), \alpha, \alpha^2\}$ with $\alpha^3 = (1)$. (See Example 2 of Section 2.3.)
    (ii) Find all subgroups of order 3 in $\mathbf{S}_3$.
11. Let $G = \{1, -1, \mathbf{i}, -\mathbf{i}\}$ be the group of four complex numbers described in Example 1 of Section 2.3.
    (i) Show that $\{1, -1\}$ is a subgroup in G.
    (ii) Show that $\{1, \mathbf{i}\}$ is not a subgroup in G.
12. Show that $H = \{(1), (12), (34), (12)(34)\}$ is a subgroup in $\mathbf{S}_4$ and that $\theta^2 = (1)$ for all θ in H. (This shows that the table for Problem 9 of Section 2.3 is the table of a group; in particular, it shows that the operation given by that table is associative.)
13. Let $G = \{\mathbf{e}, a, a^2, \ldots, a^{s-1}\}$ be a group. Is G necessarily commutative? Explain.
14. Let $G = \{x : x = a^n, n \text{ an integer}\}$ be a group. Is G necessarily abelian? Explain.
15. Let H be a subgroup in G.
    (i) If G is abelian, must H be abelian? Explain.
    (ii) If H is abelian, must G be abelian? Explain.
16. Let K be a subgroup in H and H be a subgroup in G. Explain why K must be a subgroup in G.
17. For each of the following conditions, show that the set of all permutations θ on $\{1, 2, 3, 4\}$ that satisfy the given condition is a subgroup in $\mathbf{S}_4$.
    (a) $\theta(4) = 4$.
    (b) $\theta(1)$ and $\theta(2)$ are both in $\{1, 2\}$.
    (c) $\theta(x) \leq x$ for all x in $\{1, 2, 3, 4\}$.

18. For each of the following conditions, explain why the set of all permutations θ on $\mathbf{X_n} = \{1, 2, \ldots, n\}$ that satisfy the given condition is or is not a subgroup in $\mathbf{S_n}$.
(a) $\theta(1) = 1$.
(b) $\theta(1) = 2$.
(c) $\theta(1)$ is in the set $\{1, 2\}$. (Here let $n > 2$.)
(d) $\theta(x) \geq x$ for all x in $\mathbf{X_n}$.

19. Let C consist of the elements c of a group G such that $cx = xc$ for all x in G; that is, let C be the center in G. Show that C is a subgroup in G (and thus prove Theorem 2 above).

20. Find the center in $\mathbf{S_4}$.

21. In a group G, let C be the center and C_a be the centralizer of a. Prove Theorem 3 by showing the following.
(a) C is a subgroup in C_a for all a in G.
(b) C_a is a subgroup in G for all a in G.
(c) An element g is in C if and only if g is in C_a for all a in G.
(d) An element g is in C if and only if $C_g = G$.

22. Find the centralizers of each of the following elements in $\mathbf{S_3}$.
(a) (1); (b) (12); (c) (123).

23. Let a be a fixed element of a group G and let H be a subgroup in G. Let K be the subset of G consisting of all elements of the form $a^{-1}ha$ with h in H. Prove the following.
(i) K is a subgroup in G. (The subgroup K is called the ***conjugate subgroup*** of H by a. In Section 3.2, this definition is given formally in a different context.)
(ii) H is the conjugate subgroup $\{aka^{-1} : k \in K\}$ of K by a^{-1}.

24. Let G be a group and let H be the subset of all the elements h of G such that $h^{-1} = h$.
(i) Prove that H is a subgroup in G whenever G is abelian.
(ii) If G is the group $\mathbf{S_3}$, is the subset H a subgroup? Explain.

25. Let G be a group, let s be a positive integer, and let H be the subset of all the elements of the form g^s with g in G.
(i) Prove that H is a subgroup in G whenever G is abelian.
(ii) Is H a subgroup when G is $\mathbf{S_3}$ and s is 2? Explain.
(iii) Is H a subgroup when G is $\mathbf{S_3}$ and s is 3? Explain.

26. Let G be a group, let s be a positive integer, and let H be the subset of all elements h of G such that $h^s = \mathbf{e}$.
(i) Prove that H is a subgroup in G whenever G is abelian.
(ii) Show by an example that H need not be a subgroup when G is not commutative.

27. Let G be the group $\{1, -1, \mathbf{i}, -\mathbf{i}\}$ of four complex numbers.
(i) Find all the subsets of G that are closed under multiplication.
(ii) Is each of these subsets a subgroup in G?

28. Find all the subsets of $\mathbf{S_3}$ that are closed under multiplication. Is each of these subsets a subgroup in $\mathbf{S_3}$?

29. Let γ be an s-cycle $(a_1 a_2 \ldots a_s)$ and let H consist of all the powers γ^n with n an integer. Prove that H is a subgroup of order s in any group G that has γ as an element.

30. Let a be any element of a group G and let H consist of the elements a^n for all integers n. Prove that H is a subgroup in G.

31. Let $\alpha = (12345)$ and $\beta = (12)$ be elements of a subgroup H in $\mathbf{S}_5$. Prove that $H = \mathbf{S}_5$ by showing the following.
   (i) $\alpha^{-1}\beta\alpha = (23)$, and hence (23) is in H.
   (ii) $\alpha^{-1}(23)\alpha = (34)$, and hence (34) is in H.
   (iii) (45) and (15) are in H.
   (iv) $(23)\beta(23) = (13)$, and hence (13) is in H.
   (v) (24), (35), (14), and (25) are in H.
   (vi) $H = \mathbf{S}_5$. (See Theorem 3 of Section 2.4.)

★ 32. Given that there is a 2-cycle and a 5-cycle in a subgroup H in $\mathbf{S}_5$, prove that $H = \mathbf{S}_5$.

2.6 Additive Notation, Modular Arithmetic

A group is a set with an operation that satisfies the group axioms. So far, our notation for the operation has been the same as for multiplication in our familiar number systems. For some abelian groups, it is more convenient to use additive notation. We therefore rewrite the definition of a commutative group given in Section 2.3 as follows.

DEFINITION 1 Additive Group

An **additively written group** G is a set $\hat{G}$ with an operation, called addition, that has the following five properties:

1 ***Closure*** If a and b are elements (not necessarily distinct) of $\hat{G}$, then there is a unique sum $a + b$ in $\hat{G}$.

2 ***Identity*** There is an element z in $\hat{G}$ such that, for all a in $\hat{G}$,

$$a + z = a = z + a.$$

3 ***Inverses*** For each a in $\hat{G}$ there is an element h in $\hat{G}$ such that

$$a + h = z = h + a.$$

4 ***Associativity*** If a, b, and c are elements (not necessarily distinct) in $\hat{G}$, then

$$(a + b) + c = a + (b + c).$$

5 ***Commutativity*** If a and b are in $\hat{G}$, then

$$a + b = b + a.$$

As we noted in Section 2.3, we will usually write G for $\hat{G}$.

We follow the general custom of restricting the additive notation for groups solely to abelian groups. Results that apply to all multiplicative groups also apply to all additive groups. On the other hand, results that apply to all additive groups need not apply to multiplicative groups unless they are commutative.

The identity of an additive group is generally written as 0 (***zero***). The inverse of a in an additive group is denoted by $-a$ and is called the ***negative*** (or ***additive inverse***) of a. ***Subtraction*** of b from a is defined by $a - b = a + (-b)$.

In multiplicative notation, the product $a \cdot a \cdots a$ of s equal factors a is written as a^n; we now translate this into additive notation.

NOTATION 1 **Sum of Equal Terms**

Additively, the sum $a + a + \cdots + a$ of s equal terms a is written as $s \cdot a$. We also define $0 \cdot a$ to be 0, and we let $(-s) \cdot a = -(s \cdot a)$ for all positive integers s.

Then we have

$$m \cdot a + n \cdot a = (m + n) \cdot a \quad \text{and} \quad m \cdot (n \cdot a) = (mn) \cdot a$$

for every element a of an additive group and all integers m and n. These formulas are the translations into additive notation of the rules for powers stated in Section 2.3. In the notation $m \cdot a$, the symbol a is used for an element of an additive group, and m stands for an integer (that is usually not in G); the centered dot between m and a helps to remind us that $m \cdot a$ does not mean the product of two group elements.

EXAMPLE 1 **Translation into Additive Notation**

The translation of Theorem 1 of Section 2.5 (Subgroup Conditions) into additive notation is the following:

In an additive group G, let H be a nonempty subset closed under subtraction; that is, let $h - k$ be in H whenever h and k are in H. Then H is a subgroup in G.

The special case in which G is the additive group of the integers is dealt with in Theorem 3 of Section 1.3.

The familiar number systems of elementary mathematics are a source of additive and multiplicative abelian groups. We assume some familiarity with these number systems; in particular, we assume it known that the following sets of numbers are groups under the given operation.

Groups Under Addition

1. The set **Z** of all the integers
2. The set **Q** of all the rational numbers
3. The set **R** of all the real numbers
4. The set **C** of all the complex numbers

Groups Under Multiplication

a. The subset $\{1, -1\}$ of the integers
b. The nonzero rational numbers
c. The nonzero real numbers; also, the positive real numbers
d. The nonzero complex numbers

Next we introduce "modular arithmetic." That is, we use properties of the set $\mathbf{Z}$ of all the integers to define, for each positive integer m, addition and multiplication operations on a set that is essentially the set $\{0, 1, \ldots, m-1\}$ of possible remainders in division by m. This provides a source of many abelian groups. We will see in Chapter 4 that this concept also leads to valuable rings and fields. And in Section 8.3, we describe a coding theory application that has been used in the electronic transfer of funds.

DEFINITION 2

Modular Arithmetic

For each m in $\mathbf{Z}^+$, let $\mathbf{Z}_m = \{\bar{0}, \bar{1}, \ldots, \overline{m-1}\}$. Addition in $\mathbf{Z}_m$ is given by $\bar{a} + \bar{b} = \bar{r}$, where r is the remainder in the division of $a + b$ by m. Multiplication is given by $\bar{a} \cdot \bar{b} = \bar{s}$, where s is the remainder in division of ab by m.

For example, in $\mathbf{Z}_{12}$ we have $\bar{9} + \bar{5} = \bar{2}$ and $\bar{9} \cdot \bar{5} = \bar{9}$ since $9 + 5 = 14 = 12 + 2$ and $9 \cdot 5 = 45 = 3 \cdot 12 + 9$. Addition in $\mathbf{Z}_{12}$ is related to "addition of hours on clocks"; in particular, $\bar{9} + \bar{5} = \bar{2}$ in $\mathbf{Z}_{12}$ is consistent with the fact that five hours after nine o'clock it is two o'clock. Similarly, addition of minutes (or seconds) is related to addition in $\mathbf{Z}_{60}$. And, the question of "what day of the week will it be n days from now" may be facilitated by addition in $\mathbf{Z}_7$.

EXAMPLE 2 **Addition and Multiplication in $\mathbf{Z}_4$**

Using Definition 2, one finds that addition and multiplication in $\mathbf{Z}_4 = \{\bar{0}, \bar{1}, \bar{2}, \bar{3}\}$ are as in the following tables.

$+$	$\bar{0}$	$\bar{1}$	$\bar{2}$	$\bar{3}$
$\bar{0}$	$\bar{0}$	$\bar{1}$	$\bar{2}$	$\bar{3}$
$\bar{1}$	$\bar{1}$	$\bar{2}$	$\bar{3}$	$\bar{0}$
$\bar{2}$	$\bar{2}$	$\bar{3}$	$\bar{0}$	$\bar{1}$
$\bar{3}$	$\bar{3}$	$\bar{0}$	$\bar{1}$	$\bar{2}$

$\cdot$	$\bar{0}$	$\bar{1}$	$\bar{2}$	$\bar{3}$
$\bar{0}$	$\bar{0}$	$\bar{0}$	$\bar{0}$	$\bar{0}$
$\bar{1}$	$\bar{0}$	$\bar{1}$	$\bar{2}$	$\bar{3}$
$\bar{2}$	$\bar{0}$	$\bar{2}$	$\bar{0}$	$\bar{2}$
$\bar{3}$	$\bar{0}$	$\bar{3}$	$\bar{2}$	$\bar{1}$

Clearly each of these operations is commutative. The identity for addition is $\bar{0}$, and the additive inverse (that is, negative) $-\bar{a}$ of an $\bar{a}$ in $\mathbf{Z}_4$ is $\overline{4-a}$. The identity for multiplication is $\bar{1}$. The equations $\bar{1} \cdot \bar{1} = \bar{1} = \bar{3} \cdot \bar{3}$ show that each of $\bar{1}$ and $\bar{3}$ is its own multiplicative inverse (that is, reciprocal). Neither $\bar{0}$ nor $\bar{2}$ has a multiplicative inverse in $\mathbf{Z}_4$ since $\bar{0} \cdot \bar{a} \neq \bar{1}$ and $\bar{2} \cdot \bar{a} \neq \bar{1}$ for every $\bar{a}$ in $\mathbf{Z}_4$.

DEFINITION 3 **Set V_m of Invertibles in Z_m**

An element $\bar{v}$ of $\mathbf{Z_m}$ is called an **invertible** if $\bar{v}$ has a multiplicative inverse in $\mathbf{Z_m}$. We let $\mathbf{V_m}$ denote the set of all invertibles in $\mathbf{Z_m}$.

From Example 2, we see that $\mathbf{V_4} = \{\bar{1}, \bar{3}\}$.

THEOREM 1 **Properties of Z_m**

Under addition, $\mathbf{Z_m}$ is a commutative group with $\bar{0}$ as identity and with $-\bar{a} = \overline{m - a}$ for each $\bar{a}$. Under multiplication, $\mathbf{Z_m}$ is closed and associative and has $\bar{1}$ as the identity. The subset $\mathbf{V_m}$ of invertibles is a group under multiplication.

Proof $\mathbf{Z_m}$ is closed under addition and multiplication, since the r and s in Definition 2 are remainders in division by m and hence are in $\{0, 1, \ldots, m - 1\}$. Addition is commutative (that is, $\bar{a} + \bar{b} = \bar{b} + \bar{a}$) because $a + b = b + a$ for all integers a and b, and therefore $a + b$ and $b + a$ have the same remainder in division by m. Similarly, multiplication is commutative.

Next we want to show that addition is associative; that is,

$$(\bar{a} + \bar{b}) + \bar{c} = \bar{a} + (\bar{b} + \bar{c}) \quad \text{for all } \bar{a}, \bar{b}, \bar{c} \text{ in } \mathbf{Z_m}.$$

We have $\bar{a} + \bar{b} = \bar{r}$, with r the remainder in the division of $a + b$ by m. Then $(\bar{a} + \bar{b}) + \bar{c} = \bar{r} + \bar{c} = \bar{t}$, with t the remainder in the division of $r + c$ by m. This means that there are integers q and q' with

$$a + b = qm + r, \qquad r + c = q'm + t.$$

Then

$$a + b + c = qm + r + c = qm + q'm + t = (q + q')m + t.$$

It follows that $(\bar{a} + \bar{b}) + \bar{c} = \bar{t}$, with t the (unique) remainder in the division of $a + b + c$ by m. Similarly one can show that $\bar{a} + (\bar{b} + \bar{c}) = \bar{t}$. Thus, addition in $\mathbf{Z_m}$ is associative.

The other details of the proof are straightforward and are left to the reader. ∎

The two results that follow are helpful in dealing with multiplicative properties of the complex numbers.

THEOREM 2 **Polar Form of a Complex Number**

For all real numbers a and b, the complex number $a + b\mathbf{i}$ is expressible as

$$a + b\mathbf{i} = r(\cos t + \mathbf{i} \sin t),$$

with r and t real and $r \geq 0$.

Proof Let $r = \sqrt{a^2 + b^2}$. Then

$$\left(\frac{a}{r}\right)^2 + \left(\frac{b}{r}\right)^2 = \frac{a^2 + b^2}{r^2} = \frac{r^2}{r^2} = 1.$$

Hence there is a real number t such that $\cos t = a/r$ and $\sin t = b/r$. Then

$$a + b\mathbf{i} = r\left(\frac{a}{r} + \mathbf{i}\,\frac{b}{r}\right) = r(\cos t + \mathbf{i} \sin t).$$

■

The nonnegative number $r = \sqrt{a^2 + b^2}$ is the ***absolute value*** (or ***modulus***) of the complex number $a + b\mathbf{i}$ and is denoted by $|a + b\mathbf{i}|$. A number t, such that $\cos t = a/r$ and $\sin t = b/r$, is called an ***argument*** (or ***amplitude***, or ***direction angle***) of $a + b\mathbf{i}$. (See Figure 2.3.)

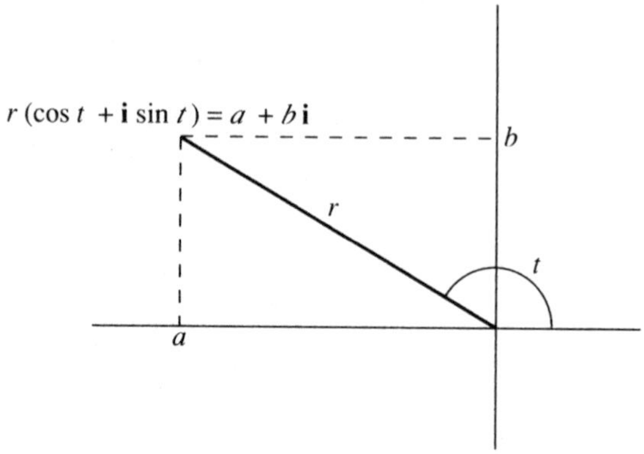

FIGURE 2.3

Since 2π is a period for the cosine and sine functions, we see that if t is an argument for $a + b\mathbf{i}$ so is $t + 2n\pi$, for every integer n.

EXAMPLE 3 **Rectangular into Polar and Vice Versa**
Here we convert $4 + 4\mathbf{i}$ into polar form and $6[\cos(\pi/3) + \mathbf{i}\sin(\pi/3)]$ into the rectangular form $a + b\mathbf{i}$. Using $a = 4$ and $b = 4$, we have $\sqrt{a^2 + b^2} = \sqrt{32} = 4\sqrt{2} = r$. An angle t with $\cos t = a/r = 4/4\sqrt{2} = 1/\sqrt{2}$ and $\sin t = b/r = 1/\sqrt{2}$ is $t = \pi/4$. Hence $4 + 4\mathbf{i} = 4\sqrt{2}[\cos(\pi/4) + \mathbf{i}\sin(\pi/4)]$.

For the second part, we start with $r = 6$ and $t = \pi/3$. Then $\cos t = \cos(\pi/3) = 1/2$ and $\sin t = \sin(\pi/3) = \sqrt{3}/2$. Thus,

$$6\left(\cos\frac{\pi}{3} + \mathbf{i}\sin\frac{\pi}{3}\right) = 6\left(\frac{1}{2} + \mathbf{i}\frac{\sqrt{3}}{2}\right) = 3 + 3\sqrt{3}\,\mathbf{i}.$$

The following theorem is illustrated in Figure 2.4.

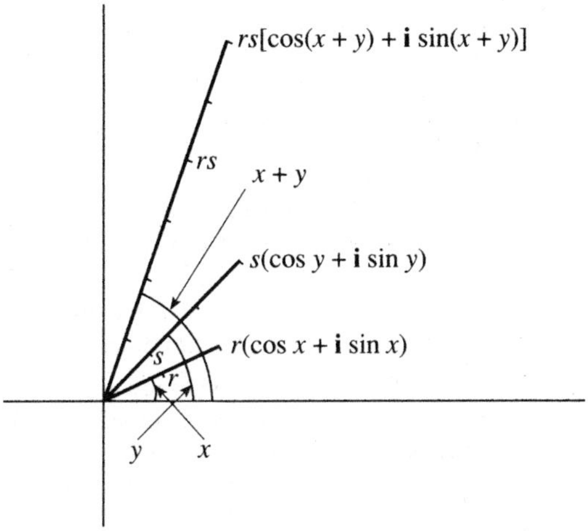

FIGURE 2.4

THEOREM 3 Polar Multiplication

The absolute value of a product of two complex numbers is the product of their absolute values; that is, $|\alpha\beta| = |\alpha| \cdot |\beta|$. Also, the sum of arguments of complex numbers α and β is an argument of $\alpha\beta$.

Proof Using the addition formulas

$$\cos(x + y) = \cos x \cdot \cos y - \sin x \cdot \sin y,$$

$$\sin(x + y) = \sin x \cdot \cos y + \cos x \cdot \sin y,$$

we obtain the desired results as follows:

$$[r(\cos x + \mathbf{i} \sin x)][s(\cos y + \mathbf{i} \sin y)]$$

$$= rs[(\cos x \cdot \cos y - \sin x \cdot \sin y) + \mathbf{i}(\sin x \cdot \cos y + \cos x \cdot \sin y)]$$

$$= rs[\cos(x + y) + \mathbf{i} \sin(x + y)]. \qquad \blacksquare$$

Problems

Problems 5 and 6 below are cited in Section 2.7.

1. For each of the following sets, tell whether it is or is not a group under addition. If it is not a group, name a group axiom it does not satisfy.
   (a) All the integral multiples of 7; that is, $\{\ldots, -7, 0, 7, 14, \ldots\}$.
   (b) All the polynomials $ax^2 + bx + c$ with a, b, and c real numbers and $a \neq 0$.
   (c) All the polynomials $ax^3 + bx^2 + cx + d$ with a, b, c, and d any real numbers.
2. Do as in Problem 1 for the following sets.
   (a) All the positive integers; that is, $\mathbf{Z}^+ = \{1, 2, 3, \ldots\}$.

(b) The nonnegative integers; that is, $\{0, 1, 2, \ldots\}$.

(c) All 2×3 real matrices; that is, all arrays of real numbers a, b, c, f, g, and h in 2 rows and 3 columns with addition defined by

$$\begin{pmatrix} a & b & c \\ f & g & h \end{pmatrix} + \begin{pmatrix} a' & b' & c' \\ f' & g' & h' \end{pmatrix} = \begin{pmatrix} a + a' & b + b' & c + c' \\ f + f' & g + g' & h + h' \end{pmatrix}.$$

3. For each of the following sets, tell whether it is or is not a group under multiplication. If it is not a group, name a group axiom it does not satisfy.

(a) $\mathbf{Z} = \{\ldots, -2, -1, 0, 1, 2, 3, \ldots\}$.

(b) The polynomials $ax + b$ with a and b real numbers.

(c) All the positive rational numbers.

4. Do as in Problem 3 for the following sets.

(a) All the rational numbers of the form $2^m 3^n$, with m and n integers.

(b) All complex numbers α with $|\alpha| \leq 1$.

(c) All 2×2 matrices of the form

$$\begin{pmatrix} \cos x & \sin x \\ -\sin x & \cos x \end{pmatrix},$$

with x a real number and multiplication defined by

$$\begin{pmatrix} a & b \\ c & d \end{pmatrix}\begin{pmatrix} f & g \\ h & k \end{pmatrix} = \begin{pmatrix} af + bh & ag + bk \\ cf + dh & cg + dk \end{pmatrix}.$$

5. Describe a subset T of the additive group $\mathbf{Z}$ of the integers such that T is closed under addition but T is not a subgroup in $\mathbf{Z}$.

6. Describe a subset S of the multiplicative group V of the nonzero complex numbers such that S is closed under multiplication but S is not a subgroup in V.

7. Restate Example 2 of Section 2.3 (without the solution) in the language of additive groups.

8. Restate Problem 30 of Section 2.5 (without the solution) in the language of additive groups.

9. Restate the multiplicative group result $(a^{-1})^{-1} = a$ in additive notation.

10. Explain why the multiplicative group formula $(ab)^{-1} = b^{-1}a^{-1}$ can be stated for additive groups as

$$-(a + b) = (-a) + (-b).$$

11. (i) Complete the rows of the following table for $\mathbf{Z}_{10}$.

$\bar{a}$	$\bar{0}$	$\bar{1}$	$\bar{2}$	$\bar{3}$	$\bar{4}$	$\bar{5}$	$\bar{6}$	$\bar{7}$	$\bar{8}$	$\bar{9}$
$-\bar{a}$	$\bar{0}$	$\bar{9}$	$\bar{8}$							
$\bar{a} + \bar{3}$	$\bar{3}$	$\bar{4}$	$\bar{5}$							
$\bar{2} \cdot \bar{a}$	$\bar{0}$	$\bar{2}$	$\bar{4}$							
$\bar{3} \cdot \bar{a}$	$\bar{0}$	$\bar{3}$	$\bar{6}$							

(ii) Find $(\bar{3})^{-1}$, that is, the reciprocal of $\bar{3}$ in $\mathbf{Z}_{10}$. [*Hint:* Use the $\bar{3} \cdot \bar{a}$ row.]

(iii) Is $\bar{2}$ an invertible in $\mathbf{Z}_{10}$? [*Hint:* Use the $\bar{2} \cdot \bar{a}$ row.]

(iv) Is there an element $\bar{b}$ in $\mathbf{Z}_{10}$ such that $\bar{b} \neq \bar{0}$ but $\bar{6} \cdot \bar{b} = \bar{0}$?

(v) Is there an element $\bar{c}$ in $\mathbf{Z}_{10}$ such that $\bar{c} \neq \bar{0}$ but $\bar{9} \cdot \bar{c} = \bar{0}$?

12. (i) Find the set $\mathbf{V}_{10}$ of invertibles in $\mathbf{Z}_{10}$, and give $(\bar{v})^{-1}$ for each $\bar{v}$ in $\mathbf{V}_{10}$.

(ii) Make the multiplication table for $\mathbf{V}_{10}$.

(iii) Is there an element $\bar{d}$ in $\mathbf{Z}_{10}$ such that $\bar{d} \neq \bar{0}$ but $\bar{7} \cdot \bar{d} = \bar{0}$?
(iv) Is there an element $\bar{f}$ in $\mathbf{Z}_{10}$ such that $\bar{f} \neq \bar{0}$ but $\bar{8} \cdot \bar{f} = \bar{0}$?

13. Find $2 \cdot M$, where M is the 2×3 real matrix

$$\begin{pmatrix} 1 & 2 & 2 \\ 4 & -6 & 3 \end{pmatrix}.$$

[Note that, in an additive group, $2 \cdot M$ denotes $M + M$. Also see Problem 2(c).]

14. Find $(-2) \cdot M$ for the M of the preceding problem.

15. Express each of the following complex numbers in the polar form $r(\cos t + \mathbf{i} \sin t)$.
(a) $-6 + 6\mathbf{i}$; (b) $9\mathbf{i}$; (c) $-5 = -5 + 0\mathbf{i}$.

16. Convert each of the following into polar form.
(a) $1 + \sqrt{3}\mathbf{i}$; (b) $\sqrt{3} - \mathbf{i}$; (c) 8.

17. Convert each of the following complex numbers from polar form into the rectangular form $a + b\mathbf{i}$.
(a) $8[\cos(\pi/6) + \mathbf{i} \sin(\pi/6)]$; (b) $\sqrt{2}[\cos(5\pi/4) + \mathbf{i} \sin(5\pi/4)]$.

18. Convert each of the following into $a + b\mathbf{i}$ form.
(a) $4[\cos(5\pi/6) + \mathbf{i} \sin(5\pi/6)]$; (b) $9[\cos(3\pi/2) + \mathbf{i} \sin(3\pi/2)]$;
(c) $6[\cos(-\pi/4) + \mathbf{i} \sin(-\pi/4)]$; (d) $7[\cos(6\pi) + \mathbf{i} \sin(6\pi)]$.

19. Express $(\sqrt{3} + \mathbf{i})^{10}(2 - 2\sqrt{3}\mathbf{i})^9$ in $a + b\mathbf{i}$ form by first converting $\sqrt{3} + \mathbf{i}$ and $2 - 2\sqrt{3}\mathbf{i}$ into polar form, then raising to powers and multiplying in polar form, and finally converting back.

20. Express $(5 + 5\mathbf{i})^{11}(-7\mathbf{i})^8$ in $a + b\mathbf{i}$ form.

21. Let $\omega = \cos(2\pi/3) + \mathbf{i} \sin(2\pi/3)$. Show the following.
(i) $\omega^3 = 1$.
(ii) $(\omega^3 - 1)/(\omega - 1) = 0$.
(iii) $\omega^2 + \omega + 1 = 0$.
(iv) $(x + 1)^3 + (x + \omega)^3 + (x + \omega^2)^3 = 3(x^3 + 1)$.

22. Let ω be the complex cube root of unity $(-1 + \mathbf{i}\sqrt{3})/2$. Use the fact that $\omega^2 + \omega + 1 = 0$ to show that, for all complex numbers α, β, and γ, one has

$$\begin{aligned}(\alpha + \beta + \gamma)(\alpha + \omega\beta + \omega^2\gamma)(\alpha + \omega^2\beta + \omega\gamma) \\ &= (\alpha + \beta + \gamma)(\alpha^2 + \beta^2 + \gamma^2 - \beta\gamma - \alpha\gamma - \alpha\beta) \\ &= \alpha^3 + \beta^3 + \gamma^3 - 3\alpha\beta\gamma.\end{aligned}$$

23. Let A consist of all nonzero complex numbers $r(\cos t + \mathbf{i} \sin t)$ with the argument t in the set $\{0, 2\pi/3, 4\pi/3\}$. Prove that A is a subgroup in the multiplicative group V of the nonzero complex numbers.

24. Let U be the set of all complex numbers α with $|\alpha| = 1$; that is, let

$$U = \{\cos t + \mathbf{i} \sin t : t \in \mathbf{R}\}.$$

Prove that U is a subgroup in the multiplicative group V of the nonzero complex numbers. (The group U is called the ***circle group*** or the ***unit circle group***.)

25. Let n be a positive integer. For $j = 0, 1, 2, \ldots, n - 1$, let

$$\alpha_j = \cos \frac{2\pi j}{n} + \mathbf{i} \sin \frac{2\pi j}{n}$$

and $U_n = \{\alpha_0, \alpha_1, \ldots, \alpha_{n-1}\}$. Show the following.
(a) $(\alpha_j)^n = 1$ for all j.
(b) U_n is a subgroup in the circle group U of Problem 24.

[By (a), each number in U_n is an nth root of 1; one can see that U_n consists of all n nth roots of unity.]

26. Let T consist of all the real numbers t such that both $\cos t$ and $\sin t$ are in $\mathbf{Q}$ (that is, both are rational). Let K consist of all complex numbers $\cos t + \mathbf{i} \sin t$ with t in T. Prove the following.
(a) T is a subgroup in the additive group of the real numbers $\mathbf{R}$.
(b) K is a subgroup in the circle group U of Problem 24.
(c) If α and β are in K, $|\alpha^2 - \beta^2|$ is in $\mathbf{Q}$.

★ **27.** Let G be a subgroup of order n in the multiplicative group V of the nonzero complex numbers. Prove that G must be the group U_n of Problem 25.

★ **28.** Let S be the set of all integers s of the form $s = m^2 + n^2$ with m and n integers. Prove the following.
(i) S is closed under multiplication.
(ii) If s is in S, so are $2s$ and $5s$.

2.7 Cyclic Groups

A mathematician who is not also something of a poet will never be a complete mathematician. —**Karl Weierstrass**

A scientist worthy of the name, above all a mathematician, experiences in his work the same expression as an artist; his pleasure is as great and of the same nature. —**Henri Poincaré**

One way to obtain subgroups H in a group G is to let a be a fixed element of G and to let H consist of the distinct elements among the integral powers

$$\ldots, a^{-2}, a^{-1}, \mathbf{e}, a, a^2, a^3, \ldots;$$

that is, we let $H = \{a^m : m \in \mathbf{Z}\}$. A subset H of this form is nonempty, since a is in H. Furthermore, H is closed under division, since the quotient $a^r(a^s)^{-1}$ of elements of H is the element a^{r-s} of H. Hence such an H is a subgroup in G by Theorem 1 of Section 2.5.

DEFINITION 1 **Cyclic Subgroup $[a]$ Generated by a in G**

Let a be in a group G. Then $\{a^m : m \in \mathbf{Z}\}$ is called the **cyclic subgroup generated by a** and this subgroup is designated as $[a]$. If G is written additively, the cyclic subgroup generated by a is $[a] = \{m \cdot a : m \in \mathbf{Z}\}$; that is, $[a]$ consists of the distinct elements among the "additive powers"

$$\ldots, -2 \cdot a, -1 \cdot a, 0, a, 2 \cdot a, 3 \cdot a, \ldots.$$

For example, in $\mathbf{S}_3$ we have $(123)^3 = (1)$, and it follows that $(123)^h = (123)^k$ when $h \equiv k \pmod 3$. Thus every integral power of (123) is equal to one of the permutations

$$(123)^0 = (1), \quad (123)^1 = (123), \quad (123)^2 = (132)$$

and hence the cyclic subgroup generated by (123) in $\mathbf{S}_3$ is

$$[(123)] = \{(1), (123), (132)\}.$$

In the additive group $\mathbf{Z}$ of the integers, the cyclic subgroup generated by 3 is

$$[3] = \{m \cdot 3 : m \in \mathbf{Z}\} = \{\ldots, -6, -3, 0, 3, 6, 9, \ldots\}.$$

In the modular arithmetic additive group $\mathbf{Z}_8 = \{\bar{0}, \bar{1}, \bar{2}, \ldots, \bar{7}\}$, the cyclic subgroup generated by $\bar{2}$ is

$$[\bar{2}] = \{m \cdot \bar{2} : m \in \mathbf{Z}\} = \{\bar{0}, \bar{2}, \bar{4}, \bar{6}\}$$

and the cyclic subgroup generated by $\bar{3}$ is

$$[\bar{3}] = \{m \cdot \bar{3} : m \in \mathbf{Z}\} = \{\bar{0}, \bar{3}, \bar{6}, \bar{1}, \bar{4}, \bar{7}, \bar{2}, \bar{5}\} = \mathbf{Z}_8.$$

Let K be a subgroup in G and a be an element of K. Then one can see that $[a]$ is a subgroup in K; in this sense, $[a]$ is the smallest subgroup in G that contains a. (This concept will be generalized in Theorem 2 of Section 3.5.)

The subgroup $[a]$ in a group G may be G itself. For example, in $\mathbf{S}_2$ one has $[(12)] = \{(1), (12)\} = \mathbf{S}_2$, and in the additive group $\mathbf{Z}$ of the integers one has

$$[1] = \{\ldots, -2 \cdot 1, -1 \cdot 1, 0, 1 \cdot 1, 2 \cdot 1, \ldots\}$$
$$= \{\ldots, -2, -1, 0, 1, 2, \ldots\} = \mathbf{Z}.$$

Also, we saw above that in $\mathbf{Z}_8$ one has $[\bar{3}] = \mathbf{Z}_8$.

DEFINITION 2

Cyclic Group, Generator of a Group

A **cyclic group** is a group G having at least one element g such that $G = [g]$; such an element g is called a **generator** of G.

A cyclic group may have several generators. For example, let $\beta = (1234)$ and $G = [\beta] = \{(1), \beta, \beta^2, \beta^3\}$. Clearly β is a generator of this G; we see that β^3 is also a generator by noting that

$$[\beta^3] = \{(1), \beta^3, (\beta^3)^2, (\beta^3)^3\} = \{(1), \beta^3, \beta^2, \beta\} = G.$$

In the additive group $\mathbf{Z}$ of the integers, one has $[-1] = \mathbf{Z} = [1]$ and hence each of 1 and -1 is a generator. In the modular arithmetic additive group $\mathbf{Z}_8$, it can be shown that each of $\bar{1}$, $\bar{3}$, $\bar{5}$, and $\bar{7}$ is a generator; that is, $\mathbf{Z}_8 = [\bar{1}] = [\bar{3}] = [\bar{5}] = [\bar{7}]$.

The symmetric group $\mathbf{S}_3$ is not cyclic, as one can show by finding all of its cyclic subgroups and noting that no one of them has order 6. (See Problem 17(i) of this section.)

THEOREM 1 **Cyclic Groups Are Abelian**

Every cyclic group $[a]$ is commutative.

Proof If x and y are elements of $[a]$, then $x = a^m$ and $y = a^n$ with m and n integers. Then

$$xy = a^m a^n = a^{m+n} = a^{n+m} = a^n a^m = yx,$$

since $m + n = n + m$ follows from the commutativity of addition of integers. Hence, $[a]$ is abelian. ■

A cyclic group may be finite or infinite. For example, $[(1234)]$ is a cyclic group of order 4, and the additive group of the integers

$$[1] = [-1] = \{\ldots, -2, -1, 0, 1, 2, 3, \ldots\}$$

is an infinite cyclic group.

We recall that the order of a group G is the number of elements in G and now present the following terminology.

DEFINITION 3 **Order of an Element of a Group**

The **order** of an element a of a group G is the order of the cyclic subgroup $[a]$ generated by a in G. If $[a]$ is infinite, one says that a has **infinite order**.

For example, in $\mathbf{S}_3$ the 3-cycle (123) has order 3 because the cyclic subgroup $[(123)] = \{(1), (123), (132)\}$ has order 3. Also, in the additive group $\mathbf{Z}$ of the integers, the cyclic subgroup

$$[2] = \{\ldots, -4, -2, 0, 2, 4, 6, \ldots\}$$

is infinite, so 2 has infinite order.

We note that only the identity $\mathbf{e}$ has order 1. If G is a finite cyclic group, it can easily be shown that an element g of G is a generator of G if and only if the order of g equals the order of G. (The proof is left to the reader as Problem 30 of this section.)

THEOREM 2 **Finding $[g]$ and the Order of g**

Let g be in a group G.

(a) If $g^n \neq \mathbf{e}$ for all n in $\mathbf{Z}^+$, then $[g] = \{\ldots, g^{-2}, g^{-1}, \mathbf{e}, g, g^2, \ldots\}$, and g has infinite order.

explain

(b) If s is the smallest positive integer such that $g^s = \mathbf{e}$, then $[g] = \{\mathbf{e}, g, g^2, \ldots, g^{s-1}\}$, and g has order s.

Proof of (a) Given that $g^n \neq \mathbf{e}$ for all n in $\mathbf{Z}^+$, we want to show that the powers

$$\dots, g^{-2}, g^{-1}, \mathbf{e}, g, g^2, g^3, \dots$$

are all distinct. We assume the contrary; that is, we assume that $g^i = g^j$ for integers i and j with $i < j$. Then

$$g^{j-i} = g^j(g^i)^{-1} = g^i(g^i)^{-1} = \mathbf{e}$$

with $j - i$ in $\mathbf{Z}^+$. This contradicts the hypothesis and thus shows that $g^i \neq g^j$ for distinct integers i and j. Hence $[g]$ is infinite and g has infinite order in this case. ∎

Proof of (b) Let s be the smallest positive integer with $g^s = \mathbf{e}$. The division algorithm expresses any integer n as $n = qs + r$ with q an integer and r in $\{0, 1, \dots, s - 1\}$. Then

$$g^n = g^{qs+r} = (g^s)^q g^r = \mathbf{e}^q g^r = \mathbf{e}g^r = g^r.$$

Thus each integral power g^n is one of the powers

$$\mathbf{e}, g, g^2, \dots, g^{s-1}.$$

These are distinct, since $g^i = g^j$ with $0 \leq i < j \leq s$ implies that $g^{j-i} = \mathbf{e}$ with $0 < j - i < s$ and this would contradict the minimal nature of s. Hence we have $[g] = \{\mathbf{e}, g, g^2, \dots, g^{s-1}\}$, and so g has order s. ∎

THEOREM 3 Finite Order Conditions

Let a be an element of a group G.

(a) If $a^m = \mathbf{e}$ with m in $\mathbf{Z}^+$, then a has finite order s and $s \mid m$.

(b) If $a^h = a^k$ with h and k in $\mathbf{Z}$ and $h > k$, then a has finite order s and $s \mid (h - k)$; that is, $h \equiv k \pmod{s}$.

The proof is similar to arguments contained in the proof of Theorem 2; the details are left to the reader as Problems 23 and 24 of this section.

EXAMPLE 1 Complex Sixth Roots of Unity

Let G be the multiplicative group of the nonzero complex numbers, and let $\alpha = (1 + \mathbf{i}\sqrt{3})/2$. Here we find $[\alpha]$, give the order of each number in $[\alpha]$, and find a generator β of $[\alpha]$ with $\beta \neq \alpha$. Calculation shows that $\alpha^2 = (-1 + \mathbf{i}\sqrt{3})/2$, $\alpha^3 = -1$, $\alpha^4 = -\alpha$, $\alpha^5 = -\alpha^2$, and $\alpha^6 = (-1)^2 = 1$. Hence 6 is the smallest positive integer s with α^s equal to the identity 1 of G; that is, α has order 6 and

$$[\alpha] = \{1, \alpha, \alpha^2, \alpha^3, \alpha^4, \alpha^5\}.$$

Similarly, one finds that

$$[\alpha^2] = \{1, \alpha^2, \alpha^4\} = [\alpha^4], \qquad [\alpha^3] = \{1, \alpha^3\}, \qquad [\alpha^5] = [\alpha].$$

Hence each of α^2 and α^4 has order 3, α^3 has order 2, and α^5 has order 6. Also, α^5 is the only other generator of $[\alpha]$.

The following result shows that it is easier to prove that a subset of a group is a subgroup when the subset is finite rather than infinite.

THEOREM 4 Finite Subgroup Conditions

Let H be a nonempty finite subset of a group G, and let H be closed under the operation of G. Then H is a subgroup in G.

Proof Since the single element subset $\{\mathbf{e}\}$ is a subgroup, we may assume that there is an a in H with $a \neq \mathbf{e}$. It then follows from closure of H under multiplication that all of the elements

$$a, a^2, a^3, a^4, \ldots \tag{1}$$

are in H. But H is finite; hence there must be repetitions in (1). This and Theorem 3(b) imply that a has finite order s.

Since $a \neq \mathbf{e}$, we have $s > 1$. Then the identity $\mathbf{e} = a^s$ and the inverse $a^{-1} = a^{s-1}$ of a are among the elements of (1), and hence they are in H. It now follows from Lemma 2 of Section 2.5 that H is a subgroup in G. ∎

An infinite subset S of a group G may be closed under the operation of G without being a subgroup in G. (In Problems 5 and 6 of Section 2.6, the reader was asked to give examples of such subsets.)

Problems

1. State the order of each of the following.
   (i) (1); (ii) (12); (iii) (123); (iv) (1234); (v) (12345).
2. What is the order of an s-cycle $(a_1 a_2 \ldots a_s)$?
3. Put (123)(34) in standard form, and find its order.
4. Put (12)(13)(14) in standard form, and find its order.
5. Find the order of each of the following.
   (i) (12)(34); (ii) (12)(345); (iii) (12)(3456); (iv) (12)(34567).
6. Find the order of each of the following.
   (i) (123)(45); (ii) (123)(456); (iii) (123)(4567);
   (iv) (123)(45678); (v) (123)(456789).
7. For each of the following, give the order of a product of disjoint cycles of the given lengths.
   (i) 2 and 3; (ii) 2 and 4; (iii) 4 and 5; (iv) 4 and 6;
   (v) 4, 5, and 6; (vi) 2, 3, 4, 5, and 6.
8. State a generalization of the answers in Problem 7.
9. Let a be an element of order 12 in a group G.
   (i) Find the smallest positive integer r such that $12 \mid (9r)$.
   (ii) Find the smallest positive integer s such that $(a^9)^s = \mathbf{e}$.
   (iii) What is the order of a^9?
   (iv) What is the order of a^8?

10. Let a be an element of order 12 in a group G.
(i) Find the smallest positive integer r such that $12 \mid (10r)$.
(ii) Find the smallest integer s such that $(a^{10})^s = \mathbf{e}$.
(iii) What is the order of a^{10}?
(iv) What is the order of a^7?

11. Let a be an element of order 6 in a group G. State the order of each of the following.
(a) a^2; (b) a^3; (c) a^4; (d) a^5; (e) a^6.

12. In $\mathbf{S}_6$, let $\alpha = (123456)$.
(i) What is the order of α?
(ii) Give the orders of the elements of the cyclic subgroup $[\alpha]$ in $\mathbf{S}_6$. [*Hint:* See Problem 11.]
(iii) List all the generators of $[\alpha]$.

13. Let a be an element of order 7 in a group G. State the order of each of the following.
(a) a^2; (b) a^3; (c) a^4; (d) a^5; (e) a^6; (f) a^7.

14. In $\mathbf{S}_7$, let $\alpha = (1234567)$.
(i) What is the order of α?
(ii) Give the orders of the elements of $[\alpha]$.
(iii) List all the generators of $[\alpha]$.

15. Let V be the multiplicative group of the nonzero complex numbers.
(i) List the elements of the cyclic subgroup $[\mathbf{i}]$ in V.
(ii) What is the order of $\mathbf{i}$ in V?
(iii) Does $\beta^4 = 1$ for every β in $[\mathbf{i}]$?
(iv) List the generators of $[\mathbf{i}]$.

16. Let V be the multiplicative group of the nonzero complex numbers, and let $\alpha = (1 + \mathbf{i})/\sqrt{2}$.
(i) List the elements of the cyclic subgroup $[\alpha]$ in V.
(ii) What is the order of α?
(iii) Does $\beta^8 = 1$ for every β in $[\alpha]$?
(iv) List all the generators of $[\alpha]$.

17. (i) For each θ in $\mathbf{S}_3$, list the elements of the cyclic subgroup $[\theta]$ in $\mathbf{S}_3$ and give the order of $[\theta]$.
(ii) Is the order of θ an integral divisor of the order $\mathbf{S}_3$ for every θ in $\mathbf{S}_3$?
(iii) Is there a positive integral divisor d of 6 such that d is not the order of any θ in $\mathbf{S}_3$? Explain why $\mathbf{S}_3$ is not cyclic.
(iv) Let $G = [a] = \{\mathbf{e}, a, a^2, a^3, a^4, a^5\}$ be a cyclic group of order 6. Is it true that a positive integer d is the order of an element of G if and only if $d \mid 6$?

18. (i) Use the fact that an element of $\mathbf{S}_4$ has a standard form of the type (1), (xy), (xyz), $(xyzw)$, or $(xy)(zw)$ to give the set of positive integers that are orders of elements of $\mathbf{S}_4$.
(ii) Is the order of θ an integral divisor of the order of $\mathbf{S}_4$ for every θ in $\mathbf{S}_4$?
(iii) Are there positive integral divisors d of 24 such that no θ in $\mathbf{S}_4$ has order d?
(iv) Let $G = [a]$ be a cyclic group of order 24. Is it true that a positive integer d is the order of an element of G if and only if $d \mid 24$?

19. Let $G = [a]$ be a cyclic group of order 30. List the elements of a subgroup H of order q for each of the following values of q.
(a) 2; (b) 3; (c) 5; (d) 6.

20. Let a be an element of finite order q in a group G. Show the following.
(a) a^{-1} also has order q.
(b) For every integer k, a^k and a^{q-k} have the same order.
(c) $[a^{-1}] = [a]$.
(d) If q is an even integer $2r$ and $r > 1$, a^r is not a generator of $[a]$.
(e) If $q > 2$, the number of generators of $[a]$ is even.

21. Let $G = [a]$ be a cyclic group of order 30.
(a) List all the elements of order 2 in G.
(b) List all the elements of order 3 in G.
(c) List all the elements of order 10 in G.
(d) List all the generators of G.

22. Let $G = [a]$ be a cyclic group of order 18.
(a) List all the elements of order 2 in G.
(b) List all the elements of order 3 in G.
(c) List all the elements of order 6 in G.
(d) List all the generators of G.

23. In a group G, let $a^m = \mathbf{e}$ with m in $\mathbf{Z}^+$. Prove that a has finite order s and that $s \mid m$. (This is the proof of Theorem 3(a) that was left to the reader.)

24. In a group G, let $a^h = a^k$ with h and k in $\mathbf{Z}$ and $h > k$. Prove that a has finite order s and that $s \mid (h - k)$. (This is the proof of Theorem 3(b) that was left to the reader.)

25. Let a^{15} equal the identity $\mathbf{e}$ of G. What are the possibilities for the order of a?

26. Let $a^m = \mathbf{e}$ in G, with m an integer. What are the possibilities for the order of a?

27. Let G have m elements g satisfying $g^2 = \mathbf{e}$.
(i) How many elements of order 1 are there in G?
(ii) How many elements of order 2 are there in G?

28. Let g be in a group G.
(i) Can g have larger order than G has? Explain.
(ii) Can g have infinite order while G is finite?

29. Is a group of order 2 or 3 necessarily cyclic? Explain.

30. Let G be a group of order m. Show that G is cyclic if and only if it has an element of order m; that is, show both of the following.
(i) If G has an element g of order m, G is cyclic.
(ii) If G is cyclic, G has an element g of order m.

31. Let a be an element of order 5 in a group. Define a function of two variables $t = F(r, s)$ by $a^r a^s = a^t$ and $0 \le t < 5$. Tabulate t for r and s ranging through $\{0, 1, 2, 3, 4\}$.

32. Do the previous problem for an a of order 4, letting r, s, and t range through $\{0, 1, 2, 3\}$.

33. Let S be a nonempty subset of the integers and let S be closed under subtraction. Show that S must be a cyclic subgroup $[m]$ in the additive group $\mathbf{Z}$ of the integers for some integer m.

34. Let H be a subgroup in the additive group $\mathbf{Z}$ of the integers. Show that H is cyclic.

35. Let a be an element of finite order in a group G. Let S be the set of all integers m such that $a^m = \mathbf{e}$. Show that S contains at least one positive integer and is closed under subtraction and hence that S consists of all the integral multiples of the order q of a; that is, show that S is the cyclic subgroup $[q]$ in the additive group of the integers.

36. (i) Explain why $[a]$ must always be a subgroup in the centralizer C_a.
(ii) Use (i) and the fact that disjoint cycles commute to find the six permutations in the centralizer C_α of $\alpha = (345)$ in $\mathbf{S}_5$.

37. Let S be a subset of the elements of a cyclic group $G = [a]$ and let T consist of all the integers m such that a^m is in S. Show that S is a subgroup in G if and only if T is a subgroup in the additive group $\mathbf{Z}$ of the integers.

38. Let H be a subgroup in a cyclic group $G = [a]$. Show that H must be cyclic. [*Hint:* Let $T = \{n : a^n \in H\}$, and use Problems 37 and 34.]

39. Let a be an element of order q in a group G, let s be an integer, and let $d = \gcd(q, s)$. Show that the order of a^s is q/d.

40. Let $[a]$ be a cyclic group of order q. Show that a^s is a generator of $[a]$ if and only if $\gcd(s, q) = 1$.

41. Let $G = [a]$ be a cyclic group of order q. For each of the following values of q, list the integers m in $\{0, 1, \ldots, q - 1\}$ such that a^m is a generator of G.
(a) $q = 3$; (b) $q = 4$; (c) $q = 5$; (d) $q = 6$; (e) $q = 7$;
(f) $q = 8$; (g) $q = 9$.

42. Let $G = [a]$ be of order q. For each of the following values of q, tell how many generators G has.
(a) $q = p$, a prime.
(b) $q = p_1 p_2$ with p_1 and p_2 distinct positive primes.
(c) $q = p^r$ with p a positive prime and r a positive integer.

43. Let $G = [a]$ be a cyclic group of order de, where d and e are positive integers. Prove that a^k has order d if and only if $k = fe$, with f a positive integer relatively prime to d.

44. Let d and m be positive integers with $d \mid m$. Prove that there are the same number of elements of order d in a cyclic group of order m as in a cyclic group of order d.

45. Let a and b be elements of a group G. Prove that ab and ba have the same order. [*Hint:* See Problems 23 and 24 of Section 2.3.]

46. Let a, b, and c be elements of a group G.
(a) Prove that abc and bca have the same order.
(b) If $ab \neq ba$, show that there is an element d in G such that dab and dba have different orders.

47. Let $\mathbf{Z}$ be the additive group of the integers. Show that each of 1 and -1 is a generator of $\mathbf{Z}$. Are there any other generators?

48. How many generators are there of an infinite cyclic group $[a]$? Explain.

49. Let V be the multiplicative group of the nonzero complex numbers and let $w = (-1 + \mathbf{i}\sqrt{3})/2$.
(i) Show that $w = \cos(2\pi/3) + \mathbf{i}\sin(2\pi/3)$.
(ii) Show that $w^2 = (-1 - \mathbf{i}\sqrt{3})/2 = \cos(4\pi/3) + \mathbf{i}\sin(4\pi/3)$.
(iii) Show that $w^3 = 1$.
(iv) List the elements of the cyclic subgroup $[w]$ in V.
(v) What is the order of w?
(vi) Does $\beta^3 = 1$ for every β in $[w]$?
(vii) List all the generators of $[w]$.

50. In the V of Problem 49, let $\alpha = (\sqrt{3} + \mathbf{i})/2$.
(i) Show that $\alpha = \cos(\pi/6) + \mathbf{i}\sin(\pi/6)$.
(ii) Show that $\alpha^6 = -1$.

(iii) List the elements of the cyclic subgroup $[\alpha]$ in V.
(iv) What is the order of α?
(v) Does $\beta^{12} = 1$ for every β in $[\alpha]$?
(vi) List all the generators of $[\alpha]$.

51. In the V of Problems 49 and 50, find the order of each of the following.
(a) $\cos(5\pi/11) + \mathbf{i}\sin(5\pi/11)$.
(b) $\cos(10\pi/11) + \mathbf{i}\sin(10\pi/11)$.
(c) $\cos(5\pi/22) + \mathbf{i}\sin(5\pi/22)$.

52. In the V of Problems 49 through 51, find the order of each of the following.
(a) $\cos(9\pi/7) + \mathbf{i}\sin(9\pi/7)$.
(b) $\cos(18\pi/7) + \mathbf{i}\sin(18\pi/7)$.
(c) $\cos(9\pi/14) + \mathbf{i}\sin(9\pi/14)$.

★ **53.** Let V be the multiplicative group of the nonzero complex numbers. Prove that the only real numbers t for which $\cos t + \mathbf{i}\sin t$ has finite order in V are those of the form $r\pi/s$ with r and s (relatively prime) integers and $s \neq 0$.

54. Let t and s be odd positive integers, with $\gcd(t, s) = 1$. In the V of Problem ★53 find the order of $\cos x + \mathbf{i}\sin x$ for each of the following.
(a) $x = t\pi/s$; (b) $x = 2t\pi/s$; (c) $x = 4t\pi/s$; (d) $x = 8t\pi/s$;
(e) $x = t\pi/2s$; (f) $x = t\pi/4s$; (g) $x = t\pi/8s$.

55. Let $n \in \mathbf{Z}^+$ and $\alpha = \cos(2\pi/n) + \mathbf{i}\sin(2\pi/n)$. Prove the following.
(a) α has order n in the multiplicative group of nonzero complex numbers.
(b) The cyclic subgroup $[\alpha]$ consists of the n complex nth roots of unity.
(c) The generators of $[\alpha]$ are the complex numbers

$$\cos(2k\pi/n) + \mathbf{i}\sin(2k\pi/n) \quad \text{with} \quad \gcd(k, n) = 1 \quad \text{and} \quad 1 \leq k \leq n.$$

(These generators are the ***primitive nth roots of unity***.)

2.8 Even and Odd Permutations

Among the symmetric groups $\mathbf{S_1}, \mathbf{S_2}, \mathbf{S_3}, \ldots$, the first one, $\mathbf{S_1} = \{(1)\}$, is exceptional in that it alone has odd order. Also $\mathbf{S_1}$ is the only $\mathbf{S_n}$ with no transpositions in it. Since 2-cycles are the basis for the material in this section, we restrict ourselves here to the $\mathbf{S_n}$ with $n > 1$.

Theorem 3 of Section 2.4 states that every permutation of $\mathbf{S_n}$ is either a transposition or a product of (not necessarily disjoint) transpositions. However, this factorization is not unique. For example,

$$(1234) = (14)(24)(34) = (12)(23)(31)(12)(14) = (12)(13)(14).$$

Although the factorization into 2-cycles is not unique, we can show that for a given permutation θ the number of transpositions in the representation has fixed parity; that is, it is either always even or always odd. This fact is needed in studying a subgroup $\mathbf{A_n}$ in $\mathbf{S_n}$, which has an essential role in the proof of the insolvability of the general fifth-degree polynomial equation. (This is discussed further in Sections 2.13 and 5.12.)

In our proof of the invariance of parity property (Theorem 1 below), we use the convenient device of considering the effect of applying a permutation θ to the subscripts i of the variables x_i in a function $F(x_1, x_2, \ldots, x_n)$ of n variables. Let $G(x_1, x_2, \ldots, x_n)$ be the result of replacing $x_1, \ldots, x_n$ in $F(x_1, \ldots, x_n)$ by

$$x_{\theta(1)}, \ldots, x_{\theta(n)},$$

respectively. Then we say that θ ***sends*** F to G or that G is the result of ***applying*** θ on F.

For example, (1234) sends $(x_1 - x_2)(x_1 - x_4)$ to $(x_2 - x_3)(x_2 - x_1)$, and (123) sends $x_1^2 + x_2^2$ to $x_2^2 + x_3^2$.

We are particularly interested in the result of applying a permutation θ on the product P_n of all the differences $x_j - x_i$ with $1 \le i < j \le n$. We note that the P_n for $n = 2$, 3, and 4 are

$$P_2 = x_2 - x_1,$$

$$P_3 = (x_2 - x_1)(x_3 - x_1)(x_3 - x_2),$$

$$P_4 = (x_2 - x_1)(x_3 - x_1)(x_3 - x_2)(x_4 - x_1)(x_4 - x_2)(x_4 - x_3).$$

EXAMPLE 1 **Applying (13) on P_4**

Here we note that the result of applying the transposition (13) on

$$(x_2 - x_1)(x_3 - x_1)(x_3 - x_2)(x_4 - x_1)(x_4 - x_2)(x_4 - x_3) = P_4$$

is

$$(x_2 - x_3)(x_1 - x_3)(x_1 - x_2)(x_4 - x_3)(x_4 - x_2)(x_4 - x_1) = -P_4$$

We generalize on this example in the following result.

LEMMA 1 **A Transposition Changes the Sign of P_n**

Every 2-cycle (uv) sends P_n to $-P_n$.

Proof Since $(vu) = (uv)$, we assume that $u < v$. As an aid in finding the effect of (uv) on P_n, we group some of the differences that are factors of P_n.

For $i < u$, we group $x_u - x_i$ and $x_v - x_i$ and note that (uv) sends

$$A = (x_u - x_i)(x_v - x_i) \quad \text{to} \quad (x_v - x_i)(x_u - x_i) = A.$$

Similarly, for $v < i$, the transposition (uv) sends

$$B = (x_i - x_u)(x_i - x_v) \quad \text{to} \quad (x_i - x_v)(x_i - x_u) = B.$$

For $u < i < v$, the transposition (uv) sends

$$C = (x_i - x_u)(x_v - x_i) \quad \text{to} \quad [-(x_v - x_i)][-(x_i - x_u)] = C.$$

Also, (uv) sends $(x_j - x_i)$ to itself if neither i nor j is u or v.

The only factor of P_n that we have not yet considered is $D = x_v - x_u$, which is sent by (uv) to

$$x_u - x_v = -(x_v - x_u) = -D.$$

Together, all of these observations show that a transposition (uv) sends P_n to $-P_n$. ∎

LEMMA 2 **θ Sends P_n to $\pm P_n$**

In $\mathbf{S_n}$, let θ be a product $\gamma_1\gamma_2 \cdots \gamma_r$ of r transpositions γ_i. Then θ sends P_n to $(-1)^r P_n$.

Proof If α sends F to G and β sends G to H, it is clear that the composite $\alpha\beta$ sends F to H. Since each γ_i changes the sign of P_n, the composite θ of the r transpositions γ_i sends P_n to $(-1)^r P_n$. ∎

THEOREM 1 **Always Even or Always Odd**

In $\mathbf{S_n}$ let $\theta = \alpha_1\alpha_2 \cdots \alpha_r = \beta_1\beta_2 \cdots \beta_s$, where each α_i and each β_j is a transposition. Then r and s are both even or both odd.

Proof It follows from Lemma 2 that, when we apply θ on P_n, we get both $(-1)^r P_n$ and $(-1)^s P_n$; thus we see that

$$(-1)^r P_n = (-1)^s P_n.$$

Hence r and s must have the same parity. ∎

DEFINITION 1 **Even Permutation, Odd Permutation**

An **even permutation** is a permutation that is expressible as a product of an even number of transpositions. A product of an odd number of transpositions is an **odd permutation.**

Theorem 1 tells us that a given permutation is either even or odd, but not both. We also note that the identity $(1) = (12)(12)$ is even in all the $\mathbf{S_n}$ with $n > 1$. In $\mathbf{S_1}$, we define the identity (1) to be even.

THEOREM 2 Permutation Parity

Let $\theta = \gamma_1\gamma_2 \cdots \gamma_r$ where $\gamma_1, \ldots, \gamma_r$ are cycles of lengths $s_1, \ldots, s_r$, respectively. Then θ is even or odd depending on whether

$$s_1 + s_2 + \cdots + s_r - r \tag{A}$$

is even or odd.

The proof is left to the reader as Problem 7 of this section.

Problems

Problem 15 below is cited in Section 3.5, and Problems 9 and 12 are cited in Section 2.10.

1. In each of the following, write the result of applying θ on $F(x_1, \ldots, x_n)$.
(a) $\theta = (123)$, $F(x_1, x_2, x_3) = x_1^2 + x_2^2 - x_3^2$.
(b) $\theta = (12)$, $F(x_1, x_2, x_3) = x_1^2 + x_2^2 - x_3^2$.
(c) $\theta = (123)$, $F(x_1, x_2, x_3, x_4) = x_1^2 + x_2^2 + x_4^2$.

2. In each of the following, write the result of applying θ on $F(x_1, \ldots, x_n)$.
(a) $\theta = (124)$, $F(x_1, x_2, x_3, x_4) = x_1^2 + x_2^2 + x_4^2$.
(b) $\theta = (23)$, $F(x_1, x_2, x_3) = (x_1 + x_2 - x_3)(x_2 + x_3 - x_1)(x_3 + x_1 - x_2)$.
(c) $\theta = (12)(34)$, $F(x_1, x_2, x_3, x_4) = x_1^2 x_2^2 x_3 x_4$.

3. Which permutations θ of $\mathbf{S_4}$ leave $x_1^2 - x_2^2 + 2x_3 + x_4^2$ invariant — that is, send it to itself?

4. Which permutations θ of $\mathbf{S_5}$ leave $x_1 + x_2 + 3(x_3 + x_4) + 2x_5$ invariant?

5. Which of the following are even permutations? Which are odd?
(a) (12345); (b) (123456); (c) (13)(12)(345); (d) (1234)(3456).

6. Show that a cycle of length s is an even permutation if s is odd and is an odd permutation if s is even.

7. Let $\theta = \gamma_1\gamma_2 \cdots \gamma_r$, where $\gamma_1, \ldots, \gamma_r$ are cycles of lengths $s_1, \ldots, s_r$, respectively. Prove that θ is even or odd depending on whether $s_1 + s_2 + \cdots + s_r - r$ is even or odd (and thus prove Theorem 2 above).

8. State a necessary and sufficient condition on the lengths of the cycles for a product $\gamma_1\gamma_2 \cdots \gamma_r$ of cycles to be even and a condition for the product to be odd.

9. Let α, α', β, and β' be in $\mathbf{S_n}$. Let α and α' be even permutations while β and β' are odd. Show the following.
(i) $\alpha\alpha'$, $\beta\beta'$, and α^{-1} are even.
(ii) $\alpha\beta$, $\beta\alpha$, and β^{-1} are odd.

10. Show that the set of even permutations in any group of permutations forms a subgroup.

11. Let $\alpha_1, \alpha_2, \ldots, \alpha_r$. be all the even permutations in $\mathbf{S_n}$ and let β be an odd permutation in $\mathbf{S_n}$. Show the following.
(i) The products $\beta\alpha_1, \beta\alpha_2, \ldots, \beta\alpha_r$ are distinct odd permutations.
(ii) The products $\alpha_1\beta, \alpha_2\beta, \ldots, \alpha_r\beta$ are distinct odd permutations.
(iii) If $n \geq 2$, there are at least as many odd permutations in $\mathbf{S_n}$ as there are even ones.

12. Let $\beta_1, \beta_2, \ldots, \beta_s$ be all the odd permutations in $\mathbf{S_n}$. Show the following.

(i) $\beta_1\beta_1, \beta_1\beta_2, \ldots, \beta_1\beta_s$ are distinct even permutations.
(ii) There are at least as many even permutations in $\mathbf{S_n}$ as odd ones.
(iii) If $n \geq 2$, there are $(n!)/2$ even permutations and the same number of odd permutations in $\mathbf{S_n}$. [See Problem 11(iii).]

13. Give an example of a group consisting entirely of even permutations.

14. Let H be a subgroup of $\mathbf{S_n}$. Show that either H consists entirely of even permutations or H has as many odd permutations as even ones.

15. Show that $\alpha^{-1}\beta^{-1}\alpha\beta$ is even for all permutations α and β in $\mathbf{S_n}$.

2.9 Groups of Symmetries

There is no branch of mathematics, however abstract, which may not someday be applied to phenomena of the real world.
—**Nicolai Ivanovich Lobachevsky**

It would seem that this statement by Lobachevsky cannot be proved or disproved. However, some proposed counterexamples have not stood the test of time. For example, the famous physicist Sir James Jeans is quoted as having stated (in 1910): "We may as well cut out group theory. That is a subject that will never be of any use in physics."

But in 1964 the existence of the omega minus particle, predicted using group theory, was confirmed by laboratory test. Thus group theory became a central theme in another cycle of restoring order to previously chaotic data on elementary particles.

In this section, we examine a very small sample of the kind of group theory that has brought new insight into geometry and physics. (See the biographical note on Felix Klein after this section.) Specifically, we introduce subgroups in $\mathbf{S_n}$ associated with rigid motions of geometric figures. We will develop this concept by considering in detail the case in which the figure is a square, and then we will look at the analogous situations for other figures, including regular polygons with n sides, isosceles triangles, and parallelograms.

Let us consider a square with vertices numbered 1, 2, 3, 4 in circular order, that is, with each of the pairs

$$\{1, 2\}, \quad \{2, 3\}, \quad \{3, 4\}, \quad \{4, 1\} \tag{1}$$

consisting of the numbers of the two vertices of a side of the square (see Figure 2.5).

After a rigid motion of the square that ends with it covering the same area as in its starting position, let vertex a_i be where vertex i was originally.

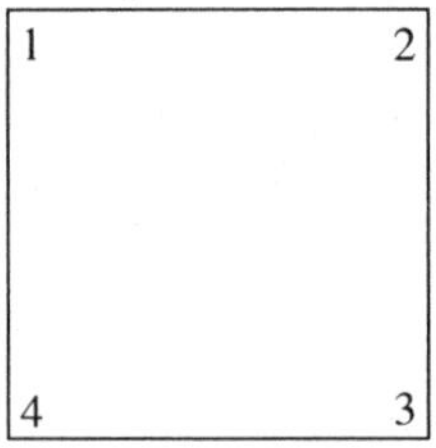

FIGURE 2.5

The rigid motion thus gives rise to a permutation

$$\theta\colon\ 1 \mapsto a_1,\quad 2 \mapsto a_2,\quad 3 \mapsto a_3,\quad 4 \mapsto a_4, \tag{2}$$

in which the pairs

$$\{a_1, a_2\},\quad \{a_2, a_3\},\quad \{a_3, a_4\},\quad \{a_4, a_1\} \tag{3}$$

are the pairs of (1), although not necessarily in the same order.

DEFINITION 1

Symmetry of a Square

The permutation θ of (2) is a **symmetry of a square** if and only if each pair of (3) is one of the pairs of (1).

How many of the permutations of the symmetric group $\mathbf{S}_4$ are symmetries of a square? We note that, in (2), a_1 may be chosen to be any number in $\{1, 2, 3, 4\}$; and then for a given a_1 there are two ways of selecting a_2 and a_4 so that $\{a_1, a_2\}$ and $\{a_4, a_1\}$ are pairs of (1). After the choice of a_1, a_2, and a_4, one must have the remaining number as a_3. Hence there are $4 \cdot 2 = 8$ symmetries of a square.

In Figure 2.6 there is a square for each of these 8 symmetries, with the new number a_i for a vertex outside the square and the old number i inside.

We now associate each of these symmetries with a rigid motion of the square. The permutations

$$\rho = (1234), \qquad \rho^2 = (13)(24), \qquad \rho^3 = (1432)$$

result from counterclockwise rotations of the square about its center through 90°, 180°, and 270°, respectively. One may associate the identity (1) with a

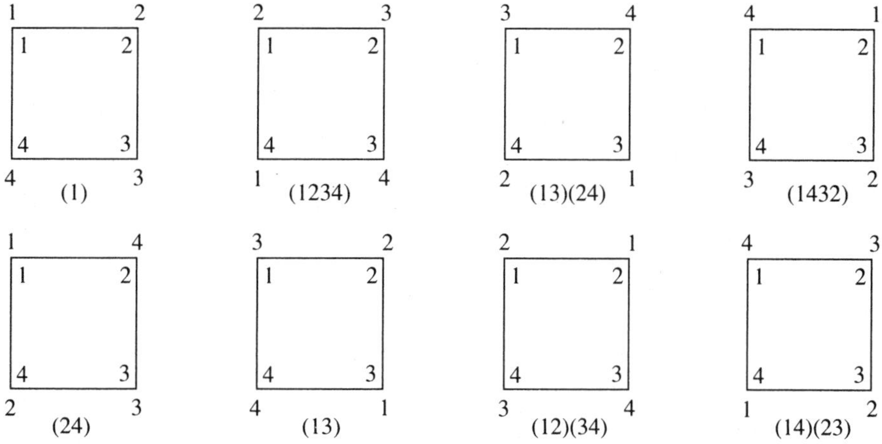

FIGURE 2.6

rotation of 360° or with any other motion that returns the square to its original position.

The symmetry $\phi = (24)$ results from a 180° flip of the square about the diagonal connecting vertices 1 and 3; the transposition (13) results from a similar flip about the other diagonal.

The symmetry (12)(34) corresponds to a 180° rotation, in 3-dimensional space, of the square about the line through the midpoints of the sides with vertices $\{1, 2\}$ and $\{3, 4\}$, respectively. The permutation (14)(23) is associated with a similar rotation about the line through the midpoints of the other two sides.

A rigid motion of a square followed by another rigid motion is equivalent to some single rigid motion. This leads us to believe that the subset of $\mathbf{S_4}$, consisting of the eight symmetries of a square, is closed under multiplication. We now prove this, using the definition above.

Let α and β be symmetries of a square, let $\gamma = \alpha\beta$, and let $\{a_1, a_2\}$ be any one of the pairs of (1). Let $\alpha(a_i) = b_i$ and $\beta(b_i) = c_i$ for $i = 1$ and 2. By the definition of a symmetry, $\{b_1, b_2\}$ is a pair of (1). Then it similarly follows that $\{c_1, c_2\}$ is a pair of (1). Since $\gamma(a_i) = c_i$, this means that the composite γ is a symmetry of a square.

THEOREM 1 Subgroup of Symmetries

The 8 symmetries of a square form a subgroup in $\mathbf{S_4}$.

Proof These permutations form a nonempty finite subset closed under the operation of $\mathbf{S_4}$. Hence this subset is a subgroup by Theorem 4 of Section 2.7. ∎

Subgroup Poset Diagram for the Octic Group

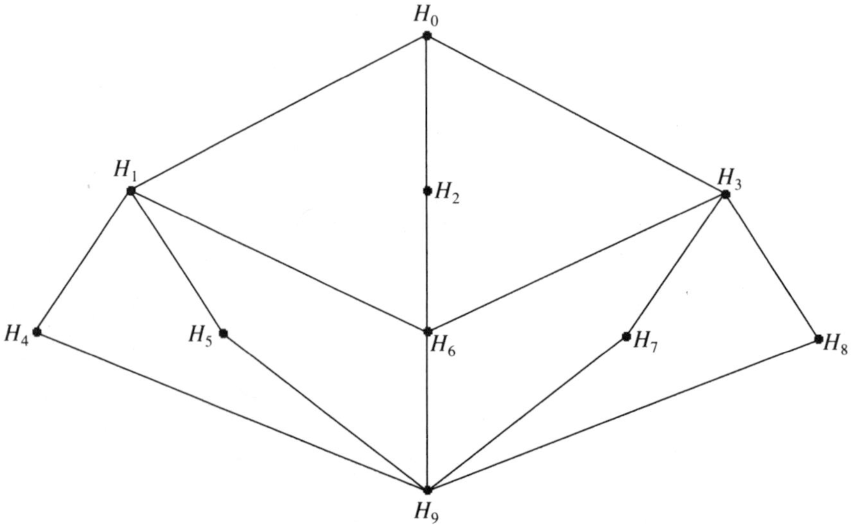

FIGURE 2.7

DEFINITION 2

Octic Group

The group of the 8 symmetries of a square is called the **octic group**.

We next use a diagram to show some of the relationships among the subgroups in the octic group.

$H_0, H_1, \ldots, H_9$, shown in Figure 2.7, stand for the subgroups in the octic group. In Problems 7 through 10 below, the reader is asked to identify the H's with the subgroups in such a way that there is a rising line or broken line from H_i to H_j if and only if H_i is a proper subgroup in H_j. This makes Figure 2.7 into a ***subgroup poset diagram*** for the octic group; the reason for the use of the word *poset* will be presented in Section 3.8.

EXAMPLE 1 Let us now consider the symmetries of a nonsquare rectangle whose vertices are numbered as in Figure 2.8.

FIGURE 2.8

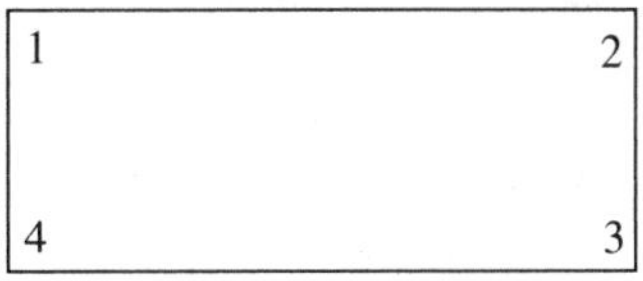

These symmetries are the permutations θ in $\mathbf{S_4}$ such that $\{\theta(1), \theta(4)\}$ and $\{\theta(2), \theta(3)\}$ are $\{1, 4\}$ and $\{2, 3\}$ in either order, and $\{\theta(1), \theta(2)\}$ and $\{\theta(3), \theta(4)\}$ are $\{1, 2\}$ and $\{3, 4\}$ in either order. It is readily confirmed that the four permutations with these properties form the subgroup $\{(1), (12)(34), (14)(23), (13)(24)\}$ in $\mathbf{S_4}$.

We now generalize some of the above material on the octic group by discussing the symmetries of a regular n-gon (that is, a regular polygon with n sides). Let the vertices of the regular n-gon be numbered in circular order as $1, 2, \ldots, n$. Then the n sides of the n-gon are determined by the pairs

$$\{n, 1\}, \quad \{1, 2\}, \quad \{2, 3\}, \quad \{3, 4\}, \ldots, \{n-1, n\}. \tag{4}$$

DEFINITION 3

Set $\mathbf{D_n}$ of All the Symmetries of a Regular n-gon

A permutation $\theta\colon 1 \mapsto a_1, 2 \mapsto a_2, \ldots, n \mapsto a_n$ in $\mathbf{S_n}$ is a **symmetry of the regular *n*-gon** if the pairs of

$$\{a_n, a_1\}, \quad \{a_1, a_2\}, \quad \{a_2, a_3\}, \quad \{a_3, a_4\}, \ldots, \{a_{n-1}, a_n\}$$

are the pairs of (4) in some order. The set of all such symmetries is denoted by $\mathbf{D_n}$.

How many of the permutations $\theta\colon 1 \mapsto a_1, \ldots, n \mapsto a_n$ in $\mathbf{S_n}$ are in the subset $\mathbf{D_n}$? We can let a_1 be any of the n numbers in $\{1, 2, \ldots, n\}$; and then, for a given a_1, there are two ways of selecting a_n and a_2 so that $\{a_n, a_1\}$ and

$\{a_1, a_2\}$ are pairs listed in (4). After the choice of a_n, a_1, and a_2, the remaining a_i are determined uniquely. Hence $\mathbf{D_n}$ consists of $2n$ of the $n!$ permutations in $\mathbf{S_n}$. One shows that $\mathbf{D_n}$ is a subgroup in $\mathbf{S_n}$ by using the arguments preceding Theorem 1.

DEFINITION 4

The Dihedral Groups $\mathbf{D_n}$

The group of the $2n$ symmetries of a regular n-gon is called the **dihedral group $\mathbf{D_n}$**.

One easily sees that the dihedral group $\mathbf{D_3}$ of symmetries of an equilateral triangle (that is, a regular 3-gon) is $\mathbf{S_3}$ itself. Also, the dihedral group $\mathbf{D_4}$ of symmetries of a regular 4-gon (that is, a square) is clearly the octic group. The 10 permutations in $\mathbf{D_5}$ are characterized in Problem 17 below, and the reader is asked in Problem 20 to give the analogous characterization of the $2n$ permutations in $\mathbf{D_n}$.

The elements of a dihedral group $\mathbf{D_n}$ describe rotations through an angle $2\pi/n$ about the center of a regular n-gon and reflections about lines connecting vertices with midpoints of opposite sides (if n is odd) or lines connecting opposite vertices and those connecting midpoints of opposite sides (if n is even). Examples are shown in Figure 2.9.

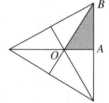

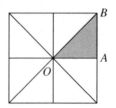

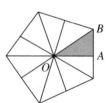

FIGURE 2.9

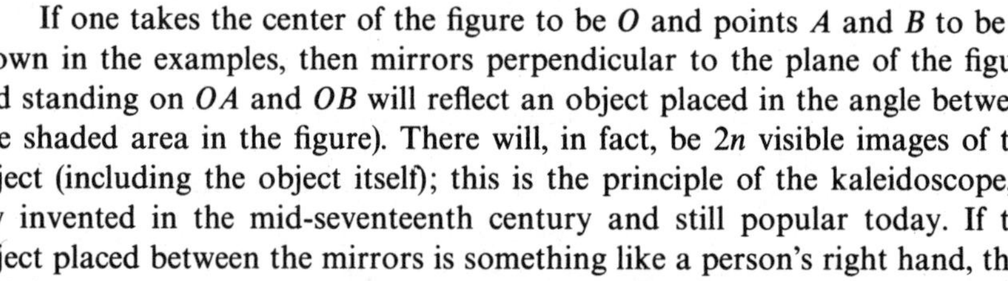

If one takes the center of the figure to be O and points A and B to be as shown in the examples, then mirrors perpendicular to the plane of the figure and standing on OA and OB will reflect an object placed in the angle between (the shaded area in the figure). There will, in fact, be $2n$ visible images of the object (including the object itself); this is the principle of the kaleidoscope, a toy invented in the mid-seventeenth century and still popular today. If the object placed between the mirrors is something like a person's right hand, then half the images will look like a right hand and half like a left hand. A full study of the geometrical properties of the dihedral groups falls more properly within the domain of a book on geometry. Nevertheless, we see here another instance in which the concept of a group arises in describing a familiar object.

Problems

1. Let $\rho = (1234)$ and $\phi = (24)$. Show the following.
(i) $\{(1), \rho, \rho^2, \rho^3, \phi, \rho\phi, \rho^2\phi, \rho^3\phi\}$ is the octic group $\mathbf{D_4}$.
(ii) $\phi\rho = \rho^3\phi$.

2. Using the notation of Problem 1, list the even permutations in the octic group $\mathbf{D}_4$.
3. Using the notation of Problem 1, make a multiplication table for the octic group $\mathbf{D}_4$.
4. Give the order of each element of the octic group $\mathbf{D}_4$.
5. List the elements of the center of the octic group $\mathbf{D}_4$.
6. (i) List all the permutations α such that $\alpha = \beta^2$ with β in the octic group.
   (ii) Show that the set H of all the α of (i) is a subgroup in the octic group.
   (iii) List all the α such that $\alpha = \beta^3$ with β in the octic group.
   (iv) Show that the set K of all the α of (iii) is a subgroup in the octic group.
7. In the octic group $\mathbf{D}_4$ find the following.
   (i) A cyclic subgroup of order 4.
   (ii) Two noncyclic subgroups of order 4.
   (iii) Five subgroups of order 2.

★ 8. Prove that the octic group $\mathbf{D}_4$ has no subgroup of order 3, 5, 6, or 7.

★ 9. Prove that the only subgroups in the octic group are $\{(1)\}$, the eight subgroups found in Problem 7, and the octic group itself.

10. Identify $H_0, H_1, \ldots, H_9$ in Figure 2.7 as the ten subgroups in the octic group in such a way that there is a rising line or broken line from H_i to H_j if and only if H_i is a proper subgroup in H_j. (Do this in one of the several possible ways. List the elements of each subgroup in standard form.)
11. Let $\rho = (1234)$ and $\phi = (24)$. Let
$$H = \{(1), \rho^2, \rho\phi, \rho^3\phi\}$$
and let K consist of the permutations $\phi^{-1}\theta\phi$ for all θ in H.
   (i) Is H a subgroup in the octic group? Explain.
   (ii) List the elements of K.
   (iii) Is K a subgroup in the octic group? Explain.
12. List the elements of the centralizer of each of the following elements in the octic group.
   (a) (24); (b) (13)(24).
13. Which permutations in $\mathbf{S}_3$ are symmetries of an isosceles triangle whose vertices are numbered 1, 2, and 3 and whose equal angles are at vertices 2 and 3?
14. Let $ABCD$ be a parallelogram whose sides AB, BC, CD, and DA have lengths 2, 1, 2, and 1, respectively. If $ABCD$ is not a rectangle, how many symmetries does it have?
15. Let α: $1 \mapsto a_1$, $2 \mapsto a_2$, $3 \mapsto a_3$, $4 \mapsto a_4$, $5 \mapsto a_5$ be a symmetry of a regular pentagon (that is, let α be in $\mathbf{D}_5$).
   (a) If $a_1 = 2$, what are the possibilities for a_2?
   (b) If $a_1 = 3$ and $a_2 = 4$, find a_3, a_4, and a_5. Does $\alpha = (12345)^2$ in this case?
   (c) If $a_1 = 3$ and $a_2 = 2$, find a_3, a_4, and a_5. Does $\alpha = \rho^3\phi$, where $\rho = (12345)$ and $\phi = (25)(34)$?
16. Let β: $1 \mapsto b_1$, $2 \mapsto b_2$, $3 \mapsto b_3$, $4 \mapsto b_4$, $5 \mapsto b_5$ be in $\mathbf{D}_5$.
   (a) If $b_1 = 5$, what are the possibilities for b_5?
   (b) If $b_1 = 4$ and $b_2 = 5$, find b_3, b_4, and b_5 and show that β is a power of $\rho = (12345)$.
   (c) If $b_1 = 4$ and $b_2 = 3$, find b_3, b_4, and b_5 and show that β is expressible in the form $\rho^r\phi$, where $\rho = (12345)$ and $\phi = (25)(34)$.

Felix Klein
1849–1925

Klein was a student at Bonn and later taught at Erlangen, Munich, Leipzig, and Göttingen. It was upon his appointment to a professorship in 1872 at Erlangen that he outlined in his inaugural address what has come to be known as his Erlanger Programm. In this address he described different geometries as the studies of properties of figures that remain invariant under different groups of transformations. For example, plane euclidean geometry is the study of the properties of figures (length, angle size, area) that are unchanged under rotation and translation, such transformations in the plane forming a group. Similarly affine geometry, projective geometry, and so on correspond to groups of transformations. This approach provided a very neat classification of various geometries in terms of the theory of groups.

Klein also provided a geometric interpretation for the problem of the solvability of the fifth-degree polynomial equation by relating it to the group of symmetries of the regular icosahedron.

He had the reputation of being a great teacher, and during his stay at Göttingen he attracted first-rate students, including many from the United States. He was a brilliant mathematician as well as teacher and provided a distinguished link in the chain of first-rank mathematicians at Göttingen in the nineteenth and early twentieth centuries: Gauss, Dirichlet, Riemann, Klein, and Hilbert. In 1893, the World's Columbian Exposition was held in Chicago to celebrate the 400th anniversary of Columbus' famous trip to America (the nineteenth-century equivalent of the fair in Seville in 1992 for the quincentennial). In conjunction with this fair, a world congress of mathematicians and astronomers was held, with a mathematical colloquium held on the campus of Northwestern University following the Exposition. Klein came from Germany to represent European mathematicians and spoke on "The Present State of Mathematics." This was a precursor of International Congresses of Mathematicians held every four years (with a few exceptions) up to the present time.

17. Let $\rho = (12345)$ and $\phi = (25)(34)$. Show the following.
(i) $\phi\rho = \rho^{-1}\phi$.
(ii) The symmetries of a regular pentagon are

$$\mathbf{D_5} = \{(1), \rho, \rho^2, \rho^3, \rho^4, \phi, \rho\phi, \rho^2\phi, \rho^3\phi, \rho^4\phi\}.$$

18. Show that every permutation of $\mathbf{S_3}$ corresponds to a symmetry of an equilateral triangle.

19. Let ρ, ϕ, and θ be in $\mathbf{D_n}$; that is, let them be symmetries of a regular n-gon (whose vertices are numbered 1, 2, ..., n in circular order). Also let $\rho(1) = 2$, $\rho(2) = 3$, $\phi(1) = 1$, and $\phi(2) = n$.
(a) Given that $\theta(1) = r$ and $\theta(2) = r + 1$, explain why $\theta = \rho^{r-1}$.

(b) Given that $\theta(1) = n$ and $\theta(2) = 1$, explain why $\theta = \rho^{n-1}$.
(c) Explain why $\phi\rho = \rho^{-1}\phi$.
(d) Given that $\theta(1) = r$ and $\theta(2) = r - 1$, explain why $\theta = \phi\rho^{r-1} = \rho^{1-r}\phi$.

20. Characterize the $2n$ symmetries of a regular n-gon in terms of a rotation ρ and a flip ϕ. [This should be analogous to our work above on the octic group $\mathbf{D_4}$ and to the work on $\mathbf{D_5}$ in Problem 17.]

2.10 The Alternating Groups $\mathbf{A_n}$

For $n > 1$, let $\mathbf{A_n}$ denote the subset of all the even permutations in the symmetric group $\mathbf{S_n}$. It is easily seen that $\mathbf{A_n}$ is closed under composition. Since $\mathbf{A_n}$ is finite and nonempty, this suffices to make $\mathbf{A_n}$ a subgroup in $\mathbf{S_n}$.

DEFINITION 1

Alternating Group

The group $\mathbf{A_n}$ of even permutations on $\mathbf{X_n} = \{1, 2, \ldots, n\}$ is the **alternating group** on $\mathbf{X_n}$.

We note from Table 2.2 that $\mathbf{A_4}$ consists of the identity (1), three elements of the form $(ab)(cd)$, and eight 3-cycles (abc). This checks with the fact that the order of $\mathbf{A_4}$ is $(4!)/2 = 24/2 = 12$. [See Problem 12(iii) of Section 2.8.]

TABLE 2.2 Multiplication Table for $\mathbf{A_4}$

In this table, the permutations of $\mathbf{A_4}$ are designated as $\alpha_1, \alpha_2, \ldots, \alpha_{12}$, and an entry k inside the table represents α_k.

	α_1	α_2	α_3	α_4	α_5	α_6	α_7	α_8	α_9	α_{10}	α_{11}	α_{12}
$(1) = \alpha_1$	1	2	3	4	5	6	7	8	9	10	11	12
$(12)(34) = \alpha_2$	2	1	4	3	8	7	6	5	11	12	9	10
$(13)(24) = \alpha_3$	3	4	1	2	6	5	8	7	12	11	10	9
$(14)(23) = \alpha_4$	4	3	2	1	7	8	5	6	10	9	12	11
$(123) = \alpha_5$	5	6	7	8	9	10	11	12	1	2	3	4
$(243) = \alpha_6$	6	5	8	7	12	11	10	9	3	4	1	2
$(142) = \alpha_7$	7	8	5	6	10	9	12	11	4	3	2	1
$(134) = \alpha_8$	8	7	6	5	11	12	9	10	2	1	4	3
$(132) = \alpha_9$	9	10	11	12	1	2	3	4	5	6	7	8
$(143) = \alpha_{10}$	10	9	12	11	4	3	2	1	7	8	5	6
$(234) = \alpha_{11}$	11	12	9	10	2	1	4	3	8	7	6	5
$(124) = \alpha_{12}$	12	11	10	9	3	4	1	2	6	5	8	7

It is left to the reader in Problem 23 below to identify $H_0, H_1, \ldots, H_9$, shown in Figure 2.10 as the subgroups in $\mathbf{A_4}$ so as to make the figure into a subgroup poset diagram for $\mathbf{A_4}$, that is, so that there is a rising line or broken line from H_i to H_j if and only if H_i is a proper subgroup in H_j.

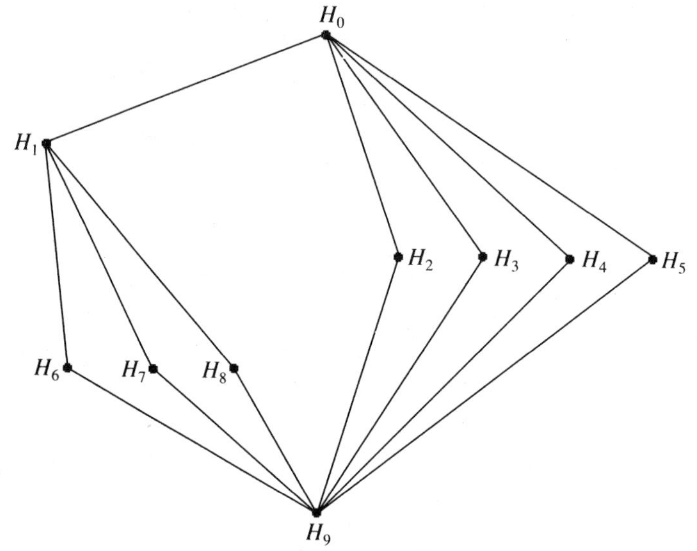

FIGURE 2.10

It is easily shown that an s-cycle $\gamma = (a_1 a_2 \ldots a_s)$ is an even permutation when s is odd and that γ is odd when s is even. (See Problems 7 and 8 of Section 2.4.) Also, the product $\alpha\beta$ of two permutations is even when α and β are both even or both odd, and the product is odd when one factor is an even permutation and the other is odd. (See Problem 9 of Section 2.8.) Thus a permutation with a standard form of the type

$$(abc)(defg)(hijk)$$

will always be even.

We next illustrate a technique for counting the permutations with a given type of standard form. Some easier counting problems of this nature occur in our investigation, in the next two sections, of the structure of $\mathbf{S_4}$ and $\mathbf{S_5}$.

EXAMPLE 1 Here we will determine how many permutations α in $\mathbf{A_{15}}$ have a standard form of the type $(abc)(defg)(hijk)$. We first note that a permutation with this type of standard form is even; hence α will be in $\mathbf{A_{15}}$ if $a, b, \ldots, k$ are 11 properly chosen elements of $\mathbf{X_{15}} = \{1, 2, 3, \ldots, 15\}$.

It is well known that one can choose a subset of r elements from a set of n elements in

$$\binom{n}{r} = \frac{n(n-1)(n-2)\cdots(n-r+1)}{1 \cdot 2 \cdot 3 \cdots r}$$

ways; therefore, the 4 elements of $\mathbf{X_{15}}$ not involved in α can be chosen in

$$\binom{15}{4} = \frac{15 \cdot 14 \cdot 13 \cdot 12}{1 \cdot 2 \cdot 3 \cdot 4} = 1365$$

ways. Then a must be the smallest of the elements of $\mathbf{X}_{15}$ that are involved in α. There are now 10 elements left from which to choose b, and after that 9 elements from which to choose c. Then d must be the smallest of the remaining elements, e can be any one of 7 elements, and so on. Thus the total number of permutations α with this type of standard form

$$1365 \cdot 10 \cdot 9 \cdot 7 \cdot 6 \cdot 5 \cdot 3 \cdot 2 = 154791000.$$

If we wanted to know how many of the permutations of $\mathbf{A}_{15}$ are products of three disjoint cycles, of which one has length 3 and the other two have length 4, we would also have to count the permutations with standard forms of the types $(abcd)(efg)(hijk)$ and $(abcd)(efgh)(ijk)$.

Problems

1. Give the standard forms of all the elements of order two in $\mathbf{A}_4$.
2. List in standard form all the elements of order three in $\mathbf{A}_4$.
3. Are there any elements of order greater than three in $\mathbf{A}_4$?
4. List all the subgroups of order one or two in $\mathbf{A}_4$.
5. List all the subgroups of order three in $\mathbf{A}_4$.
6. Does $\mathbf{A}_4$ have a cyclic subgroup of order four or six? Explain.
7. Find the smallest subgroup in $\mathbf{A}_4$ that contains (12)(34) and (123).
8. Find the smallest subgroup in $\mathbf{A}_4$ that contains (14)(23) and (123).
9. Let $\alpha_1 = (1)$, $\alpha_5 = (123)$, and $\alpha_9 = (132)$.
   (i) For each θ of $\mathbf{A}_4$, find the set $\{\alpha_1\theta, \alpha_5\theta, \alpha_9\theta\}$.
   (ii) How many different subsets of $\mathbf{A}_4$ are there in the sets of (i)?
10. Let $\alpha_1 = (1)$, $\alpha_5 = (123)$, and $\alpha_9 = (132)$. Show that the sets

$$\{\alpha_1\theta, \alpha_5\theta, \alpha_9\theta\} \quad \text{and} \quad \{\theta\alpha_1, \theta\alpha_5, \theta\alpha_9\}$$

    are different when $\theta = (124)$ and are the same when $\theta = (123)$.
11. Let $\alpha_1 = (1)$, $\alpha_2 = (12)(34)$, $\alpha_3 = (13)(24)$, and $\alpha_4 = (14)(23)$.
    (i) Find the subset $\{\alpha_1\theta, \alpha_2\theta, \alpha_3\theta, \alpha_4\theta\}$ for each θ in $\mathbf{A}_4$.
    (ii) How many different subsets are there in (i)?
    (iii) Show that

$$\{\alpha_1\theta, \alpha_2\theta, \alpha_3\theta, \alpha_4\theta\} = \{\theta\alpha_1, \theta\alpha_2, \theta\alpha_3, \theta\alpha_4\}$$

    for every θ in $\mathbf{A}_4$.
12. Which of the elements of $\mathbf{A}_4$ are squares of elements of $\mathbf{A}_4$? Which are cubes? Which are fifth powers?
13. How many elements of $\mathbf{A}_5$ are products of two disjoint 2-cycles? How many elements of $\mathbf{A}_5$ are 3-cycles? How many are 5-cycles?
14. Are there any other elements of $\mathbf{A}_5$ besides those counted in the parts of Problem 13?
15. Give the set of positive integers that are orders of elements of $\mathbf{A}_5$.

16. How many of the elements of $\mathbf{A}_{10}$ are products $(abcd)(efgh)$ of two disjoint 4-cycles?

17. Let $\alpha_2 = (12)(34)$. Complete the following table, and note that $\theta^{-1}\alpha_2\theta$ is a product of two disjoint transpositions for every θ in $\mathbf{A}_4$.

θ	α_1	α_2	α_3	α_4	α_5	α_6	$\cdots$	α_{12}
$\theta^{-1}\alpha_2\theta$	α_2	α_2			α_4			

18. Let $\alpha_5 = (123)$. Show that $\theta^{-1}\alpha_5\theta$ is a 3-cycle for every θ in $\mathbf{A}_4$.

19. Find the centralizer of $\alpha_2 = (12)(34)$ in $\mathbf{A}_4$.

20. Find the centralizer of $\alpha_5 = (123)$ in $\mathbf{A}_4$.

21. (i) Is $(1234)(56)$ in $\mathbf{A}_6$?
(ii) Is $(1234)(56)$ the square of an element of $\mathbf{S}_6$? Explain.

22. Let a, b, c, and d be distinct positive integers. Use the fact that

$$(ab)(ac) = (abc) \quad \text{and} \quad (ab)(cd) = (abc)(adc)$$

to prove that every α in $\mathbf{A}_n$ is the product of 3-cycles when $n \geq 3$.

23. Identify $H_0, H_1, \ldots, H_9$, shown in Figure 2.10, as the subgroups in $\mathbf{A}_4$ so that there is a rising line or broken line from H_i to H_j if and only if H_i is a proper subgroup in H_j. (Do this in one of the possible ways.)

2.11 Cosets of a Subgroup

Let us consider the orders of subgroups in the symmetric group $\mathbf{S}_4$. The cyclic subgroups $[(1)]$, $[(12)]$, $[(123)]$, and $[(1234)]$ have orders 1, 2, 3, and 4, respectively. The subset of all permutations θ in $\mathbf{S}_4$ such that $\theta(4) = 4$ is a subgroup of order 6. The octic group, the alternating group $\mathbf{A}_4$, and $\mathbf{S}_4$ itself are subgroups of orders 8, 12, and 24, respectively. These orders

$$1,\ 2,\ 3,\ 4,\ 6,\ 8,\ 12,\ 24$$

of subgroups in $\mathbf{S}_4$ are all the positive integral divisors of the order 24 of $\mathbf{S}_4$.

In the alternating group $\mathbf{A}_4$, which has order 12, we have found subgroups with

$$1,\ 2,\ 3,\ 4,\ 12$$

as their orders. We shall prove that these are the only possibilities for the order of a subgroup in $\mathbf{A}_4$.

These examples, and all the others that we have encountered so far, might lead one to conjecture that the order of a subgroup H in a finite group G must be an integral divisor of the order of G but that a positive integral divisor of the order of G need not be the order of a subgroup in G. This conjecture will be proved to be correct with the help of the following concept.

DEFINITION 1

Coset for a of H in G

Let H be a subgroup in G and let a be a fixed element of G. The set of all products ah with h in H is called the **left coset** for a of H in G and is denoted by aH. The set Ha of all ha with h in H is the **right coset** for a of H in G.

Symbolically, $aH = \{ah : h \in H\}$ and $Ha = \{ha : h \in H\}$.

EXAMPLE 1 **Cosets of $[(12)]$ in $\mathbf{S}_3$**

Let H be the cyclic subgroup $[(12)] = \{(1), (12)\}$ in $\mathbf{S}_3$. Here we find all the left cosets and all the right cosets of H in $\mathbf{S}_3$. We see that

$$\begin{aligned}(123)H &= \{(123)h : h \in H\} = \{(123)(1), (123)(12)\}\\ &= \{(123), (23)\}, \quad \text{and}\\ H(123) &= \{h(123) : h \in H\} = \{(1)(123), (12)(123)\}\\ &= \{(123), (13)\}.\end{aligned}$$

Similarly, we find that

$$(1)H = (12)H = H = H(1) = H(12),$$

$$(23)H = (123)H = \{(23), (123)\}, \qquad (13)H = (132)H = \{(13), (132)\},$$

$$H(23) = H(132) = \{(23), (132)\}, \qquad H(13) = H(123) = \{(13), (123)\}.$$

We note that, in Example 1, $\theta H = H\theta$ for $\theta = (1)$ or (12) but that $\theta H \neq H\theta$ for $\theta = (123), (132), (23)$, or (13). We also see from $(13)H = (132)H$ that one can have $aH = bH$ with $a \neq b$.

EXAMPLE 2 **Cosets of $[(123)]$ in $\mathbf{S}_3$**

Let $N = [(123)] = \{(1), (123), (132)\}$. Now we find all the left cosets and all the right cosets of N in $\mathbf{S}_3$. We have

$$\begin{aligned}(23)N &= \{(23)(1), (23)(123), (23)(132)\}\\ &= \{(23), (12), (13)\}, \quad \text{and}\\ N(23) &= \{(1)(23), (123)(23), (132)(23)\}\\ &= \{(23), (13), (12)\}.\end{aligned}$$

Similarly, we find that

$$\theta N = N\theta = \{(1), (123), (132)\} = N \quad \text{for } \theta = (1), (123), \text{ and } (132);$$

$$\theta N = N\theta = \{(23), (13), (12)\} \quad \text{for } \theta = (23), (13), \text{ and } (12).$$

We see that $gH = Hg$ may be true for all g in G, as in Example 2, or it may be true for some but not all elements of G, as in Example 1. One can easily show that $hH = Hh = H$ for each h in H. [See Problems 17(iii) and 18 of this section.]

Since the identity $\mathbf{e}$ of a group G is in every subgroup H, an element $a = a\mathbf{e} = \mathbf{e}a$ is always in both its left coset aH and its right coset Ha.

The following lemmas are part of the proof of the important Theorem of Lagrange on the orders of subgroups of finite groups. (See Theorem 1 below; also see the biographical note on Lagrange after this section.)

LEMMA 1 Number of Elements in a Coset

Let H be a finite subgroup in G. Then the number of elements in any left coset aH (or any right coset Ha) is equal to the order of H.

Proof Let $H = \{h_1, h_2, \ldots, h_s\}$. The cancellation theorem tells us that either $ah_i = ah_j$ or $h_i a = h_j a$ implies that $h_i = h_j$. This shows that there are exactly s distinct elements in each of

$$aH = \{ah_1, \ldots, ah_s\} \quad \text{and} \quad Ha = \{h_1 a, \ldots, h_s a\}. \qquad \blacksquare$$

LEMMA 2 Distinct Left (Right) Cosets Are Disjoint

Let H be a subgroup in G and let a and b be elements of G. Then the left cosets aH and bH either have no elements in common or are identical subsets of G. The same is true of the right cosets Ha and Hb.

Proof Let c be in both aH and bH; that is, let $c = ah = bh_1$ with h and h_1 in H. Then $a = bh_1h^{-1}$.

Now any element ah_2 of aH may be expressed as

$$ah_2 = (bh_1h^{-1})h_2 = b(h_1h^{-1}h_2) = bh_3,$$

where $h_3 = h_1h^{-1}h_2$ is a product of elements of H and hence is in H by group closure. Therefore, $ah_2 = bh_3$ is in bH; and we have shown that every element of aH is in bH. Similarly, every element of bH is in aH and so $aH = bH$. This completes the proof for left cosets; the proof for right cosets is similar. $\blacksquare$

THEOREM 1 Lagrange's Theorem

Let G be a finite group with order r. Then the order of each subgroup H in G and the order of each element a of G is an integral divisor of r. Also, $g^r = \mathbf{e}$ for every g in G.

Proof Let H be a subgroup in G. Also, let ord $G = r$ and ord $H = s$. Since G is finite and each left coset of H in G has s elements with no overlapping, there are a finite number, say t, of left cosets of H in G. Every a of G is in its left coset aH.

Together, the t nonoverlapping left cosets, each with s elements, have all r elements of G; hence $r = st$ and (ord H) | (ord G).

Let a be an element of order u in G. Then the cyclic subgroup $[a]$ has order u and hence $u \mid r$. Also, $r = uv$, with v an integer, and

$$a^r = a^{uv} = (a^u)^v = \mathbf{e}^v = \mathbf{e}.$$

Thus we see that $g^r = \mathbf{e}$ for all g in G. ∎

COROLLARY **Groups of Prime Order**

Let the order of G be a prime p. Then each element of G, except the identity, has order p; G is cyclic; G has $p - 1$ generators; and G is abelian.

DEFINITION 2 **Index of a Subgroup**

The number of right cosets of a subgroup H in G is called the **index** of H in G.

In the proof of Lagrange's Theorem, we have shown that the index of a subgroup H in a finite group G is the quotient of the order of G by the order of H; that is,

$$(\text{index of } H \text{ in } G) = \frac{\text{ord } G}{\text{ord } H}, \qquad \text{ord } G = (\text{ord } H)(\text{index of } H \text{ in } G).$$

Hence the index of H in G is an integral divisor of the order of G.

For example, the subgroup $[(12)] = \{(1), (12)\}$ has order 2 and index 3 in the symmetric group $\mathbf{S}_3$ of order 6.

Beginning with the next section, we shall show that an especially important role in group theory is played by the type of subgroup characterized in the following definition.

DEFINITION 3 **Normal Subgroup**

Let N be a subgroup in G such that $aN = Na$ for each a in G. Then N is a **normal subgroup** in G.

Examples 1 and 2 show that in $\mathbf{S}_3$ the subgroup [(123)] is normal but the subgroup [(12)] is not.

If one wishes to show that a given subgroup H is not normal in a group G, it is not enough just to produce elements h in H and g in G such that $hg \neq gh$. One must show that there are elements h in H and g in G such that the element hg of the right coset Hg is not in the left coset gH or such that gh is not in Hg. A very useful variation of this technique is included in the following result.

THEOREM 2 Equivalent Conditions for Normality

A subgroup H is normal in a group G if and only if $g^{-1}hg$ is in H for every h in H and g in G.

The proof of Theorem 2 is left to the reader as Problem 20 of this section.

This result frequently provides the easiest way to show that a given subgroup H is not normal in G, as we next illustrate.

EXAMPLE 3 Nonnormal Subgroup in $\mathbf{S}_3$

Using Theorem 2, we see that $H = \{(1), (12)\}$ is not a normal subgroup in $\mathbf{S}_3$, since (12) is in H, (13) is in $\mathbf{S}_3$, and $(13)^{-1}(12)(13) = (23)$ is not in H.

A normal subgroup may be called an ***invariant subgroup***, or a ***self-conjugate subgroup***, in some texts; this terminology will conform with the definition of *conjugate subgroup* in Section 3.2.

It is not necessary to distinguish between left and right cosets in an abelian group; this includes all additively written groups. Hence every subgroup in an abelian group is a normal subgroup. However, a subgroup H in G need not be normal when H is abelian and G is not. (See Example 1.)

In an additive group G, the coset for a of N in G is denoted by $a + N$. For example, let H be the cyclic subgroup [3] in the additive group $\mathbf{Z}$ of the integers. Then we easily see that there are exactly 3 cosets of H in $\mathbf{Z}$ and that they are

$$0 + H = 3 + H = -3 + H = \cdots = H = \{\ldots, -6, -3, 0, 3, 6, \ldots\},$$
$$1 + H = 4 + H = -2 + H = \cdots = \{\ldots, -5, -2, 1, 4, 7, \ldots\},$$
$$2 + H = 5 + H = -1 + H = \cdots = \{\ldots, -4, -1, 2, 5, 8, \ldots\}.$$

If we let N be the subgroup [2], in $\mathbf{Z}$, then the two cosets of N in $\mathbf{Z}$ are $0 + N = N$, which is the set of the even integers, and $1 + N$, which is the set of the odd integers.

Problems

Problem 3 below is cited in Section 2.13.

1. Let $H = \{(1), (12)(34)\}$ and $K = \{(1), (12)(34), (13)(24), (14)(23)\}$.

(i) Find two left cosets of H in the alternating group $\mathbf{A_4}$ that are also right cosets of H in $\mathbf{A_4}$.
(ii) Show that H is not a normal subgroup in $\mathbf{A_4}$.
(iii) Show that H is a normal subgroup in K.

2. Let $H = [(123)] = \{(1), (123), (132)\}$.
(i) What is the index t of H in $\mathbf{A_4}$?
(ii) Find the t left cosets of H in $\mathbf{A_4}$. (See Table 2.2 in Section 2.10.)
(iii) Find the t right cosets of H in $\mathbf{A_4}$.
(iv) Is H a normal subgroup in $\mathbf{A_4}$? Explain.

3. Let $N = \{(1), (12)(34), (13)(24), (14)(23)\}$.
(i) Does N contain every α of $\mathbf{A_4}$ with $\alpha^2 = (1)$?
(ii) Is the cube β^3 of every β of $\mathbf{A_4}$ in N?
(iii) What is the index t of N in $\mathbf{A_4}$?
(iv) Find the t left cosets and the t right cosets of N in $\mathbf{A_4}$.
(v) Is N a normal subgroup in $\mathbf{A_4}$? Explain.

4. (i) Find all the cosets of $N = [a^3]$ in a cyclic group $G = [a]$ of order 12.
(ii) Find all the cosets of $H = [\bar{3}]$ in the modular arithmetic additive group $\mathbf{Z_{12}} = \{\bar{0}, \bar{1}, \bar{2}, \ldots, \overline{11}\}$.

5. (a) Explain why a subgroup in the center of G is always a normal subgroup in G.
(b) Is the trivial subgroup $[\mathbf{e}] = \{\mathbf{e}\}$ normal in every group G? Explain.

6. Show that a subgroup $H = \{\mathbf{e}, h\}$ of order 2 is a normal subgroup in G if and only if H is contained in the center of G.

7. (a) For $n > 1$, describe the left cosets and the right cosets of the alternating subgroup $\mathbf{A_n}$ in the symmetric group $\mathbf{S_n}$.
(b) Give the left cosets and the right cosets of $[\bar{2}]$ in the $\mathbf{Z_{12}}$ of Problem 4(ii).

8. Prove that a subgroup of order m in a group of order $2m$ is always a normal subgroup.

9. (i) Use Lagrange's Theorem to show that a group G of order 4 is either cyclic or such that $g^2 = \mathbf{e}$ for all g in G.
(ii) Explain why every group of order 4 must be abelian.

10. Show that a group G of order 27 is either cyclic or such that $g^9 = \mathbf{e}$ for all g in G. Generalize on this and Problem 9(i).

11. Explain why the following are all the subgroups in the symmetric group $\mathbf{S_3}$: $\{(1)\}$, three cyclic subgroups of order 2, one cyclic subgroup of order 3, and $\mathbf{S_3}$ itself.

12. (a) Explain why the octic group $\mathbf{D_4}$ has no subgroup of order 3, 5, 6, or 7.
(b) Find all the normal subgroups in the octic group. [*Hint:* See Problem ★9 of Section 2.9 and Problems 5, 6, and 8 of this section.]

13. Show the following.
(i) A cyclic group of order 22 has one element of order 2 and ten elements of order 11.
(ii) A noncyclic group of order 22 has 21 elements, each of which has order 2 or 11.
(iii) A group of order 110 must have a cyclic subgroup of order 2, 5, or 11.

14. Generalize on part (iii) of Problem 13.

15. (i) Give the set of orders of elements of the alternating group $\mathbf{A_6}$.
(ii) What is the smallest positive integer m such that $\alpha^m = (1)$ for all α in $\mathbf{A_6}$?

Comte Joseph Louis de Lagrange
1736–1813

Lagrange, though of French parentage and generally considered to be a French mathematician, was born in Turin, Italy, and spent his early years there. In 1766 he succeeded Euler as mathematician at the Berlin Academy. He was appointed to the post by Frederick the Great, who suggested in the invitation that the greatest of European geometers should be close to the greatest of kings!

Lagrange made important contributions to analysis (*Théorie des Fonctions Analytiques*, 1797) and the calculus of variations. In these areas he introduced levels of rigor previously unknown in analysis. Much of his work parallels that of Euler, but Euler's style was less concerned with rigor. Lagrange was the first to prove that any positive integer can be expressed as the sum of at most four squares. This is the first case of Waring's Problem, a problem to which Euler had made contributions. He also was the first to prove (in 1770) the so-called Wilson's Theorem, a result first announced by Waring in his *Meditationes Algebraicae* (1770) as being the work of Waring's student, John Wilson.

Also in 1770 he published a method for solving algebraic equations, which essentially reduced the problem of solving a given equation to that of solving a related equation called a resolvent. In the case of the quadratic, cubic, and quartic equations, the resolvent was of lower degree and this provided a systematic technique for solving polynomial equations of degree 2, 3, or 4. However, the Lagrange resolvent for the fifth-degree polynomial has degree 6. This may have helped to dispel the misguided confidence of those who expected all polynomial equations to be solvable by radicals. However, it was still possible to search for a solution of the general quintic by other methods until Ruffini, Abel, and Galois settled this question completely.

The work of Lagrange on permutations of the roots of a polynomial equation provided a foundation for the group-theoretic approach of Galois.

16. Do both parts of Problem 15 with $\mathbf{A_6}$ replaced by a cyclic group of order 360 [and (1) replaced by $\mathbf{e}$].

17. Let H be a subgroup in G.
 (i) Show that $aH = bH$ if and only if $a^{-1}b$ is in H.
 (ii) If $aH = bH$, does $b^{-1}a$ have to be in H?
 (iii) Show that $hH = \mathbf{e}H = H$ for each h in H.

18. Do the analogue of Problem 17 for right cosets.

19. Let H be a subgroup in G. Prove that H is normal in G if and only if, for every g in

G and h in H, there exist h_1 and h_2 in H such that

$$hg = gh_1 \quad \text{and} \quad gh = h_2 g.$$

20. Let H be a subgroup in G. Prove that H is normal in G if and only if, for every g in G and h in H, the element $g^{-1}hg$ is in H. (This proves Theorem 2.)

21. Let N be a normal subgroup in G. Let a and b be in G and let n_1 and n_2 be in N. Prove that there is an n_3 in N such that

$$an_1bn_2 = abn_3.$$

22. Let $n \geq 3$ and let every 3-cycle (abc) of $\mathbf{S_n}$ be in a subgroup H in $\mathbf{S_n}$. Prove the following.
(i) Every even permutation of $\mathbf{S_n}$ is in H.
(ii) H is either $\mathbf{A_n}$ or $\mathbf{S_n}$.

23. Let H be the subgroup $\{(1), (123), (132)\}$ in $\mathbf{A_4}$. Show that there are 9 distinct products $\alpha\alpha'$ in which α is in H and α' is in the left coset $(12)(34)H$. (Table 2.2 in Section 2.10 may be helpful.)

24. Let N be the subgroup $\{(1), (12)(34), (13)(24), (14)(23)\}$ in $\mathbf{A_4}$. Show that there are only 4 distinct permutations among the 16 products $\alpha\alpha'$ in which α is in N and α' is in the left coset $(123)N$.

25. Let K be a subgroup in H and let H be a subgroup in G. What are the possibilities for ord H if ord $K = 60$ and ord $G = 4200$?

26. Let K be a subgroup in H and let H be a subgroup in G. What are the possibilities for ord H if ord $K = 6$ and ord $G = 288$?

27. In a nonabelian group G, let a be an element that is not in the center C and let $C_a = \{g : g \in G \text{ and } ga = ag\}$ be the centralizer of a in G. Explain why C must be a proper subgroup in C_a and C_a must be a proper subgroup in G.

28. Use Lagrange's Theorem to prove that the center C cannot have prime index in a finite group G.

29. In a group G, let H be a subgroup and a an element of order 2. Explain why the left coset aH consists of the inverses of the elements of the right coset Ha.

30. Let H and K be subgroups in G. Show the following.
(a) The intersection $I = H \cap K$ is a subgroup in G.
(b) If H and K are normal subgroups in G, so is I.

31. Give an example of groups H, K, and G such that H is normal in K, K is normal in G, but H is not normal in G.

32. Let s be a fixed positive integer, G be a group, and $H = \{a^s : a \in G\}$ be a subgroup in G. Prove that H is normal in G. [*Hint:* Use Theorem 2.]

2.12 Quotient Groups

We show below that the collection of cosets of a normal subgroup N in a group G becomes a group when the product of such cosets is defined appropriately. This is an important tool for the study of groups, and it also enables us to create new algebraic structures from previously known ones.

Let us look at two examples. In the first one, let A denote the alternating subgroup $\mathbf{A_n}$ in a symmetric group $\mathbf{S_n}$ with $n > 1$. Also, let B consist of the permutations of $\mathbf{S}_n$ not in A, namely, the odd permutations. The product of two elements of A or of two elements of B is in A; the product of an element of A and an element of B (in either order) is in B. (See Problem 9 of Section 2.8.) These statements may be represented by the following table:

	A	B
A	A	B
B	B	A

This clearly is the table for a group $\{A, B\}$ of order 2 with A as the identity.

Next let N be the normal subgroup

$$[3] = \{\ldots, -9, -6, -3, 0, 3, 6, 9, \ldots\}$$

in the additive group $\mathbf{Z}$ of the integers. We use the notation

$$P = 1 + N = \{\ldots, -8, -5, -2, 1, 4, 7, 10, \ldots\},$$

$$Q = 2 + N = \{\ldots, -7, -4, -1, 2, 5, 8, 11, \ldots\}$$

for the other cosets of N in $\mathbf{Z}$. Then the table

$+$	N	P	Q
N	N	P	Q
P	P	Q	N
Q	Q	N	P

for a group $\{N, P, Q\}$ of order 3, with N as identity, may be interpreted as summarizing certain statements about the three cosets N, P, and Q. One of these statements is that the sum of an integer in P and an integer in Q is always an integer in N.

We now generalize on these examples. In the following definition and lemma, we use multiplicative notation for the operation in G and leave it to the reader to make the appropriate changes for additive groups.

DEFINITION 1 **Product of Left Cosets**

Let H be a subgroup in G. Then the **product** $aH \cdot bH$ of left cosets aH and bH is the subset of G consisting of all products uv with u in aH and v in bH.

LEMMA 1 **Product of Cosets of a Normal Subgroup**

Let N be a normal subgroup in G. Then the product $aN \cdot bN$ of cosets aN and bN is the coset $(ab)N$.

Proof Since the identity $\mathbf{e}$ of G is in N, the product $ab \cdot n = a\mathbf{e} \cdot bn$ is in the coset product $aN \cdot bN$ for every n in N; thus every element of $(ab)N$ is in $aN \cdot bN$. For the converse, we let an and bn' be any elements of aN and bN, respectively. Since N is normal in G, $Nb = bN$ and hence $nb = bn_1$ for some n_1 in N. Then

$$an \cdot bn' = a(nb)n' = a(bn_1)n' = (ab)(n_1 n') = (ab)n_2,$$

where $n_2 = n_1 n'$ is in N. This shows that each element of the coset product $aN \cdot bN$ is in the coset $(ab)N$. Since we have already shown that every element of $(ab)N$ is in $aN \cdot bN$, we have the desired multiplication formula

$$aN \cdot bN = (ab)N$$

for the cosets of a normal subgroup N in G. ∎

It should be kept in mind that if H is not normal in G, then the product $aH \cdot bH$ (of left cosets of H and G) need not be a coset of H in G. (For an example see Problem 1 of this section.)

If N is a normal subgroup in G, it follows from Lemma 1 that the product AB, of cosets A and B of N, can be determined by performing one multiplication ab of elements a and b of A and B, respectively, and noting that AB is the coset $(ab)N$ containing ab. Thus the lemma shows that the coset product is well defined by the formula

$$aN \cdot bN = (ab)N.$$

THEOREM 1 **Quotient Group G/N**

Let N be normal in G and let G/N denote the collection of all the cosets of N in G. Then the operation of coset multiplication makes G/N into a group. (This group is called the **quotient group**, or **factor group**, of G by N.) The identity element in the quotient group G/N is the coset N.

Proof Let $\mathbf{e}$ be the identity of G. Then

$$\mathbf{e}N \cdot bN = (\mathbf{e}b)N = bN = (b\mathbf{e})N = bN \cdot \mathbf{e}N$$

shows that $\mathbf{e}N = N$ is the identity of G/N. Also,

$$aN \cdot a^{-1}N = (aa^{-1})N = \mathbf{e}N = (a^{-1}a)N = a^{-1}N \cdot aN$$

tells us that each aN in G/N has an inverse $a^{-1}N$ in G/N.

For associativity in G/N, we note that $(aN \cdot bN)cN$ and $aN(bN \cdot cN)$ consist, respectively, of all products $(uv)w$ and all products $u(vw)$ with u in aN, v in bN, and w in cN. Since $(uv)w = u(vw)$ by associativity in G, the triple products of cosets are equal and G/N is a group. ∎

We will see in the next section that the following simple result can give us much information about the possibilities for normal subgroups in a group.

THEOREM 2 Normal Subgroups of Index *m* Contain All *m*th Powers

Let N be a normal subgroup of index m in G. Then:

(a) In G/N, $C^m = N$ for all cosets C of N in G.

(b) For all g in G, g^m is in N.

Proof The order of the quotient group G/N is the number of cosets of N in G, that is, the index of N in G. It therefore follows from Lagrange's Theorem that C^m is the identity coset N for all cosets C.

Since C^m consists of all products $c_1c_2 \cdots c_m$ with each c_i in C, we have as a special case that c^m is in N for each c in every coset C of N. But each g of G is in some coset C; hence g^m is in N for all g in G. ■

We will use the following result in our characterization of all groups of order 6 and of all groups of order 8.

THEOREM 3 Finite Subgroup with Index 2

Let H be a subgroup of order t in a group G of order $2t$. Let K consist of the t elements of G not in H. Then.

(a) H is normal in G, and $G/H = \{H, K\}$.

(b) g^2 is in H for every g in G.

Proof (a) For each g in G, the coset gH (or the coset Hg) consists of t distinct elements of G, by Lemma 1 of Section 2.11. If g is in H, these t elements are the t elements of H, because H is closed under the operation of G. That is, $gH = H = Hg$ if g is in H.

If g is not in H, then gH is not identical to H, since g is in gH but is not in H; it then follows from Lemma 2 of Section 2.11 that gH and H are disjoint. Hence gH consists of the t elements of G not in H; that is, $gH = K$ when g is in K. Similarly, $Hg = K$ when g is in K.

Thus $gH = Hg$ for every g in G, and H is normal in G. It is also clear that $G/H = \{H, K\}$.

(b) Since H is normal in G, it follows from Theorem 2 with $m = 2$ that g^2 is in H for every g in G. ■

Lagrange's Theorem and the properties of cosets furnish powerful tools for, among other things, describing the possibilities for groups whose orders are fairly small.

Any group of prime order is cyclic, and it is easy to make its operation table; this takes care of groups whose order is in $\{2, 3, 5, 7, 11, \ldots\}$.

If ord $G = 4$, then Lagrange's Theorem tells us that each element in G has order 1, 2, or 4. Hence, either G is cyclic or $g^2 = \mathbf{e}$ for every g in such a group. The only possible operation table for a noncyclic G with order 4 is given in the answer to Problem 9(i) of Section 2.3.

In Examples 1, 2, and 3, which follow, we give all possibilities for groups of order 6; the similar but more difficult task for groups of order 8 is considered in Problems 17 through 21 below.

EXAMPLE 1 **Groups of Order 6, Part I**

Let ord $G = 6$. By Lagrange's Theorem, each element of G has order 1, 2, 3, or 6. Here we show that G must have an element with order 3.

If a is an element of order 6 in G, then a^2 has order 3. Hence we need only show that it is impossible for the orders of all elements of G to be 1 or 2.

Assume that G consists of the identity $\mathbf{e}$ and five elements with order 2; then $g^2 = \mathbf{e}$ for every g in G. This would make G abelian, as we saw in Problem 11 of Section 2.3, and G would have a subgroup $H = \{\mathbf{e}, a, b, ab\}$ of order 4, by Problem 5 of Section 2.5. But this would contradict Lagrange's Theorem, since 4 is not an integral divisor of 6. Therefore, G must have an element with order 3.

EXAMPLE 2 **Groups of Order 6, Part II**

Let ord $G = 6$. Here we show that there must be an element with order 2 in G. If a is an element of order 6 in G, then a^3 has order 2. Hence we need only show that it is impossible for the order of each g in G to be 1 or 3.

If b is an element of order 3, then b^2 (which is the other generator of $[b]$) also has order 3 and $b^2 \neq b$. Also, if b and c have order 3, the sets $\{b, b^2\}$ and $\{c, c^2\}$ are either identical or disjoint. Hence G has an even number of elements with order 3, and G cannot consist of the identity and 5 elements with order 3.

EXAMPLE 3 **Groups of Order 6, Part III**

Let ord $G = 6$. By Examples 1 and 2, there is an element a with order 3 and an element b with order 2 in G. Then $N = [a] = \{\mathbf{e}, a, a^2\}$ is a normal subgroup in G, by Theorem 3. Clearly b is not in N, since the elements of N have order 1 or 3. Hence the cosets of N in G are N and $Nb = \{b, ab, a^2b\}$. This means that

$$G = \{\mathbf{e}, a, a^2, b, ab, a^2b\}.$$

Also, ba is in Nb, since ba is in bN and $bN = Nb$ by the normality of N in G. We cannot have $ba = b$, since this would lead to $a = \mathbf{e}$. Thus ba equals either ab or a^2b.

If $ba = ab$, we see that G is cyclic with ab as a generator, since

$$(ab)^2 = abab = a^2b^2 = a^2, \quad (ab)^3 = a^3b^3 = b, \quad (ab)^4 = a^4b^4 = a,$$

$$(ab)^5 = a^5b^5 = a^2b, \quad \text{and} \quad (ab)^6 = a^6b^6 = \mathbf{e}.$$

If $ba = a^2b$, one finds that the multiplication table for G is the same as for $\mathbf{S_3} = \{(1), \rho, \rho^2, \phi, \rho\phi, \rho^2\phi\}$ if the permutation $\rho = (123)$ is replaced by a and $\phi = (23)$ is replaced by b.

Problems

The tables of the inside back cover of the book may be helpful in some of the following problems.

1. Let H be the subgroup $\{(1), (12)(34)\}$ in the alternating group $\mathbf{A_4}$. Let $\alpha_3 = (13)(24)$ and $\alpha_5 = (123)$. Show that there are four distinct permutations in the product

$$\alpha_3 H \cdot \alpha_5 H$$

of left cosets and hence that this product is not a coset of H in $\mathbf{A_4}$. (This should not be surprising, since H is not a normal subgroup in $\mathbf{A_4}$.)

2. Let $G = \{(1), (12)(34), (13)(24), (14)(23)\}$ and $H = \{(1), (12)(34)\}$. Use the definition of the product of left cosets and Table 2.2 of Section 2.10 to show that the product of any two left cosets of H in G is also a left coset of H in G. (This should not be surprising, since here H is a normal subgroup.)

3. Let N be the normal subgroup $\{(1), (12)(34), (13)(24), (14)(23)\}$ in $\mathbf{A_4}$. Let P and Q be the cosets $\alpha_5 N$ and $\alpha_9 N$, respectively, where $\alpha_5 = (123)$ and $\alpha_9 = (132)$. Make the multiplication table for the quotient group $A_4/N = \{N, P, Q\}$.

4. Let G be the octic group $\{\varepsilon, \rho, \rho^2, \rho^3, \phi, \rho\phi, \rho^2\phi, \rho^3\phi\}$, where $\rho = (1234)$ and $\phi = (24)$, and let $C = \{\varepsilon, \rho^2\}$ be the center of G.
   (i) Explain why C is normal in G and hence why the quotient group G/C exists.
   (ii) Is G/C cyclic? Explain.
   (iii) Is G/C abelian? Explain.

5. For a fixed positive integer m, let $[m]$ be the cyclic group generated by m in the additive group $\mathbf{Z}$ of the integers; that is, let $[m]$ consist of the integral multiples of m.
   (i) Explain why $[m]$ is a normal subgroup in $\mathbf{Z}$.
   (ii) What is the index of $[m]$ in $\mathbf{Z}$?
   (iii) Explain why the cosets of $[4]$ in $\mathbf{Z}$ are

$$0 + [4], \quad 1 + [4], \quad 2 + [4], \quad \text{and} \quad 3 + [4].$$

   (iv) Tabulate the operation of the quotient group $\mathbf{Z}/[4]$, using $\bar{0}$, $\bar{1}$, $\bar{2}$, and $\bar{3}$ to denote the cosets listed in (iii).
   (v) Is the table of part (iv) the same as for the modular arithmetic additive group $\mathbf{Z_4} = \{\bar{0}, \bar{1}, \bar{2}, \bar{3}\}$?

6. Let m be a positive integer and let $\bar{k}$ denote the coset $k + [m]$ of the normal subgroup $[m]$ in $\mathbf{Z}$ for $k = 0, 1, \ldots, m - 1$. With this notation, is the table for the quotient group $\mathbf{Z}/[m]$ the same as that for the modular arithmetic additive group $\mathbf{Z_m}$?

7. Assume that there exists a subgroup H of order 6 in the alternating group $\mathbf{A_4}$ (which has order 12). Explain why each of the following is true.
   (i) α^2 would be in H for every α in $\mathbf{A_4}$.
   (ii) Each of the eight 3-cycles (abc) would be in H.
   (iii) There is no such H.

8. Let N be a normal subgroup with index 2 in the symmetric group $\mathbf{S_4}$. Explain why each of the following is true.
   (i) θ^2 is in N for every θ in $\mathbf{S_4}$.
   (ii) Each of the 3-cycles $\alpha_5, \alpha_6, \ldots, \alpha_{12}$ is in N.
   (iii) $\alpha_2 = \alpha_5 \alpha_{10}$ is in N, and so are α_3 and α_4.
   (iv) N must be $\mathbf{A_4}$.

9. Let G be a finite group and S consist of the elements with order greater than 2 in G. Explain why each of the following is true.
   (i) If s is in S, then $s \neq s^{-1}$ and s^{-1} is in S.
   (ii) If s and t are in S, the pairs $\{s, s^{-1}\}$ and $\{t, t^{-1}\}$ are either identical or disjoint.
   (iii) S has an even number of elements.

(iv) If ord G is even, G has an even number of elements g satisfying $g^2 = \mathbf{e}$, and hence G has an odd number of elements of order 2.
(v) G has at least one element of order 2 if and only if ord G is even.

10. (a) Let ord $G = 2p$, with p in the set $\{3, 5, 7, 11, \ldots\}$ of odd primes. Prove that G has an element of order p. [*Hint:* See Example 1.]
(b) Let ord $G = 10$. Explain why there are a and b in G such that $G = \{\mathbf{e}, a, a^2, a^3, a^4, b, ab, a^2b, a^3b, a^4b\}$ and $a^5 = \mathbf{e} = b^2$.

11. Let N be a normal subgroup of prime index p in a group G. Explain why the quotient group G/N is cyclic.

12. Let C be the center in a group G. Explain why each of the following is true.
(i) C is normal in G and thus G/C exists.
(ii) If G/C is cyclic, that is, $G/C = [bC]$ for some b in G, then every x in G is expressible as $b^n c$ with n in $\{0, 1, \ldots\}$ and c in C.
(iii) If G/C is cyclic, $xy = yx$ for all x and y in G, and hence $G = C$.
(iv) The index of C in G cannot be a prime.

13. Describe a cyclic normal subgroup N with index 2 in the dihedral group $\mathbf{D_n}$.

14. Let N be a normal subgroup in G. Show that G/N is abelian if and only if $a^{-1}b^{-1}ab$ is in N for all a and b in G.

15. Let $\mathbf{R}$ be the additive group of the real numbers, let N be its cyclic subgroup $[2\pi]$ (where π is the number 3.14159 ...), and let T be the quotient group $\mathbf{R}/N$. In T, let $\bar{a}$ denote the coset $a + N$. Explain why each of the following is true.
(a) $\overline{2m\pi} = \bar{0}$ for all integers m.
(b) For every $\bar{a}$ in T, there is a real number b with $-\pi < b \le \pi$ and $\bar{a} = \bar{b}$.
(c) The order of $\overline{\pi/12}$ in T is 24.
(d) $\bar{1}$ has infinite order in T.

16. Let $\mathbf{R}$ be the additive group of the real numbers, $\mathbf{Z}$ be its cyclic subgroup $[1] = \{\ldots, -2, -1, 0, 1, 2, \ldots\}$, and W be the quotient group $\mathbf{R}/\mathbf{Z}$.
(a) What is the order of the coset $(-2/5) + \mathbf{Z}$ in W?
(b) Use the fact that $\sqrt{3}$ is irrational to show that the coset $\sqrt{3} + \mathbf{Z}$ does not have finite order in W.

17. Let ord $G = 8$. Explain why each of the following is true.
(i) Each g in G has order 1, 2, 4, or 8.
(ii) Either G has an element of order 4 or $g^2 = \mathbf{e}$ for each g in G.
(iii) If $g^2 = \mathbf{e}$ for each g in G, then G is abelian and has the form

$$G = \{\mathbf{e}, a, b, c, ab, ac, bc, abc\},$$

with $a^2 = b^2 = c^2 = \mathbf{e}$. [One such group is the subgroup in $\mathbf{S_6}$ obtained by letting $\mathbf{e} = (1)$, $a = (12)$, $b = (34)$, and $c = (56)$.]

18. Let ord $G = 8$. Let a be an element of order 4 in G and let b be in G but not in $N = [a] = \{\mathbf{e}, a, a^2, a^3\}$. Explain why each of the following is true.
(i) N is normal in G; that is, $gN = Ng$ for every g in G.
(ii) $G = \{\mathbf{e}, a, a^2, a^3, b, ab, a^2b, a^3b\}$.
(iii) b^2 is in N. [*Hint:* See Theorem 3(b).]
(iv) If $b^2 = a$, then $G = [b]$; that is, G is cyclic with b as a generator.
(v) If $b^2 = a^3$, then $G = [b]$.
(vi) If b has order 4, then $b^2 = a^2$. [*Hint:* Use part (iii) and note the order of b^2.]

19. Let a, b, G, and N be as in Problem 18. Explain why each of the following is true.
(i) $ba^k \ne b$ for k in $\{1, 2, 3\}$.
(ii) $ba^kb^{-1} \ne \mathbf{e}$ for k in $\{1, 2, 3\}$.

(iii) $(bab^{-1})^k = ba^kb^{-1}$ for k in $\mathbf{Z}^+$.
(iv) bab^{-1} has order 4, $bab^{-1} \neq a^2$, and $ba \neq a^2b$.
(v) ba is either ab or a^3b.

20. Make the operation table for a group

$$G = \{\mathbf{e}, a, a^2, a^3, b, ab, a^2b, a^3b\}$$

under each of the following assumptions.
(a) $a^4 = \mathbf{e} = b^2$ and $ba = a^3b$.
(b) $a^4 = \mathbf{e}$, $b^2 = a^2$, and $ba = a^3b$.

21. Let ord $G = 8$. Explain why G must be as in Problem 17(iii), or as in Problem 18(iv), or have the form

$$G = \{\mathbf{e}, a, a^2, a^3, b, ab, a^2b, a^3b\}$$

and satisfy one of the following conditions.
(i) $a^4 = \mathbf{e} = b^2$ and $ba = ab$. [One such G is the subgroup in $\mathbf{S}_6$ with $\mathbf{e} = (1)$, $a = (1234)$, and $b = (56)$.]
(ii) $a^4 = \mathbf{e} = b^2$ and $ba = a^3b$. [One such G is the octic group $\mathbf{D}_4$.]
(iii) $a^4 = \mathbf{e}$, $b^2 = a^2$, and $ba = a^3b$. [One such G is described in Section 4.7; it is called the ***quaternion group***.]

22. Let $H = \{h_1, h_2, \ldots, h_s\}$ be a finite subgroup in G. For every a and b in G, let $(ab)H = aHb$, where $aHb = \{ahb : h \in H\}$. Prove that H is normal in G.

2.13 Solvable Groups

"There is no use trying," she said: "one can't believe impossible things." "I daresay you haven't had much practice," said the Queen. "When I was your age, I always did it for half-an-hour a day. Why, sometimes I've believed as many as six impossible things before breakfast."
—Charles Lutwidge Dodgson (Lewis Carroll)

Techniques for solving second-degree polynomial equations,

$$ax^2 + bx + c = 0,$$

go back at least to the times of the ancient Greeks. In the first half of the sixteenth century, Italian mathematicians found formulas expressing the roots of the general third-degree and fourth-degree polynomial equations,

$$ax^3 + bx^2 + cx + d = 0,$$
$$ax^4 + bx^3 + cx^2 + dx + e = 0,$$

in terms of the coefficients, using addition, subtraction, multiplication, division, and extraction of roots. (Techniques for solving polynomial equations of degree 2, 3, or 4 are given in Section 5.11.)

Then for approximately 300 years, some of the greatest mathematicians of all time tried to find a formula of this type for the roots of the general fifth-degree polynomial equation.

But no such formula is possible, as was proved by Niels Henrik Abel in a paper published in 1824. Paolo Ruffini gave a prior demonstration, but it was not generally accepted as a convincing proof. (See the biographical note on Ruffini at the end of Section 2.3.)

Modern treatments of solvability of polynomial equations are based on the work of Évariste Galois, in which group theory plays a major role. We shall go into this matter further in Section 5.12. A key concept is the following.

DEFINITION 1

Solvable Finite Group

A finite group G is **solvable** if there exists a chain

$$G_0, G_1, G_2, \ldots, G_r$$

of groups such that $G_0 = G$, G_{i+1} is a normal subgroup with prime index in G_i for $0 \leq i \leq r - 1$, and $G_r = \{\mathbf{e}\}$.

Note that, in Definition 1, the ratio (ord G_i)/(ord G_{i+1}) must be a prime number.

In Example 1 we shall show that the symmetric group $\mathbf{S}_4$ is solvable. Theorem 2 proves that the alternating group $\mathbf{A}_5$ is not solvable. It is left to the reader, in the problems for this section, to show that $\mathbf{A}_4$ is solvable and $\mathbf{S}_5$ is not.

EXAMPLE 1 **$\mathbf{S}_4$ Is Solvable**

Here we show that the symmetric group $\mathbf{S}_4$ is solvable. Let $G_0 = \mathbf{S}_4$, $G_1 = \mathbf{A}_4$,

$$G_2 = \{(1), (12)(34), (13)(24), (14)(23)\},$$

$G_3 = \{(1), (12)(34)\}$, and $G_4 = \{(1)\}$. For $i = 0$, 2, and 3, G_{i+1} has index 2 in G_i and hence is normal in G_i by Theorem 3 of Section 2.12. Also, G_2 is a normal subgroup with index 3 in G_1. (See Problem 3 of Section 2.11.) Since G_{i+1} is a normal subgroup with prime index in G_i for $0 \leq i \leq 3$, we have proved that $\mathbf{S}_4$ is solvable.

DEFINITION 2

Simple Group

A group G is **simple** if it has only itself and $\{\mathbf{e}\}$ as normal subgroups and $G \neq \{\mathbf{e}\}$.

For example, it follows from Lagrange's Theorem that every group of prime order is simple.

THEOREM 1 **Unsolvable Simple Group**

Let G be a finite simple group whose order is not prime. Then G is not solvable.

Proof This follows readily from Definitions 1 and 2. ■

It can be shown that the alternating groups $\mathbf{A_n}$ are simple for $n \geq 5$ and hence that these groups are not solvable. We content ourselves here with the outline of a proof, using a combinatorial technique, that $\mathbf{A_5}$ is not solvable. The proof that $\mathbf{A_n}$ is not solvable for $n \geq 5$ is asked for in Problem 17 of Section 3.5.

Our proofs of nonsolvability are based on Theorem 2 of Section 2.12 (which states that a normal subgroup N of index m in a group G has the property that g^m is in N for every g in G) and on the counting of the number of squares, cubes, or fifth powers in the group under consideration.

THEOREM 2 **Unsolvable Alternating Group**

$\mathbf{A_5}$ is not solvable.

Proof It follows from Definition 1 that a finite solvable group G must have a normal subgroup of prime index in G. Since the order of $\mathbf{A_5}$ is $60 = 2^2 \cdot 3 \cdot 5$, a subgroup with prime index would have to have index 2, 3, or 5. It therefore suffices to show that $\mathbf{A_5}$ has no normal subgroup with index 2, 3, or 5. With the help of Theorem 2 of Section 2.12, we deal with these cases in the following lemmas. ■

LEMMA 1 **$\mathbf{A_5}$ Has No Subgroup of Order 30**

There is no subgroup with index 2 in $\mathbf{A_5}$.

Proof By Theorem 3 of Section 2.12, such a subgroup would contain α^2 for every α in $\mathbf{A_5}$. The formulas $(abc) = (acb)^2$ and $(abcde) = (adbec)^2$ show that the twenty 3-cycles (abc) and the twenty-four 5-cycles $(abcde)$ in $\mathbf{A_5}$ are squares. With the identity, this makes a total of 45 squares in $\mathbf{A_5}$, too many to fit into a subgroup of order 30. ■

We leave the proofs of the following two results to the reader as Problems 5 and 6 of this section.

LEMMA 2 $\mathbf{A_5}$ has no normal subgroup with index 3.

LEMMA 3 $\mathbf{A_5}$ has no normal subgroup with index 5.

We also leave the proof of the following result to the reader as Problem 11 of this section.

THEOREM 3 **Unsolvable Symmetric Group**

$\mathbf{S_5}$ is not solvable.

Lagrange's Theorem states that the order of a subgroup in a finite group G is an integral divisor of the order of G. Paolo Ruffini was the first to show that the converse is not true; that is, there exist finite groups G for which some positive integral divisor of the order of G is not the order of a subgroup in G. (See Problem 7 of Section 2.12 and Lemma 1 of this section for such groups and such divisors.)

Problems

1. Show that the octic group $\mathbf{D_4}$ is solvable, by producing a chain of subgroups with the necessary properties.
2. Show that the symmetric group $\mathbf{S_3}$ is solvable.
3. Show that the dihedral group $\mathbf{D_6}$ is solvable.
4. (i) Which permutations are cubes of elements of $\mathbf{A_4}$?
   (ii) What is the only subgroup of order 4 in $\mathbf{A_4}$?
5. Explain why the following are true.
   (i) Every permutation of $\mathbf{A_5}$ is a 1-cycle, 3-cycle, 5-cycle, or product of two disjoint 2-cycles.
   (ii) 40 of the 60 elements of $\mathbf{A_5}$ are cubes of elements of $\mathbf{A_5}$.
   (iii) There is no normal subgroup with index 3 in $\mathbf{A_5}$. (This is the proof of Lemma 2 left to the reader.)
6. Prove that there is no normal subgroup with index 5 in $\mathbf{A_5}$ (and thus prove Lemma 3).
7. Let N be a subgroup with index 2 in $\mathbf{S_5}$. Prove that $N = \mathbf{A_5}$ by showing the following.
   (i) Every 3-cycle (xyz) of $\mathbf{S_5}$ is in N.
   (ii) Every 5-cycle $(abcde)$ of $\mathbf{S_5}$ is in N.
   (iii) Every product of two disjoint transpositions of $\mathbf{S_5}$ is in N. [*Hint:* Use $(ab)(cd) = (abc)(adc)$.]
8. Prove that there is no normal subgroup with index 3 in $\mathbf{S_5}$.
9. Prove that there is no normal subgroup with index 5 in $\mathbf{S_5}$.
10. Explain why the only normal subgroup with prime index in $\mathbf{S_5}$ is $\mathbf{A_5}$. [*Hint:* Use Problems 7–9.]
11. Use Problem 10 and Theorem 2 to prove that $\mathbf{S_5}$ is not solvable (and thus prove Theorem 3).
12. Prove the following.
    (i) No one of the three subgroups with index 6 in $\mathbf{A_4}$ is a normal subgroup in $\mathbf{A_4}$.

William Burnside
1852–1927

Burnside, an English mathematician, conjectured early in the twentieth century that all groups of odd order are solvable. This remained an open question until 1963 when a paper "Solvability of Groups of Odd Order" appeared in the *Pacific Journal of Mathematics.* In this paper the authors, two Americans, **John Thompson** (1932–) of the University of Chicago and **Walter Feit** (1930–) of Yale, proved that all groups of odd order are solvable. Thus nonabelian finite simple groups have even order; the proof required over 250 pages and took up a whole issue of the journal. In 1965 they were awarded the Cole Prize of the American Mathematical Society for this achievement, and in 1970 Thompson received the Fields Medal at the International Congress of Mathematicians in Nice.

The Fields Medal is taken to be the equivalent in mathematics of a Nobel Prize. It is often reported — and there may be no basis for the report — that there is no Nobel Prize in mathematics because Alfred Nobel felt a personal animosity toward the Swedish mathematician G. M. Mittag-Leffler; and since, if there had been a Nobel Prize in mathematics, Mittag-Leffler might have won it, Nobel decreed that there would be no Nobel Prize in mathematics. The Fields Medal is restricted to the work of mathematicians under the age of forty. (For more information on the Fields Medals, see Albers et al., *International Mathematical Congresses/An Illustrated History, 1893–1986,* in the Annotated Bibliography.)

If a finite group G is not simple, it has a nontrivial proper normal subgroup N, and much information concerning the structure of G is obtainable from the structures of N and G/N, which have lower orders than G. If N or G/N is not simple, the process can be continued. Hence simple groups are in a sense the fundamental building blocks for a theory of the structure of finite groups. The determination of all finite simple groups was generally considered to be practically impossible until Thompson and Feit proved that the order of any such group is either a prime or an even integer. Then an enormous effort was begun by a number of first-class algebraists to finish the so-called classification problem for all simple groups. This reached a successful conclusion in 1980. The contributors are too numerous to mention here, but the list includes **Michael Aschbacher** (1944–) of the California Institute of Technology and **Daniel Gorenstein** (1923–1992) of Rutgers University. There are various categories of simple groups; some of the groups are of astounding size and complexity. The so-called "monster" group F_1 has order equal to $2^{46} \cdot 3^{20} \cdot 5^9 \cdot 7^6 \cdot 11^2 \cdot 13^3 \cdot 17 \cdot 19 \cdot 23 \cdot 29 \cdot 31 \cdot 41 \cdot 47 \cdot 59 \cdot 71$, a large number indeed! Mathematicians in this field are currently trying to shorten the lengthy proofs for this classification problem.

(ii) $\mathbf{A_4}$ has no normal subgroup with index 4.
(iii) $\mathbf{A_4}$ has exactly three normal subgroups. (See Problem 4.)

13. Let $G = [a]$ be a cyclic group of order 25. Find a subgroup G_1 of index 5 in G, explain why G_1 is a normal subgroup, and show that G is solvable.

14. Let $G = [a]$ be a cyclic group of order 1001. Find a subgroup G_1 of index 7 in G, a subgroup G_2 of index 11 in G_1, and a subgroup G_3 of index 13 in G_2. Then show that G is solvable.

15. Let $G = [a]$ be a cyclic group of order m and let

$$m = p_1 p_2 \cdots p_r,$$

where the p_i are positive primes. Show that G is solvable.

16. Use Problem 15 and a factorization theorem of Section 1.7 to explain why every finite cyclic group is solvable.

17. Let N be a finite cyclic normal subgroup of prime index in G. Explain why G is solvable.

18. Prove that the dihedral group $\mathbf{D_n}$ is solvable for $n \geq 3$.

★ **19.** Prove that $\mathbf{A_5}$ is simple.

2.14 More on Symmetry

We have observed in Section 2.9 that certain groups of permutations are associated with symmetries of regular polygons. Groups do, in fact, describe symmetries of geometric forms in much more general ways. For example, one can develop groups that describe algebraically the symmetries that one observes in patterns in wallpaper, or in the elaborate illustrations of the Dutch artist, M. C. Escher. These groups are made up of elements that represent translations, rotations and other transformations of the geometric figures. George Pólya and the Swiss mineralogist, Paul Niggli, in 1924 classified all the "symmetries" in the plane — there are seventeen in all — in a pair of papers published in a crystallographic journal. This essentially categorizes all the wallpaper patterns that are possible. The methods used were group-theoretic and the motivation for their joint work was a problem in chemistry, although Pólya admitted later that he had spent a lot of time in the architecture library in Zürich studying the Moorish decorations in the Alhambra in Granada, Spain. It turns out that the result had been discovered in 1891 by a Russian crystallographer, E. S. Fedorov, and, in fact, sixteen of the seventeen symmetries had been found even earlier and appear in work of Camille Jordan from 1869. Nevertheless Pólya's mathematical treatment was the one that caught the attention of the mathematical community and, as it turns out, the artistic community as well. Recent scholarship has shown that Escher was actually influenced by this work; he meticulously copied Pólya's 1924 paper and used some of the symmetries he found there in subsequent designs.

The groups associated with these symmetries in the plane are infinite and of somewhat less interest to us in the context of this chapter. Here we have largely concentrated on finite groups. We shall therefore look now at some groups associated with geometric figures in three dimensions. These too are

FIGURE 2.11

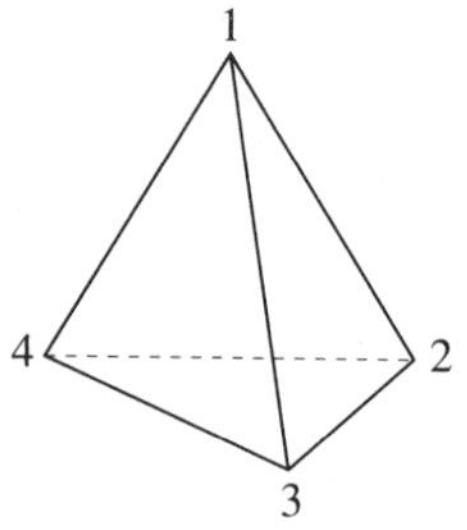

related to questions in the study of crystals and molecular structure. The latter problem, involving the benzene molecule, led to Pólya's developing a famous combinatorial theorem that bears his name.

As you recall, in the case of the dihedral groups, we considered rotations of the regular polygons in the plane and flips about axes. These flips involved moving out of the plane in which the figure sits and into three-dimensional space. This makes sense as long as we think of the figures as essentially being in three-dimensional space all the time. But if we tried to do something analogous to a figure like a cube in three-space, we would find ourselves moving out of the familiar three-space into four-dimensional space. This is too hard to visualize, so we shall restrict ourselves here to rotations in three-dimensional space.

Let us consider all the rotations of the regular tetrahedron (shown in Figure 2.11) that will bring the figure back into its original position but permute the vertices. If we label the vertices 1, 2, 3, and 4 as indicated in the figure, we can identify the rotations with permutations just as we did with the dihedral groups. Of course, if we do not rotate the figure at all, we get the identity, (1). There are three 180° rotations about axes connecting midpoints of opposite edges — for example, about the line connecting the midpoint of edge $\{1, 4\}$ and the midpoint of edge $\{2, 3\}$. This particular rotation is (14)(23). The other two are (12)(34) and (13)(24).

There are also rotations about the altitudes. Let us consider first the altitude connecting vertex 1 with the centroid of the opposite face having vertices 2, 3, and 4. The figure will remain in the same position if we rotate the figure about this line through 120° or 240°, but, of course, the vertices will be in new locations. These two rotations correspond to the permutations (234) and (243). Of course, one more rotation about 120° gives a rotation through the angle 360°, and this corresponds to the identity. Now there are three more altitudes; those from vertices 2, 3, and 4. They yield, respectively, the permutations (134), (143); (124), (142); and (123), (132). So the set of symmetries of the regular tetrahedron consists of the identity, three products of two 2-cycles, and eight 3-cycles. Of course, we recognize this as an old friend, the alternating group, $\mathbf{A}_4$, of Section 2.10.

EXAMPLE 1 Let us consider the set of rotations of the cube that leave the position of the cube fixed but permute the vertices. This will be a group of higher order than we have been dealing with up to this time; but since the cube has eight ver-

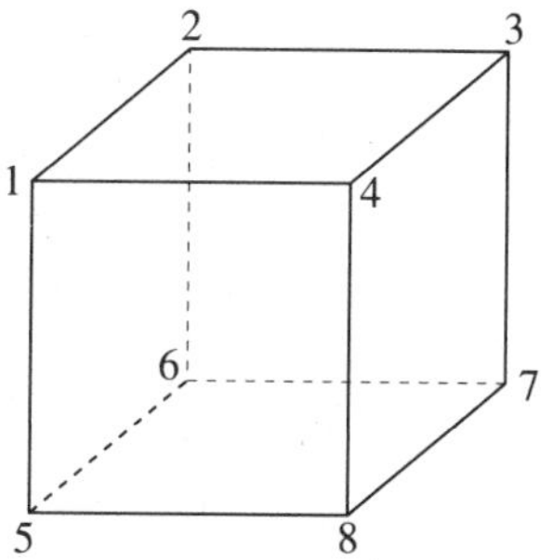

FIGURE 2.12

tices, we should be able to identify this set of rotations with a subset (and, in fact, it will be a subgroup) of $\mathbf{S}_8$. First we identify the vertices as shown in Figure 2.12.

There will be four rotations associated with the line connecting the centroid of the upper face (with vertices 1, 2, 3, and 4) and the centroid of the lower face (with vertices 5, 6, 7, and 8). Rotations about multiples of 90° yield, with clockwise rotation, viewed from above, the following permutations: (1), (1234)(5678), (13)(24)(57)(68), and (1432)(5876). Similarly, two other such lines connect centroids of opposite faces, and these yield the following additional permutations: (1265)(3784), (16)(25)(38)(47), (1562)(3487), (1485)(2376), (18)(27)(36)(45), (1584)(2673).

There are further rotations through 180° about lines connecting midpoints of opposite edges — for example, the midpoint of edge $\{1, 2\}$ and edge $\{7, 8\}$: (12)(35)(46)(78). (It helps in seeing these if one actually has in hand a small cube that one can rotate to see what happens to the vertices.) There are six such lines and, correspondingly, six rotations; the others are as follows:

Connecting midpoints of $\{2, 3\}$, $\{5, 8\}$: (17)(23)(46)(58)

Connecting midpoints of $\{3, 4\}$, $\{5, 6\}$: (17)(28)(34)(56)

Connecting midpoints of $\{1, 4\}$, $\{6, 7\}$: (14)(28)(35)(67)

Connecting midpoints of $\{1, 5\}$, $\{3, 7\}$: (15)(28)(37)(46)

Connecting midpoints of $\{2, 6\}$, $\{4, 8\}$: (17)(26)(35)(48).

Of course, some rotations remain uncounted: those about lines connecting opposite vertices. The easiest way to see the permutations these lead to is to look at the cube along one of the diagonals. Say we look at the vertex 4 and down the diagonal $\{4, 6\}$. Then we see that we shall have new symmetries here corresponding to rotations through 120° and 240°. The permutations will involve three cycles of the vertices that are connected to vertices 4 and 6 by one edge, namely, (138)(275) and (183)(257). There are four such diagonals, each giving a permutation that can be written as a product of two disjoint 3-cycles. The other three are as follows:

Diagonal $\{3, 5\}$: (186)(247) and (168)(274)

Diagonal $\{2, 8\}$: (136)(475) and (163)(457)

Diagonal $\{1, 7\}$: (245)(386) and (254)(368).

In summary, the following represent rotations of the cube that leave the cube in a fixed position:

$\beta_1 = (1)$

$\beta_2 = (1234)(5678)$

$\beta_3 = (1432)(5876)$

$\beta_4 = (1265)(3784)$

$\beta_5 = (1562)(3487)$

$\beta_6 = (1485)(2376)$

$\beta_7 = (1584)(2673)$

$\beta_8 = (13)(24)(57)(68)$

$\beta_9 = (16)(25)(38)(47)$

$\beta_{10} = (18)(27)(36)(45)$

$\beta_{11} = (12)(35)(46)(78)$

$\beta_{12} = (17)(23)(46)(58)$

$\beta_{13} = (17)(28)(34)(56)$

$\beta_{14} = (14)(28)(35)(67)$

$\beta_{15} = (15)(28)(37)(46)$

$\beta_{16} = (17)(26)(35)(48)$

$\beta_{17} = (138)(275)$

$\beta_{18} = (183)(257)$

$\beta_{19} = (186)(247)$

$\beta_{20} = (168)(274)$

$\beta_{21} = (136)(475)$

$\beta_{22} = (163)(457)$

$\beta_{23} = (245)(386)$

$\beta_{24} = (254)(368)$

We note in the example above that there are 24 rotations of the cube. In fact, it can be shown that, were we to form a group table for this set of rotations (and they do, indeed, form a group), we would have a group with the same structure as $\mathbf{S}_4$. The rotations of a cube or an octahedron have the same structure as $\mathbf{S}_4$; this is called the octahedral group.

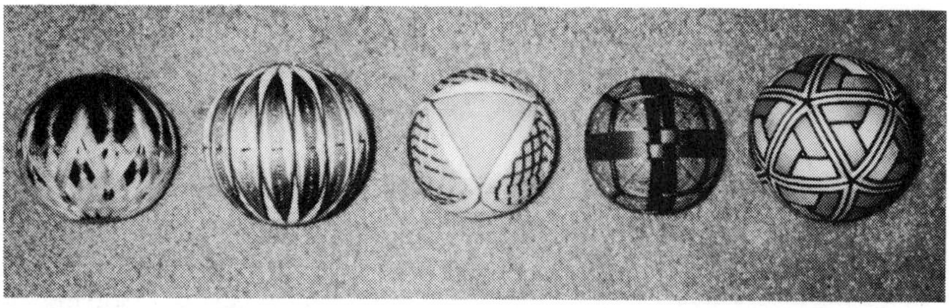

Temari balls constructed by Kasuko Yamamoto and illustrating (from left to right) a cyclic group symmetry, a dihedral group symmetry, tetrahedral, octahedral, and icosahedral symmetries.

It is rather surprising, but there are relatively few finite groups of rotations in three-dimensional space: the cyclic groups of order n; the dihedral groups $\mathbf{D_n}$; $\mathbf{A_4}$ (the tetrahedral group; the set of rotations of a tetrahedron); $\mathbf{S_4}$ (the octahedral group; the set of rotations of a cube or of an octahedron), and $\mathbf{A_5}$ (the icosahedral group; the set of rotations of a dodecahedron or of an icosahedron).

Problems

1. (a) What are the orders of the various elements of the octahedral group (the group of Example 1)?
   (b) How many elements are there of each order?
2. Check the elements of $\mathbf{S_4}$. What are the orders of the elements, and how many are there of each order? Do the results agree with those of Problem 1?
3. Find a subgroup of order 12 in the octahedral group, and list its elements.
4. Does the group of Problem 3 have a familiar group structure? What is it?
5. (a) Find the group of rotations of a pyramid with a square base, as shown in the accompanying figure, by listing the permutations of the vertices.

FIGURE 2.13

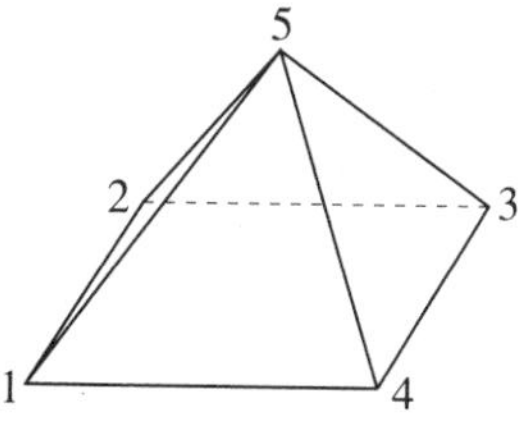

   (b) Set up the multiplication table for the elements of part (a).
6. What are the orders of the elements of the group in Problem 5? Does the group have a familiar structure? What is it?
7. (a) Give the permutations of the vertices for the rotations of a triangular prism, as shown in the accompanying figure.

George Pólya
1887–1985

Pólya contributed primarily to the field of mathematical analysis, but he also made major contributions in mathematical probability (most significantly in the theory of random walk), in geometry (where he described one of the famous space-filling curves), in combinatorics (where he proved the Pólya Enumeration Theorem, in which symmetry groups play an essential role), and in number theory (where the Pólya conjecture had a close relationship to the famous Riemann hypothesis). Born in Budapest, he went on to study not only in Budapest, but also in Göttingen and Paris before becoming a professor at the Swiss Federal Institute of Technology in Zürich and later at Stanford University. Throughout his career he was interested in mathematical problem solving, an area that led to a number of his most influential books. The most popular of these is *How To Solve It*, which has appeared in 19 languages and has sold well over a million copies.

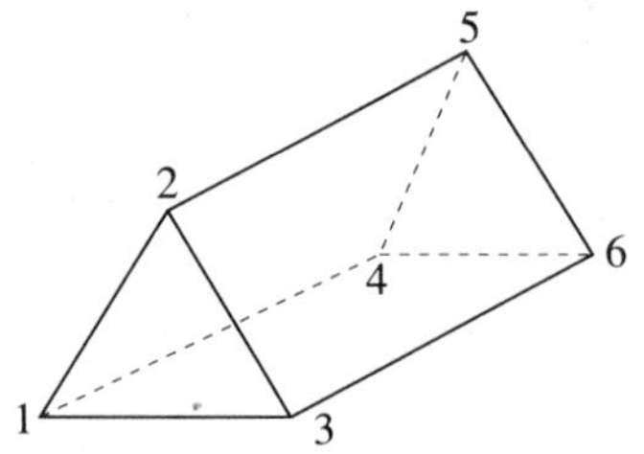

FIGURE 2.14

(b) Set up a multiplication table for the group elements in part (a).
(c) Does the group structure look familiar? What is it?

8. (a) Give the permutations of the vertices for the rotations of the box (prism, with square base), shown in the accompanying figure.

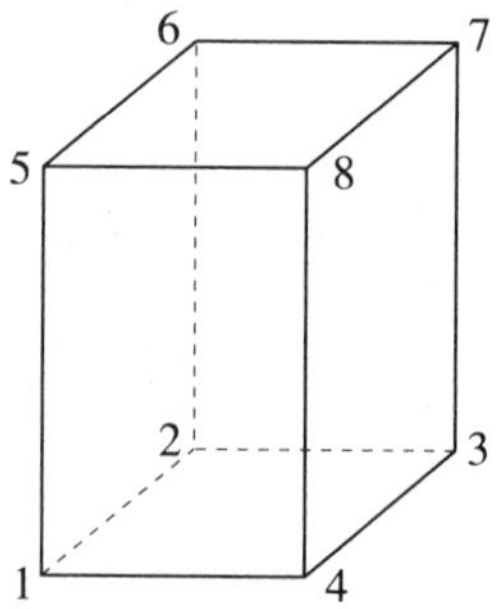

FIGURE 2.15

(b) Set up the multiplication table for the permutations of part (a). What is the order of the group? What are the orders of the elements?
(c) Does the group structure look familiar? What is it?

9. What is the group of rotations of a prism whose cross section is a regular polygon with n sides?

10. What is the group of rotations of a pyramid whose base is a regular polygon with n sides?

★ 2.15 The Sylow Theorems (Without Proofs)

In this section, G is always a finite group with order m. By Lagrange's Theorem, $(\text{ord } H) \mid m$ for every subgroup H in G. However, there is not necessarily a subgroup with order d for every positive integral divisor d of m. (See Problem 7 of Section 2.12.) But there are partial converses of Lagrange's Theorem; we state here without proof a theorem of this nature and some related results, all due to the Norwegian mathematician Ludwig Sylow (1832–1918).

THEOREM 1 Sylow's Theorem, Part I

Let p and n be positive integers with p a prime. Then G has a subgroup of order p^n if and only if $p^n \mid (\text{ord } G)$.

The case $n = 1$ of this theorem is known as ***Cauchy's Theorem***. For each positive prime divisor p of $m = \text{ord } G$, the case with the largest possible exponent n is especially noteworthy.

DEFINITION 1 Sylow p-Subgroup

Let p and n be positive integers with p a prime. Let the order of G be an integral multiple of p^n but not of p^{n+1}. Then a subgroup of order p^n in G is a **Sylow *p*-subgroup** in G.

THEOREM 2 Sylow's Theorem, Part II

Let H and K be any two Sylow p-subgroups in G for the same prime p. Then there is an element a in G such that

$$K = \{a^{-1}ha : h \in H\}, \qquad H = \{aka^{-1} : k \in K\};$$

that is, H and K are conjugate subgroups. (See Problem 23 of Section 2.5.)

THEOREM 3 Sylow's Theorem, Part III

For each positive prime divisor p of $m = \text{ord } G$, the number of Sylow p-subgroups in G is an integral divisor d of m having the form $d = 1 + kp$, with k a nonnegative integer.

In Section 3.16, the structure of finite commutative groups is characterized completely in terms of the Sylow p-subgroups.

Problems

1. Do the following for the alternating group $\mathbf{A}_4$.
   (a) Find a Sylow 2-subgroup, and explain why it is the only Sylow 2-subgroup.
   (b) Show that the Sylow 3-subgroups $H = [(123)]$ and $K = [(124)]$ are conjugate subgroups, and find two other Sylow 3-subgroups.
2. Describe a Sylow 2-subgroup in the symmetric group $\mathbf{S}_4$.
3. Show that $\mathbf{S}_5$ has six Sylow 5-subgroups.
4. Show that $\mathbf{S}_5$ has ten Sylow 3-subgroups.
5. Describe a Sylow 2-subgroup in $\mathbf{S}_5$.

★ 6. How many Sylow 2-subgroups are there in $\mathbf{S}_5$?

7. Let p and q be primes with $p > q > 0$. Use Theorem 3 to show that a group G of order pq has only one Sylow p-subgroup and that this subgroup is normal.
8. Show that the number of Sylow 5-subgroups in a group G of order 55 must be 1 or 11.

Review Problems

1. Let a, b, and c be in a group and $b \neq c$. Show that $ab^{-1} \neq ac^{-1}$.
2. Find the standard form of each of the following products.
   (a) (123)(145).
   (b) (1234)(1567).
   (c) (12345)(16789).
   (d) (14)(24)(34)(12)(24)(23).
3. Let $\alpha = (13)(15)(16)(21)(24)(26)$ and $\beta = (125)(326)(14)$. Find the standard forms of α, α^{-1}, β, and β^{99}, and give the order of each of these four permutations.
4. (a) Express (132)(154)(123)(145) in standard form.
   (b) Let $\alpha = (axb)$ and $\beta = (bcy)$. Show that $\alpha^{-1}\beta^{-1}\alpha\beta$ is a 3-cycle.
5. Let $G = [a]$, a cyclic group of order 80.
   (a) What is the order of a^{36}? Explain.
   (b) List the elements of a subgroup H of order 5.
   (c) List the elements of a subgroup N of index 10 in G.
   (d) List all the elements of order 10 in G.
6. If a is in a group G and a^{52} is the identity, what are the possibilities for the order of a?
7. Let $G = [a]$, a cyclic group of order 6. Give all the subsets of G that are closed under the operation of G. Is each of these subsets a subgroup in G? Explain.
8. (a) Give the set of orders of elements of $\mathbf{A}_7$.
   (b) Give the set of orders of elements of $\mathbf{S}_7$.
   (c) How many of the elements of $\mathbf{S}_{10}$ are products $(abcd)(efghi)$ of two disjoint cycles, one of length 4 and the other of length 5?

9. In the multiplicative group V of the nonzero complex numbers, find the order of $\cos t + \mathbf{i} \sin t$, where
(a) $t = 11\pi/9$; (b) $t = 10\pi/9$; (c) $t = 9\pi/10$.

10. (i) Find four elements each of which is a generator of the modular arithmetic additive group $\mathbf{Z}_{12}$.
(ii) Give the multiplicative inverse of each of the generators listed for (i).
(iii) Does $\bar{2}$ have a multiplicative inverse in $\mathbf{Z}_{12}$?

11. Let H be a subgroup in $\mathbf{A}_4$ and let the order of H be at least 7. Must H be $\mathbf{A}_4$? Explain.

12. (a) Let G be a group of order 64 and let G contain an element a such that a^{32} is not the identity. Must G be cyclic? Explain.
(b) Show that a group of order 64 must have an element of order 2.
(c) Let G be a group of order 96 and a be an element of G such that neither a^{48} nor a^{32} is the identity. Explain why G must be cyclic.

13. Let G be a finite group. Show the following.
(a) G has an even number of elements of order 3.
(b) The number of elements of order 5 is a multiple of 4.
(c) G has an even number of elements x such that x^2 is not the identity.

14. Let H be a subgroup of order s in a group G of order t and let $s < t$. Explain why $2s \leq t$.

15. Let r and f be elements of order 5 and 2, respectively, in a group G. Let $fr = r^4 f$. Show that the smallest subgroup in G containing r and f is

$$H = \{\mathbf{e}, r, r^2, r^3, r^4, f, rf, r^2f, r^3f, r^4f\}.$$

16. Show that the group H of Problem 15 is solvable.

17. Let H be a subgroup in the symmetric group $\mathbf{S}_5$ and let each of (123), (1234), and (12345) be in H. Show that $H = \mathbf{S}_5$.

18. (a) What is the smallest positive integer s such that there is a nonabelian group of order s?
(b) What is the smallest positive integer t such that there is a noncyclic group of order t?

19. What is the smallest positive integer m such that $\alpha^m = (1)$ for every permutation α of $\mathbf{A}_7$?

20. In the real numbers, the equation $x^2 = x$ has 0 and 1 as solutions. How many solutions are there of $x^2 = x$ in a group G? Explain.

21. (a) Let a be an element of order 3 in a group G. Show that a is a square in G; that is, show that there is an element b in G such that $b^2 = a$.
(b) Show that every element of order 5 in G is a square.

22. Generalize on parts (a) and (b) of Problem 21.

23. (a) Let a be an element of order 2 in a group G. Show that there exist elements b and c in G with $a = b^3 = c^5$.
(b) Let d have order 3 in G. Show that $d = f^5$ for some f in G.
(c) Let g have order 5 in G. Show that $g = h^3$ for some h in G.

24. Generalize on Problems 21–23.

25. Let m be the order of a finite cyclic group G. Let g be any element of G and let s be a positive integer such that $\gcd(m, s) > 1$. Explain why g^s is not a generator of G.

26. Let H be a subgroup in G. Prove the following.

(a) If $aH = Hb$, then $aH = Ha = bH = Hb$.
(b) H is normal in G if and only if for every g in G there is a g' in G such that $gH = Hg'$ (or such that $Hg = g'H$).
(c) H is normal in G if and only if for every h in H and g in G there is an h' in H such that $hg = gh'$ (or such that $gh = h'g$).

CHAPTER 3

SETS AND MAPPINGS

To see what is general in what is particular and what is permanent in what is transitory is the aim of scientific thought.
— **Alfred North Whitehead**

The *function* concept is an indispensable tool in all of modern mathematics. We are accustomed to using it in calculus, of course, where we deal with real valued functions of a real variable — that is, functions from a set of real numbers to a set of real numbers. The permutations treated in the previous chapter are special functions from the set $\mathbf{X}_n = \{1, 2, \ldots, n\}$ to itself.

In this chapter we generalize these examples and define a function from any set X to any set Y. We usually substitute the synonym *mapping* for the word *function*, to lessen the possibility of confusing the general concept with the previously familiar special cases.

As we have seen in our study of abstract groups, a given group table can represent groups of very different things; a group with elements that are complex numbers, for example, may have essentially the same group table as a group consisting of geometric rotations. A natural question arises: when are two groups "the same"? And for a given order, how many "different" groups are there? We shall use the idea of a mapping to help us answer such questions. And along the way we'll encounter Cayley's Theorem, which at first encounter seems quite surprising but then, when one understands the proof, seems very natural. We shall also see how to construct new groups out of old ones, using direct products.

In 1854, George Boole, an Irish mathematician-logician, published a book entitled *The Laws of Thought*. It attracted little interest in the nineteenth century. But in this landmark book Boole developed an algebra of sets, or of logic, and this has had a profound effect on twentieth-century mathematics. Not only did the theory of sets, and hence Boole's algebra, have an effect on the way we look at mathematics, it provided a logical tool essential to the design of computers. We shall investigate this algebra, very different from group theory, late in this chapter.

3.1 Mappings

DEFINITION 1

Mapping, Image

A **mapping** θ from a set X to a set Y is an assignment to each element x in X of a unique element y in Y. The element y in Y assigned to a given x in X is called the **image** of x under θ and may be denoted by $\theta(x)$.

DEFINITION 2

Domain, Codomain, Image Set

Let θ be a mapping from X to Y. Then X is the **domain** and Y is the **codomain** of θ. The **image set** of θ is the subset of Y consisting of the images y of all the elements x of X.

Symbolically, the image set of a mapping θ from X to Y is

$$\{\theta(x) : x \in X\}.$$

Many elementary calculus texts use *range* for *codomain*; many others use *range* for *image set*. Because of this ambiguity, we shall avoid the word *range* in what follows.

NOTATION **Mapping Arrows**

One indicates that θ is a mapping from X to Y by writing

$$\theta : X \to Y.$$

An alternative way of indicating that $y = \theta(x)$ is to write

$$\theta : x \mapsto y$$

or to write

$$x \overset{\theta}{\mapsto} y.$$

Either of these may be read as "θ sends x to y" or as "x goes to y under θ."

Both kinds of arrows are used when a mapping f from X to Y is given by

$$f : X \to Y; \qquad x \mapsto f(x),$$

where $f(x)$ is a rule for obtaining the image of x under f. For example,

$$f : \mathbf{Z} \to \{0, 1, 2, \ldots\}; \qquad n \mapsto n^2$$

designates the mappings in which the image of each integer n is n^2.

EXAMPLE 1 Let $X = \{1, 2, 3, 4\}$ and $Y = \{a, b, c\}$. Let $\theta : X \to Y$ be such that

$$\theta : 1 \mapsto a,\ 2 \mapsto c,\ 3 \mapsto a,\ 4 \mapsto a.$$

A pictorial representation of the mapping θ is given in Figure 3.1.

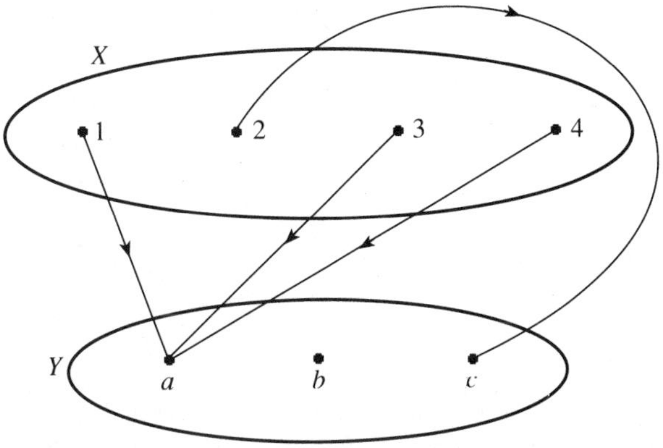

FIGURE 3.1

As is required by the definition of a mapping, each x of X has a unique image. The definition does not require that every element of Y be the image of some x in X; we see here that no element of X has b as its image. Also, an element of Y may be the image of more than one element of X. For example, the element a of Y is the image of three elements of X under this mapping. Since the element b of the codomain is not an image, the image set $\{a, c\}$ is a proper subset of the codomain $\{a, b, c\}$.

EXAMPLE 2 Let $G = [a]$ and $G' = [b]$ be cyclic groups of orders 6 and 3, respectively. Let θ be the mapping from G to G' with

$$\mathbf{e} \mapsto \mathbf{e}',\quad a \mapsto b,\quad a^2 \mapsto b^2,\quad a^3 \mapsto \mathbf{e}',\quad a^4 \mapsto b,\quad a^5 \mapsto b^2.$$

Here the image set of θ is the same set as the codomain $G' = \{\mathbf{e}', b, b^2\}$. This mapping is illustrated in Figure 3.2.

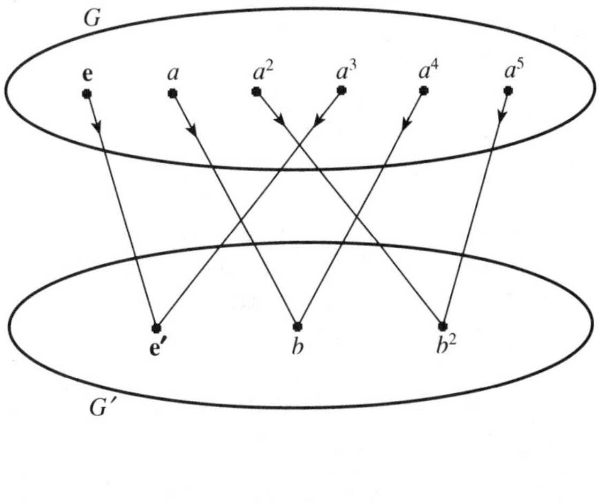

FIGURE 3.2

EXAMPLE 3 Let X be a proper subset of Y and let θ be the mapping from X to Y with $\theta(x) = x$ for all x in X. The image set is X, a proper subset of the codomain Y. A special case of this type of mapping is depicted in Figure 3.3.

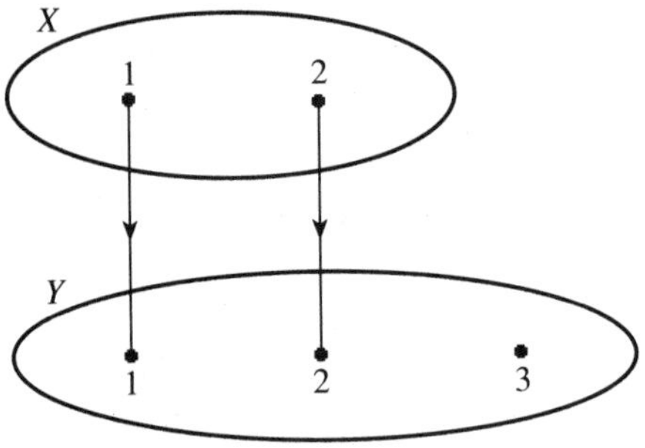

FIGURE 3.3

EXAMPLE 4 $\theta : \mathbf{R} \to \mathbf{R}; \quad r \mapsto 2^r$
Let θ be the mapping from the set $\mathbf{R}$ of the real numbers to itself with $\theta(r) = 2^r$. The image set of θ consists of the positive numbers and is a proper subset of the codomain $\mathbf{R}$. Figure 3.4 is a partial representation of this mapping.

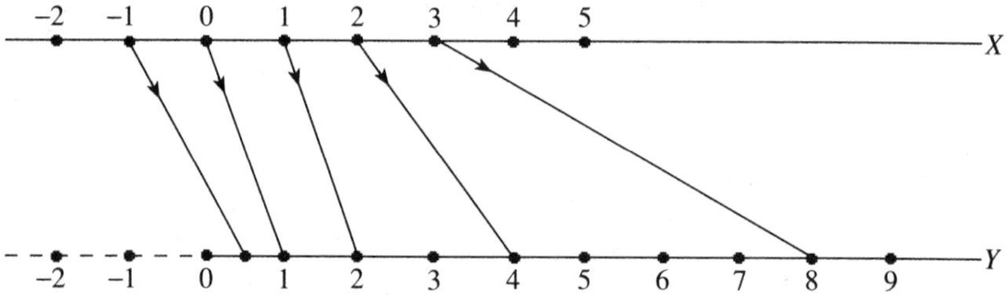

FIGURE 3.4

EXAMPLE 5 $\theta : \mathbf{R} \to \mathbf{R}^+; \quad r \mapsto 2^r$
Let θ be the mapping from the real numbers $\mathbf{R}$ to the positive real numbers $\mathbf{R}^+$ given by $\theta(r) = 2^r$. Here we have modified the mapping of Example 4 so that the codomain $\mathbf{R}^+$ is the same as the image set. Figure 3.4, excluding the dashed part of the line, depicts this θ.

EXAMPLE 6 $\theta : G \to G; \quad x \mapsto xa$
Let a be a fixed element of a group G and let θ be the mapping from G to itself with $\theta(x) = xa$ for all x in G. If b is in G, $\theta(ba^{-1}) = ba^{-1}a = b$; this shows that every element of the codomain G is in the image set of this θ.

Let θ be a fixed mapping from X to Y. By definition of a mapping, a given element x of X always determines one and only one image $\theta(x)$. However, if $\theta(x)$ is given, one may not be able to determine x uniquely. In Example 1, if $\theta(x)$ is known to be a, then all one can deduce is that x is 1, 3, or 4.

The following terminology is applied to mappings θ such that each $\theta(x)$ in the image set determines x uniquely.

DEFINITION 3

Injection

Let θ be a mapping such that distinct elements x_1 and x_2 of the domain always have distinct images $\theta(x_1)$ and $\theta(x_2)$; that is, let $x_1 \neq x_2$ imply $\theta(x_1) \neq \theta(x_2)$. Then θ is said to be an **injection**, or an **injective mapping**.

Equivalently, θ is an injecton if $[\theta(x_1) = \theta(x_2)] \Rightarrow (x_1 = x_2)$.

An injection is also called a ***one-to-one mapping***. The mapping of Example 1 is not injective, since $1 \neq 3$ but $\theta(1) = \theta(3)$. (See Figure 3.1.) In Example 2, $\mathbf{e} \neq a^3$ and $\theta(\mathbf{e}) = \theta(a^3)$; hence, this θ is also not an injection. (See Figure 3.2.) The mappings of Examples 3, 4, 5, and 6 are injections (one-to-one mappings), since $\theta(x_1) = \theta(x_2)$ implies $x_1 = x_2$ for these mappings. (See Figures 3.3 and 3.4.)

DEFINITION 4

Surjection

If the image set of a mapping θ from X to Y is the same set as the codomain Y, the mapping is said to be **surjective**, or a **surjection**.

In other words, a mapping is surjective when every element of the codomain is in the image set. A surjection from X to Y is also called a mapping from X ***onto*** Y.

The mapping of Example 1 is not a surjection (mapping onto the codomain), since the element b of the codomain is not in the image set. (See Figure 3.1.) The mappings of Examples 3 and 4 are also not surjective. The mappings of Examples 2, 5, and 6 are surjections. (See Figure 3.2 and Figure 3.4 without the dashed part.)

DEFINITION 5

Bijection, Permutation

A mapping that is both injective and surjective is said to be **bijective**, or a **bijection**. A bijection from X to itself is a **permutation** on X.

We note that this definition of a permutation on a general set X generalizes the definition of a permutation on $\{1, 2, \ldots, n\}$ given in Section 2.1.

Of the examples above, only the mappings in Examples 5 and 6 are bijections (one-to-one mappings onto the codomain). The mapping in Example 6 is a permutation.

One may note that Figure 3.1 represents a mapping that is neither surjective nor injective; Figure 3.2, a surjection that is not injective; Figure 3.3, an injection that is not surjective; and Figure 3.4 (without the dashed part), a bijection.

DEFINITION 6

Pre-image, Complete Inverse Image

Let θ be a mapping from X to Y. If m' is an element of the image set of θ, an element m with $\theta(m) = m'$ is called a **pre-image** or an **antecedent** of m'. For every y in Y, $\theta^{-1}(y)$ denotes the set of all x in X such that $\theta(x) = y$; the subset $\theta^{-1}(y)$ of X is called the **complete inverse image** of y.

Symbolically, $\theta^{-1}(y) = \{x \in X : \theta(x) = y\}$.

We note that the set $\theta^{-1}(y)$ of all antecedents of y is empty when y is not in the image set of θ, and hence θ is surjective if and only if $\theta^{-1}(y)$ is nonempty for all y in the codomain Y. Also, θ is injective if and only if $\theta^{-1}(y)$ never has more than one element. Hence θ is bijective if and only if $\theta^{-1}(y)$ has exactly one element for every y in Y. One may think of θ^{-1} as a mapping from Y to the class of subsets of X.

If α is a bijection from X to Y, there is a mapping β from Y to X with $\beta(y)$ the unique pre-image of y; it can be seen that this β is a bijection from Y to X.

DEFINITION 7

Inverse of a Bijection

For a bijection α from X to Y, the mapping β from Y to X with $\beta(y)$ the unique pre-image of y in X is called the **inverse** of α and may be denoted by α^{-1}.

The existence of an inverse mapping for every bijection α sometimes leads us to use the two-headed arrow notation

$$x \overset{\alpha}{\leftrightarrow} y$$

to indicate that α sends x to y and α^{-1} sends y to x.

We note that $\alpha^{-1}(y)$ may have either of two meanings when α is a bijection, and we leave it to the context to tell us whether $\alpha^{-1}(y)$ is the unique pre-image x or is the complete inverse image $\{x\}$ under the bijection α.

A finite sequence $a_1, a_2, \ldots, a_n$ of elements from a set A characterizes the mapping α from $\{1, 2, \ldots, n\}$ to A with $\alpha(i) = a_i$ for $1 \le i \le n$. An infinite sequence $b_1, b_2, b_3, \ldots$ of elements from a set B is a way of looking at the mappings $\beta : \mathbf{Z}^+ \to B$ with $\beta(i) = b_i$ for each i in $\mathbf{Z}^+$.

When the elements of X and Y are denoted by Greek letters α, β, and so on, we shall generally use Latin letters f, g, and so on for mappings from X to Y.

Some synonyms for the word *mapping* are ***map***, ***function***, ***correspondence***, and ***transformation***. We repeat that an injection may be referred to as a ***one-to-one mapping*** and that a surjection from X to Y may be called a mapping from X ***onto*** Y, or an ***onto mapping***. A bijection from X to Y is also called a ***one-to-one correspondence*** between X and Y, or a ***one-to-one mapping*** from X ***onto*** Y.

Problems

Problem 12 below is cited in Section 3.14.

1. Describe a bijection θ from the set $\mathbf{Z}^+ = \{1, 2, 3, \ldots\}$ to its proper subset $Y = \{2, 3, 4, \ldots\}$.

2. Describe a bijection θ from the set $\mathbf{Z} = \{\ldots, -2, -1, 0, 1, 2, \ldots\}$ to its proper subset $E = \{\ldots, -4, -2, 0, 2, 4, \ldots\}$.

3. (i) Tabulate one of the injections (one-to-one mappings) θ from $X = \{1, 2, 3\}$ to $Y = \{1, 2, 3, 4\}$.
(ii) Is θ surjective (a mapping onto Y)? Explain.

4. (i) Describe an injection θ from a set $X = \{x_1, x_2, \ldots, x_n\}$ with n elements to the set $\mathbf{Z}$ of the integers.
(ii) Is θ surjective?

5. Let θ be the mapping from the set $\mathbf{R}$ of the real numbers to the set N of the nonnegative real numbers given by $\theta(x) = |x|$.
(i) Is θ injective? Explain.
(ii) Is θ surjective? Explain.
(iii) List the elements of the complete inverse image $\theta^{-1}(5)$.

6. Let $G = [a]$ be a cyclic group of order 12.
(i) Tabulate the mapping θ from G to G with $\theta(x) = x^3$.
(ii) List the elements of the image set of θ.
(iii) Is θ injective? Explain.
(iv) Is θ surjective? Explain.
(v) List the elements of the complete inverse images $\theta^{-1}(a^9)$ and $\theta^{-1}(a^{10})$.

7. Let $X = \{x_1, x_2, x_3\}$ be a set with 3 elements and let θ be a mapping from X to itself.
(i) Can θ be injective without being surjective?
(ii) Can θ be surjective without being injective?

8. (i) Describe an injection α, from the set $\mathbf{Z}$ of the integers to itself, such that α is not a surjection. [*Hint*: See Problem 2.]
(ii) Describe a surjection β from $\mathbf{Z}$ to $\mathbf{Z}$ that is not an injection.

9. How many mappings are there from a set X with 2 elements to a set Y with 3 elements?

10. How many mappings are there from a set X with m elements to a set Y with n elements?

11. Let θ be a mapping from a group G to a set Y. An element p of G, such that $\theta(xp) = \theta(x)$ for all x in G, is called a ***right period*** of θ. Prove that the set of all right periods of θ is a subgroup in G.

12. Let a be a fixed element of a group G, and let θ be the mapping from G to itself with $\theta(x) = xa$ for all x in G. In Example 6 above we showed that θ is a surjection. Explain why θ is also an injection and hence is a bijection.

13. Let $\rho = (1234)$ and $\phi = (24)$. Let G be the octic group

$$\{\varepsilon, \rho, \rho^2, \rho^3, \phi, \rho\phi, \rho^2\phi, \rho^3\phi\};$$

N be its center $\{\varepsilon, \rho^2\}$; and N, $B = N\rho$, $C = N\phi$, and $D = N\rho\phi$ be the cosets of the normal subgroup N. Let f be the mapping from G to the quotient group G/N such that the image of each element α of G is its coset $N\alpha$.

(i) Does $f(\alpha\beta) = f(\alpha)f(\beta)$ for all α and β in G? Explain.
(ii) Tabulate the mapping f.
(iii) Is $f(\varepsilon)$ the identity of G/N? Explain.
(iv) Does $f(\alpha^{-1}) = [f(\alpha)]^{-1}$ for all α in G? Explain.
(v) Which elements of G are in the complete inverse image $f^{-1}(N)$?

14. Let N be a normal subgroup in G and θ be the mapping from G to G/N with $\theta(a) = aN$.

(i) Does $\theta(ab) = \theta(a)\theta(b)$ for all a and b in G? Explain.
(ii) Does θ send the identity of G to the identity of G/N? Explain.
(iii) Does $\theta(a^{-1}) = [\theta(a)]^{-1}$ for all a in G? Explain.
(iv) What elements of G are in the complete inverse image $\theta^{-1}(N)$?
(v) Is θ surjective? Explain.
(vi) What must be true of N for θ to be injective?

15. Let r_n be the remainder when the integer n is divided by 3 and $\mathbf{Z}_3$ be the modular arithmetic additive group $\{\bar{0}, \bar{1}, \bar{2}\}$.

(i) Complete the following partial table for

$$f : \mathbf{Z} \to \mathbf{Z}_3; \quad n \mapsto \bar{r}_n.$$

n	$\cdots$	-2	-1	0	1	2	3	4	5	6	7
$\bar{r}_n$	$\cdots$	$\bar{1}$	$\bar{2}$	$\bar{0}$	$\bar{1}$	$\bar{2}$	$\bar{0}$				

(ii) Does $f(m + n) = f(m) + f(n)$ for all m and n in $\mathbf{Z}$?

16. Let $g : \mathbf{Z} \to \mathbf{Z}_5 = \{\bar{0}, \bar{1}, \bar{2}, \bar{3}, \bar{4}\}$; $n \mapsto \bar{s}_n$, where s_n is the remainder when n is divided by 5. Does $g(mn) = g(m)g(n)$ for all integers m and n?

17. Let α and β be bijections from A to B and from B to C, respectively. Describe a bijection γ from A to C.

18. Let an injection α: $A \to B$ have C as its image set. Use α to characterize a bijection $\beta : A \to C$, that is, a one-to-one mapping from A onto C.

3.2 Group Isomorphisms

Mathematicians do not study objects, but relations among objects; they are indifferent to the replacement of objects by others as long as relations do not change. Matter is not important, only form interests them. — **Henri Poincaré**

There are many groups of order 3, for example, the groups

$$G_1 = \{(1), (123), (132)\}, \qquad G_2 = \{1, (-1 + \mathbf{i}\sqrt{3})/2, (-1-\mathbf{i}\sqrt{3})/2\},$$

with composition of permutations the operation for G_1 and multiplication of complex numbers the operation for G_2. However, the differences between these groups are superficial in that one can obtain the table for the operation of G_2 by replacing (1) with 1, (123) with $(-1 + \mathbf{i}\sqrt{3})/2$, and (132) with $(-1 - \mathbf{i}\sqrt{3})/2$ in the row headings, column headings, and entries of the table for G_1. Actually, the construction of the table for any group $G = \{\mathbf{e}, b, c\}$ of order 3, given in Example 2 of Section 2.3, shows that all groups of order 3 have essentially the same structure.

On the other hand, two groups of order 4 may differ in more than the names for their elements and the names for their operations; for example, one can be cyclic and the other noncyclic.

We shall be able to make these statements more precise after introducing the terminology that follows.

DEFINITION 1

Group Homomorphism

A **group homomorphism** from a group G to a group G' is a mapping $\theta : G \to G'$ such that

$$\theta(ab) = \theta(a)\theta(b) \qquad \text{for all } a \text{ and } b \text{ in G.} \qquad \text{(P)}$$

The property of θ given in formula (P) is called ***preservation of the operations*** of G and G'; it states that the image of a product is the product of the images. As we see in Figure 3.5, the multiplication ab is performed in G, while $\theta(a)\theta(b)$ denotes a product in G'.

DEFINITION 2

Group Isomorphism

A bijective homomorphism is called an **isomorphism**; that is, a group isomorphism θ from G to G' is a bijection θ such that $\theta(ab) = \theta(a)\theta(b)$ for all a and b in G.

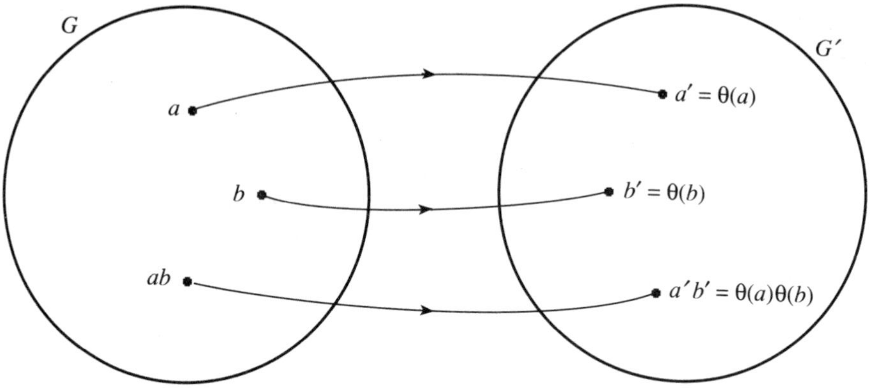

FIGURE 3.5

We study isomorphisms in this section and take up the more general concept of homomorphisms in the next section.

One says that G is ***isomorphic*** to G' if there exists an isomorphism θ from G to G'. The inverse mapping θ^{-1} of such a θ is easily seen to be an isomorphism from G' to G. Hence the groups in an isomorphism play interchangeable roles, and we may replace *G is isomorphic to G′* with *G and G′ are isomorphic.*

Preservation of the operations implies that whenever elements a, b, and c of a group G are such that $ab = c$, their images a', b', and c' under an isomorphism θ from G to G' must satisfy $a'b' = c'$ in G'. If the operation of G is specified by a table, it follows that replacement of the row and column headings and the entries of this table by their images under the isomorphism θ gives one the table for the operation of G'. This means that G and G' have the same structure; the only possibilities for differences are in the names for their elements and for their operations.

EXAMPLE 1 **Noncyclic Groups of Order 4**

Let $G = \{\mathbf{e}, b, c, d\}$ and $G' = \{\mathbf{e}', u, v, w\}$ be any two noncyclic groups of order 4. Here we show that G and G' are isomorphic. Since G and G' are noncyclic, it follows from Lagrange's Theorem that each x in G or G' has order 1 or 2 and hence that x^2 is the identity. Then one shows easily that the tables for the operations of G and G' are as follows:

	e	b	c	d
e	**e**	b	c	d
b	b	**e**	d	c
c	c	d	**e**	b
d	d	c	b	**e**

	e′	u	v	w
e′	**e′**	u	v	w
u	u	**e′**	w	v
v	v	w	**e′**	u
w	w	v	u	**e′**

We see that the table on the left goes over into the one on the right, by applying the bijection θ with

$$\mathbf{e} \mapsto \mathbf{e}', \quad b \mapsto u, \quad c \mapsto v, \quad d \mapsto w.$$

Hence this bijection satisfies $\theta(x)\theta(y) = \theta(xy)$ for all x and y in G and is an isomorphism. There are 6 isomorphisms from G onto G'. (See Problem 3 of this section.)

This example shows that there is essentially only one noncyclic group of order 4; such a group is called a ***Klein 4-group*** or a ***4-group*** and is often denoted by V, the initial letter of "Viergruppe," which is "4-group" in German. (See the biographical note on Felix Klein after Section 2.9.) Note that in a Klein 4-group all squares equal the identity; that is, each entry on the main diagonal of the group table is an $\mathbf{e}$.

The following result will help us to set up isomorphisms, when they exist, and to prove that certain pairs of groups are not isomorphic.

THEOREM 1 Properties of Group Isomorphisms

Let θ be a group isomorphism from G to G'. Then:

(a) $\mathbf{e} \overset{\theta}{\mapsto} \mathbf{e}'$, where $\mathbf{e}$ and $\mathbf{e}'$ are the identities of G and G'.

(b) If $a \overset{\theta}{\mapsto} a'$, then $a^n \overset{\theta}{\mapsto} (a')^n$ for all integers n. In particular, $a^{-1} \overset{\theta}{\mapsto} (a')^{-1}$.

(c) If $a \overset{\theta}{\mapsto} a'$, then a and a' have equal orders.

Proof Using preservation of the operations and properties of the identities, we have

$$\theta(\mathbf{e})\theta(\mathbf{e}) = \theta(\mathbf{ee}) = \theta(\mathbf{e}) = \mathbf{e}'\theta(\mathbf{e}).$$

Then the equality $\theta(\mathbf{e})\theta(\mathbf{e}) = \mathbf{e}'\theta(\mathbf{e})$ and right cancellation give us $\theta(\mathbf{e}) = \mathbf{e}'$; this proves (a).

For (b) we start with $aa^{-1} = \mathbf{e}$. Preservation of the operations and the result in (a) give us

$$\theta(a)\theta(a^{-1}) = \theta(aa^{-1}) = \theta(\mathbf{e}) = \mathbf{e}'.$$

The equality $\theta(a)\theta(a^{-1}) = \mathbf{e}'$ shows that $\theta(a^{-1}) = [\theta(a)]^{-1}$, that is, $a^{-1} \mapsto (a')^{-1}$. It is left to the reader to complete a proof of (b) by mathematical induction.

For (c), we let $a \mapsto a'$ and denote the orders of a and a' by q and q', respectively. From (a) we have $\mathbf{e} \mapsto \mathbf{e}'$, while $\mathbf{e} = a^q \mapsto (a')^q$ by (b). Hence $(a')^q = \mathbf{e}'$. Since q is a positive integer and q' is the least positive integer m with $(a')^m = \mathbf{e}'$, we have $q \geq q'$. Similarly, $q' \geq q$. Together, these inequalities imply that $q = q'$. ■

In the proof of Theorem 1, we used the fact that a group isomorphism θ preserves products to show that it also preserves the identity, (integral) powers, and the inverse and order of all elements.

If groups G and G' are isomorphic and either one has finite order, the existence of a bijection from G to G' implies that the two groups have the same order. The following result will help us see that the converse is not true; that is, groups of the same order need not be isomorphic.

LEMMA 1 Cyclic and Noncyclic Groups of the Same Order

Let ord $G = q =$ ord G' and let $G = [a]$ be cyclic. Then G and G' are isomorphic if and only if G' is cyclic.

Proof If $G' = [b]$ is cyclic, then the mapping

$$\theta : G \to G';\ a^j \mapsto b^j \qquad \text{for } j = 0, 1, \ldots, q - 1$$

is a bijection that preserves the operations, since elements x and y of $[a]$ can be written as $x = a^j$ and $y = a^k$, and then

$$\theta(xy) = \theta(a^j a^k) = \theta(a^{j+k}) = b^{j+k} = b^j b^k = \theta(a^j)\theta(a^k) = \theta(x)\theta(y).$$

Hence θ is an isomorphism.

If g' is not cyclic, it has no element of order q and there is no element in G' to serve as the image of the generator a of G under an isomorphism. [See Theorem 1(c).] In this case, G and G' are not isomorphic. ∎

EXAMPLE 2 Classification of Groups of Order 4

Let $G_1 = [(1234)]$, $G_2 = \{(1), (12), (34), (12)(34)\}$, and G be any group of order 4. If G is cyclic, it is isomorphic to G_1 by Lemma 1. If G is not cyclic, then Example 1 tells us that G is isomorphic to G_2. Also, we see from Lemma 1 that G_1 and G_2 are not isomorphic. Thus there are essentially only two groups of order 4. One deduction from this classification result is that all groups of order 4 are abelian.

EXAMPLE 3 Classification of Groups of Order 6

It follows from Lemma 1 that $G_1 = [(123456)]$ and $G_2 = \mathbf{S}_3$ are not isomorphic. Also, Example 3 of Section 2.12 shows that every group of order 6 is isomorphic either to G_1 or to G_2.

A classification of all groups of order 8 can be based on Problem 21 of Section 2.12.

The groups G and G' of an isomorphism need not be distinct. For example, if $G = \{\mathbf{e}, b, c\}$ is a group of order 3, then $G = [b] = [c]$, and it follows from the proof of Lemma 1 that the mapping θ with

$$\mathbf{e} \mapsto \mathbf{e}, \quad b \mapsto c, \quad c \mapsto b$$

is an isomorphism from G onto itself.

DEFINITION 3

Group Automorphism

A group isomorphism from G to itself is a **group automorphism** of G.

If G is a group, the mapping θ with $\theta(g) = g$ for all g in G is always an automorphism of G; this is called the ***trivial***, or ***identity***, automorphism of G. A nontrivial automorphism θ of G is one with $\theta(a) \neq a$ for at least one a in G.

The following result provides both a source of automorphisms and a tool for studying normal subgroups.

THEOREM 2 **Conjugation**

Let a be a fixed element of a group G and let θ be the mapping from G to itself with $\theta(x) = a^{-1}xa$ for all x in G. Then θ is an automorphism of G. Moreover, $a^{-1}xa$ has the same order as x for each x in G.

Proof We see that θ is surjective by noting that for every b in G one has $\theta(aba^{-1}) = a^{-1}(aba^{-1})a = b$. Also, θ is injective since $a^{-1}xa = a^{-1}ya$ implies that $x = y$. Hence, θ is bijective. Further,

$$\theta(x)\theta(y) = (a^{-1}xa)(a^{-1}ya) = a^{-1}(xy)a = \theta(xy)$$

shows that θ preserves the operations. Hence, θ is an automorphism. Finally, $a^{-1}xa$ and x have the same order by Theorem 1(c). ■

DEFINITION 4

Inner Automorphism, Conjugate

An **inner automorphism** of a group G is a mapping $\theta : x \mapsto a^{-1}xa$ from G onto itself, where a is a fixed element of G. This θ is called **conjugation** by a. The image $\theta(b) = a^{-1}ba$ is the **conjugate** of b by a. Also $\{g^{-1}bg : g \in G\}$ is the **set of conjugates** of b in G.

Let H be a subgroup in G, and θ be the inner automorphism of conjugation by a. Then one can show [see Problem 23(i) of Section 2.5] that

$$K = \{a^{-1}ha : h \in H\} = \{\theta(h) : h \in H\} \tag{1}$$

is also a subgroup in G.

DEFINITION 5

Conjugate Subgroup

If H is a subgroup in G and a is an element of G, the subgroup K of (1) is the **conjugate subgroup** of H by a.

We note that, if K is the conjugate subgroup of H by a, then H is the conjugate subgroup of K by a^{-1} [see Problem 23(ii) of Section 2.5]; that is, H and K are conjugate subgroups of each other.

THEOREM 3 Tests for a Subgroup to Be Normal

A subgroup H is normal in G if and only if any one of the following is true:
(a) The conjugate $g^{-1}hg$ is in H for every h in H and every g in G.
(b) The conjugate subgroup $\{g^{-1}hg : h \in H\}$, of H by g, is contained in H for every g in G.
(c) The conjugate subgroup of H by g is H for every g in G.

Proof Part (a) is Problem 20 of Section 2.11, and (b) is a rewording of (a). We leave (c) to the reader as Problem 29 of this section. ∎

Because of the result in Theorem 3(c), a normal subgroup in G is frequently called a ***self-conjugate subgroup*** or an ***invariant subgroup***.

Problems

1. Let $G = [a]$ and $G' = [b]$ be cyclic groups of order 5.
(i) Complete the following table so that each θ_i is an isomorphism from G to G'.

x	e	a	a^2	a^3	a^4
$\theta_1(x)$		b			
$\theta_2(x)$		b^2			
$\theta_3(x)$		b^3			
$\theta_4(x)$		b^4			

[Note that $\theta(a^2) = \theta(a)\theta(a) = [\theta(a)]^2$, $\theta(a^3) = [\theta(a)]^3$, and so on, for an isomorphism θ.]
(ii) Explain why there are exactly 4 isomorphisms from G to G'.

2. Let $G = [a]$ and $G' = [b]$ be cyclic groups of order 6.
(i) Complete the following table so that α and β are isomorphisms from G to G'.

x	e	a	a^2	a^3	a^4	a^5
$\alpha(x)$		b				
$\beta(x)$		b^5				

(ii) Explain why there are just 2 isomorphisms from G to G'.

3. Let $G = \{\mathbf{e}, b, c, d\}$ and $G' = \{\mathbf{e}', u, v, w\}$ be noncyclic groups of order 4. The isomorphism from G to G' with

$$\mathbf{e} \mapsto \mathbf{e}', \quad b \mapsto u, \quad c \mapsto v, \quad d \mapsto w$$

is given in Example 1. Tabulate five other isomorphisms from G to G'.

4. In an isomorphism between groups of permutations, is the image of an odd permutation necessarily odd? [*Hint:* Consider the groups $G = \{(1), (12), (34), (12)(34)\}$ and $G' = \{(1), (12)(34), (13)(24), (14)(23)\}$.]

5. Explain why the following are true.
   (a) The only automorphism of a group of order 2 is the trivial automorphism.
   (b) A group of order 3 has exactly 2 automorphisms.

6. (a) Write the formula that expresses preservation of the operations in a mapping θ from a multiplicative group G to an additive group G'.
   (b) Is the mapping θ with $\theta(x) = 10^x$ an isomorphism from the additive group $\mathbf{R}$ of the real numbers to the multiplicative group $\mathbf{R}^+$ of the positive numbers? Explain.

7. Let G be isomorphic to G' and let G be abelian. Does G' have to be abelian? Explain.

8. Let θ be an isomorphism from G to G'. Let C and C' be the centers of G and G', respectively. Are C and C' isomorphic? Explain.

9. Complete the following table so that f, g, and h are automorphisms of the symmetric group $\mathbf{S}_3$.

α	(1)	(123)	(132)	(23)	(13)	(12)
$f(\alpha)$		(123)		(23)		
$g(\alpha)$		(123)		(13)		
$h(\alpha)$		(123)		(12)		

[Use preservation of the operations. For example, if $\rho = (123)$ and $\phi = (23)$, then under the mapping g,

$$(13) = (123)(23) = \rho\phi \mapsto g(\rho\phi) = g(\rho)g(\phi) = (123)(13) = (12).]$$

10. Complete the following table so that f, g, and h are automorphisms of the symmetric group $\mathbf{S}_3$.

α	(1)	(123)	(132)	(23)	(13)	(12)
$f(\alpha)$		(132)		(23)		
$g(\alpha)$		(132)		(13)		
$h(\alpha)$		(132)		(12)		

11. Explain why every automorphism of $\mathbf{S}_3$ must be one of the 6 automorphisms of Problems 9 and 10.

12. Explain why $\mathbf{S}_3$ is not isomorphic to [(123456)].

13. For the automorphism g of Problem 9, find a β in $\mathbf{S}_3$ such that $g(\theta) = \beta^{-1}\theta\beta$ for all θ in $\mathbf{S}_3$.

14. For the automorphism h of Problem 9, find a γ in $\mathbf{S}_3$ such that $h(\theta) = \gamma^{-1}\theta\gamma$ for all θ in $\mathbf{S}_3$.

15. Let $G = [a]$ and $G' = [b]$ be cyclic groups of the same order. Show that, among the isomorphisms θ from G to G', there is exactly one with $\theta(a) = c$ if and only if c is a generator of G'.

16. Let G and G' be cyclic groups of order 8. How many isomorphisms are there from G to G'?

17. Explain why any two groups of the same prime order must be isomorphic.

18. How many automorphisms are there of a group of prime order p?

19. Describe two automorphisms of the additive group of the integers.

20. Let G and G' be infinite cyclic groups. How many isomorphisms are there from G to G'? Explain.

21. Let g be an element of a group G. Show the following.
(i) Conjugates $a^{-1}ga$ and $b^{-1}gb$ of g are equal if and only if ba^{-1} is in the centralizer C_g. (The definition of *centralizer* is in Section 2.5.)
(ii) $a^{-1}ga = b^{-1}gb$ if and only if a and b are in the same right coset of the centralizer C_g in G.
(iii) If G is finite, the number n_g of conjugates of g in G equals the index of the centralizer C_g in G; that is, $n_g = (\text{ord } G)/(\text{ord } C_g)$. Also, $n_g = 1$ if and only if g is in the center in G.

22. Show that the set of conjugates of (12) in $\mathbf{S}_4$ is

$$\{(12), (13), (14), (23), (24), (34)\},$$

and hence that the centralizer of (12) in $\mathbf{S}_4$ has index 6 (and order 4). (See Problem 21.)

23. Let α and β be the inner automorphisms of a group G with $\alpha(x) = a^{-1}xa$ and $\beta(x) = b^{-1}xb$.
(i) Show that $\alpha = \beta$ [that is, $\alpha(x) = \beta(x)$ for all x in G] if and only if ba^{-1} is in the center C in G.
(ii) Show that $\alpha = \beta$ if and only if $Ca = Cb$.
(iii) Let C have index q in G. Explain why there are exactly q inner automorphisms of G.

24. Explain why the only inner automorphism of an abelian group is the trivial automorphism θ with $\theta(x) = x$ for all x in G.

25. How many inner automorphisms are there of the octic group?

26. How many inner automorphisms are there of the alternating group $\mathbf{A}_4$?

27. Let M be the multiplicative group of 2×2 matrices of the form

$$m(x) = \begin{pmatrix} \cos x & \sin x \\ -\sin x & \cos x \end{pmatrix}$$

with x a real number. Let U be the multiplicative group of complex numbers of the form

$$\cos x + \mathbf{i} \sin x,$$

with x real. [See Problems 4(c) and 24 of Section 2.6. U is called the circle group.] Prove that M and U are isomorphic.

28. Let U be as in Problem 27. Let T be the quotient group $\mathbf{R}/[2\pi]$ of the additive group $\mathbf{R}$ of the real numbers by its cyclic subgroup $[2\pi]$. Show that U and T are isomorphic.

29. Let N be a normal subgroup in G. Show that every element m of N is a conjugate $g^{-1}ng$ of an element n of N by an element g of G. Then show that the previous statement and Theorem 3(b) imply Theorem 3(c).

30. Show that the subgroup $H = \{(1), (12)\}$ is not normal in the symmetric group $\mathbf{S}_3$, by producing a conjugate subgroup K of H that is different from H.

31. Let H be a subgroup in G and a be an element of G.

(a) Show that $Ha = aK$, where K is the conjugate subgroup of H by a.
(b) Show that $aH = La$ for some subgroup L in G.

32. Describe a set $\{G_1, G_2, G_3, G_4, G_5\}$ of five groups, each of order 8 and no two isomorphic, such that every group of order 8 is isomorphic to one of these five groups. [Assume, as will be noted in Section 3.4, that

$$G_5 = \{(1), \alpha, \alpha^2, \alpha^3, \beta, \alpha\beta, \alpha^2\beta, \alpha^3\beta\},$$

with $\alpha = (1234)(5876)$ and $\beta = (1537)(2648)$, is a group.]

★ **33.** Prove that the octic group $\mathbf{D_4}$ has 8 automorphisms, of which 4 are inner automorphisms.

★ **34.** Prove that the alternating group $\mathbf{A_4}$ has 24 automorphisms, of which 12 are inner automorphisms.

35. Let G be a finite abelian group of order n and let θ be the mapping from G to itself with $\theta(g) = g^m$ for all g in G, where m is an integer such that $\gcd(m, n) = 1$. Prove the following.

(i) There exists a fixed integer r such that $g^{rm} = g$ for all g in G.
(ii) θ is injective and surjective; that is, θ is a one-to-one mapping from G onto itself.
(iii) θ is an automorphism of G.

36. Let θ be a bijection from a set S to a group G. Prove that S becomes a group isomorphic to G when multiplication is defined by

$$ab = \theta^{-1}[\theta(a)\theta(b)]$$

for all a and b in S.

3.3 Group Homomorphisms

We recall from the previous section that a group homomorphism θ from G to G' is a mapping from G to G' with $\theta(ab) = \theta(a)\theta(b)$ for all a and b in G, and we note that the multiplication ab is performed in G while $\theta(a)\theta(b)$ is performed in G'.

The preservation of operations property of a homomorphism from G to G' helps in deducing properties of one of the groups from properties of the other. The work here with group homomorphisms will be extended and applied to rings in Chapters 4 and 5.

In this section, we shall see that group homomorphisms are closely related to normal subgroups and quotient groups.

DEFINITION 1

Kernel of a Homomorphism

The **kernel** of a group homomorphism θ from G to G' is the set of all k in G such that $\theta(k) = \mathbf{e}'$, where $\mathbf{e}'$ is the identity of G'.

In other words, the kernel of a group homomorphism θ from G to G' is the complete inverse image $\theta^{-1}(\mathbf{e}')$.

DEFINITION 2

Natural Map

Let N be a normal subgroup in G. Then the **natural map** from G to the quotient group G/N is the mapping θ with $a \mapsto aN$, that is, the mapping that sends each element a of G to the coset for a of N in G.

We are now ready for our first result.

THEOREM 1 **The Natural Map Is a Surjective Homomorphism**

Let N be a normal subgroup in G. Then the natural map θ from G to G/N is a surjective homomorphism with N as its kernel.

Proof Using the formula $aN \cdot bN = (ab)N$ of Section 2.12 for multiplication of cosets of a normal subgroup, we see that

$$\theta(ab) = (ab)N = aN \cdot bN = \theta(a)\theta(b);$$

that is, the natural map θ is a homomorphism. It is surjective because a given coset gN in G/N is the image of g under θ.

The identity of G/N is the coset N. Also, $\theta(g) = N$ if and only if g is in N. These two statements tell us that the kernel of θ is N. ■

EXAMPLE 1 The subgroup $N = [(123)] = \{(1), (123), (132)\}$ has index 2 in the symmetric group $\mathbf{S}_3$, and hence it follows from Theorem 3 of Section 2.12 that N is a normal subgroup in $\mathbf{S}_3$. Let $\phi = (23)$. Then the natural map f from $\mathbf{S}_3$ to $\mathbf{S}_3/N$ has the table

α	(1)	(123)	(132)	(23)	(13)	(12)
$f(\alpha)$	N	N	N	ϕN	ϕN	ϕN

EXAMPLE 2 Let g be the mapping fron $\mathbf{S}_3$ to $\mathbf{S}_4$ given by the table

α	(1)	(123)	(132)	(23)	(13)	(12)
$g(\alpha)$	(1)	(1)	(1)	(23)	(23)	(23)

Here we see that g is a group homomorphism.

The image set of g is the subgroup $\{(1), (23)\}$ of order 2 in the codomain $\mathbf{S}_4$. Each of the tables in Examples 1 and 2 may be considered to be the table for a mapping from $\mathbf{S}_3$ to a group $\{\mathbf{e}, b\}$ of order 2, with $\mathbf{e}$ and b given different names in the two examples. Since the natural map f is a homomorphism, this implies that g is also a homomorphism.

We could also verify preservation of the operations by noting that $g(\alpha) = (1)$ when α is an even permutation and $g(\alpha) = (23)$ when α is odd.

The homomorphism g of Example 2 is not surjective; for example, the element $\rho = (1234)$ of the codomain $\mathbf{S}_4$ is not in the image set $\{\varepsilon, \phi\}$, where $\phi = (23)$. Hence the complete inverse image $g^{-1}(\rho)$ is the empty set. We note that, for the elements of the image set, the complete inverse images are the cosets

$$g^{-1}(\varepsilon) = \{(1), (123), (132)\}, \qquad g^{-1}(\phi) = \{(23), (13), (12)\}$$

of the kernel $K = [(123)]$ of g.

The parts of the following result generalize on this observation and other properties of the homomorphisms of Examples 1 and 2. (One might compare this result with Theorem 1 of Secton 3.2.)

THEOREM 2 Properties of Group Homomorphisms

Let θ be a group homomorphism from G to G'. Then:

(a) $\theta(\mathbf{e}) = \mathbf{e}'$; that is, the identity of G is sent by θ to the identity of G'. (This also implies that $\mathbf{e}$ is in the kernel K and $\mathbf{e}'$ is in the image set M of θ.)
(b) $\theta(g^n) = [\theta(g)]^n$ for every integer n and each g in G.
(c) The image set M of θ is a subgroup in G'.
(d) The kernel K of θ is a normal subgroup in G.
(e) If $\theta(a) = a'$, the complete inverse image $\theta^{-1}(a')$ is the coset aK. Also, the order of a' is an integral divisor of the order of a.
(f) G/K is isomorphic to the image group M.
(g) If θ is injective, G is isomorphic to M.

Proof The proofs of (a) and (b) are similar to the analogous results for isomorphisms in Theorem 1 of Section 3.2.

For (c), let a' and b' be in the image set M. Then there are a and b in G such that $\theta(a) = a'$ and $\theta(b) = b'$. Now

$$\theta(ab^{-1}) = \theta(a)\theta(b^{-1}) = \theta(a)[\theta(b)]^{-1} = a'(b')^{-1},$$

using preservation of the operations and (b). This means that $a'(b')^{-1}$ is in M. Hence, M is closed under division. Also, M is nonempty, since $\mathbf{e}'$ is in M by (a). The last two statements and Theorem 1 of Section 2.5 tell us that M is a subgroup in G'.

For (d), let j and k be in the kernel K. Then

$$\theta(jk^{-1}) = \theta(j)\theta(k^{-1}) = \theta(j)[\theta(k)]^{-1} = \mathbf{e}'(\mathbf{e}')^{-1} = \mathbf{e}'$$

shows that jk^{-1} is in K; that is, K is closed under division. Also, K is nonempty by (a). Hence, K is a subgroup in G.

Next we see that, for every g in G and k in the kernel K,

$$\theta(g^{-1}kg) = \theta(g^{-1})\theta(k)\theta(g) = [\theta(g)]^{-1}\mathbf{e}'\theta(g)$$
$$= [\theta(g)]^{-1}\theta(g) = \mathbf{e}'.$$

This means that $g^{-1}kg$ is an element k_1 of K. Then $kg = gk_1$, which shows that the right coset Kg is contained in the left coset gK. Similarly, one sees that gK is contained in Kg. Hence $Kg = gK$ for all g in G; that is, K is a normal subgroup in G.

For (e), we first let b be in aK. Then $b = ak$ with k in K, and

$$\theta(b) = \theta(ak) = \theta(a)\theta(k) = \theta(a)\mathbf{e}' = \theta(a) = a';$$

that is, b is in $\theta^{-1}(a')$. Conversely, let $\theta(b) = a'$. Then

$$\theta(a^{-1}b) = \theta(a^{-1})\theta(b) = [\theta(a)]^{-1}\theta(b) = (a')^{-1}a' = \mathbf{e}'.$$

Hence, $a^{-1}b$ is an element k of K, and $b = ak$ is in aK.

Now let $\theta(a) = a'$ and let the orders of a and a' be q and q', respectively. Using (a) and (b) we have

$$\mathbf{e}' = \theta(\mathbf{e}) = \theta(a^q) = (a')^q.$$

The equality $\mathbf{e}' = (a')^q$ implies that $q' \mid q$, as desired.

For (f), we note that (e) implies that $\theta(a) = \theta(b)$ if and only if $aK = bK$. This means that the mapping α from G/K to M with $\alpha(gK) = \theta(g)$ is well defined and is injective. Also, for any c' in M there is a c in G with $\theta(c) = c'$. Then $\alpha(cK) = \theta(c) = c'$; that is, α is surjective. Finally,

$$\alpha(aK \cdot bK) = \alpha(abK) = \theta(ab) = \theta(a)\theta(b) = \alpha(aK)\alpha(bK)$$

shows that α preserves the operations. Together, these statements tell us that α is an isomorphism.

For (g), we let β be the mapping from G to M with $\beta(g) = \theta(g)$ for all g in G; that is, β is the modification of θ in which one ignores the elements of the codomain G' that are not in the image set M. This changes the injection θ into the bijection β. Also, β preserves the operations, since it has the same effect as the homomorphism θ. Hence, β is an isomorphism from G to M. ∎

THEOREM 3 Normal Subgroups ↔ Kernels

A subset S of a group G is a normal subgroup in G if and only if S is the kernel of a homomorphism from G to some group G'.

Proof Theorem 1 tells us that every normal subgroup N in G is the kernel of a homomorphism, the natural map, from G to the quotient group G/N. Conversely, Theorem 2(d) states that the kernel of a homomorphism from G to G' is a normal subgroup in G. ∎

We next present some notation and preliminary results concerning the quotient group $\mathbf{Z}/[m]$ of the additive group $\mathbf{Z}$ of the integers by a cyclic subgroup $[m]$; this will facilitate our applications of the concept of natural maps to $\mathbf{Z}$. The material given here will be very helpful in Chapter 4.

Let m be a fixed positive integer. Since $\mathbf{Z}$ is abelian, the cyclic subgroup

$$[m] = \{\ldots, -3m, -2m, -m, 0, m, 2m, 3m, \ldots\}$$

is normal in $\mathbf{Z}$, and the quotient group $\mathbf{Z}/[m]$ exists. This quotient group consists of cosets

$$a + [m] = \{\ldots, a - 2m, a - m, a, a + m, a + 2m, \ldots\}.$$

In Definition 2 (Modular Arithmetic) of Section 2.6, we described addition and multiplication operations on the set $\mathbf{Z_m} = \{\bar{0}, \bar{1}, \ldots, \overline{m-1}\}$ whose elements $\bar{a}$ are related to the remainders in division by the positive integer m. Now we interpret the symbol $\bar{a}$ as a notation for a coset of the cyclic subgroup $[m]$. We will see that the new interpretation does not change addition and multiplication in $\mathbf{Z_m}$.

NOTATION **Coset $\bar{a}$**

Let the positive integer m be known from the context, and let

$$[m] = \{\ldots, -2m, -m, 0, m, 2m, 3m, \ldots\}.$$

For each integer a, we let $\bar{a}$ denote the coset $a + [m]$; that is,

$$\bar{a} = \{\ldots, a - 2m, a - m, a, a + m, a + 2m, a + 3m, \ldots\}.$$

In our previous modular arithmetic work, the symbol $\bar{a}$ for an element of $\mathbf{Z_m} = \{\bar{0}, \bar{1}, \ldots, \overline{m-1}\}$ was used only for integers a satisfying $0 \leq a < m$. From now on, a can be any integer in the symbolism $\bar{a}$. For example, with $m = 4$ one has

$$\bar{1} = 1 + [4] = \{\ldots, -7, -3, 1, 5, 9, \ldots\},$$

and it is easily seen that

$$\cdots \overline{-7} = \overline{-3} = \bar{1} = \bar{5} = \bar{9} = \ldots.$$

In general, elements $\bar{a}$ and $\bar{b}$ of the quotient group $\mathbf{Z}/[m]$ are equal when b is in $\bar{a} = a + [m]$.

It is easily shown that $\{\bar{0}, \bar{1}, \ldots, \overline{m-1}\}$ is the set of distinct cosets of $[m]$ in $\mathbf{Z}$ and that, in the quotient group

$$\mathbf{Z}/[m] = \{\bar{0}, \bar{1}, \ldots, \overline{m-1}\},$$

addition is given by $\bar{a} + \bar{b} = \bar{r}$, where r is the remainder in the division of $a + b$ by m. (See Problem 5 of this section.) Since $\mathbf{Z}/[m]$ and the modular arithmetic additive group $\mathbf{Z_m}$ now have the same elements and the same operation, we can use $\mathbf{Z_m}$ to denote $\mathbf{Z}/[m]$.

The natural map θ from $\mathbf{Z}$ to the quotient group $\mathbf{Z}/[m] = \mathbf{Z_m}$ has $\theta(n) = \bar{n}$ for all integers n. Since the homomorphism θ preserves addition,

$$\overline{a+b} = \theta(a+b) = \theta(a) + \theta(b) = \bar{a} + \bar{b}.$$

Hence, $\bar{2} = \overline{1+1} = \bar{1} + \bar{1} = 2 \cdot \bar{1}$. Similarly, $\bar{3} = 3 \cdot \bar{1}$, $\bar{4} = 4 \cdot \bar{1}$, and so on. Also, $m \cdot \bar{1} = \bar{m} = \bar{0}$, since m is in $\bar{0} = [m]$. Thus we see that $\mathbf{Z_m}$ is a cyclic group $[\bar{1}]$ of order m.

We next give some further terminology for special kinds of homomorphisms.

DEFINITION 3

Homomorphic

If there exists a surjective homomorphism from G to G', one says that G is **homomorphic** to G'.

In Problem 25 below, the reader is asked to show that G can be homomorphic to G' without G' being homomorphic to G.

DEFINITION 4

Endomorphism, Monomorphism, Epimorphism

A homomorphism from G to itself is an **endomorphism.** A homomorphism θ from G to G' may be called a **monomorphism** if θ is injective and an **epimorphism** if θ is surjective.

We see that a bijective endomorphism is an automorphism.

Problems

1. Let $G = [a]$ be a cyclic group of order 9 and let $G' = [(123)]$.
   (i) Complete the following table so as to make θ a homomorphism from G to G'. [*Hint*: Use $\theta(a^2) = \theta(a)\theta(a)$, $\theta(a^3) = \theta(a^2)\theta(a)$, etc.]

g	e	a	a^2	a^3	a^4	a^5	a^6	a^7	a^8
$\theta(g)$	(1)	(123)							

   (ii) What is the kernel of θ?
   (iii) Give the complete inverse image of (123) and of (132).
2. Show that the f of the following table is not a homomorphism from $\mathbf{S_3}$ to $[(123)]$, by finding β_1 and β_2 in $\mathbf{S_3}$ such that

$$f(\beta_1\beta_2) \neq f(\beta_1)f(\beta_2).$$

β	(1)	(123)	(132)	(23)	(13)	(12)
$f(\beta)$	(1)	(123)	(132)	(132)	(123)	(1)

3. Let $G = [a]$ and $G' = [b]$ be cyclic groups of orders 3 and 2, respectively. Prove that there is no homomorphism θ from G to G' with $\theta(a) = b$.

4. Let $\mathbf{D}_4$ be the octic group and $V = \{\mathbf{e}, b, c, d\}$ be a noncyclic group of order 4 (Klein 4-group) with $b^2 = c^2 = d^2 = \mathbf{e}$. Also, let $\rho = (1234)$ and $\phi = (24)$.
   (i) Complete the following table for a homomorphism f from $\mathbf{D}_4$ to V.

α	ε	ρ	ρ^2	ρ^3	ϕ	$\rho\phi$	$\rho^2\phi$	$\rho^3\phi$
$f(\alpha)$		b	$\mathbf{e}$		c			

   (ii) Is f injective? Is f surjective? Explain each answer.
   (iii) Give the complete inverse image for each element of the codomain.

5. Let m be a fixed positive integer. In $\mathbf{Z}/[m]$, show the following.
   (i) $\bar{a} = \bar{b}$ if and only if $a \equiv b \pmod{m}$.
   (ii) $\bar{n} = \bar{r}$, where r is the remainder in the division of n by m.
   (iii) The set of distinct cosets is $\{\bar{0}, \bar{1}, \ldots, \overline{m-1}\}$.
   (iv) $\bar{a} + \bar{b} = \bar{r}$, where r is the remainder in the division of $a + b$ by m.
   (v) $\mathbf{Z}/[m] = [\bar{1}]$; that is, $\mathbf{Z}/[m]$ is cyclic with $\bar{1}$ as a generator.

6. Let G be the quotient group $\mathbf{Z}/[6] = \{\bar{0}, \bar{1}, \bar{2}, \bar{3}, \bar{4}, \bar{5}\}$ and let $G' = \{\mathbf{e}, b\}$ be a group of order 2.
   (i) Tabulate a homomorphism θ from G to G' with $\theta(\bar{5}) = b$.
   (ii) What is the kernel of θ?
   (iii) What set is the complete inverse image $\theta^{-1}(b)$?

7. Let $\mathbf{R}^{\pm}$ and $\mathbf{R}^{+}$ denote the multiplicative groups of the nonzero real numbers and the positive real numbers, respectively. Let θ be the mapping from $\mathbf{R}^{\pm}$ to $\mathbf{R}^{+}$ with $\theta(x) = |x|$.
   (i) Explain why θ is a group homomorphism.
   (ii) What is the kernel of θ?
   (iii) List the elements of the complete inverse image $\theta^{-1}(6)$.

8. Let $\mathbf{R}^{\pm}$ and V be the multiplicative groups of the nonzero real numbers and the nonzero complex numbers, respectively. Let θ be the mapping from V to $\mathbf{R}^{\pm}$ with $\theta(z) = |z|$; that is, let θ send $r(\cos x + \mathbf{i} \sin x)$ to r.
   (i) Explain why θ is a group homomorphism.
   (ii) What is the image set of θ?
   (iii) What is the kernel of θ?
   (iv) Explain why $\theta^{-1}(2)$ consists of all complex numbers $a + b\mathbf{i}$ with $a^2 + b^2 = 4$.
   (v) What is $\theta^{-1}(-2)$?

9. Let G and G' be groups with $\mathbf{e}'$ the identity of G'. Let θ be the mapping from G to G' such that $\theta(g) = \mathbf{e}'$ for all g in G.
   (i) Show that θ is a homomorphism.
   (ii) What is the kernel of θ?

10. (i) Let $f\colon G \to G'$ be a group homomorphism with kernel K. Show that f is injective if and only if $K = \{\mathbf{e}\}$.
   (ii) Let H be a subgroup in G. Describe an injective group homomorphism from H to G.

11. Let N be a normal subgroup in G. Let q be the order of an element a of G and let q' be the order of the coset aN in the quotient group G/N. Explain why $q' \mid q$.

12. Let K and M be the kernel and image set, respectively, of a group homomorphism θ from G to G'. Let the orders of G, G', K, and M be r, s, t, and u, respectively. Explain why the following are true.
(i) $u \mid s$. [See Theorem 2(c).]
(ii) $r = tu$. [See Theorem 2(f).]
(iii) $r \mid (st)$.

13. Let θ be a group homomorphism from G to G'. Let a have order q in G and let the image set of θ be a group M of order m. Let $\gcd(q, m) = 1$. Use Theorem 2(e) above and Lagrange's Theorem to prove that a is in the kernel of θ.

14. Let f be a homomorphism from the alternating group $\mathbf{A_4}$ to $[(123)]$. Use Problem 13 to show that $f(\alpha) = (1)$ for each α in the subset $\{(12)(34), (13)(24), (14)(23)\}$ of $\mathbf{A_4}$.

15. Let s be a fixed integer and let θ be the mapping from a group G to itself with $\theta(g) = g^s$ for all g in G.
(i) Show that θ is a homomorphism when G is abelian.
(ii) Show by an example that θ need not be a homomorphism when G is not commutative.

16. Let θ be the mapping from a group G to itself with $\theta(g) = g^{-1}$ for all g in G. Prove that θ is an isomorphism if and only if G is abelian.

17. Let $\mathbf{Z}$ be the additive group of the integers and let a be a fixed element of a group G. Let θ be the mapping from $\mathbf{Z}$ to G with $\theta(n) = a^n$ for all integers n.
(i) Show that θ is a homomorphism.
(ii) What is the kernel of θ when a has order q?
(iii) What is the kernel of θ when a has infinite order?

18. Let θ be a homomorphism from a cyclic group $G = [a]$ to a group G'. Let $\theta(a) = a'$.
(i) Explain why the image set M is the cyclic group $[a']$.
(ii) Explain why the kernel K is cyclic. (See Problem 38 of Section 2.7.)
(iii) Explain why the quotient group G/K is cyclic. [*Hint:* Use the natural map.]

19. Let G be a cyclic group $[a]$. Let b' be any element of a group G'.
(i) Show that there is at most one homomorphism from G to G' with $\theta(a) = b'$.
(ii) Show that there is a homomorphism θ from G to G' with $\theta(a) = b'$ if and only if the order of b' is an integral divisor of the order of a.
(iii) State a condition on the orders of a and b' for the homomorphism of (ii) to be injective.

20. Let each of G and G' have order 4 with G cyclic and G' noncyclic. Explain why there are exactly 4 homomorphisms from G to G'.

21. Let $G = [a]$ be a cyclic group of order q. Explain why there are exactly q endomorphisms of G, that is, homomorphisms from G to itself.

22. Let G be a Klein 4-group (noncyclic group of order 4). Show that there are exactly 16 endomorphisms of G, that is, homomorphisms from G to itself.

23. Let G consist of all 2×2 matrices $\alpha = \begin{pmatrix} a & b \\ c & d \end{pmatrix}$ with a, b, c, and d complex numbers and with $ad - bc \neq 0$. Let multiplication in G be defined by

$$\begin{pmatrix} a & b \\ c & d \end{pmatrix}\begin{pmatrix} a' & b' \\ c' & d' \end{pmatrix} = \begin{pmatrix} aa' + bc' & ab' + bd' \\ ca' + dc' & cb' + dd' \end{pmatrix}.$$

Let V be the multiplicative group consisting of all the complex numbers except 0 and let f be the mapping from G to V given by $f(\alpha) = ad - bc$.

(a) Show that G is an infinite nonabelian group.

(b) Show that f is a homomorphism from G to V.

(c) Let H be the subset of all α in G with $f(\alpha)$ in $\{1, -1\}$. Show that H is a subgroup in G, and describe a subgroup K of index 2 in H.

24. Let θ be a homomorphism from $\mathbf{S_5}$ to $\mathbf{S_2}$. Explain why the following are true.

(i) The image set of θ has either 1 or 2 elements.

(ii) The kernel K of θ has index 1 or 2 in $\mathbf{S_5}$. [See (i) of this problem and (ii) of Problem 12.]

(iii) K is either $\mathbf{S_5}$ or $\mathbf{A_5}$. (See Problem 7 of Section 2.13.)

(iv) There are only two homomorphisms from $\mathbf{S_5}$ to $\mathbf{S_2}$.

25. Give an example of groups G and G' such that G is homomorphic to G' but G' is not homomorphic to G.

3.4 Cayley's Theorem

In this section we prove Cayley's Theorem, which states that every group of order n is isomorphic to a subgroup in the symmetric group $\mathbf{S_n}$. (See the biographical note on Cayley on p. 160.) But first we consider a problem related to Cayley's Theorem.

EXAMPLE 1 **The Quaternion Group**

In Problem 20(b) of Section 2.12, we assumed that there exists a group

$$G = \{\mathbf{e}, a, a^2, a^3, b, ab, a^2b, a^3b\}$$

of order 8 in which $a^4 = \mathbf{e}$, $b^2 = a^2$, and $ba = a^3b$, and we asked for its operation table. If we use the notation

$$q_1 = \mathbf{e}, \quad q_2 = a, \quad q_3 = a^2, \quad q_4 = a^3,$$

$$q_5 = b, \quad q_6 = ab, \quad q_7 = a^2b, \quad q_8 = a^3b,$$

the requested table turns out to be Table 3.1. For example, we find that

$$q_6 q_7 = aba^2b = a(ba)ab = a(a^3b)ab = a^4(ba)b = \mathbf{e}a^3bb$$
$$= a^3a^2 = a^4a = a = q_2.$$

Does such a group exist? In other words, does the operation given by this table make G into a group? We show next that it does. (In Section 4.7, we will see why it is called the *quaternion group*.)

It is easy to see that G is closed under the operation, has an identity, and has inverses for all of its elements. To complete the proof that Table 3.1 is a group table, we need to show associativity; that is, we need to show that $(xy)z = x(yz)$ for all x, y, and z in G. There are 8 choices for each of x, y, and z and hence a total of $8^3 = 512$ cases. Verification of this number of cases does not appeal to us. Instead, we establish associativity by setting up a bijection θ, which preserves the operations, from G to a subgroup G' in the symmetric group $\mathbf{S_8}$. This transfers associativity from $\mathbf{S_8}$ to G.

TABLE 3.1 In this table, an entry k inside the table stands for q_k.

	q_1	q_2	q_3	q_4	q_5	q_6	q_7	q_8
q_1	1	2	3	4	5	6	7	8
q_2	2	3	4	1	6	7	8	5
q_3	3	4	1	2	7	8	5	6
q_4	4	1	2	3	8	5	6	7
q_5	5	8	7	6	3	2	1	4
q_6	6	5	8	7	4	3	2	1
q_7	7	6	5	8	1	4	3	2
q_8	8	7	6	5	2	1	4	3

To each element q_j of G, we associate the permutation β_j on $\mathbf{X_8} = \{1, 2, \ldots, 8\}$ such that β_j sends x to y when $q_x q_j = q_y$. This means that $\theta(q_j) =$

$$\beta_j = \begin{pmatrix} 1 & 2 & 3 & \ldots & 8 \\ b_1 & b_2 & b_3 & \ldots & b_8 \end{pmatrix},$$

where the second row $b_1, \ldots, b_8$ is taken from the entries (from top to bottom) in the column headed by q_j in Table 3.1. For example,

$$q_2 \overset{\theta}{\mapsto} \beta_2 = \begin{pmatrix} 1 & 2 & 3 & 4 & 5 & 6 & 7 & 8 \\ 2 & 3 & 4 & 1 & 8 & 5 & 6 & 7 \end{pmatrix} = (1234)(5876).$$

One can also see that the images of q_1, q_3, q_4, and so on are $\beta_1 = (1)$, $\beta_3 = (13)(24)(57)(68)$, $\beta_4 = (1432)(5678)$, $\beta_5 = (1537)(2648)$, $\beta_6 = (1638)(2745)$, $\beta_7 = (1735)(2846)$, and $\beta_8 = (1836)(2547)$. We will see in the following theorem that the bijection $\theta : q_i \mapsto \beta_i$ preserves the operations. This transfers the associativity property of the permutations β_i to the q_i. Hence, G is a group and θ is a group isomorphism from G to $G' = \{\beta_1, \ldots, \beta_8\}$.

THEOREM 1 Cayley's Theorem

Every group G of order n is isomorphic to a subgroup in the symmetric group $\mathbf{S_n}$.

Proof Let $G = \{g_1, g_2, \ldots, g_n\}$. For $1 \le j \le n$, let β_j be the permutation on $\mathbf{X_n} = \{1, 2, \ldots, n\}$ such that

$$\beta_j(x) = y \qquad \text{when } g_x g_j = g_y.$$

Let α be the mapping from G to $\mathbf{S_n}$ with $\alpha(g_j) = \beta_j$ and M be the image set of α. We next show that α is injective.

Let β_a and β_b be in the image set M of α. Let $\beta_a(1) = c$ and $\beta_b(1) = d$; that is, let $g_1 g_a = g_c$ and $g_1 g_b = g_d$. If $\beta_a = \beta_b$, then 1 has the same image under β_a and under β_b; that is, $c = d$. It follows that

$$g_c = g_d, \qquad g_1 g_a = g_1 g_b, \qquad \text{and} \qquad g_a = g_b.$$

The fact that $\beta_a = \beta_b$ implies $g_a = g_b$ shows that α is injective.

To prove that β preserves the operations, we assume that $g_a g_b = g_c$ and show (as follows) that this implies $\beta_a \beta_b = \beta_c$. Let x be any integer in $\mathbf{X_n} = \{1, 2, \ldots, n\}$. Also, let

$$g_x g_a = g_y \quad \text{and} \quad g_y g_b = g_z. \tag{1}$$

Then

$$g_x g_c = g_x g_a g_b = g_y g_b = g_z. \tag{2}$$

The equations in (1) and (2) and the definition of the β's tell us that

$$x \overset{\beta_a}{\mapsto} y \overset{\beta_b}{\mapsto} z \quad \text{and} \quad x \overset{\beta_c}{\mapsto} z.$$

Since x is any integer in $\mathbf{X_n}$, this means that $\beta_a \beta_b = \beta_c$. Hence α is a homomorphism. Finally, it follows from parts (c) and (g) of Theorem 2 of Section 3.3 that the image set M of α is a subgroup in $\mathbf{S_n}$ and that G is isomorphic to M. ∎

The reader may have observed in the above example and proof that each element of the group G is associated with the permutation characterized by the corresponding column of the group table. Thus, given a group table, one can easily furnish the isomorphic permutation group.

In Problem 8 below, the reader is asked to identify the H_i in Figure 3.6 so as to make it the subgroup poset diagram for the group whose operation is

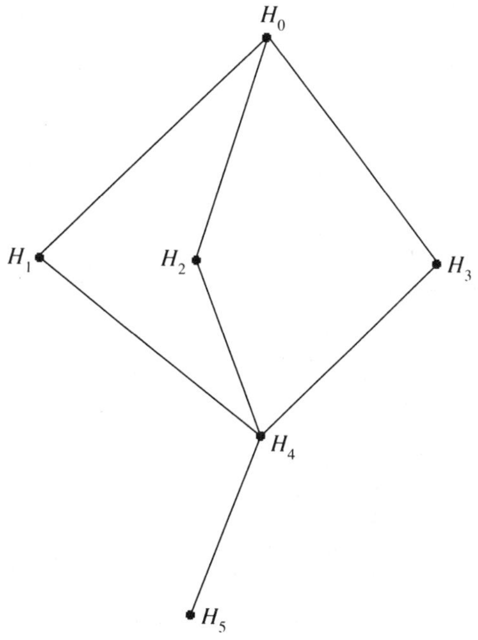

FIGURE 3.6

Arthur Cayley
1821–1895

Cayley, like Fermat, was a lawyer as well as a mathematician. He was educated at Trinity College, Cambridge, a college that, incidentally, has produced or been home to an extraordinary number of great mathematicians and physicists, including Sir Isaac Newton, Isaac Barrow, Roger Cotes, Augustus De Morgan, James Clerk-Maxwell, Sir J. J. Thomson, Lord Rayleigh, Lord Rutherford, Sir Arthur Eddington, Sir James Jeans, G. H. Hardy, J. E. Littlewood, and, briefly, the Indian Srinivasa Ramanujan.

Cayley is generally credited with the first discussion of abstract groups, although earlier references appear in the work of Cauchy and formal definitions were given only later by Kronecker, Weber, and Frobenius. Prior to Cayley's (and probably Cauchy's) work, groups were thought of as permutation groups and were considered only in connection with specific applications.

Today Cayley is largely remembered for his contributions to the theory of matrices — notably his definition of matrix multiplication — and for his work on invariants. Much of his work in algebra was done with J. J. Sylvester, another English mathematician, who also worked in the United States on two separate occasions. Cayley was very prolific, rivaling Cauchy and Euler in the total number of his published works.

given by Table 3.1. This group, or any group isomorphic to it, is called a ***quaternion group***. (Another description is given in Theorem 7 of Section 4.7.)

Problems

1. Use the technique in Cayley's Theorem to obtain an isomorphism f from $\mathbf{S_3}$ to a subgroup in $\mathbf{S_6}$. Number the elements of $\mathbf{S_3}$ as $\theta_1 = (1)$, $\theta_2 = (123)$, $\theta_3 = \theta_2^2$, $\theta_4 = (23)$, $\theta_5 = \theta_2\theta_4$, and $\theta_6 = \theta_3\theta_4$.

2. Use the technique of Cayley's Theorem to obtain an isomorphism α from a Klein 4-group $G = \{g_1, g_2, g_3, g_4\}$, with g_1 the identity, to a subgroup in $\mathbf{S_4}$.

3. List the orders q for which a cyclic group of order q is isomorphic to a subgroup in $\mathbf{S_5}$.

4. Give some examples of noncyclic groups G such that G is isomorphic to a subgroup in $\mathbf{S_n}$ with ord $G > n$.

5. Let $G = \{q_1, q_2, \ldots, q_8\}$ be the quaternion group of Table 3.1. Find the following.
(i) All the elements of order 2.
(ii) All the elements of order 4.
(iii) A subgroup of order 2.
(iv) Three subgroups of order 4.

6. Show that the quaternion group G has only one subgroup of order 2.

7. Show that the quaternion group G has only three subgroups of order 4.

8. Identify $H_0, H_1, \ldots, H_5$ in Figure 3.6 as the six subgroups in the quaternion group so that there is a rising line or broken line from H_i to H_j if and only if H_i is a proper subgroup in H_j.

9. Find the center of the quaternion group.

10. Explain why every subgroup is normal in the quaternion group G.

3.5 Unions, Intersections, Partitions

A set whose elements are themselves sets will be called a ***collection of sets***. In this section, we extend the terminology and notations given on the inside front cover of the book for unions and intersections of finite collections of sets to include infinite collections.

These concepts are applied to groups here and to rings and fields in later chapters.

DEFINITION 1

Union of the Sets of a Collection

Let Γ be a collection of sets. Then the set consisting of those objects that are in at least one of the sets A of Γ is the **union** of the sets of Γ and is denoted by

$$\bigcup_{A \in \Gamma} A$$

DEFINITION 2

Intersection of the Sets of a Collection

The set consisting of those objects that are in all of the sets of Γ is the **intersection** of the sets of Γ and is denoted by

$$\bigcap_{A \in \Gamma} A$$

For example, it follows from Theorem 3 of Section 2.5 that

$$C = \bigcap_{a \in G} C_a,$$

where G is any group, C is the center in G, and C_a denotes the centralizer of a in G.

DEFINITION 3

Smallest Set of a Collection

If the intersection D of the sets of Γ is one of the sets of Γ, one calls D the **smallest set** of Γ.

EXAMPLE 1 Let $G = \{\varepsilon, \rho, \rho^2, \rho^3, \phi, \rho\phi, \rho^2\phi, \rho^3\phi\}$ be the octic group $\mathbf{D}_4$ of symmetries of the square. We recall that $\varepsilon = (1)$, $\rho = (1234)$, and $\phi = (24)$. Let

$$H = \{\varepsilon, \rho, \rho^2, \rho^3\}, \qquad K = \{\varepsilon, \phi, \rho^2, \rho^2\phi\},$$
$$L = \{\varepsilon, \rho\phi, \rho^2, \rho^3\phi\}.$$

Each of H, K, and L is a finite subset of G that is closed under multiplication; hence each of these subsets is a subgroup in G. Then we easily see that the union and intersection of the sets of the collection $\{H, K, L\}$ are

$$H \cup K \cup L = G \quad \text{and} \quad H \cap K \cap L = \{\varepsilon, \rho^2\} = [\rho^2],$$

respectively. We also note that

$$[\rho] \cup [\phi] \cup [\rho\phi] = \{\varepsilon, \rho, \rho^2, \rho^3, \phi, \rho\phi\}$$

and that this union of subgroups is not a subgroup.

Example 1 shows that the union of a collection of subgroups in a group G may or may not be a subgroup in G. We can be more definite about the intersection of subgroups.

THEOREM 1 **The Intersection of Subgroups Is a Subgroup**

Let Γ be a nonempty collection of subgroups in a group G and let D be the intersection of the subgroups of Γ. Then D is also a subgroup in G.

The proof of Theorem 1 is left to the reader as Problem 6(a) of this section.

THEOREM 2 **Smallest Subgroup Containing a Subset**

Let T be a subset of a group G, let Γ be the collection of all the subgroups H in G such that $T \subseteq H$, and let D be the intersection of the subgroups of Γ. Then D is a subgroup in G and hence is the smallest subgroup of Γ.

Proof Since $T \subseteq G$, the collection Γ is nonempty; and it follows from Theorem 1 that D is a subgroup in G. Also, $T \subseteq D$. Then D is the smallest subgroup of Γ, by definition of the smallest set of a collection. ∎

Theorem 2 enables us to introduce some terminology.

DEFINITION 4

Subgroup Generated by a Subset

Let T be a subset of a group G. The **subgroup generated** by T in G is the intersection D of all subgroups H in G such that $T \subseteq H$. If $D = G$, we say that the subset T **generates** the group G.

Clearly, a cyclic subgroup $[a]$ is the subgroup generated by the singleton subset $\{a\}$. It is left to the reader as Problem 41 below to show that the subgroup generated by T in G consists of all products $x_1 x_2 \cdots x_n$ in which each x_i is either some t in T or is t^{-1} for some t in T and n is a varying positive integer.

EXAMPLE 2 Here we show that $\mathbf{A_4}$ is the subgroup generated by the subset $T = \{(123), (143)\}$ in the group $\mathbf{S_4}$.

Let Γ consist of all subgroups H in $\mathbf{S_4}$ with $T \subseteq H$. Since $(123)(143) = (12)(34)$, $(143)(123) = (14)(23)$, and $(12)(34) \cdot (14)(23) = (13)(24)$, every H in Γ must have the groups $[(123)]$ and

$$\{(1), (12)(34), (14)(23), (13)(24)\}$$

of orders 3 and 4, respectively, as subgroups. Then it follows from Lagrange's Theorem that the order of every H in Γ is a multiple of 3 and 4, and hence of 12. Since 12 and 24 are the only positive integers that are both multiples of 12 and divisors of the order 24 of $\mathbf{S_4}$, such an H must be $\mathbf{S_4}$ or have index 2 in $\mathbf{S_4}$. But $\mathbf{A_4}$ is the only subgroup with index 2 in $\mathbf{S_4}$. (See Problem 8 of Section 2.12.) Hence, $\mathbf{A_4}$ and $\mathbf{S_4}$ are the only possibilities for subgroups H of Γ. Now we note that $T \subseteq \mathbf{A_4}$ and $T \subseteq \mathbf{S_4}$. It then follows that $\Gamma = \{\mathbf{A_4}, \mathbf{S_4}\}$ and that the subgroup generated by T is the smallest subgroup $\mathbf{A_4}$ of Γ.

One can easily show that $ab = ba$ in a group G if and only if $a^{-1}b^{-1}ab$ is the identity of G. For this reason, elements of the form $a^{-1}b^{-1}ab$ play an important role in noncommutative groups such as the symmetric groups $\mathbf{S_n}$ with $n > 2$.

DEFINITION 5

Commutator, Commutator Subgroup

If $c = a^{-1}b^{-1}ab$ with a and b in a group G, then c is a **commutator** of G. The **commutator subgroup** in G is the subgroup generated by the subset of all the commutators of G; that is, the commutator subgroup is the smallest subgroup containing all the commutators.

EXAMPLE 3 **Commutator Subgroup in an Abelian Group**

If G is an abelian group, $a^{-1}b^{-1}ab = a^{-1}ab^{-1}b = \mathbf{e}$ for all a and b in G, and it follows that the commutator subgroup in G is the trivial subgroup $\{\mathbf{e}\}$.

EXAMPLE 4 **3-Cycles as Commutators**

Here we show that every 3-cycle $\gamma = (abc)$ is a commutator of the alternating group $\mathbf{A_n}$ (or the symmetric group $\mathbf{S_n}$) for $n \geq 5$. For $n \geq 5$, there are numbers d and e in $\mathbf{X_n} = \{1, 2, \ldots, n\}$ that differ from a, b, and c. Then $\alpha = (acd)$ and $\beta = (abe)$ are in $\mathbf{A_n}$, and one easily verifies that γ is the commutator $\alpha^{-1}\beta^{-1}\alpha\beta$.

It can be shown that every 3-cycle of $\mathbf{S_4}$ is a commutator of $\mathbf{S_4}$ but that no 3-cycle is a commutator of $\mathbf{A_4}$. (See Example 6 and Problem 12 of this section.)

EXAMPLE 5 **Commutators in $\mathbf{S_n}$ Are Even**

Now we see that every commutator in $\mathbf{S_4}$ is an even permutation and hence the commutator subgroup in $\mathbf{S_n}$ is a subgroup in $\mathbf{A_n}$. First we note that the inverse of an even permutation is even and the inverse of an odd permutation is odd. Then a consideraton of cases in which each of α and β is either even or odd shows that a commutator $\alpha^{-1}\beta^{-1}\alpha\beta$ is always even. (See Problem 15 of Section 2.8.) Now $\mathbf{A_n}$ is a subgroup, in $\mathbf{S_n}$, that contains the commutators, and hence $\mathbf{A_n}$ contains the smallest subgroup with this property.

EXAMPLE 6 **Commutator Subgroup in $\mathbf{S_4}$**

Next we show that $\mathbf{A_4}$ is the commutator subgroup in $\mathbf{S_4}$. We easily verify that

$$(13)^{-1}(124)^{-1}(13)(124) = (123) \quad \text{and} \quad (13)^{-1}(142)^{-1}(13)(142) = (143);$$

that is, (123) and (143) are commutators in $\mathbf{S_4}$. Then it follows from Example 2 that every permutation of $\mathbf{A_4}$ is in the commutator subgroup in $\mathbf{S_4}$. Conversely, Example 5 tells us that every element of the commutator subgroup is in $\mathbf{A_4}$. Together, the last two statements give us the desired result.

For examples of multiplicative groups G of 3 by 3 matrices with polynomial entries such that the commutator subgroup in G contains noncommutators, see "Products of commutators are not always commutators: an example" by Phyllis Joan Cassidy in the *American Mathematical Monthly* 86 (1979), 772. (We will deal with matrices in Section 4.7 and with polynomials in Chapter 5.)

We next present some additional terminology.

DEFINITION 6

Disjoint Sets

Sets A and B are **disjoint** if they have no element in common — that is, if $A \cap B = \varnothing$.

DEFINITION 7

Partition of a Set

Let X be a nonempty set and let Γ be a collection of subsets of X. Then Γ is a **partition** of X if the following three conditions are met:

(a) The empty set $\varnothing$ is not a member of Γ.
(b) The union of the sets of Γ is X.
(c) Any two distinct sets A and B of Γ are disjoint; that is, if A and B are in Γ, then either $A \cap B = \varnothing$ or $A = B$.

EXAMPLE 7 **Coset Partition**

Let H be a subgroup in G and let Γ consist of all the left cosets (or of all the right cosets) of H in G. It follows from Lemma 2 of Secton 2.11 that Γ is a partition of G.

EXAMPLE 8 **Nonpartition**

Is the collection $\Gamma = \{H, K, L\}$ of subgroups of the octic group $\mathbf{D_4}$, given in Example 1, a partition of $\mathbf{D_4}$? No, because distinct members of Γ have the identity $\mathbf{e}$ in common and hence are not disjoint. (The conditions that the empty set not be in Γ and that $H \cup K \cup L = \mathbf{D_4}$ are met.)

EXAMPLE 9 For every cyclic subgroup H in a group G, let $g(H)$ denote the set of generators of H. Then the collection Γ of all these subsets $g(H)$ of G is a partition of G. The proof of this statement is straightforward and is left to the reader. (See Problems 19, 20, and 21 of this section.)

EXAMPLE 10 For every element a of a group G, let $M(a)$, denote the set of all conjugates $g^{-1}ag$ of a by elements g of G. Then the collection Γ of all these $M(a)$ is a partition of G.

The special case of Example 10, in which G is the alternating group $\mathbf{A}_4$, is dealt with in Problem 33 below.

EXAMPLE 11 Let $\alpha : X \to Y$ and let Γ be the collection of complete inverse images $\alpha^{-1}(y)$ for all y in the image set of α. Then Γ is a partition of X.

EXAMPLE 12 Let S be any nonempty set and Γ be the collection of all the singleton subsets $\{s\}$ of S. Then Γ is a partition of S.

In Section 3.7, we shall reformulate partitions of a set X, using the concept of "equivalence relations" on X.

Problems

1. Let $G = \{\mathbf{e}, b, c, d\}$ be a Klein 4-group — that is, a noncyclic group of order 4. Show that G is the union of its three subgroups that have order 2.

2. Let $\rho = (1234)$ and $\phi = (24)$.
(a) Show that the octic group $\mathbf{D}_4 = \{(1), \rho, \rho^2, \rho^3, \phi, \rho\phi, \rho^2\phi, \rho^3\phi\}$ is the union of five cyclic subgroups.
(b) Find the subgroup in $\mathbf{D}_4$ generated by $\{\rho^2, \phi\}$.

3. Let $\rho = (123456)$ and $\phi = (26)(35)$. Find the subgroup in $\mathbf{S}_6$ generated by each of the following.
(a) $\{\rho, \phi\}$; (b) $\{\rho^2, \phi\}$; (c) $\{\rho^3, \phi\}$; (d) $\{\rho^2, \rho\phi\}$;
(e) $\{\rho^3, \rho\phi\}$; (f) $\{\rho^3, \rho^2\phi\}$; (g) $\{\phi, \rho\phi\}$; (h) $\{\phi, \rho^2\phi\}$.

4. For each of the following choices of G and T, find the subgroup in G generated by the subset T.
(a) $G = \mathbf{S}_3$ and $T = \{(23), (13)\}$.
(b) $G = \mathbf{S}_5$ and $T = \{(12), (345)\}$.
(c) $G = \mathbf{S}_6$ and $T = \{(12), (34), (56)\}$.
(d) $G = \mathbf{S}_5$ and $T = \{(25)(34), (12345)\}$.
(e) $G = \mathbf{S}_5$ and $T = \{(25)(34), (13)(45)\}$.
(f) $G = \mathbf{S}_6$ and $T = \{(16)(25)(34), (26)(35)\}$.

5. (a) Let a be in G. Explain why $[a]$ is the subgroup generated by $\{a\}$.
(b) Let T be a subset of a subgroup H in G. Explain why T generates the same subgroup in H as it does in G.

6. Let Γ be a nonempty collection of subgroups in G. Let D and U be the intersection and union, respectively, of the sets of Γ.
(a) Prove that D is a subgroup in G (and thus prove Theorem 1).
(b) Let H be the subgroup in G generated by U. Prove that every group K of Γ is a subgroup in H.

7. Let G have subgroups H and K of orders r and s, respectively. Let $d = \gcd(r, s)$ and $m = \operatorname{lcm}[r, s]$.
(a) Explain why the order of $H \cap K$ must be a positive integral divisor of d.
(b) Let L be the subgroup in G generated by $H \cup K$. Explain why $m \mid (\operatorname{ord} L)$.

8. Prove that a quotient group G/N is abelian if and only if the commutator subgroup in G is a subgroup in N.

9. Let H be a subgroup in G such that every commutator of G is an element of H. Show the following.

(i) H is normal in G. [*Hint:* Use $ha = a(a^{-1}hah^{-1})h$ to show that $Ha \subseteq aH$ for every a in G.]

(ii) The quotient group G/H is abelian. (See Problem 8 above or Problem 14 of Section 2.12.)

(iii) Let θ be the natural map from G to G/H. Then $\theta(a)\theta(b) = \theta(b)\theta(a)$ for all a and b in G.

(iv) Let $a_1, a_2, \ldots, a_s$ be any elements of G and m be a product of the $2s$ elements

$$a_1, a_2, \ldots, a_s, a_1^{-1}, a_2^{-1}, \ldots, a_s^{-1}$$

in any order. Then $\theta(m)$ is the identity of G/H; that is, m is in the kernel H of the natural map θ.

10. Let C be the commutator subgroup in G. Explain why the following are true.

(a) C is normal in G.

(b) C is the smallest normal subgroup N in G such that G/N is abelian.

11. Find the commutator subgroup in the octic group, $\mathbf{D_4}$.

12. Show that the commutator subgroup in $\mathbf{A_4}$ has order 4.

13. Let $n \geq 3$ and let T consist of all the 3-cycles (abc) of $\mathbf{S_n}$. Prove that the subgroup generated by T in $\mathbf{S_n}$ is $\mathbf{A_n}$.

14. Show that $\mathbf{A_n}$ is the commutator subgroup in $\mathbf{S_n}$ for $n = 2$ and 3.

15. (a) Prove that $\mathbf{A_n}$ is the commutator subgroup in $\mathbf{S_n}$ for $n > 1$. (Use Problem 13 and Examples 4 and 6 above.)

(b) Prove that $\mathbf{A_n}$ is its own commutator subgroup for $n \geq 5$.

16. Let G be the subgroup in $\mathbf{S_8}$ generated by the subset $\{\alpha, \beta\}$, where $\alpha = (1234)(5876)$ and $\beta = (1537)(2648)$. Show the following.

(i) $\beta^2 = \alpha^2$, $\beta\alpha = \alpha^3\beta$, and β is not in $[\alpha] = \{(1), \alpha, \alpha^2, \alpha^3\}$.

(ii) $G = \{(1), \alpha, \alpha^2, \alpha^3, \beta, \alpha\beta, \alpha^2\beta, \alpha^3\beta\}$.

17. Use Problems 8 and 15(b) to prove that $\mathbf{A_n}$ is not solvable for $n \geq 5$.

18. Let $G = \{a_1, a_2, \ldots, a_{2t+1}\}$ be a finite group of odd order. Let m be a product of the $2t + 1$ elements of G in any order. Prove that m is in the commutator subgroup of G. [*Hint:* First explain why no a_j has order 2, and then use Problem 9(iv).]

19. Let S be the subset of all elements of order 7 in a group G. For every s in S, let $A(s) = \{s, s^2, \ldots, s^6\}$.

(i) Show that the collection Γ of all the sets $A(s)$ is a partition of S. (Do not use the result in Example 9.)

(ii) If S has a finite number m of elements, explain why $6 \mid m$.

20. Do the analogue of Problem 19(ii) in which 7 is replaced by a general prime p.

21. For every cyclic subgroup A in G let $g(A)$ denote the set of generators in A. Prove that the collection of all these subsets $g(A)$ is a partition of G (and thus prove the statement in Example 9).

22. Let Γ consist of the centralizers C_a for all the elements a in a group G. Explain why the intersection of the sets of Γ is the center C in G.

23. Explain why a collection Γ of subgroups in G is not a partition of G unless $\Gamma = \{G\}$.

24. Show that a cyclic group $G = [a]$ is the union of a collection Γ of subgroups in G if and only if G itself is a member of Γ.

25. Let K be a subgroup in H and H be a subgroup in G. Show that every left coset of H in G is the union of a collection of left cosets of K in G.

26. State the analogue for right cosets of Problem 25.

27. Let H be a subgroup in G and U be the union of the left cosets of H in G that are also right cosets. Prove that U is a subgroup in G.

28. Let H, G, and U be as in Problem 27. Prove that H is a normal subgroup in U.

29. Let H and K be subgroups of G. Let neither one of H and K be a subset of the other; that is, let there be an h that is in H but is not in K and a k that is in K but is not in H. Explain why the following are true.
    (i) hk is not in H and is not in K.
    (ii) $H \cup K$ is not closed under the operation of G and therefore is not a subgroup in G.

30. Show that the union $U = H \cup K$ of subgroups H and K in G is a subgroup in G if and only if one of H and K is a subgroup in the other.

31. Let $G_1, G_2, G_3, \ldots$ be an infinite sequence of groups such that G_i is a subgroup in G_{i+1} for all i. Let G be the union of all the groups of this sequence. Define multiplication in G so that G becomes a group with each G_i as a subgroup.

32. Let S be the set of standard forms of permutations in all the symmetric groups $\mathbf{S_n}$. Motivated by Problem 31, define multiplication in S so as to make it into an infinite noncommutative group.

33. Give the partition described in Example 10 for the special case in which G is the alternating group $\mathbf{A_4}$. [The partition is of the form $\{B, C, D, E\}$, where each of C and D contains four conjugate 3-cycles and E has three permutations.]

34. Let there be a finite number $N(q)$ of elements having order q in a group G. Let $\phi(q)$ be the number of generators of a cyclic group of order q. Use Problem 21 above to prove that $\phi(q) \mid N(q)$.

35. Let T be the subset $\{(123), (12)(34)\}$ of $\mathbf{S_4}$. Show that the subgroup in $\mathbf{S_4}$ generated by T is $\mathbf{A_4}$.

36. Let U be the subset $\{(12), (12345)\}$ of $\mathbf{S_5}$. Use Problem 31 of Section 2.5 to show that the subgroup in $\mathbf{S_5}$ generated by U is $\mathbf{S_5}$ itself.

37. Give the partition Γ of Example 7 in the special case in which $G = \mathbf{A_4}$ and $H = \{(1), (12)(34), (13)(24), (14)(23)\}$.

★ 38. Let H and K be two different subgroups each of index 2 in a group G. Prove that

$$D = H \cap K$$

is a normal subgroup of index 4 in G and that G/D is not cyclic.

★ 39. Let H and K be subgroups in G and let H have index m in G. Prove that the index of $H \cap K$ in K must be in the set $\{1, 2, \ldots, m\}$.

★ 40. Let a group G be generated by a subset T. Let α and β be homomorphisms from G to a group G' such that $\alpha(t) = \beta(t)$ for all t in T. Prove that $\alpha(g) = \beta(g)$ for all g in G.

41. Prove that the subgroup generated by T in G consists of all products $x_1 x_2 \cdots x_n$ in which each x_i is either some t in T or is t^{-1} for some t in T and in which n is a varying positive integer.

42. In a finite group G, let a, b, and ab be elements whose orders are 2, 2, and m, respectively. Let $a \neq b$ and H be the subgroup in G generated by $\{a, b\}$. Show the following.

(i) If $ab = ba$, then H is a Klein 4-group.
(ii) If $ab \neq ba$, then H is isomorphic to the dihedral group $\mathbf{D_m}$.
(iii) ord $H = 2m$.

43. What is the order of the subgroup H in $\mathbf{S_9}$ generated by

$$\{(12)(34)(56)(78), (19)(28)(37)(46)\}?$$

3.6 Cartesian Products, Direct Products

We assume that the reader is familiar with ordered pairs from coordinate geometry. Here we use this concept as an aid in constructing new groups from known groups.

DEFINITION 1 Cartesian Product

If S and T are sets, the set of all ordered pairs (s, t) with s in S and t in T is called the **cartesian product** of S and T (in that order) and is denoted by $S \times T$ (read as "S cross T").

The adjective *cartesian* is in honor of René Descartes, one of the founders of analytic geometry. (See the biographical note near the close of this section.)

We note that ordered pairs (a, b) differ from pairs $\{a, b\}$ in the two following ways:

1. A pair $\{a, b\}$ is a set with two distinct elements, and hence the notation $\{a, b\}$ should only be used when $a \neq b$. However, the ordered pair (a, a) is in $S \times T$ when a is in both S and T.
2. If $a \neq b$, $(a, b) \neq (b, a)$ but $\{a, b\} = \{b, a\}$.

In $S \times T$, $(a, b) = (c, d)$ if and only if both $a = c$ and $b = d$. One can also show that $S \times T \neq T \times S$ unless $S = T$, $S = \varnothing$, or $T = \varnothing$. (See Problems 1 and 2 of this section.)

THEOREM 1 Cartesian Product of Groups

Let G_1 and G_2 be groups and let G be the cartesian product $G_1 \times G_2$ of their sets. Then G is a group under the operation given by

$$(a_1, a_2)(b_1, b_2) = (a_1 b_1, a_2 b_2).$$

The proof is left to the reader as Problem 7 of this section.

DEFINITION 2

Direct Product of Groups

If G_1 and G_2 are groups, the group $G_1 \times G_2$ of Theorem 1 is called the **direct product** of G_1 and G_2.

Now we extend the concepts of cartesian product and direct product as follows. If $S_1, S_2, \ldots, S_n$ are sets, the set of all ordered n-tuples

$$(x, x_2, \ldots, x_n), \quad x_j \text{ in } S_j$$

is called the ***cartesian product*** of $S_1, S_2, \ldots, S_n$ and is designated as

$$S_1 \times S_2 \times \cdots \times S_n.$$

An ordered n-tuple $(x_1, \ldots, x_n)$ differs from a set $\{y_1, \ldots, y_n\}$ with n elements in the same ways that an ordered pair differs from a pair. Specifically, the same element may be used more than once among the coordinates x_j of $(x_1, \ldots, x_n)$; also,

$$(a_1, \ldots, a_n) = (b_1, \ldots, b_n)$$

if and only if $a_1 = b_1, a_2 = b_2, \ldots, a_n = b_n$.

If $G_1, G_2, \ldots, G_n$ are groups, then the cartesian product

$$G = G_1 \times G_2 \times \cdots \times G_n$$

of their sets is a group under the operation given by

$$(a_1, a_2, \ldots, a_n)(b_1, b_2, \ldots, b_n) = (a_1 b_1, a_2 b_2, \ldots, a_n b_n);$$

this group is called the ***direct product*** of $G_1, \ldots, G_n$.

The mapping α_i from a cartesian product

$$P = S_1 \times S_2 \times \cdots \times S_n$$

to S_i with $\alpha_i(s_1, s_2, \ldots, s_n) = s_i$ is called the ***projection*** of P on S_i.

Many mathematicians characterize a mapping θ from S to T using the subset U of $S \times T$ consisting of all the ordered pairs (s, t) with $\theta(s) = t$. Such a subset U of $S \times T$ has the property that for every s in S there is exactly one ordered pair (x, y) in U with $x = s$. If one is given a subset U of $S \times T$ with this property, one can define the associated mapping from S to T as the θ with $\theta(a) = b$ whenever (a, b) is in U.

Problems

1. Let $S = \{1, 2\}$ and $T = \{1, 3, 4\}$.

(i) List the six ordered pairs of $S \times T$.

(ii) List the six ordered pairs of $T \times S$.
(iii) Does $S \times T = T \times S$ for these sets S and T?

2. Explain why $S \times T = T \times S$ if and only if $S = T$, $S = \varnothing$, or $T = \varnothing$.

3. How many elements are there in $S \times T$ when S has m elements and T has n elements?

4. Describe a bijection from $(S \times T) \times U$ to $S \times (T \times U)$.

5. Let $G_1 = [a] = \{\mathbf{e}, a, a^2\}$ and $G_2 = [b] = \{\mathbf{e}', b\}$ be cyclic groups of order 3 and 2, respectively.
(i) Make the multiplication table for the direct product $G = G_1 \times G_2$.
(ii) Is G cyclic? Explain.
(iii) Is G isomorphic to the symmetric group $\mathbf{S_3}$? Explain.
(iv) Tabulate a surjective homomorphism from G onto G_1.

6. Show that every noncyclic group of order 4 (that is, every Klein 4-group) is isomorphic to a direct product $G_1 \times G_2$, with G_1 and G_2 each of order 2. (See Example 1 of Section 3.2.)

7. Give the details of showing that the operation

$$(a_1, a_2)(b_1, b_2) = (a_1 b_1, a_2 b_2)$$

makes a cartesian product of groups into a group. (This is the proof of Theorem 1.)

8. (i) Explain why the direct product of abelian groups is also abelian.
(ii) Is the octic group (of symmetries of a square) isomorphic to a direct product of two groups having 2 and 4 as their orders? Explain.

9. Is the projection of a direct product $G_1 \times G_2$ on G_1 a homomorphism? Explain.

10. Let $G = G_1 \times G_2 \times G_3$, where each of G_1, G_2, and G_3 has order 2. Find the smallest positive integer m such that g^m is the identity of G for all g in G.

11. Let $\mathbf{R}^{\pm}$, $\mathbf{R}^{+}$, and M denote the multiplicative groups of the nonzero real numbers, the positive real numbers, and the invertibles $\{1, -1\}$ of $\mathbf{Z}$, respectively. Explain why $\mathbf{R}^{\pm}$ is isomorphic to the direct product $\mathbf{R}^{+} \times M$.

12. Let $\mathbf{R}^{\pm}$ and M be as in Problem 11 and let $\mathbf{R}$ denote the additive group of real numbers. Explain why $\mathbf{R}^{\pm}$ is isomorphic to the direct product $\mathbf{R} \times M$.

13. Let $\mathbf{C}$ and $\mathbf{R}$ denote the additive groups of the complex and of the real numbers, respectively. Explain why $\mathbf{C}$ is isomorphic to $\mathbf{R} \times \mathbf{R}$.

14. Let V, $\mathbf{R}^{+}$, and U denote the multiplicative groups of the nonzero complex numbers, the positive real numbers, and the complex numbers of absolute value 1, respectively. Explain why V is isomorphic to $\mathbf{R}^{+} \times U$.

15. Let H and K be groups isomorphic to H' and K', respectively. Explain why $H \times K$ is isomorphic to $H' \times K'$.

16. Let V and $\mathbf{R}^{+}$ be as in Problem 14. Let T denote the quotient group $\mathbf{R}/[2\pi]$ of the additive group $\mathbf{R}$ of the real numbers by its cyclic subgroup $[2\pi]$. Explain why V is isomorphic to $\mathbf{R}^{+} \times T$.

17. Let $M = \{1, -1\}$ be a group of order 2 with 1 as identity and let $G = \mathbf{S_n} \times M$. Let H be the subset of G consisting of all (α, m) such that $m = 1$ when α is even and $m = -1$ when α is odd. Show that H is a subgroup in G and that H is isomorphic to $\mathbf{S_n}$.

18. Let $V = \{\mathbf{e}, b, c, d\}$ be a Klein 4-group; that is, let $g^2 = \mathbf{e}$ for every g in V. Let $\mathbf{Z_2}$ be the modular arithmetic additive group $\{\bar{0}, \bar{1}\}$. Let α be a bijection from V to $\mathbf{Z_2} \times \mathbf{Z_2}$ for which $\alpha(\mathbf{e}) = (\bar{0}, \bar{0})$. Show that α is an isomorphism.

René Descartes
1596–1650

Descartes, the French philosopher, has a place in the history of mathematics largely due to his work *La Géométrie*, which was published as part of his masterpiece, *Discours de la Méthode* (1637), and which contained the elements of analytic geometry. To say that Descartes invented analytic geometry is, as with most such statements, an oversimplification. One can indeed trace the ideas of analytic geometry back as far as Apollonius (c. 250 B.C.) and the application of algebra to geometrical problems was certainly in the works of Viète (1540–1603) and Oughtred (1574–1660). Fermat wrote a treatise on analytic geometry at roughly the same time as *La Géométrie*. However, it was not published until later, so Descartes receives the credit for the discovery. *La Géométrie* is not a systematic treatment of analytic geometry; instead, it is a series of problems in which the techniques of analytic geometry are used — but not as those of us familiar with modern analytic geometry might expect. Descartes just took as origin a point in an existing figure and measured off distances along two existing lines. Axes are not necessarily perpendicular, and the choice of the origin was arbitrary.

In the third book of *La Géométrie*, Descartes points out that if a cubic equation with rational coefficients has a rational root, then its roots can be constructed by straightedge and compass. He derives the cubic equation on which the angle trisection problem is based (see Section 6.3); however, he does not go on to consider the nonconstructability with straightedge and compass, but instead he proceeds to perform the trisection with the aid of a parabola and circle. It was not until the early nineteenth century that Wantzel used methods introduced by Galois to settle these nonconstructability problems once and for all.

Descartes made contributions to all of the traditional sciences except perhaps chemistry. He wrote on mathematics, and physics, geology, and even the theory of music. Physics claimed much of his time during his most productive years, 1629–1649, which he spent in Holland. In 1649, he accepted an invitation from Queen Christina of Sweden to come to the Swedish court. He did not long withstand the rigors of the Stockholm winter and the perverse conviction of the Queen that 5 o'clock in the morning was the proper time to receive philosophy lessons from Descartes. He died one year after going to Sweden.

Seventeen years after his death, his bones were returned to Paris for entombment in the Panthéon. Because his political and philosophical views were potentially embarrassing to the government, there was no public oration. The mathematician Carl G. J. Jacobi commented: "It is often more convenient to possess the ashes of great men than to possess the men themselves during their lifetime."

19. Let $G = \{\mathbf{e}, a, b, c, ab, ac, bc, abc\}$ be a group of order 8 in which $g^2 = \mathbf{e}$ for every g in G. Show that there exists an isomorphism β from G to $\mathbf{Z_2} \times \mathbf{Z_2} \times \mathbf{Z_2}$ with

$$\beta(a) = (\bar{1}, \bar{0}, \bar{0}), \quad \beta(b) = (\bar{0}, \bar{1}, \bar{0}), \quad \beta(c) = (\bar{0}, \bar{0}, \bar{1}).$$

20. Let $G = \{\mathbf{e}, a, a^2, a^3, b, ab, a^2b, a^3b\}$ be a group of order 8 in which $a^4 = \mathbf{e} = b^2$ and $ba = ab$. Show that G is isomorphic to $\mathbf{Z}_4 \times \mathbf{Z}_2$ and that G has four elements of order 4.

★ 21. Let s and t be positive integers and let $G = [a]$ be a cyclic group of order st. Show that G is isomorphic to $[a^s] \times [a^t]$ if and only if s and t are relatively prime.

3.7 Relations

In the paragraph preceding the problems for Section 3.6, we gave the characterization of a mapping from S to T as a special kind of subset of the cartesian products $S \times T$. Here we study general subsets of the cartesian product of a set with itself.

DEFINITION 1

Relation

A **relation** on a set S is a subset of the cartesian product $S \times S$.

In some other texts, what we call "a relation on S" is called "a binary relation on S."

NOTATION 1 If R stands for a relation on S, aRb denotes that the ordered pair (a, b) is in the subset R of $S \times S$, and $a\bar{R}b$ means that (a, b) is not in R. The letter R may be replaced by another letter or by an appropriate symbol such as $|$, $<$, $>$, $\leq$, $\geq$, $\subset$, $\subseteq$, $\supset$, or $\supseteq$.

One may read aRb as "a is related to b (under R)" and $a\bar{R}b$ as "a is not related to b (under R)."

EXAMPLE 1 Let S be the set of integers $\{1, 2, 3\}$. Then the $<$ relation on S is the subset $\{(1, 2), (1, 3), (2, 3)\}$ of $S \times S$, and the $\leq$ relation on S is

$$\{(1, 1), (1, 2), (1, 3), (2, 2), (2, 3), (3, 3)\}.$$

If P stands for $\leq$, then $2P2$ tells us that 2 is less than or equal to itself and $3\bar{P}2$ tells us that 3 is not less than or equal to 2.

A mapping θ from a set S to itself may be thought of as a relation F on S with the special property that for every a in S there is exactly one b in S with aFb. This unique b is of course $\theta(a)$.

DEFINITION 2

Reflexive, Transitive, Symmetric, Antisymmetric

Let R be a relation on a set S. Then:

(i) R is **reflexive** if aRa for all a in S;
(ii) R is **transitive** if aRb and bRc together imply aRc;
(iii) R is **symmetric** if aRb implies bRa;
(iv) R is **antisymmetric** if aRb and bRa together imply $a = b$.

EXAMPLE 2 Let A, B, C, D, and E be the relations on the integers $\mathbf{Z}$ in which xAy means that $x < y$, xBy means that $x \leq y$, xCy means that $x - y$ is an even integer, xDy means that $x \mid y$, and xEy means that $x = y$. All five of these relations are transitive. All but A are reflexive. (For D, this implies that we have defined 0 to be an integral divisor of itself.) C and E are symmetric while A, B, and E are antisymmetric. (A is vacuously antisymmetric, since $x < y$ and $y < x$ are contradictory.) Thus E is both symmetric and antisymmetric. We note that D is neither symmetric nor antisymmetric: D is not symmetric, since $3D6$ but $6\bar{D}3$; and D is not antisymmetric, since $(-3)D3$ and $3D(-3)$ but 3 and -3 are not equal.

The relations in Example 2 show that "antisymmetric" has a different meaning from "not symmetric."

DEFINITION 3

Equivalence Relation

A relation that is reflexive, symmetric, and transitive is an **equivalence relation.**

EXAMPLE 3 Let H be a subgroup in G and let R be the relation on G for which aRb means that a and b are the same left coset of H in G; that is, $aH = bH$. This relation is easily seen to be reflexive, symmetric, and transitive; hence it is an equivalence relation. The special case in which G is the additive group $\mathbf{Z}$ of the integers and H is a cyclic subgroup $[m]$ is called the relation of "congruence modulo m."

THEOREM 1 **Partitions and Equivalence Relations**

Let Γ be a partition of a set S and let E be the relation on S for which aEb means that a and b are the same member set C of the partition Γ. Then E is an equivalence relation.

The proof is left to the reader as Problem 8(a) of this section.

THEOREM 2 **Equivalence Classes and Partitions**

Let E be an equivalence relation on a nonempty set S. For every a in S, let $C(a)$ be the subset of all elements x of S satisfying aEx. Then the collection Γ of all these subsets $C(a)$ is a partition of S. [The $C(a)$ are called the **equivalence classes** of the equivalence relation E.]

We leave the proof to the reader as Problem 8(b) of this section.

The transitive relation L on the integers $\mathbf{Z}$ in which xLy means $x < y$ has the property that, if a and b are in $\mathbf{Z}$, then exactly one of the following holds:

$$\text{(i) } a = b; \qquad \text{(ii) } aLb; \qquad \text{(iii) } bLa. \tag{T}$$

We next generalize on this relation.

DEFINITION 4 **Trichotomy**

A relation L on a set S has the **trichotomy** property if, whenever a and b are in S, then one and only one of the statements (i), (ii), (iii) of display (T) above is true.

DEFINITION 5 **Linear Ordering**

If a relation R on a set S is transitive and has the trichotomy property, R is a **linear ordering** and (S, R) is a **linearly ordered set**.

If S is any set of real numbers (for example, $\mathbf{Z}^+$, $\mathbf{Z}$, $\mathbf{Q}^+$, $\mathbf{Q}$, $\mathbf{R}^+$, or $\mathbf{R}$ itself), then clearly $<$ is a linear ordering on S and $(S, <)$ is a linearly ordered set; also $(S, >)$ is a linearly ordered set.

NOTATION 2 If $<$ is a linear ordering on a set S and x and y are in S, $x \leq y$ means that either $x < y$ or $x = y$.

DEFINITION 6 **Reverse of a Relation**

Let R be a relation on a set S. Then the relation V on S, such that xVy if and only if yRx is called the **reverse** of the relation R.

For example, on the set of real numbers, $<$ and $>$ are reverse relations of each other; and the same is true of $\leq$ and $\geq$.

Problems

1. Let H be a subgroup in G. Let L be the relation on G for which aLb means that a is in the left coset for b of H in G. Without using Theorem 1, show that L is reflexive, symmetric, and transitive and hence is an equivalence relation.
2. Let E be the relation on a group G for which aEb means that a and b have the same order. Show that E is an equivalence relation.
3. Let E be the relation on a group G for which aEb means that either $b = a$ or $b = a^{-1}$. Show that E is an equivalence relation.
4. Let C be the relation on a group G such that aCb means that $a = g^{-1}bg$ for some g in G; that is, aCb means that a is a conjugate of b. Show that C is an equivalence relation.
5. Let E be the relation on a group G for which aEb means that $[a] = [b]$. Show that E is an equivalence relation.
6. Let L be the relation on a group G such that aLb means that $ab = ba$. Is L always an equivalence relation? Explain.
7. Let S be any nonempty set of real numbers.
   (i) Is $=$ an equivalence relation on S?
   (ii) If so, what are the equivalence classes $C(a)$ of this equivalence relation?
8. (a) Prove Theorem 1 of this section.
   (b) Prove Theorem 2 of this section.
9. Consider the relation $\leq$ on a set X of real numbers.
   (a) Is $\leq$ an antisymmetric, reflexive, and transitive relation on X?
   (b) Let there be more than one number in X. Is the relation $\leq$ on X an equivalence relation? Is it a linear order relation? Explain.
10. Consider the relation $\subseteq$ on any collection Γ of sets.
    (a) Is $\subseteq$ an antisymmetric, reflexive, and transitive relation on Γ?
    (b) Give an example in which $\subseteq$ is not an equivalence relation on Γ.
    (c) Give an example in which $\subseteq$ is not a linear ordering of Γ.
11. How many distinct relations are there in a set S with two elements?
12. How many distinct relations are there in a set S with three elements?
13. (a) Let $|$ be the "is an integral divisor of" relation on the set $\mathbf{Z}$ of the integers. Is $|$ reflexive? Is it symmetric? Is it transitive? Is it a linear order relation?
    (b) Answer the questions of part (a) with $|$ replaced by the relation P on $\mathbf{Z}$ such that aPb means that $\gcd(a, b) = 1$.
14. Let P consist of all the subsets of a set S and let I be the relation on P for which $\alpha I \beta$ means that $\alpha \subset \beta$. Is I reflexive? Symmetric? Transitive? Is P a linear order relation when S has more than one element?
15. Let A be an antisymmetric relation on a set X. Let s and t be elements of X such that sAx and tAx for all x in X. Prove that $s = t$.
16. Let A be the relation on $\mathbf{Z}$ such that mAn means that both $m | n$ and $n | m$. Is A an equivalence relation?

17. Let L be the relation on the rational numbers $\mathbf{Q}$ for which aLb means that $a > b$. Is L a linear order relation?

18. Let V be the reverse of a relation R on a set S. (This means that xVy if and only if yRx.)
(a) Show that V is antisymmetric, reflexive, and transitive whenever R is antisymmetric, reflexive, and transitive.
(b) If R is symmetric, is V also symmetric?
(c) If R is a linear order relation, is V also a linear order relation?

19. Show that there are exactly 6 linear order relations on a set $S = \{a, b, c\}$ with 3 elements.

20. How many linear order relations are there on a set with n elements?

21. Let α be a mapping from S to itself, let R be the relation on S for which xRy means that $y = \alpha(x)$, and let V be the reverse relation of R. Show that α is bijective if and only if there exists a mapping β from S to itself such that sVt means that $t = \beta(s)$.

22. Let R be a relation on a set S. Prove that R is its own reverse if and only if R is symmetric.

23. Let L be a relation on a set S such that L is symmetric and transitive but is not reflexive.
(i) Prove that there is an element a in S such that $a\bar{L}s$ for all s in S.
(ii) Give an example to show that there exists such a relation L.

24. Let Γ be a collection of sets and let ρ be the relation on Γ such that $A\rho B$ means that there exists a bijection from A to B. Prove that ρ is an equivalence relation.

25. Let Γ be the collection of all subgroups in a group G. Also let ρ be the relation on Γ such that $H\rho K$ if and only if K is the conjugate subgroup of H by some g in G. Show that ρ is an equivalence relation.

26. Let ord $G = p^2$, where p is a positive prime. Let I be the relation on G in which aIb means that a and b are conjugates; that is, $c^{-1}ac = b$ for some c in G. Let $C(a)$ consist of all g in G such that aIg. Show the following.
(i) I is an equivalence relation.
(ii) If G is abelian, each $C(a)$ is just the singleton set $\{a\}$.
(iii) If G is not cyclic, then G consists of $p^2 - 1$ elements having order p and the identity.
(iv) If G is not abelian, then $C(\mathbf{e}) = \{\mathbf{e}\}$ and $C(a)$ has p elements when $a \neq \mathbf{e}$.
(v) G must be abelian.

27. Let ord $G = p^2$, where p is a positive prime. Prove that G is either cyclic or isomorphic to a direct product $G_1 \times G_2$ of cyclic groups each of order p.

★ **28.** Let P be the set of all nonnegative real numbers and let R be the relation on P for which aRb means that

$$a - \sqrt{a} \le b - \tfrac{1}{4} \le a + \sqrt{a}.$$

Show that R is symmetric but not transitive.

3.8 Partially Ordered Sets

The relation $\le$ on any set X of real numbers is antisymmetric, reflexive, and transitive, and the same is true of the relation $\subseteq$ on any collection Γ of subsets of a set S. (See Problems 9 and 10 of Section 3.7.) We next introduce notation and terminology that will enable us to study simultaneously all relations having these three properties.

DEFINITION 1

Partial Ordering $\preccurlyeq$, Poset $(S, \preccurlyeq)$

A **partial ordering** on a set S is an antisymmetric, reflexive, and transitive relation on S. The notation $\preccurlyeq$ is reserved for partial orderings. A **partially ordered set**, or (for short) **poset**, is an ordered pair $(S, \preccurlyeq)$ with $\preccurlyeq$ a partial ordering on S.

It follows from the answers to Problems 9 and 10 of Section 3.7 that $(X, \leq)$ is a poset for every set X of real numbers and that $(\Gamma, \subseteq)$ is a poset for any collection Γ of sets.

NOTATIONS $\prec, \succcurlyeq, \succ$

If $(S, \preccurlyeq)$ is a poset and x and y are in S, then $x \prec y$ means that $x \preccurlyeq y$ but $x \neq y$. Also $y \succcurlyeq x$ and $y \succ x$ are alternate notations for $x \preccurlyeq y$ and $x \prec y$, respectively; that is, $\preccurlyeq$ and $\succcurlyeq$ are reverses of each other, and the same is true of $\prec$ and $\succ$.

In Problem 18(a) of Section 3.7, the reader was asked to prove that $(S, \succcurlyeq)$ is a poset for every poset $(S, \preccurlyeq)$. Thus $(X, \geq)$ is a poset for every subset X of the real numbers, and $(\Gamma, \supseteq)$ is a poset for every collection Γ of sets.

DEFINITION 2

Poset Diagram

Let $(S, \preccurlyeq)$ be a poset, with S finite. Let D be a diagram having a vertex for every element of S and having a rising line or broken line from the vertex for an element s to that for an element s' if and only if $s \prec s'$. Then D is a **poset diagram** for $(S, \preccurlyeq)$. (Some of the vertices may be isolated — that is, may have no lines emanating from them. It is customary to use as few segments as possible while satisfying this definition.)

EXAMPLE 1 Let G be a group and Γ be the collection of all subgroups in G. Then $(\Gamma, \subseteq)$ is a poset; there are poset diagrams for such posets in Sections 2.9, 2.10, and 3.4. Similarly, $(\Gamma, \supseteq)$ is a poset. If $H \lhd K$ denotes that H is a normal subgroup in K, $(\Gamma, \lhd)$ is generally not a poset, since $\lhd$ is not necessarily transitive.

DEFINITION 3

Least Element, Greatest Element

Let $(X, \preccurlyeq)$ be a poset. A **least** element in X is an element s in X such that $s \preccurlyeq x$ for all x in X. A **greatest** element in X is an element g in X such that $x \preccurlyeq g$ for all x in X.

We will see in Problems 9 and 10 below that there is at most one least element and at most one greatest element in a given poset. The poset $(\mathbf{Z}^+, \leq)$ has 1 as its least element and has no greatest element. If S is the set of all real numbers x satisfying $3 < x \leq 5$, then the poset $(S, \leq)$ has 5 as its greatest element but has no least element. The poset $(\mathbf{Z}, \leq)$ has neither a least nor a greatest element. If $S = \{a, b\}$ and $P = \{\varnothing, \{a\}, \{b\}, S\}$, the poset $(P, \subseteq)$ has $\varnothing$ as its least element and S as its greatest element.

EXAMPLE 2 Let $S = \{1, 2, 3, 4, 6, 12\}$ and D be the relation on S for which xDy means that x is an integral divisor of y. Then (S, D) is a poset with 1 as the least element and 12 as the greatest element.

Problems

1. Let D be the relation on $S = \{2, 3, 4, 6, 12\}$ for which xDy means that $x \mid y$. Which ordered pairs of $S \times S$ are in D?
2. Do Problem 1 with $\{2, 3, 4, 6, 12\}$ replaced by $\{1, 2, 3, 4, 6\}$.
3. Does the poset (S, D) of Problem 1 have a least element? A greatest element?
4. Does the poset (S, D) of Problem 2 have a least element? A greatest element?
5. Let m be a least element for a poset $(S, \preccurlyeq)$. What is the role of m for the poset $(S, \succcurlyeq)$?
6. Do Problem 5 with m replaced by a greatest element M for $(S, \preccurlyeq)$.
7. In a poset $(X, \preccurlyeq)$, let $x \preccurlyeq y$ and $y \preccurlyeq x$. Explain why one must have $x = y$.
8. In a poset $(X, \preccurlyeq)$, is it possible to have $x \prec y$ and $y \prec x$ simultaneously? [*Hint:* See the meaning of the notation $x \prec y$ in $(X, \preccurlyeq)$.]
9. Prove that there is at most one least element in a poset $(X, \preccurlyeq)$.
10. Do Problem 9 with "least" replaced by "greatest."
11. (i) Construct a poset diagram for the poset (S, D) of Problem 1.
    (ii) Why does the diagram of part (i) have fewer segments than there are ordered pairs in D?
12. Give an example of a poset $(S, \preccurlyeq)$ in which S is infinite and there is both a least element and a greatest element.
13. Let M be the relation on the set $\mathbf{Z}^+$ of positive integers for which aMb means $b \mid a$; that is, a is an integral multiple of b. Is M reflexive? Is it symmetric? Is it antisymmetric? Is it transitive? Is it a linear ordering? Is it a partial ordering?

14. Let P consist of all the subsets of a set S and let I be the relation on P for which $\alpha I \beta$ means that $\alpha \subseteq \beta$; that is, α is a subset of β. Is I antisymmetric? Is it reflexive? Is it transitive? Is it a partial ordering?

15. Is the relation I of Problem 14 a linear ordering when S has more than one element? Explain.

3.9 Power Sets

The essence of mathematics is its freedom. — **Georg Cantor**

No one can expel us from the paradise which Cantor created for us. — **David Hilbert**

The main topic of this section has applications to fields such as computer science and logic. We will also see that it furnishes important examples for the other topics of this chapter.

DEFINITION 1

Power Set, Universe, Complement

The collection of all the subsets of a set S is called the **power set** for S and is denoted by $P(S)$. S is called the **universe** for $P(S)$. If A is an element of $P(S)$, that is, a subset of S, then the subset consisting of the elements of S not in A is the **complement** of A (in S) and is denoted by $\bar{A}$ (or by $S \backslash A$ when one wishes to show the role of the universe explicitly).

EXAMPLE 1 Let S be a set $\{s_1, s_2, s_3\}$ with three elements. Then

$$P(S) = \{\varnothing, \{s_1\}, \{s_2\}, \{s_3\}, \{s_1, s_2\}, \{s_1, s_3\}, \{s_2, s_3\}, S\}.$$

THEOREM 1 **Power Set Poset**

For any set S, $(P(S), \subseteq)$ is a poset with $\varnothing$ as the least element and S as the greatest element.

Proof Since $A \subseteq A$ for all sets A, the relation $\subseteq$ on $P(S)$ is reflexive. If $A \subseteq B$ and $B \subseteq C$, one has $A \subseteq C$; hence, $\subseteq$ is transitive. Also, $A \subseteq B$ and $B \subseteq A$ together imply that $A = B$; hence, $\subseteq$ is antisymmetric. Thus $(P(S), \subseteq)$ is a poset. Since $\varnothing \subseteq A \subseteq S$ for all A in $P(S)$, $\varnothing$ is the least element and S is the greatest element for this poset. ∎

We next introduce a few concepts that are somewhat related to power sets.

DEFINITION 2

Bit, n-bit String, Set $\mathbf{W_n}$

A **bit** is a digit in the set $\{0, 1\}$ of binary digits. An ***n*-bit string** has the form $a_1a_2 \ldots a_n$, with each a_i in $\{0, 1\}$. We use $\mathbf{W_n}$ to denote the set of all n-bit strings.

For example, $\mathbf{W_3} = \{000, 001, 010, 011, 100, 101, 110, 111\}$ is the set of all 3-bit strings.

NOTATION **2^S, Base 2 Numeral**

If S is a set, $\mathbf{2^S}$ denotes the set of all mappings from S into $\{0, 1\}$. If $S = \{s_1, s_2, \ldots, s_n\}$ has n elements, the **binary numeral** for a mapping α in 2^S is the n-bit string $a_1a_2 \ldots a_n$ with $a_i = \alpha(s_i)$ for $1 \leq i \leq n$.

For example, 011 is the binary numeral for the mapping

$$\alpha : s_1 \mapsto 0, \quad s_2 \mapsto 1, \quad s_3 \mapsto 1.$$

DEFINITION 3

Characteristic Function, Binary Numeral

Let A be a subset of S. Then the **characteristic function** for A is the mapping α in 2^S with $\alpha(s) = 1$ when s is in A and $\alpha(s) = 0$ when s is not in A. If S is finite, the binary numeral for the characteristic function α is also said to be the **binary numeral** for the set A.

EXAMPLE 2 Let A be the subset $\{s_2, s_4, s_5\}$ of $S = \{s_1, s_2, s_3, s_4, s_5\}$. Then the characteristic function for A is the mapping α from S into $\{0, 1\}$ with $\alpha(s_2) = \alpha(s_4) = \alpha(s_5) = 1$ and $\alpha(s_1) = \alpha(s_3) = 0$. Hence 01011 is the binary numeral for A or for α.

EXAMPLE 3 Let $S = \{s_1, s_2, \ldots, s_n\}$. For any subset A of S, that is, for any element A of the power set $P(S)$, let $f(A)$ be the binary numeral for A. Then f is a bijection from $P(S)$ onto $\mathbf{W_n}$. If we fix n as 3, the table for this bijection is

A	$\varnothing$	$\{s_3\}$	$\{s_2\}$	$\{s_2, s_3\}$	$\{s_1\}$	$\{s_1, s_3\}$	$\{s_1, s_2\}$	S
$f(A)$	000	001	010	011	100	101	110	111

,

and the poset diagram for $(P(S), \subseteq)$ is as shown in Figure 3.7, with the numeral for A as the label of the vertex A.

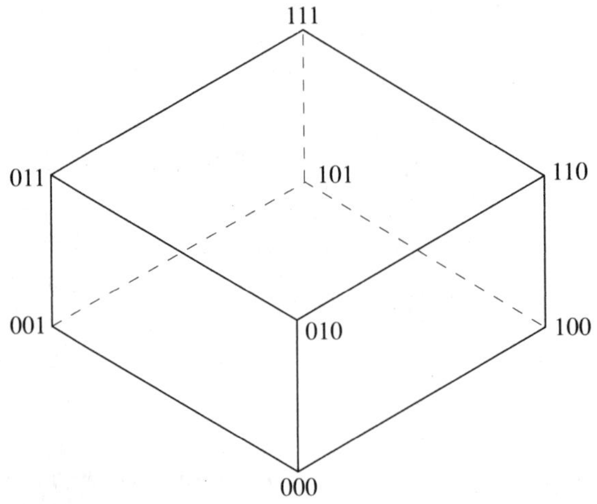

FIGURE 3.7

Problems

1. Let $S = \{s_1, s_2, s_3, s_4, s_5\}$ and P be its power set.
   (a) What is the binary numeral for $A = \{s_1, s_4\}$? Also, give the complement $\bar{A}$ and its binary numeral.
   (b) Give the element B of P with 01110 as its binary numeral and also its complement $\bar{B}$.
   (c) Give the numeral for the least element of $(P, \subseteq)$.
   (d) Tell how to obtain the numeral $c_1c_2c_3c_4c_5$ for $A \cup B$ from the numerals $a_1a_2a_3a_4a_5$ and $b_1b_2b_3b_4b_5$ for A and B.
   (e) Does $A \cup \bar{A} = S$ for all A in P?
2. Let P be the power set for $S = \{s_1, s_2, \ldots, s_n\}$.
   (a) Describe the binary numeral for the greatest element of $(P, \subseteq)$.
   (b) Tell how to obtain the numeral $c_1c_2 \ldots c_n$ for $\bar{A}$ from the numeral $a_1a_2 \ldots a_n$ for A.
   (c) Tell how to obtain the numeral $d_1d_2 \ldots d_n$ for $A \cap B$ from the numerals $a_1a_2 \ldots a_n$ and $b_1b_2 \ldots b_n$ for A and B.
   (d) Does $A \cap \bar{A} = \emptyset$ for all A in P?
3. Give the answer to Problem 2(d) when S is an infinite set.
4. Give the answer to Problem 1(e) when S is an infinite set.
5. In a power set $P(S)$, let $B = \bar{A}$. Does $\bar{B} = A$?
6. Let A and B be in $P(S)$, $A \cup B = S$, and $A \cap B = \emptyset$. Is each of A and B the complement of the other in S?
7. Is $(P(S), \supseteq)$ a poset for every set S? If so, what is its least element?
8. What is the greatest element for $(P(S), \supseteq)$?
9. How many mappings are there from $\{s_1, s_2, \ldots, s_n\}$ to $\{0, 1, 2\}$?
10. How many mappings are there from $\{s_1, s_2, \ldots, s_n\}$ to $\{1, 2, \ldots, b\}$?

11. Let S be a set. Is every mapping α in 2^S the characteristic function for some subset A of S?

12. If S is a set with n elements, how many elements are there in $P(S)$?

13. Let $\le$ be the relation on the set P_n of n-digit binary numerals for which $a_1a_2 \ldots a_n \le b_1b_2 \ldots b_n$ means that $a_i \le b_i$ for $i = 1, 2, \ldots, n$.
   (a) Explain why $(P_n, \le)$ is a poset.
   (b) What is the least element of this poset?
   (c) What is the greatest element?

14. Let T be a subset of P_n and $\le$ be as in Problem 13.
   (a) Explain why $(T, \le)$ is a poset.
   (b) Does it always have a least element? Explain.
   (c) Does it always have a greatest element? Explain.

15. Let S be a set and T consist of some of the subsets of S in $P(S)$. Explain why $(T, \subseteq)$ is a poset.

16. Give an example in which the $(T, \subseteq)$ of Problem 15 has neither a least nor a greatest element.

3.10 Operations

Some material concerning operations has been presented in Chapter 2; we now give a more formal and more general treatment.

DEFINITION 1

Operation on a Set

Let S^n be the cartesian product $S \times S \times \cdots \times S$ of n copies of a set S. An ***n*-ary operation** on S is a mapping F from S^n to S. In particular, a **unary operation** on S is a mapping from S to itself, and a **binary operation** on S is a mapping from $S \times S$ to S.

By this definition, S must be closed under the operation on S.

NOTATION **Image Under an Operation**

The image of an ordered n-tuple $(s_1, s_2, \ldots, s_n)$ in S^n under an n-ary operation F on S may be denoted by $F(s_1, s_2, \ldots, s_n)$. The image $B(x, y)$ under a binary operation may also be written as $x + y$ when B represents addition, as $x \cdot y$ or xy when B is multiplication, as $x - y$ when B is subtraction, and as x/y or $x \div y$ when B is division. The symbols $\circ$, $\square$, and $\triangle$ are used frequently to represent abstract binary operations.

It is clear from the preceding definition and notation that an n-ary operation on the set **R** of real numbers is the same as a real valued function of n real variables.

EXAMPLE 1 **Some Unary Operations**

The mapping f with $f(x) = x^2$ is a unary operation on the set $\mathbf{Z}^+$ of the positive integers (or on any of the sets $\mathbf{Z}$, $\mathbf{Q}$, $\mathbf{R}$, and $\mathbf{C}$). If c is a fixed element of a group G, then each of the mappings

$$\alpha : x \mapsto x^{-1}, \quad \beta : x \mapsto cx, \quad \gamma : x \mapsto xc, \quad \delta : x \mapsto c^{-1}xc$$

is a unary operation on G.

EXAMPLE 2 **Some Binary Operations**

The mapping f from $\mathbf{Q} \times \mathbf{Q}$ to $\mathbf{Q}$ with $(x, y) \mapsto (2x + y)/3$ is a binary operation on the set $\mathbf{Q}$ of rational numbers. On the set $\mathbf{Z}^+$ of positive integers, the binary operations lcm and gcd are the mappings from $\mathbf{Z}^+ \times \mathbf{Z}^+$ to $\mathbf{Z}^+$ with

$$(x, y) \mapsto \mathrm{lcm}[x, y] \quad \text{and} \quad (x, y) \mapsto \gcd(x, y),$$

respectively.

EXAMPLE 3 **An *n*-ary Operation**

Let $A(x_1, x_2, \ldots, x_n) = (x_1 + x_2 + \cdots + x_n)/n$ whenever the x_i are in $\mathbf{Q}$. Clearly, this averaging function is an n-ary operation on $\mathbf{Q}$.

DEFINITION 2

Associativity, Commutativity, Identity, Inverse

Let θ be a binary operation on S. Then θ is **associative** if $\theta(\theta(x, y), z) = \theta(x, \theta(y, z))$ for all x, y, z in S. θ is **commutative** if $\theta(x, y) = \theta(y, x)$ for all x, y in S. An element e of S is an **identity** under the operation θ if $\theta(x, e) = x = \theta(e, x)$ for all x in S. If e is an identity under θ and s is in S, then an **inverse** of s with respect to θ and e is an element s^{-1} of S such that $\theta(s, s^{-1}) = e = \theta(s^{-1}, s)$.

Using the symbol $\circ$ for an abstract binary operation on a set S, we can rewrite these definitions as follows.

Associativity If $(x \circ y) \circ z = x \circ (y \circ z)$ for all x, y, z in S, then $\circ$ is associative.

Commutativity If $x \circ y = y \circ x$ for all x, y in S, then $\circ$ is commutative.

Identity If $x \circ e = x = e \circ x$ for all x in S, then e is an identity under $\circ$.

Inverse If e is an identity under $\circ$ and $s \circ t = e = t \circ s$, then t is an inverse of s with respect to $\circ$.

The binary operation f on $\mathbf{Q}$ with $f(x, y) = (2x + y)/3$ is not commutative, since, for example, $f(3, 0) = 2$ and $f(0, 3) = 1$. This f is not associative, since

$$f(f(6, 9), 0) = f(7, 0) = 14/3, \quad f(6, f(9, 0)) = f(6, 6) = 6.$$

Each of the binary operations lcm and gcd on $\mathbf{Z}^+$ is commutative, since

$$\text{lcm}[x, y] = \text{lcm}[y, x] \quad \text{and} \quad \gcd(x, y) = \gcd(y, x)$$

for all x and y in $\mathbf{Z}^+$. Each of lcm and gcd is also associative, since

$$\text{lcm}[\text{lcm}[x, y], z] = \text{lcm}[x, \text{lcm}[y, z]] \text{ for all } x, y, z \text{ in } \mathbf{Z}^+,$$

$$\gcd(\gcd(x, y), z) = \gcd(x, \gcd(y, z)) \text{ for all } x, y, z \text{ in } \mathbf{Z}^+.$$

The identity under lcm is 1, since $\text{lcm}[n, 1] = n = \text{lcm}[1, n]$ for all n in $\mathbf{Z}^+$, but there is no identity under gcd, since there is no e in $\mathbf{Z}^+$ with $\gcd(n, e) = n$ for all n in $\mathbf{Z}^+$. However, on the set $X = \{1, 2, 3, 6\}$, gcd is a binary operation with 6 as its identity. These assertions are justified by the following table, which shows that $\gcd(x, y)$ is in X for all x and y in X and $\gcd(x, 6) = x = \gcd(6, x)$ for each x in X.

gcd	1	2	3	6
1	1	1	1	1
2	1	2	1	2
3	1	1	3	3
6	1	2	3	6

THEOREM 1 **Uniqueness of the Identity**

There is at most one identity under a binary operation $\circ$ on a set S.

The proof is the same as that for Lemma 1 of Section 2.3.

On the basis of this result, we can now write "the identity" instead of "an identity."

EXAMPLE 4 **Operations on Power Sets**

Let S be a nonempty set. The operation , of taking complements with respect to S, is a unary operation on the power set $P(S)$. The binary operation $\cup$ on $P(S)$ given by the mapping

$$P(S) \times P(S) \to P(S), \qquad (A, B) \mapsto A \cup B$$

is commutative and has $\varnothing$ as the identity, since

$$A \cup B = B \cup A \quad \text{and} \quad A \cup \varnothing = A = \varnothing \cup A$$

for all subsets A and B of S. This operation is also associative, since

$$(A \cup B) \cup C = A \cup (B \cup C) \quad \text{for all } A, B, C \text{ in } P(S).$$

The binary operation $\cap$ on $P(S)$ is also commutative and associative, and it has S as its identity. In the special case in which S is a singleton $\{s\}$, the power set is $P(S) = \{\varnothing, S\}$, and tables for $^{-}$, $\cup$, and $\cap$ as are follows:

A	$\varnothing$	S
$\bar{A}$	S	$\varnothing$

$\cup$	$\varnothing$	S
$\varnothing$	$\varnothing$	S
S	S	S

$\cap$	$\varnothing$	S
$\varnothing$	$\varnothing$	$\varnothing$
S	$\varnothing$	S

.

DEFINITION 3

Min and Max Operations

Let $(S, <)$ be a linearly ordered set. For x and y in S, let **min**(x, y) be x if $x \leq y$ and be y if $y < x$. Also, let **max**(x, y) be x if $y \leq x$ and be y if $x < y$.

Clearly, Definition 3 makes each of min and max into a binary operation on any linearly ordered set.

EXAMPLE 5 If $(S, <)$ is a linearly ordered set, each of the binary operations min and max on S is commutative, since

$$\min(x, y) = \min(y, x) \quad \text{and} \quad \max(x, y) = \max(y, x)$$

for all x and y in S. Each of these operations is also associative, since

$$\min(\min(x, y), z) = \min(x, \min(y, z)),$$

$$\max(\max(x, y), z) = \max(x, \max(y, z))$$

for all x, y, z in S. If there is a least element t in S (that is, if $t \leq x$ for all x in S), then t is the identity under max, since $\max(x, t) = x = \max(t, x)$ for all x in S. Similarly, if g is the greatest element in S, then g is the identity under min.

Problems

1. Let P be the power set for $S = \{1, 2\}$.
(a) On P, tabulate the unary operation $A \mapsto \bar{A}$ of complementation.
(b) Tabulate the binary operation $\cup$ on P.
(c) What is the identity for $\cup$?

2. Do as in Problem 1(b) and (c) with $\cup$ replaced by $\cap$.

3. Let P be the power set for $S = \{s_1, s_2, s_3\}$.
(a) In terms of the numerals of

$$\mathbf{W}_3 = \{000, 001, 010, 100, 011, 101, 110, 111\},$$

tabulate the binary operation $\cap$ on P.
(b) What is the identity under the operation of (a)?

4. Do Problem 3 with $\cap$ replaced by $\cup$.

5. (i) Give the operation table for lcm on the set $X = \{1, 2, 3, 6\}$.
(ii) What is the identity in X under lcm?
(iii) Does each x in X have an inverse in X under lcm? Explain.

6. (i) Give the operation tables for lcm and gcd on $Y = \{1, 2, 4, 8\}$.

(ii) What is the identity in Y under lcm?
(iii) Does each y in Y have an inverse in Y under lcm? Explain.
(iv) What is the identity in Y under gcd?
(v) Does each y in Y have an inverse in Y under gcd? Explain.

7. Is there an identity for the binary operation lcm on $X = \{2, 3, 6\}$?

8. Is there an identity for gcd on $X = \{1, 2, 3\}$?

9. (i) Give the operation table for min on the set $S = \{2, 3, 4\}$.
(ii) Name the identity in S under min.
(iii) Does each s in S have an inverse under min? Explain.
(iv) Does $\mathbf{Z}^+$ have an identity under min? If so, name it.
(v) Does $\mathbf{Z}$ have an identity under min? If so, name it.

10. Repeat the parts of Problem 9 with min replaced by max.

11. Verify that $\min(x, \max(y, z)) = \max(\min(x, y), \min(x, z))$ for each of the following choices of (x, y, z).
(i) (2, 3, 4); (ii) (3, 2, 4); (iii) (4, 2, 3).

12. Verify that $\max(x, \min(y, z)) = \min(\max(x, y), \max(x, z))$ for each of the cases of Problem 11.

13. Let $a = (p_1)^{e_1}(p_2)^{e_2} \cdots (p_r)^{e_r}$ and $b = (q_1)^{f_1}(q_2)^{f_2} \cdots (q_s)^{f_s}$ be the standard factorizations for positive integers a and b. Write an expression for $\gcd(a, b)$ using the min operation.

14. Let a and b have the standard factorization of Problem 13. Write an expression for $\text{lcm}[a, b]$ using the max operation.

15. Let S have 3 elements.
(a) How many unary operations are there on S?
(b) How many binary operations are there on S?
(c) How many of the binary operations are commutative?
(d) How many n-ary operations are there on S?

16. Repeat the parts of Problem 15 for a set S having m elements.

17. Let P be the power set for $S = \{s_1, s_2, \ldots, s_n\}$. In P, let $A \cup B = C$ and let $a_1a_2 \ldots a_n$, $b_1b_2 \ldots b_n$, and $c_1c_2 \ldots c_n$ be the binary numerals for A, B, and C. Tell how to find c_i directly from a_i and b_i. [*Hint:* Use one of the operations min and max of Definition 3 above.]

18. Do Problem 17 with $\cup$ replaced by $\cap$.

19. Let M be the set of nonzero real numbers and let $C = M \times M$. Let θ be the operation in M (mapping from C to M) such that the image of (a, b) under θ is the real number product of a and $|b|$.
(i) Explain why θ is associative but not commutative.
(ii) Show that M has two "right identities" under θ, that is, two elements j such that, for all a in M,

$$(a, j) \overset{\theta}{\mapsto} a$$

(iii) Show that M has no left identity — that is, no element k such that, for all b in M,

$$(k, b) \overset{\theta}{\mapsto} b.$$

(iv) Let j be either one of the right identities. Show that every element b of M has a left inverse (with respect to j) under θ; that is, show the existence of a b^{-1} with

$$(b^{-1}, b) \overset{\theta}{\mapsto} j.$$

20. Let P be the power set for a nonempty set S. For each member A in P (that is, subset A of S), let $\bar{A}$ be the complement of A in S.
(i) If $\bar{A} = C$, what is $\bar{C}$?
(ii) If $\bar{A} = \bar{B}$, does $A = B$?
(iii) For every Y in P, is there an X in P such that $\bar{X} = Y$?
(iv) Is the mapping from P to P with $A \mapsto \bar{A}$ a bijection?
(v) Does $\overline{A \cup B} = \bar{A} \cap \bar{B}$ for all A and B in P?
(vi) Does $\overline{A \cap B} = \bar{A} \cup \bar{B}$ for all A and B in P?

3.11 Algebraic Structures

An ***algebraic structure*** is a set S, called the ***carrier*** of the structure, with one or more operations on S. The type of structure depends on the axioms satisfied by these operations. By the formal definition that an n-ary operation on S is a mapping from S^n to S, the set S is automatically closed under any operation on S. Binary operations are most common, and associativity is the most useful axiom; therefore we begin with a type of structure that involves a single binary operation and has associativity as the sole axiom.

DEFINITION 1

Semigroup

A **semigroup** is an ordered pair $(S, \circ)$, with S a set and with $\circ$ an associative binary operation on S.

We note that every group is a semigroup, since associativity is one of the axioms for the binary operation of a group. A semigroup may or may not be commutative; thus a ***commutative semigroup*** is an ordered pair $(S, \circ)$ such that $\circ$ is a commutative and associative binary operation on S.

EXAMPLE 1 Let E be the set of even integers. Since the product $x \cdot y$ of even integers is always an even integer, $\cdot$ is a binary operation on E. As this operation is associative, $(E, \cdot)$ is a semigroup; it is not a group. Since $\cdot$ is also commutative, $(E, \cdot)$ is a commutative semigroup.

EXAMPLE 2 $(\mathbf{Z}, -)$ is not a semigroup, since the binary operation of subtraction is not associative, as we see in the example

$$(10 - 6) - 3 = 4 - 3 = 1, \qquad 10 - (6 - 3) = 10 - 3 = 7.$$

$(\mathbf{Z}, \div)$ is not a semigroup, since $\mathbf{Z}$ is not closed under division; and hence $\div$ is not an operation on $\mathbf{Z}$. Let $\mathbf{Q}^+$ be the set of positive rational numbers; then $\div$ is an operation on $\mathbf{Q}^+$ but $(\mathbf{Q}^+, \div)$ is not a semigroup, since $\div$ is not associative.

DEFINITION 2

Subsemigroup

If $(T, \square)$ and $(S, \circ)$ are semigroups with T a subset of S and with $x \square y = x \circ y$ for all x and y in T, then $(T, \square)$ is a **subsemigroup** in $(S, \circ)$.

If $(T, \square)$ is a subsemigroup in $(S, \circ)$, we will usually denote the operation of the subsemigroup by the same symbol as for the semigroup, since $x \square y = x \circ y$ for all x and y in the subset T of S.

DEFINITION 3

Powers

Let $(S, \circ)$ be a semigroup and $a \in S$. Then a^2 denotes $a \circ a$; a^3 denotes $a \circ a^2$; and so on. Also, $\langle a \rangle$ denotes the set $\{a^n : n \in \mathbf{Z}^+\}$ of positive integral powers of a.

THEOREM 1 **Principal Subsemigroup Generated by an Element**

Let $(S, \circ)$ be a semigroup and $a \in S$. Then $(\langle a \rangle, \circ)$ is a subsemigroup in $(S, \circ)$.

The result follows readily from the definitions; $\langle a \rangle$ is called the ***principal subsemigroup generated by*** a.

EXAMPLE 3 **Semigroup $(\mathbf{Z}_m, *)$**

Let $*$ denote the operation of multiplication in $\mathbf{Z}_m = \{\bar{0}, \bar{1}, \ldots, \overline{m-1}\}$. That is, $\bar{a} * \bar{b} = \bar{r}$, where r is the remainder in the division of ab by m. Then $(\mathbf{Z}_m, *)$ is a semigroup. In the special case with $m = 100$, we note that the powers $\bar{6}$, $\bar{6}^2$, $\bar{6}^3, \ldots$ form the sequence $\bar{6}, \overline{36}, \overline{16}, \overline{96}, \overline{76}, \overline{56}, \overline{36}, \overline{16}, \overline{96}, \ldots$ in which the first term $\bar{6}$ does not reappear and the block of the next five terms repeats endlessly. Hence, in $(\mathbf{Z}_{100}, *)$, the principal subsemigroup $\langle \bar{6} \rangle$ is $\{\bar{6}, \overline{36}, \overline{16}, \overline{96}, \overline{76}, \overline{56}\}$.

DEFINITION 4

Monoid

A **monoid** is a semigroup $(S, \circ)$ having an identity under $\circ$.

Thus a monoid is an ordered triple $[S, \circ, \mathbf{e}]$ in which $\circ$ is an associative binary operation on S and $\mathbf{e}$ is the identity for the operation $\circ$. The two axioms for this type of algebraic structure are as follows:

M_1 ***Associativity*** $(x \circ y) \circ z = x \circ (y \circ z)$ for all x, y, z in S.

M_2 ***Identity*** $x \circ \mathbf{e} = x = \mathbf{e} \circ x$ for all x in S.

Since these axioms are included among the group axioms, every group is a monoid. In fact, a group is a monoid $[G, \circ, \mathbf{e}]$ such that every element g of G has an inverse (with respect to $\circ$).

EXAMPLE 4 $[\mathbf{Z}^+, \cdot, 1]$ is a monoid but is not a group, since, for example, the positive integer 2 does not have a multiplicative inverse in $\mathbf{Z}^+$. Let $N = \{0, 1, 2, \ldots\}$; then $[N, +, 0]$ is a monoid and is not a group, since, for example, 1 does not have an inverse under $+$ in N.

EXAMPLE 5 Let $P(S)$ be the power set for a nonempty set S. Then $[P(S), \cup, \varnothing]$ and $[P(S), \cap, S]$ are commutative monoids, since $\cup$ and $\cap$ are commutative and associative binary operations on $P(S)$ with $\varnothing$ the identity under $\cup$ and S the identity under $\cap$. Neither of these monoids is a group, since only $\varnothing$ has an inverse under $\cup$ and only S has an inverse under $\cap$.

Problems

1. (a) Is $(\mathbf{Z}^+, +)$ a semigroup? Is it a monoid? Is it a group?
(b) Do part (a) with $(\mathbf{Z}^+, +)$ replaced by $(\mathbf{Z}^+, \cdot)$.

2. Let $\mathbf{Q}^+$ be the set of positive rational numbers.
(a) Is $(\mathbf{Q}^+, +)$ a semigroup? Is it a monoid? Is it a group?
(b) Do part (a) with $(\mathbf{Q}^+, +)$ replaced by $(\mathbf{Q}^+, \cdot)$.

3. Let $\mathbf{R}$ be the real numbers and $-$ be the binary operation of subtraction. Is $(\mathbf{R}, -)$ a semigroup? Is it a monoid? Is it a group?

4. Let $\mathbf{R}^+$ be the positive real numbers. Is $(\mathbf{R}^+, \div)$ a semigroup? Is it a monoid? Is it a group?

5. (a) Give an example of a semigroup that is not a monoid.
(b) Does there exist a monoid that is not a semigroup? Explain.

6. (a) Give an example of a monoid that is not a group.
(b) Does there exist a group that is not a monoid? Explain.

7. Let $*$ be the multiplication operation of $\mathbf{Z}_{100}$. List the elements of the principal subsemigroup $\langle \bar{2} \rangle$ in $(\mathbf{Z}_{100}, *)$.

8. Do as in Problem 7, with $\langle \bar{2} \rangle$ replaced by each of the following. (a) $\langle \bar{8} \rangle$; (b) $\langle \bar{5} \rangle$.

9. Is $(\mathbf{Z}^+, \text{gcd})$ a semigroup? Explain. If so, does it have an identity?

10. (i) Is $(\mathbf{Z}^+, \text{lcm})$ a semigroup?
(ii) Is $[\mathbf{Z}^+, \text{lcm}, 1]$ a monoid? If so, is it a group? Explain.

11. Let $X = \{1, 2, 3, 5, 6, 10, 15, 30\}$.
(i) Is $\text{lcm}[x, y]$ in X for all x and y in X?
(ii) Is lcm a binary operation on X?
(iii) Is (X, lcm) a semigroup? If so, does it have an identity?

12. Repeat Problem 11 with lcm replaced by gcd.

13. Repeat Problem 11 with X now $\{2, 3, 5, 6, 10, 15, 30\}$.

14. Repeat Problem 11 with lcm replaced by gcd and X now $\{1, 2, 3, 5, 6, 10, 15\}$.

15. Since $(\mathbf{Z}^+, <)$ is a linearly ordered set, min is a binary operation on $\mathbf{Z}^+$.
   (i) Is $(\mathbf{Z}^+, \text{min})$ a semigroup?
   (ii) Is $[\mathbf{Z}^+, \text{min}, 1]$ a monoid?
   (iii) Is $[\mathbf{Z}^+, \text{min}, 1]$ a group? Explain.

16. Is $(\mathbf{Z}^+, \text{max})$ a semigroup? If so, does it have an identity?

17. Is $(\mathbf{Z}, \text{max})$ a semigroup? If so, does it have an identity? Explain.

18. Is $(\mathbf{Z}, \text{min})$ a semigroup? If so, does it have an identity?

19. Let X be a nonempty finite subset of $\mathbf{Z}$. Which element of X is the identity $\mathbf{e}$ of the monoid $[X, \text{min}, \mathbf{e}]$?

20. Repeat Problem 19 with min replaced by max.

★ 21. Let $(X, \square)$ be a semigroup in which there is an element g such that $x \square g = x$ for all x in X. Also, for each x in X let there be an element y_x in X such that $x \square y_x = g$. Prove the following.
   (i) If h is in X and $h \square h = h$, then $h = g$.
   (ii) $y_x \square x = g$ for each x in X.
   (iii) $g \square x = x$ for each x in X.
   [*Note:* These parts show that a semigroup $(X, \square)$ is a group if it has a right identity g and a right inverse y_x with respect to g for every x in X.]

3.12 Boolean Algebra

The structures studied in this section were introduced by the British mathematician George Boole (1815–1864) in his effort to provide an algebraic basis for logic. These structures have applications in situations featuring dichotomy. In logic the dichotomy is that of "true" or "false." In electrical circuits it is that of switches being in the "on" or "off" position and of the current "flowing" or "not flowing." In set theory it is that of an element being "in" or "out" of a set.

We start by considering an algebraic structure $(S, \square, \triangle)$ in which $\square$ and $\triangle$ are binary operations on the carrier set S.

DEFINITION 1

Distributivity

If $x \square (y \triangle z) = (x \square y) \triangle (x \square z)$ for all x, y, z in S, then $\square$ is **left distributive** over $\triangle$. When each of $\square$ and $\triangle$ is commutative, the word left can be omitted.

EXAMPLE 1 For the structure $(\mathbf{Z}, +, \cdot)$, we know that multiplication is distributive over addition; that is, $x(y + z) = xy + xz$ for all x, y, z in $\mathbf{Z}$. However, addition is not distributive over multiplication, since, for example, $1 + (2 \cdot 3)$ does not equal $(1 + 2) \cdot (1 + 3)$.

EXAMPLE 2 Let P be the power set for a general universe S. We saw in Example 5 of Section 3.11 that $[P, \cup, \varnothing]$ and $[P, \cap, S]$ are commutative monoids. The unary operation $^{-}$ of forming the complement is related to the identities of these monoids by the properties that

$$A \cup \bar{A} = S \quad \text{and} \quad A \cap \bar{A} = \varnothing \quad \text{for all } A \text{ in } P.$$

In the structure $(P, \cup, \cap)$, each of the binary operations $\cup$ and $\cap$ is distributive over the other; that is,

$$A \cup (B \cap C) = (A \cup B) \cap (A \cup C),$$

$$A \cap (B \cup C) = (A \cap B) \cup (A \cap C),$$

for all A, B, C in P.

EXAMPLE 3 Let $Y = \{1, 2, 3, 5, 6, 10, 15, 30\}$ and $y' = 30/y$ for each y in Y. Then the algebraic structure

$$[Y, \text{lcm}, \text{gcd}, ', 1, 30]$$

has the following properties:

(1) Each of $[Y, \text{lcm}, 1]$ and $[Y, \text{gcd}, 30]$ is a commutative monoid.
(2) Each of lcm and gcd is distributive over the other.
(3) The identities 1 and 30 for the monoids of (1) are distinct.
(4) $\text{lcm}[y, y'] = 30$ and $\text{gcd}(y, y') = 1$ for all y in Y.

Examples 2 and 3 state that the algebraic structures

$$[P(S), \cup, \cap, ^{-}, \varnothing, S],$$

$$[Y, \text{lcm}, \text{gcd}, ', 1, 30]$$

share certain properties. The following definition enables us to study simultaneously all structures having these properties.

DEFINITION 2

Boolean Algebra, Join, Meet, Complement

A **boolean algebra** is an ordered 6-tuple $[X, \vee, \wedge, ', O, I]$ in which O and I are special elements of the carrier X, $\vee$ and $\wedge$ are binary operations and $'$ is a unary operation on X, and the following four axioms are satisfied:

(B_1) Each of $[X, \vee, O]$ and $[X, \wedge, I]$ is a commutative monoid.
(B_2) Each of $\vee$ and $\wedge$ is distributive over the other.
(B_3) $O \neq I$.
(B_4) $x \vee x' = I$ and $x \wedge x' = O$ for each x in X.

One calls $x \wedge y$ the **meet** of x and y (or "x wedge y"), $x \vee y$ the **join** of x and y (or "x vee y"), and x' the **complement** (or **negation**) of x.

The statements in Example 2 imply that $[P(S), \cup, \cap, ^{-}, \varnothing, S]$ is a boolean algebra for every nonempty set S. The assertions (1), (2), (3), and (4) in Example 3 tell us that

$$[Y, \text{lcm}, \text{gcd}, ', 1, 30]$$

is a boolean algebra.

DEFINITION 3

Power Set Boolean Algebras

We call $[P(S), \cup, \cap, ^{-}, \varnothing, S]$ the ***P(S)* boolean algebra.**

The power set boolean algebras furnish examples in which the carrier $P(S)$ is infinite (when S is infinite) and in which the carrier has 2^n elements (when S has n elements). In particular, when S has only one element, the carrier consists just of the identities $\varnothing$ and S under $\cup$ and $\cap$, respectively. For this doubleton carrier, we can replace $\varnothing$ by O and S by I, as we do in the following definition.

DEFINITION 4

Doubleton Boolean Algebra

The **doubleton boolean algebra** is the boolean algebra whose carrier is the doubleton $\{O, I\}$.

THEOREM 1 **Complements of O and *I***

In a boolean algebra, $O' = I$ and $I' = O$.

Proof Since O is the identity under $\vee$, $O' = O \vee O'$. Also $O \vee O' = I$ by Axiom (B_4). Hence, $O' = I$. It is left to the reader as Problem 9 below to give the similar proof that $I' = O$. ∎

DEFINITION 5

Dual of a Statement of Boolean Algebras

A theorem applying to all boolean algebras $[X, \vee, \wedge, ', O, I]$ is a true statement involving the elements O and I, letters representing arbitrary elements of the carrier, and the operations $'$, $\vee$, and $\wedge$. The **dual** of such a statement is obtained by replacing each appearance of O, I, $\vee$, and $\wedge$ in the statement by I, O, $\wedge$, and $\vee$, respectively.

For example, the equations $x \vee x' = I$ and $x \wedge x' = O$ in Axiom (B_4) are duals of each other. The equation

$$x \vee (y \wedge z) = (x \vee y) \wedge (x \vee z) \quad \text{for all } x, y, z \text{ in } X,$$

which states that $\vee$ is left distributive over $\wedge$, and the equation

$$x \wedge (y \vee z) = (x \wedge y) \vee (x \wedge z) \quad \text{for all } x, y, z \text{ in } X$$

(which is the condition for $\wedge$ to be left distributive over $\vee$) are duals of one another. The statements "$[X, \vee, O]$ is a commutative monoid" and "$[X, \wedge, I]$ is a commutative monoid" in Axiom (B_1) are duals of each other. The dual of the inequality $O \neq I$ of Axiom (B_3) is $I \neq O$, which is obviously equivalent to Axiom (B_3). Taken together, these facts tell us that the dual of each of the four axioms for boolean algebras is the same axiom, slightly rewritten.

THEOREM 2 Principle of Duality

The dual of any theorem on boolean algebras is also a theorem.

Outline of Proof The dual of each statement in the axioms for boolean algebras is also part of these axioms. Hence one can organize the statements and proofs of all the theorems presented so that a theorem that is not self-dual is followed by its dual, and with the proof of the original result made into the proof of its dual by replacing each step and its justification (that is, reference to an axiom or previous theorem) with the dual step and justification. (This procedure is illustrated in the solutions to Problems 9 through 12 of this section.) ■

Each of the four parts of the following theorem consists of a pair of equations that are duals of one another.

THEOREM 3 Some Properties of Boolean Algebras

Let $B = [X, \vee, \wedge, ', O, I]$ be a boolean algebra. Then:

(a) $x \vee x = x$ and $x \wedge x = x$ for all x in X. (These are the **idempotent laws** for $\vee$ and $\wedge$, respectively.)
(b) $x \vee I = I$ and $x \wedge O = O$ for all x in X.
(c) $x \vee (x \wedge y) = x$ and $x \wedge (x \vee y) = x$ for all x in X. (These are the **absorption laws** for boolean algebras.)
(d) $(x \vee y)' = x' \wedge y'$ and $(x \wedge y)' = x' \vee y'$ for all x and y in X. (These are **DeMorgan's Laws.**)

Proof For (a), one sees that

$$x = x \vee O = x \vee (x \wedge x') = (x \vee x) \wedge (x \vee x') = (x \vee x) \wedge I = x \vee x,$$

using the fact that O is the identity under $\vee$, Axiom (B_4), distributivity of $\vee$ over $\wedge$, Axiom (B_4), and the fact that I is the identity under $\wedge$. Hence, $x = x \vee x$. Then $x = x \wedge x$ follows by duality (that is, by Theorem 2).

For (b), let x be any element of X. One sees that

$$I = x \vee x' = (x \vee x) \vee x' = x \vee (x \vee x') = x \vee I,$$

using Axiom (B_4), part (a) above, associativity of $\vee$, and Axiom (B_4). Hence $I = x \vee I$. Using duality (Theorem 2), we also get $O = x \wedge O$.

For (c), let x and y be any elements of X. We have

$$x \vee (x \wedge y) = (x \wedge I) \vee (x \wedge y) = x \wedge (I \vee y) = x \wedge I = x,$$

using the fact that I is the identity under $\wedge$, distributivity of $\wedge$ over $\vee$, part (b), and the fact that I is the identity under $\wedge$. Hence $x \vee (x \wedge y) = x$. By duality (Theorem 2), we also have $x \wedge (x \vee y) = x$.

Part (d) is left to the reader as Problems 23 and 24 below. ■

DEFINITION 6

Boolean Homomorphisms and Isomorphisms

Let $A = [X, \vee, \wedge, ', O, I]$ and $B = [Y, \Box, \Delta, \bar{\ }, a, b]$ be boolean algebras. A **boolean homomorphism** from A to B is a mapping f from X to Y such that $f(s') = \overline{f(s)}$, $f(s \vee t) = f(s) \Box f(t)$, and $f(s \wedge t) = f(s) \Delta f(t)$ for all s and t in X. If such an f is a bijection from X onto Y, then f is a **boolean isomorphism** from A onto B.

Thus a boolean homomorphism from A to B is a mapping from the carrier of A to the carrier of B that preserves the operations involved in the definition of a boolean algebra.

EXAMPLE 4 **A Boolean Isomorphism**

Let $S = \{a, b, c\}$ and $A = [P(S), \cup, \cap, \bar{\ }, \varnothing, S]$ be the boolean algebra of Example 2 for this particular set S. Let $Y = \{1, 2, 3, 5, 6, 10, 15, 30\}$ and $B = [Y, \text{lcm}, \text{gcd}, ', 1, 30]$ be the boolean algebra of Example 3. Then the mapping f from $P(S)$ to X given by the accompanying table

T	$\varnothing$	$\{a\}$	$\{b\}$	$\{c\}$	$\{a, b\}$	$\{a, c\}$	$\{b, c\}$	S
$f(T)$	1	2	3	5	6	10	15	30

is a boolean isomorphism from A onto B, since f is a bijection such that $f(\bar{T}) = [f(T)]'$ for all members T in $P(S)$ and

$$f(T_1 \cup T_2) = \text{lcm}[f(T_1), f(T_2)] \quad \text{and} \quad f(T_1 \cap T_2) = \text{gcd}(f(T_1), f(T_2))$$

for all T_1 and T_2 in $P(S)$.

DEFINITION 7 **Boolean Cartesian Product Operations**

Let B_i be a boolean algebra $(Y_i, \vee, \wedge, ', O, I)$ for $i = 1, 2, \ldots, n$. Then the operations $\vee$, $\wedge$, and $'$ are defined on the cartesian product $Y = Y_1 \times Y_2 \times \cdots \times Y_n$ as follows:

$$(a_1, a_2, \ldots, a_n) \vee (b_1, b_2, \ldots, b_n) = (a_1 \vee b_1, a_2 \vee b_2, \ldots, a_n \vee b_n)$$

$$(a_1, a_2, \ldots, a_n) \wedge (b_1, b_2, \ldots, b_n) = (a_1 \wedge b_1, a_2 \wedge b_2, \ldots, a_n \wedge b_n)$$

$$(a_1, a_2, \ldots, a_n)' = (a_1', a_2', \ldots, a_n').$$

THEOREM 4 **Boolean Direct Product**

Under the operations of Definition 7, the cartesian product of the carriers of n boolean algebras is also a boolean algebra. (We call it the **boolean direct product.**)

The proof is straightforward, and the details are left to the reader.

Problems

1. Let $B = [X, \vee, \wedge, ', O, I]$ be a boolean algebra. Justify each of the following.
   (i) $O \vee O = O$ and $O \vee I = I$.
   (ii) $I \wedge O = O$ and $I \wedge I = I$.
   (iii) $I \vee I = I$ and $O \wedge O = O$.
   (iv) $O' = I$ and $I' = O$.
2. Use the equations in Problem 1 and commutativity of $\vee$ and $\wedge$ to tabulate the operations $'$, $\vee$, and $\wedge$ for the doubleton boolean algebra (whose carrier is $\{O, I\}$).
3. Tabulate the operations $'$, lcm, and gcd for the boolean algebra $A = [X, \text{lcm}, \text{gcd}, ', 1, 15]$ in which $X = \{1, 3, 5, 15\}$ and $x' = 15/x$ for each x in X.
4. Tabulate the operations $\bar{\ }$, $\cup$, and $\cap$ for the boolean algebra $B = [P(S), \cup, \cap, \bar{\ }, \varnothing, S]$ in which $S = \{a, b\}$.
5. Given that f is a boolean isomorphism from the A of Problem 3 onto the B of Problem 4 and that $f(3) = \{a\}$, find $f(5)$, $f(1)$, and $f(15)$.
6. Repeat Problem 5 with $f(3) = \{a\}$ replaced by $f(3) = \{b\}$.
7. Tabulate the unary operation $'$ and the binary operation lcm for the boolean algebra $[Y, \text{lcm}, \text{gcd}, ', 1, 30]$ of Example 4 with $Y = \{1, 2, 3, 5, 6, 10, 15, 30\}$.
8. Tabulate the operation gcd for the boolean algebra of Problem 7.

In Problems 9–24, 31, and 32 below, B denotes an arbitrary boolean algebra $[X, \vee, \wedge, ', O, I]$ *with X as its carrier.*

9. Prove that $I' = O$. [This part of Theorem 1, which was left to the reader, follows from Theorem 2 (Principle of Duality).] Instead of using Theorem 2, dualize each step of the proof that $O' = I$ in Theorem 1.
10. Prove that $x \wedge x = x$ for all x in X by giving the dual of each step in the proof that $x \vee x = x$ in Theorem 3(a). Do not use the Principle of Duality.

11. Prove that $x \wedge O = O$ for all x in X by giving the dual of each step in the proof that $x \vee I = I$ in Theorem 3(b). Do not use the Principle of Duality.

12. Prove that $x \wedge (x \vee y) = x$ for all x and y in X by giving the dual of each step in the proof that $x \vee (x \wedge y) = x$ in Theorem 3(c). Do not use the Principle of Duality.

13. Let S be a nonempty subset of the carrier X and let S be closed under $'$, $\vee$, and $\wedge$. Prove that O is in S.

14. Prove that I is in the S of Problem 13.

15. Given that a and b are elements in X with $a \vee b = I$ and $a \wedge b = O$, prove that $b = a'$ by justifying the following steps:

$$\begin{aligned} b &= b \vee O = b \vee (a \wedge a') = (b \vee a) \wedge (b \vee a') \\ &= (a \vee b) \wedge (b \vee a') = I \wedge (b \vee a') = (a \vee a') \wedge (b \vee a') \\ &= (a \wedge b) \vee a' = O \vee a' = a'. \end{aligned}$$

16. Prove that $x'' = x$ for all x in X. [*Hint:* Use Problem 15. Also note that $x'' = (x')'$.]

17. Prove that $x \neq x'$ for each x in X by assuming that $a = a'$ for some a in X and justifying the following steps. [They show that this assumption implies $O = I$, which contradicts Axiom (B_3).]

$$O = a \wedge a' = a \wedge a = a = a \vee a = a \vee a' = I.$$

18. Prove that X cannot be a set $\{O, a, I\}$ with three elements. [*Hint:* Each choice of an element of this set to be a' leads to a contradiction. For example, if $a' = O$, then $a = a'' = O' = I$, using Problem 16.]

19. Prove that the mapping f with $x \mapsto x'$ is a bijection from X onto itself.

20. When X is finite, explain why it must have an even number of elements. [*Hint:* The doubleton subsets $\{x, x'\}$ form a partition of X.]

21. Prove that $(x \vee y) \wedge (x' \wedge y') = O$ for all x and y in X by justifying the following steps:

$$\begin{aligned} (x \vee y) \wedge (x' \wedge y') &= [x \wedge (x' \wedge y')] \vee [y \wedge (x' \wedge y')] \\ &= [(x \wedge x') \wedge y'] \vee [x' \wedge (y \wedge y')] \\ &= (O \wedge y') \vee (x' \wedge O) \\ &= O \vee O = O. \end{aligned}$$

22. Prove that $(x \vee y) \vee (x' \wedge y') = I$ for all x and y in X.

23. Explain why $(x \vee y)' = x' \wedge y'$. [This is part of the proof of Theorem 3(d).]

24. Explain why $(x \wedge y)' = x' \vee y'$. [This equation and the equation in Problem 23 are known as ***DeMorgan's Laws.***]

25. Let f be a boolean homomorphism from $[X, \vee, \wedge, ', O, I]$ to $[Y, \square, \triangle, ^{-}, a, b]$. Prove that $f(O) = a$.

26. For the f of Problem 25, prove that $f(I) = b$.

27. Let $S = \{s_1, s_2, s_3\}$. Using binary numerals for the members of $P(S)$, complete the following table for an isomorphism from the power set boolean algebra $[P(S), \cup, \cap, ^{-}, \varnothing, S]$ onto the boolean algebra $[X, \text{lcm}, \text{gcd}, ', 1, 30]$ with $X = \{1, 2, 3, 5, 6, 10, 15, 30\}$ and $x' = 30/x$.

x	000	001	010	011	100	101	110	111
$f(x)$		2	3		5			

28. (a) Do Problem 27 with "isomorphism" replaced by "homomorphism" and the table replaced with the following one.

x	000	001	010	100	011	101	110	111
$f(x)$		1	3	5				

(b) Repeat Problem 27 with X now $\{1, 3, 5, 7, 15, 21, 35, 105\}$ and with $f(001) = 3$, $f(010) = 5$, and $f(100) = 7$.

29. Let D be the doubleton boolean algebra whose carrier is $\{O, I\}$ and let D^n be the direct product of n copies of D. Let $S = \{s_1, s_2, \ldots, s_n\}$ be a set with n elements. Describe an isomorphism f from the power set boolean algebra

$$[P(S), \cup, \cap, ^{-}, \varnothing, S]$$

onto D^n.

30. Tabulate the f of Problem 29 in the special case of $n = 3$.

31. Let the relation $\leq$ on the carrier X of B be defined so that $x \leq y$ if and only if $x \vee y = y$. Prove that $\leq$ is a partial ordering of X.

32. Let the relation $\leq$ on the carrier X of B be such that $x \leq y$ if and only if $x \wedge y = x$. Prove that $\leq$ is a partial ordering of X.

33. What are the least and greatest elements for the poset $(X, \leq)$ of Problem 31?

34. What are the least and greatest elements for the poset of Problem 32?

35. What is the meaning of the relation $\leq$ of Problem 31 in the special case of the boolean algebra $[P(S), \cup, \cap, ^{-}, \varnothing, S]$ of Example 2?

36. Do Problem 35 with the $\leq$ of Problem 31 replaced by the $\leq$ of Problem 32.

37. Verify that $\min(x, \max(y, z)) = \max(\min(x, y), \min(x, z))$ for elements x, y, z of a linearly ordered set X for each of the following cases. [These are all of the cases with $y \leq z$.]
(i) $x \leq y \leq z$; (ii) $y < x \leq z$; (iii) $y \leq z < x$.

38. Let X be a linearly ordered set. Explain how the previous problem helps show that the binary operation min on X is distributive over max. [*Hint:* Write the inequalities on x, y, z for the other possible cases, and use commutativity for min and max to relate these cases to those in Problem 37.]

39. Prove that $\max(x, \min(y, z)) = \min(\max(x, y), \max(x, z))$ for all elements x, y, z of a linearly ordered set X.

★ **40.** Show that the poset $(X, <)$ with $X = \{1, 2, 4, 8\}$ cannot be made into a boolean algebra whose binary operations $\vee$ and $\wedge$ are given by max and min, respectively. [*Hint:* Show that every choice of the unary operation $'$ leads to a contradiction.]

★ **41.** In a boolean algebra, let $a \vee b = a \vee c$ and $a \wedge b = a \wedge c$. Prove that $b = c$.

★ **42.** Show that in Definition 6 the condition "$f(s') = \overline{f(s)}$" can be replaced with "$f(O) = a$ and $f(I) = b$" without changing the meaning of homomorphism.

3.13 Boolean Functions and Their Applications

In this section, D stands for the doubleton boolean algebra whose carrier is $\{0, 1\}$.

George Boole
1815–1864

Boole was born in Lincoln, England, the son of a shopkeeper. But though born in England he was claimed by the Irish and became professor of mathematics at Queen's College, Cork, Ireland, in 1849, after spending some years teaching in preparatory schools. He was largely self-educated, which may help explain his strikingly original work in mathematics. Five years after his appointment to Queen's College, he published the work for which he is best known. The full title of the work is *An Investigation of the Laws of Thought, on which are founded the Mathematical Theories of Logic and Probabilities*, usually referred to succinctly as *The Laws of Thought.* Boole's ambition in writing this book is clear from some lines from the text: "The design of the following treatise is to investigate the fundamental laws of those operations of the mind by which reasoning is performed; to give expression to them in the symbolical language of a Calculus, and upon this foundation to establish the science of Logic and construct its method; ... and, finally, to collect from the various elements of truth brought to view in the course of these inquiries some probable intimations concerning the nature and constitution of the human mind." That's no mean task to set for oneself.

Of course the algebra that is developed here is the algebra of sets and logic and is now referred to as boolean algebra. It is the basis for the whole discipline called *lattice theory* and has played an important role in the design of computers. Boolean algebra may not tell us all that much about how the human mind works — the human mind is much too complex to be reduced to these simple operations — but it does tell us something about how a computer works, since such a "mind" is fast but basically rather simple. Boole lived to write on differential equations and finite differences, but nothing in these efforts matched his work that led to boolean algebra. He died relatively young.

DEFINITION 1 **Boolean Function**

An ***n*-ary boolean function** is an n-ary operation on the doubleton boolean algebra D that can be expressed in terms of the operations $\vee$, $\wedge$, and $'$.

EXAMPLE 1 $\alpha(x, y, z) = x \wedge (y \vee z)$, $\beta(x, y, z) = (x \wedge y) \vee (x \wedge z)$, and $\gamma(x, y, z) = x \wedge [(x' \vee y) \vee z]$ are ternary boolean functions.

We will see shortly that a boolean function may represent such diverse things as an electrical circuit or an insurance policy. Thus the ability to prove that two boolean functions are equal may help one to design a simpler (and thus more reliable and less expensive) circuit that is equivalent to a complicated circuit, or it may help one to rewrite a legal document in more understandable form.

When a boolean function is applied to an electrical circuit, each of the variables x, y, and so on stands for one or more switches; and the possible values 0 and 1 for a variable denote the open and closed positions, respectively, for its switches. Also, $\theta \vee \phi$ indicates that the θ and ϕ parts of the circuit are in parallel; $\theta \wedge \phi$ indicates that these parts are in series; and θ' denotes a part of the circuit that is open (in other words, a part through which current cannot flow) when the θ part is closed and that is closed when θ is open. (Here θ or ϕ may represent a single switch or a portion of the circuit.)

The circuits depicted in Figures 3.8a, 3.8b, and 3.8c are associated with the boolean functions α, β, and γ of Example 1. If one of the switches labeled with x in Figure 3.8b is open, the other one labeled with x must also be open. In Figure 3.8c, the x' switch must be open when the x switch is closed, and it must be closed when the x switch is open. Looking at any one of the figures, one can easily see that current will flow only when the x switch is closed (or in Figure 3.8b, when the x switches are closed) and at least one of the y and z switches is closed. That is, current will flow only when the positions of the switches are given by $(x, y, z) = (1, 0, 1)$, $(1, 1, 0)$, or $(1, 1, 1)$. Clearly, the three circuits are equivalent, and this tells us that α, β, and γ are equal to each other as boolean functions. The circuit of Figure 3.8a seems to be the simplest of these equivalent circuits.

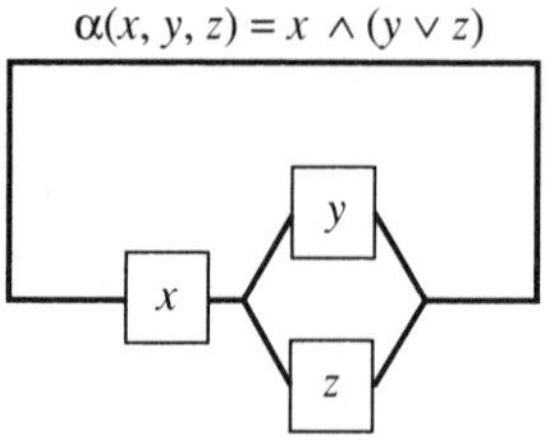

FIGURE 3.8a

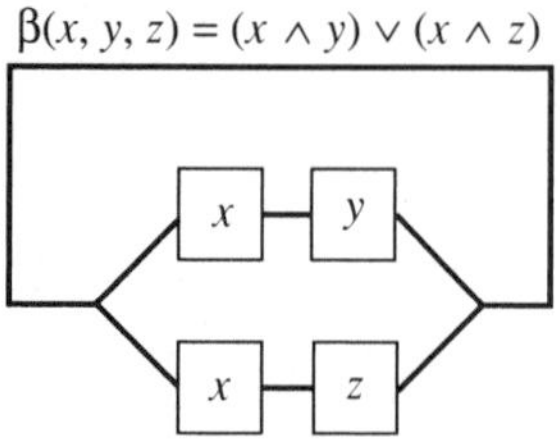

FIGURE 3.8b

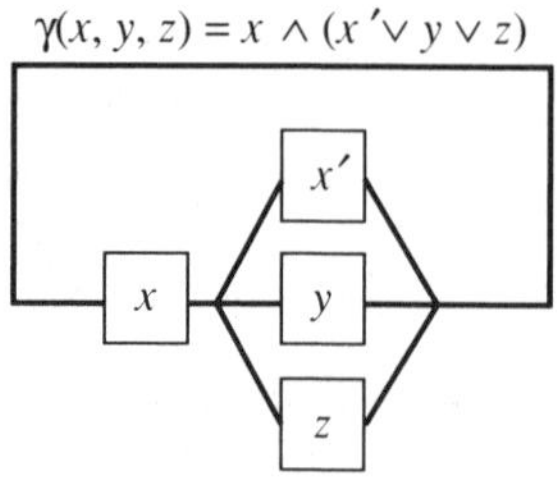

FIGURE 3.8c

When a circuit is complicated, it may be useful to analyze it by tabulating its boolean function. The tables for the operations $\vee$, $\wedge$, and $'$ on D are as shown.

$\vee$	0	1
0	0	1
1	1	1

$\wedge$	0	1
0	0	0
1	0	1

x	0	1
x'	1	0

Using these tables, one can tabulate the boolean function γ of Example 1 (and Figure 3.8c), as given in Table 3.2.

Similarly one could tabulate the boolean functions α and β of Example 1 and Figures 3.8a and 3.8b. This would again establish that these functions are

TABLE 3.2

(x, y, z)	x'	$x' \vee y$	$(x' \vee y) \vee z$	$\gamma = x \wedge [(x' \vee y) \vee z]$
(0, 0, 0)	1	1	1	0
(0, 0, 1)	1	1	1	0
(0, 1, 0)	1	1	1	0
(0, 1, 1)	1	1	1	0
(1, 0, 0)	0	0	0	0
(1, 0, 1)	0	0	1	1
(1, 1, 0)	0	1	1	1
(1, 1, 1)	0	1	1	1

all equal; that is, $\alpha(x, y, z) = \beta(x, y, z) = \gamma(x, y, z)$ for each of the eight possible choices of the ordered triple (x, y, z).

Boolean functions may also be helpful in formal logic or in the analysis of compound statements in important documents. We now introduce meanings for $\vee$, $\wedge$, and $'$ in such applications.

A statement that is either true or false (but not both) will be called a ***proposition*** or an ***assertion***. One manufactures compound propositions from simpler ones in the following way.

If p and q are propositions, let p' denote the assertion that p is false, let $p \vee q$ denote the assertion that at least one of p and q is true, and let $p \wedge q$ denote the assertion that both p and q are true.

Note that $\vee$ has the meaning of "and/or"; it is called the ***logical or*** and also is called the ***inclusive or***. One may read $\wedge$ as "and." Also, p' is called the ***negation*** of p. Here O denotes a false proposition such as "$3 + 4 = 5$," and I denotes some fixed true proposition.

EXAMPLE 2 Let p stand for the assertion that $3 + 4 = 5$ and q represent the assertion that $3^2 + 4^2 = 5^2$. Then $p \vee q$ is the true assertion that at least one of p and q is true, and $p \wedge q$ is the false assertion that p and q are both true. Also p' and q' are the assertions that p and q, respectively, are false. One can translate p' as "3 + 4 is not equal to 5." Since p is false, p' is true; since q is true, q' is false.

EXAMPLE 3 Valid checks for the account of a certain organization must be signed by the president and also by either the secretary or the treasurer of the organization. Thus a check is valid only if $\alpha(p, s, t) = 1$, where $\alpha(p, s, t) = p \wedge (s \vee t)$ and p, s, or t has the value 1 when the check is signed by the president, secretary, or treasurer, respectively, and has the value 0 if the respective signature is not present.

EXAMPLE 4 For a fixed employee of some university, let g and n denote the employee's age and years of service at the institution. The employee is eligible for retirement benefits if and only if at least one of the following conditions is satisfied:

(i) $g + n \geq 75$;

(ii) $g \geq 65$ and $n \geq 5$.

Letting x be the assertion that $g + n \geq 75$, y be the assertion that $g \geq 65$, z be the assertion that $n \geq 5$, and $\alpha(x, y, z) = x \vee (y \wedge z)$, one sees that eligibility occurs if and only if $\alpha(x, y, z) = 1$. We note that when $(g, n) = (62, 11)$ one has $(x, y, z) = (0, 0, 1)$ and $\alpha(0, 0, 1) = 0 \vee (0 \wedge 1) = 0 \vee 0 = 0$; hence the employee is not eligible. However, when $(g, n) = (66, 6)$, one has $(x, y, z) = (0, 1, 1)$ and $\alpha(0, 1, 1) = 0 \vee (1 \wedge 1) = 0 \vee 1 = 1$; in this case the employee is eligible. The employee is also eligible when $(g, n) = (50, 25)$, since then $(x, y, z) = (1, 0, 1)$ and $\alpha(1, 0, 1) = 1 \vee (0 \wedge 1) = 1 \vee 0 = 1$.

Next we introduce a symbol that is useful in applying boolean functions to formal logic and to the statements and proofs of theorems.

DEFINITION 2

Implication

If P and Q are assertions (or are boolean functions), $P \Rightarrow Q$ is an alternate symbolism for the assertion (or boolean function) $P' \vee Q$. One calls $\Rightarrow$ the **implication sign**. An assertion $P \Rightarrow Q$ is called an **implication**, with P as the **hypothesis** or **premise** and Q as the **conclusion**; $P \Rightarrow Q$ is read in any of the following ways:

(i) If P, then Q.
(ii) P only if Q.
(iii) P implies Q.
(iv) P is a sufficient condition for Q.
(v) Q is a necessary condition for P.

EXAMPLE 5 Let a, b, and c be elements of a group. Let P, Q, and R be the assertions that $ab = ac$, that $ba = ca$, and that $b = c$, respectively. Then the Cancellation Theorem of Section 2.3 can be expressed in the form $(P \vee Q) \Rightarrow R$. By definition of implication, this has the same meaning as $(P \vee Q)' \vee R$, which in turn can be rewritten as $(P' \wedge Q') \vee R$ by using one of DeMorgan's Laws. (See Problem 23 of Section 3.12.)

DEFINITION 3

Equivalence

For assertions (or boolean functions) P and Q, $P \Leftrightarrow Q$ denotes the assertion (or boolean function) $[P \Rightarrow Q] \wedge [Q \Rightarrow P]$. The **equivalence** $P \Leftrightarrow Q$ is read in any of the following ways:

(i) P is equivalent to Q.
(ii) P if and only if Q.
(iii) P is a necessary and sufficient condition for Q.

EXAMPLE 6 For elements a and b of a group, let P be the assertion that $(ab)^{-1} = a^{-1}b^{-1}$ and Q be the assertion that $ab = ba$. Problem 5 of Section 2.3 asks the reader to "show that $(ab)^{-1} = a^{-1}b^{-1}$ if and only if $ab = ba$." This can be rewritten as "show that $P \Leftrightarrow Q$."

DEFINITION 4

Converse, Contrapositive

Let P and Q be assertions. The **converse** of the implication $P \Rightarrow Q$ is the implication $Q \Rightarrow P$. The **contrapositive** of $P \Rightarrow Q$ is $Q' \Rightarrow P'$.

Let $P \Rightarrow Q$ be a true implication (that is, a theorem). Then its converse may be true, as in Example 6 above, or its converse may be false, as in Example 7 below. Theorem 1 below shows that $P \Rightarrow Q$ is true if and only if its contrapositive $Q' \Rightarrow P'$ is true.

EXAMPLE 7 For a finite group G of order r, let P be the assertion that "d is the order of a subgroup in G" and Q be the assertion that "d is a positive integral divisor of r." Then Lagrange's Theorem in Section 2.11 tells us that $P \Rightarrow Q$; that is, "if d is the order of a subgroup in G, then d is a positive integral divisor of r." The converse $Q \Rightarrow P$ of $P \Rightarrow Q$ is the false statement that "if d is a positive integral divisor of r, then d is the order of a subgroup in G"; see Problem 7 of Section 2.12 for a counterexample demonstrating that this particular converse is false. The contrapositive $Q' \Rightarrow P'$ of $P \Rightarrow Q$ is the true statement that "if d is not a positive integral divisor of r, then d is not the order of a subgroup in G." Thus it follows from the contrapositive that a subset with 5 elements is not a subgroup in the octic group $\mathbf{D}_4$.

DEFINITION 5

Tautology, Contingency, Contradiction

An n-ary boolean function α is called a **tautology** if $\alpha(x_1, x_2, \ldots, x_n) = 1$ for each of the 2^n choices of $(x_1, x_2, \ldots, x_n)$. Also, α is a **contradiction** if $\alpha(x_1, x_2, \ldots, x_n) = 0$ for all choices of the x_i; and α is a **contingency** if it has 1 as a value for some choice of the x_i and has 0 as a value for another choice of the x_i.

We note that a given boolean function is a tautology or a contingency or a contradiction, but cannot be two of these.

EXAMPLE 8 Let $\alpha(x, y) = (x \Rightarrow y) \Leftrightarrow (x' \vee y)$, $\beta(x, y) = x \wedge y$, and $\gamma(x) = x \wedge x'$. Using the definitions of implication and equivalence, one sees that α is a tautology. Since $\beta(1, 1) = 1$ and $\beta(1, 0) = 0$, β is a contingency. Since $\gamma(x) = 0$ for $x = 1$ and for $x = 0$, γ is a contradiction. Also, $(x \wedge x') \wedge y$ is a contradiction.

THEOREM 1 Contrapositive

(i) $(x \Rightarrow y)$ and $(y' \Rightarrow x')$ are equal boolean functions.
(ii) $[(x \Rightarrow y) \Leftrightarrow (y' \Rightarrow x')]$ is a tautology.
The proof is left to the reader as Problems 5 and 10 of this section.

Theorem 1 enables us to prove a theorem $P \Rightarrow Q$ and then to apply its contrapositive $Q' \Rightarrow P'$ when it is convenient to do so; see Example 7 above for a specific illustration. This technique will also be applied in Problems 29 and 30 of Section 5.3.

Problems

1. Suppose that the functioning of a circuit with three switches is described by the boolean function $x \vee (y \wedge y')$. Is there an equivalent circuit with just one switch? Explain.

2. Let the functioning of a circuit with four switches be characterized by $(x \wedge y) \vee (x \wedge z)$. Is there an equivalent circuit with three switches? Explain.

3. Let $\alpha(x, y, z) = x \wedge (y \vee z)$ and $\beta(x, y, z) = (x \wedge y) \vee z$. Are α and β equal boolean functions? Explain.

4. Let $\gamma(x, y, z) = z \vee (y \wedge x)$ and $\beta(x, y, z) = (x \wedge y) \vee z$. Are γ and β equal boolean functions? Explain.

5. Let $\alpha(x, y) = (x \Rightarrow y)$ and $\beta(x, y) = (y' \Rightarrow x')$. Show that α and β are equal boolean functions by tabulating each for the four possible ordered pairs (x, y). [This is the proof of Theorem 1(i).]

6. Let $\alpha(x, y) = (x \Rightarrow y)$ and $\gamma(x, y) = (y \Rightarrow x)$.
 (i) Find an ordered pair (x_1, y_1) such that $\alpha(x_1, y_1) = 1$ and $\gamma(x_1, y_1) = 0$.
 (ii) Find an ordered pair (x_2, y_2) such that $\alpha(x_2, y_2) = 1 = \gamma(x_2, y_2)$.
 (iii) Is $[(x \Rightarrow y) \Rightarrow (y \Rightarrow x)]$ a tautology, a contingency, or a contradiction?

7. Let $\alpha(x, y) = x \Rightarrow y$ and $\beta(x, y) = x' \vee y$.
 (i) Evaluate $\beta(0, 0)$.
 (ii) Use the answer to part (i) and the definition of implication to evaluate $\alpha(0, 0)$.
 (iii) Use the answer to part (ii) to tell whether or not the following assertion is true: "If n^2 is negative for some integer n, then $\sqrt{2}$ is a rational number."

8. For each of the following implications, tell whether it is true or false.
 (a) If $3 = -3$, then $3^2 = -3^2$.
 (b) If $3 = -3$, then $3^2 = (-3)^2$.
 (c) If $3 = 1 + 2$, then $3^2 = -3^2$.
 (d) If $3 = 1 + 2$, then $3^2 = (-3)^2$.

9. Let $\alpha(x, y) = (x \Leftrightarrow y)$ and $\beta(x, y) = [(x \wedge y) \vee (x' \wedge y')]$. Are α and β equal boolean functions? Explain.

10. Prove that boolean functions α and β are equal if and only if $\alpha \Leftrightarrow \beta$ is a tautology (and thus complete the proof of Theorem 1).

11. Let $\alpha(0, 0) = 0 = \alpha(1, 1)$ and $\alpha(1, 0) = 1 = \alpha(0, 1)$. Give a formula for $\alpha(x, y)$ that uses only symbols chosen from the set $\{x, y, ', \Leftrightarrow\}$.

12. Let $\beta(0, 0) = \beta(0, 1) = \beta(1, 0) = 1$ and $\beta(1, 1) = 0$. Give a formula for $\beta(x, y)$ that uses only symbols chosen from the set $\{x, y, ', \Rightarrow\}$.

13. Let $\alpha(p, s, t) = p \wedge (s \vee t)$. Evaluate $\alpha(0, 1, 1)$ and use this to tell whether in Example 3 a check is valid when signed by the secretary and the treasurer but not by the president.

14. Using the α of Problem 13, evaluate $\alpha(1, 0, 0)$ and then translate that result into a statement concerning the validity of a check in Example 3.

15. In Example 4, would an employee with 20 years of service be eligible for retirement benefits at age 50? Explain.

16. In Example 4, what is the lowest age at which an employee would be eligible for retirement with 4 years of service?

17. Let a, b, and p be integers with p prime. Assign assertions involving a, b, and p as meanings of R, S, and T so that Lemma 1 of Section 1.7 is expressible as $R \Rightarrow (S \vee T)$.

18. Assign assertions as meanings for $P, Q_1, Q_2, \ldots, Q_n$ so that Lemma 2 of Section 1.7 is expressible as $P \Rightarrow (Q_1 \vee Q_2 \vee \cdots \vee Q_n)$.

19. For each of the following boolean functions, tell whether it is a tautology, a contingency, or a contradiction, and explain your answer.
(a) $(x \vee y) \Rightarrow (x \wedge y)$.
(b) $[(x \Rightarrow y) \wedge (z \Rightarrow w)] \Rightarrow [(x \wedge z) \Rightarrow (y \wedge w)]$.
(c) $(x \vee y) \wedge (x \vee y') \wedge x'$.
(d) $[(x \Rightarrow y) \wedge (y \Rightarrow z)] \Rightarrow (x \Rightarrow z)$.
(e) $[(x \wedge y) \Rightarrow (z \wedge w)] \Rightarrow (x \Rightarrow z)$.

20. Do as in Problem 19 for each of the following.
(a) $(x \wedge y) \Rightarrow (x \vee y)$.
(b) $(x \vee y) \wedge (x \vee y') \wedge (x' \vee y) \wedge (x' \vee y')$.
(c) $[(x \wedge y) \Rightarrow z] \Rightarrow (x \Rightarrow z)$.
(d) $[x \Rightarrow (y \wedge z)] \Rightarrow (x \Rightarrow y)$.
(e) $[(x \wedge z) \Rightarrow (y \wedge z)] \Rightarrow (x \Rightarrow y)$.

3.14 Composition of Mappings, Groups of Bijections

Here we generalize on composition of permutations by defining the composite of a mapping from S to T and a mapping from T to U, where S, T, and U are any (finite or infinite) sets.

We then generalize on the symmetric groups $\mathbf{S_n}$ by showing that the set of bijections from any (finite or infinite) set X to itself becomes a group $\mathbf{B}(X)$ under composition of mappings. Also, it is noted that the automorphisms of a group G form a subgroup in $\mathbf{B}(G)$.

Let S, T, and U be any sets, not necessarily distinct. The ***composite*** of a mapping α from S to T and a mapping β from T to U is the mapping $\gamma = \alpha\beta$ with

$$s \overset{\gamma}{\mapsto} u \quad \text{whenever } s \overset{\alpha}{\mapsto} t \overset{\beta}{\mapsto} u.$$

The composite $\beta\alpha$ in the other order does not have meaning unless $S = U$, and it need not equal $\alpha\beta$ even when it has meaning.

In Theorem 1, we shall use the special case of the following lemmas in which all the sets are the same.

LEMMA 1 Surjections, Injections, Bijections

Let α and β be mappings from S to T and from T to U, respectively. Then:
(a) If α and β are surjections, so is $\alpha\beta$.
(b) If α and β are injections, so is $\alpha\beta$.
(c) If α and β are bijections, so is $\alpha\beta$.

Proof For (a), let u be any element of U. If β and α are surjections, there is an element t in T with $\beta(t) = u$, and there is an element s in S with $\alpha(s) = t$. Now the composite $\alpha\beta$ sends s to u; that is, every u of U is in the image set of $\alpha\beta$.

For (b), let s_1 and s_2 be elements of S with $s_1 \neq s_2$. Let $\alpha(s_i) = t_i$ and $\beta(t_i) = u_i$ for $i = 1$ and 2. If α and β are injections, it follows that $t_1 \neq t_2$ and then that $u_1 \neq u_2$. That is, distinct elements s_1 and s_2 of S have distinct images u_1 and u_2 under $\alpha\beta$.

Putting the results of (a) and (b) together, we have (c). ∎

LEMMA 2 Composition Is Associative

Let α, β, and γ be mappings from S to T, T to U, and U to V, respectively. Then $(\alpha\beta)\gamma = \alpha(\beta\gamma)$.

Proof Let s be any element of S and let

$$s \overset{\alpha}{\mapsto} t \overset{\beta}{\mapsto} u \overset{\gamma}{\mapsto} v.$$

Then

$$s \xmapsto{\alpha\beta} u, \qquad t \xmapsto{\beta\gamma} v, \qquad s \xmapsto{(\alpha\beta)\gamma} v, \qquad s \xmapsto{\alpha(\beta\gamma)} v.$$

That is, every s of S has the same image under both $(\alpha\beta)\gamma$ and $\alpha(\beta\gamma)$. ∎

THEOREM 1 Group B(X) of Bijections of X

Let X be any (finite or infinite) set and let $\mathbf{B}(X)$ consist of all the bijections from X to itself. Then $\mathbf{B}(X)$ is a group under composition of mappings.

Proof Lemma 1(c) implies that $\mathbf{B}(X)$ is closed under composition, and Lemma 2 tells us that this operation is associative. The identity of $\mathbf{B}(X)$ is the bijection ε with $\varepsilon(x) = x$ for all x in X.

In Section 3.1, we noted that a bijection α from S to S has an inverse mapping α^{-1} that is also a bijection from S to S. Also, $\alpha\alpha^{-1} = \varepsilon = \alpha^{-1}\alpha$. Hence composition of mappings in **B**(X) satisfies the group axioms. ■

THEOREM 2 Subgroup of Automorphisms A(G)

Let **A**(G) consist of all the automorphisms of a group G. Then **A**(G) is a subgroup in the group of bijections **B**(G).

The proof is left to the reader as Problem 3 of this section.

The following result states that the structure of **B**(X) is determined by the number of elements in the set X.

THEOREM 3 Condition for B(X) and B(Y) to Be Isomorphic

Let there exist a bijection β from X to Y. Then for every bijection θ from X to itself, the mapping $\beta^{-1}\theta\beta$ is a bijection from Y to itself. Also, the mapping f with $f(\theta) = \beta^{-1}\theta\beta$ is a group isomorphism from **B**(X) to **B**(Y).

The proof is left to the reader as Problem 7 of this section.

THEOREM 4 Generalized Cayley Theorem

Let G be a (finite or infinite) group. Then the group **B**(G) of bijections of G has a subgroup M that is isomorphic to G.

Proof For a fixed element a of G, let β_a be the mapping from G to G with $\beta_a(x) = xa$ for all x in G. We noted in Secton 3.1 that β_a is a bijection from G to itself. (See Example 6 and Problem 12 of Section 3.1.)

Hence the mapping θ with $\theta(a) = \beta_a$ is a mapping from G to **B**(G). We next show that θ is an injective homomorphism.

For any x in G,

$$x \overset{\beta_a}{\mapsto} xa \overset{\beta_b}{\mapsto} (xa)b = x(ab).$$

This shows that the $\beta_a\beta_b = \beta_{ab}$. Then

$$\theta(ab) = \beta_{ab} = \beta_a\beta_b = \theta(a)\theta(b).$$

Thus θ is a homomorphism.

Now let $\theta(a) = \theta(b)$; that is, let $\beta_a = \beta_b$. Then for all x in G, we have $\beta_a(x) = \beta_b(x)$, or $xa = xb$. Clearly this implies $a = b$. Hence, θ is injective. Now it follows from parts (g) and (c) of Theorem 2 of Section 3.3 that G is isomorphic to the image set M of θ and that M is a subgroup in **B**(G). ■

If G is a group of finite order n, Theorem 3 tells us that $\mathbf{B}(G)$ is isomorphic to the symmetric group $\mathbf{S_n}$. Then Theorem 4 implies that G is isomorphic to a subgroup in $\mathbf{B}(G)$, and hence to a subgroup in $\mathbf{S_n}$. This shows that Theorems 3 and 4 contain Cayley's Theorem as a special case.

Problems

1. Let G_1, G_2, G_3 be cyclic groups $[a]$, $[b]$, $[c]$ of orders 12, 6, and 2, respectively. Let α be the mapping from G_1 to G_2 with $\alpha(a^s) = b^s$ for $s = 0, 1, \ldots, 11$ and let β be the mapping from G_2 to G_3 with $\beta(b^t) = c^t$ for $t = 0, 1, \ldots, 5$. Tabulate the composite mapping $\gamma = \alpha\beta$ from G_1 to G_3, and tell which of α, β, and γ are homomorphisms.
2. Let G_1, G_2, G_3 be any three groups. Let α and β be homomorphisms from G_1 to G_2 and from G_2 to G_3, respectively. Prove that the composite $\alpha\beta$ is a homomorphism from G_1 to G_3.
3. Prove Theorem 2; that is, prove that the set $\mathbf{A}(G)$ of automorphisms of a group G is a subgroup in the group $\mathbf{B}(G)$ of bijections from G to G.
4. Prove that the set of inner automorphisms of a group G is a subgroup in $\mathbf{A}(G)$ and hence is also a subgroup in $\mathbf{B}(G)$.
5. Prove that $\mathbf{B}(X)$ is nonabelian when X has at least three elements.
6. Give an example of an infinite noncommutative group.
7. Prove Theorem 3; that is, prove that $\mathbf{B}(X)$ and $\mathbf{B}(Y)$ are isomorphic if there is a bijection β from X to Y.
8. Let S be the set consisting of all the real numbers except 0 and 1. Let $\theta_1, \theta_2, \theta_3, \theta_4, \theta_5$, and θ_6 be the mappings from S to S such that
$$\theta_1(x) = x, \quad \theta_2(x) = 1/x, \quad \theta_3(x) = 1 - x,$$
$$\theta_4(x) = 1/(1 - x), \quad \theta_5(x) = (x - 1)/x,$$
$$\theta_6(x) = x/(x - 1)$$
for all x in S. Let T be the set of these six mappings. Prove the following.
   (i) Each of the θ_i, $1 \le i \le 6$, is a bijection from S to S.
   (ii) The set T is closed under the operation of composition of mappings.
   (iii) T is isomorphic to the symmetric group S_3.
9. Let C be the center in a group G.
   (i) Explain why C is normal in G and hence G/C exists.
   (ii) Prove that G/C is isomorphic to the group of inner automorphisms of G.
10. For each element g of a group G, let σ_g be the inner automorphism of G given by $\sigma_g(x) = g^{-1}xg$ for all x in G. Let F be the mapping from G to $\mathbf{A}(G)$ given by $F(g) = \sigma_g$ for all g in G. Show the following.
   (i) F is a homomorphism.
   (ii) The kernel of F is the center C in G.
11. ★ Let $\mathbf{B}(\mathbf{Z})$ be the group of bijections of the set $\mathbf{Z}$ of integers. Let β be the bijection in $\mathbf{B}(\mathbf{Z})$ with $\beta(n) = n + 1$ for all n in $\mathbf{Z}$. Is the cyclic subgroup $[\beta]$ isomorphic to the additive group of $\mathbf{Z}$? Explain.
12. Let X and Y be nonempty sets. Let α and β be mappings from X to Y and from Y to X, respectively, such that $\alpha\beta$ is a bijection from X to itself. Prove that α is an injection and that such a β is a surjection.

13. Let α be an injection from X to Y and let u be a known element of X. Show that there is at least one mapping β from Y to X such that $\alpha\beta$ is the identity mapping in $\mathbf{B}(X)$ and that β must be a surjection.

★ **14.** Let β be a surjection from Y to X. Show that there is at least one (injective) mapping α from X to Y such that $\alpha\beta$ is the identity of $\mathbf{B}(X)$. (This problem is starred because its solution requires something called the "Axiom of Choice.")

★ 3.15 Conjugacy Classes and the Class Equation

Here we use a method of partitioning groups and a counting technique to study the structure of finite noncommutative groups.

If a is an element of a group G, the set $\{g^{-1}ag : g \in G\}$, of all conjugates of a by elements of G, is called the ***conjugacy class*** of a in G and we denote this set by CC(a). Since a is its own conjugate by the identity **e**, it is clear that one always has a in CC(a). It can also be shown that, for any elements a and b in G, CC(a) and CC(b) are either identical or disjoint. These facts imply the following result.

THEOREM 1 Partition by Conjugacy

In each finite group G, there exist elements $g_1, g_2, \ldots, g_r$ such that the conjugacy classes CC(g_i) for $i = 1, 2, \ldots, r$ form a partition of G.

We have outlined a proof above and also refer the reader to Problems 4 and 8(b) of Section 3.7 for another approach.

For any finite set S, let $|S|$ denote the number of elements in S. In particular, $|G| = \text{ord } G$ for any finite group G.

THEOREM 2 Number of Conjugates

Let C_a and CC(a) be the centralizer and conjugacy class, respectively, of a in a finite group G.

(a) $\text{CC}(a)| = |G|/|C_a|$; that is, the number of conjugates of a in G is the index of the centralizer of a in G.

(b) $\text{CC}(a)| = 1$, or $\text{CC}(a) = \{a\}$, if and only if a is in the center C in G.

The proof is left to the reader. [See Problem 21(iii) of Section 3.2.]

THEOREM 3 The Class Equation

Let $g_1, \ldots, g_r$ be elements of a finite group G whose conjugacy classes CC(g_i) form a partition of G. Let n_i be the index of the centralizer of g_i in G. Then

$$|G| = n_1 + n_2 + \cdots + n_r. \tag{1}$$

Proof Since each element of G is in one and only one of the $CC(g_i)$,

$$|G| = |CC(g_1)| + |CC(g_2)| + \cdots + |CC(g_r)|.$$

Using Theorem 2(a), we substitute n_i for $|CC(g_i)|$ and thus obtain (1). ∎

EXAMPLE 1 Let $\mathbf{D}_4$ be the octic group $\{\varepsilon, \rho, \rho^2, \rho^3, \phi, \rho\phi, \rho^2\phi, \rho^3\phi\}$, with $\rho = (1234)$ and $\phi = (24)$. Using the middle table on the inside back cover of the book, one can show that the conjugacy classes in $\mathbf{D}_4$ are as follows:

$$CC(\varepsilon) = \{\varepsilon\}, \quad CC(\rho) = \{\rho, \rho^3\} = CC(\rho^3), \quad CC(\rho^2) = \{\rho^2\},$$

$$CC(\phi) = \{\phi, \rho^2\phi\} = CC(\rho^2\phi), \quad CC(\rho\phi) = \{\rho\phi, \rho^3\phi\} = CC(\rho^3\phi).$$

Thus we have both of the following:

$$\mathbf{D}_4 = CC(\varepsilon) \cup CC(\rho) \cup CC(\rho^2) \cup CC(\phi) \cup CC(\rho\phi),$$

$$8 = |\mathbf{D}_4| = |CC(\varepsilon)| + |CC(\rho)| + |CC(\rho^2)| + |CC(\phi)| + |CC(\rho\phi)|$$

$$= 1 + 2 + 1 + 2 + 2.$$

We illustrate the use of the class equation in the proof of the following result.

THEOREM 4 Groups with Prime Power Order

Let ord $G = p^m$, with p and m positive integers and p prime. Let C be the center in G. Then $p \mid (\text{ord } C)$, and hence ord $C > 1$.

Proof The only conjugate of an element c in the center C is c itself; in other words, the only conjugacy class that contains c is $CC(c) = \{c\}$. Therefore, if $g_1, \ldots, g_r$ are elements of G such that the $CC(g_i)$ partition G, each element of C must be one of these g_i. Let ord $C = s$; then we may assume that $g_1, \ldots, g_s$ are elements of C and that $g_{s+1}, g_{s+2}, \ldots, g_r$ are those g_i that are not in C. Then it follows from Theorem 2(b) that in the class equation

$$|G| = n_1 + n_2 + \cdots + n_s + n_{s+1} + \cdots + n_r$$

we have $n_1 = n_2 = \cdots = n_s = 1$ and $n_i > 1$ for $i = s + 1, \ldots, r$.

Using the fact that $n_i = 1$ for $1 \le i \le s$ and the hypothesis $|G| = p^m$, we then have

$$p^m = s + n_{s+1} + n_{s+2} + \cdots + n_r. \tag{2}$$

Each n_i is the index of a subgroup in G, and hence each n_i is one of the positive integral divisors $1, p, p^2, \ldots, p^m$ of $p^m = \text{ord } G$. For $i > s$, we have $n_i > 1$; and thus these n_i are of the form p^k, with k a positive integer. From (2), we have

$$\text{ord } C = s = p^m - n_{s+1} - n_{s+2} - \cdots - n_r.$$

Since $p \mid n_i$ for $i > s$, this implies that $p \mid (\text{ord } C)$ and hence that ord $C > 1$, as desired. ■

Theorem 4 implies that, in a group of order $8 = 2^3$, the order of the center is 2, 4, or 8. Actually, only 2 and 8 are possible orders for the center in a group of order 8. Example 1 of this section shows that the center in the octic group has order 2 and that a cyclic (or other abelian) group of order 8 has a center of order 8. In Problem 3 below, the reader is asked to prove that no group of order p^3, with p prime, has a center of order p^2.

Problems

1. Let the alternating group $\mathbf{A_4}$ have $\alpha_1, \ldots, \alpha_{12}$ as its elements, with the α_i as in the row headings for the table on the inside rear cover of the book. In $\mathbf{A_4}$, show the following.
(a) $CC(\alpha_1) = \{\alpha_1\}$.
(b) $CC(\alpha_2) = \{\alpha_2, \alpha_3, \alpha_4\}$.
(c) $CC(\alpha_5) = \{\alpha_5, \alpha_6, \alpha_7, \alpha_8\}$.
(d) $CC(\alpha_9) = \{\alpha_9, \alpha_{10}, \alpha_{11}, \alpha_{12}\}$.

2. Show that in the symmetric group $\mathbf{S_4}$, the conjugacy class of (123) consists of all eight 3-cycles.

3. Let ord $G = p^3$, with p prime. Use Theorem 4 of this section and Problem 27 of Section 2.11 to prove that the center C in G must have order p or p^3.

4. Let G be a nonabelian group of order p^3, with p prime. Prove that the center C in G must have order p.

5. Let a be an element different from the identity $\mathbf{e}$ of a noncyclic group G of order 9. Explain why the following are true.
(i) The order of a is 3, and hence a commutes with at least 3 elements of G.
(ii) The centralizer C_a has index 1 or 3 in G.
(iii) Under the assumption that C_a is not G, a would have 3 conjugates (that is, there would be 3 distinct elements of the form $g^{-1}ag$ with g in G).
(iv) $C_a = G$. [*Hint:* A set of 8 elements cannot be partitioned into subsets each having 3 elements.]
(v) Every group of order 9 is abelian.

6. Let G be a group of order p^2, where p is a prime.
(i) Prove that G is abelian.
(ii) Prove that either G is cyclic or G is isomorphic to the direct product of two groups of order p.

★ 3.16 The Fundamental Theorem on Abelian Groups (Without Proof)

Let G be a finite group and let the standard factorization of ord G be

$$\text{ord } G = m = (p_1)^{r_1}(p_2)^{r_2} \cdots (p_s)^{r_s}. \tag{1}$$

We now state without proof the following result.

Fundamental Theorem on Finite Abelian Groups

(a) Let G be a finite commutative group with ord G as in (1). For each p_i in (1), there exists a (Sylow) subgroup G_i of order $(p_i)^{r_i}$ in G, and G is isomorphic to the direct product $G_1 \times G_2 \times \cdots \times G_s$.
(b) Every finite commutative group is isomorphic to the direct product of a finite number of its cyclic subgroups.

For example, an abelian group G of order 360 is isomorphic to the direct product $G_1 \times G_2 \times G_3$ of subgroups G_1, G_2, and G_3 with ord $G_1 = 2^3$, ord $G_2 = 3^2$, and ord $G_3 = 5$. Also, the subgroup G_2 is either cyclic or isomorphic to the direct product of two cyclic groups of order 3. There are three possibilities for G_1. One is that G_1 is cyclic; a second is that it is isomorphic to a direct product of two cyclic groups whose orders are 2 and 4; and the third is that it is isomorphic to a direct product of three cyclic groups of order 2.

Review Problems

1. Let $G = \{\mathbf{e}, a, a^2, a^3\}$ be a cyclic group of order 4 and G' be the multiplicative group $\{1, -1, \mathbf{i}, -\mathbf{i}\}$ of 4 complex numbers.
(i) Tabulate two isomorphisms from G to G'.
(ii) Explain why there are no other isomorphisms from G to G'.

2. What is the smallest positive integer m such that there exist nonisomorphic groups of order m?

3. Let $\rho = (1234)$, $\phi = (24)$, and G be the octic group

$$G = \{(1), \rho, \rho^2, \rho^3, \phi, \rho\phi, \rho^2\phi, \rho^3\phi\}.$$

Let f and g be the mappings from G to itself with $f(\theta) = \theta^2$ and $g(\theta) = \theta^3$ for all θ in G.
(a) Is f injective? Is f surjective? Is f a homomorphism? Give reasons for the answers.
(b) Is g injective? Is g surjective? Is g a homomorphism? Give reasons for the answers.

4. Let G be the octic group (with its elements as in Problem 3) and $G' = \{\mathbf{e}, b\}$ be a group of order 2.
(i) Tabulate a homomorphism f from G to G' with $f(\rho) = b$ and $f(\rho\phi) = \mathbf{e}$.
(ii) What is the kernel of f?
(iii) What is the complete inverse image $f^{-1}(b)$?
(iv) What is the image set M of f?

5. Let f be the mapping from the octic group G to the symmetric group $\mathbf{S_4}$ with $f(\theta) = (12) \cdot \theta \cdot (12)$ for all θ in G. Show that f is an injective homomorphism.

6. Let $\mathbf{A_4} = \{\alpha_1, \alpha_2, \ldots, \alpha_{12}\}$, with the α_i as in Table 2.2 of Section 2.10 and let $G = \{\mathbf{e}, b, b^2\}$ be a cyclic group of order 3. Given that f is a homomorphism from $\mathbf{A_4}$ to G with $f(\alpha_2) = \mathbf{e}$ and $f(\alpha_5) = b$, find the following complete inverse images.
(i) $f^{-1}(\mathbf{e})$; (ii) $f^{-1}(b)$; (iii) $f^{-1}(b^2)$.

7. Let G have subgroups H and K, each of order 43. Explain why either $H = K$ or $H \cap K = \{\mathbf{e}\}$.

8. Explain why the union of all the subgroups in a group G is a subgroup in G.

9. Let $G = \{\mathbf{e}, b, c\}$ be a cyclic group of order 3. Tabulate the products $(b, c)(x, y)$ for all the ordered pairs (x, y) in the direct product $G \times G$.

10. Let L be the relation on $\mathbf{Z}$ for which aLb means that $a^2 = b^2$. Is L symmetric? Reflexive? Transitive? An equivalence relation? A linear order relation? Explain.

11. How many antisymmetric relations are there on a set $\{a, b\}$ with 2 elements? Of these, how many are also reflexive?

12. Do Problem 11 for a set $\{a, b, c\}$ with 3 elements.

13. Let D be the relation on $\mathbf{Z}^+$ for which aDb means $a \mid b$. Is D reflexive? Symmetric? Antisymmetric? Transitive? A partial ordering?

14. Do Problem 13 with $\mathbf{Z}^+$ replaced by the nonzero integers.

15. Let $S = \{1, 2, 3, 4\}$.
   (a) How many subsets are there in the power set $P(S)$?
   (b) How many mappings are there from $P(S)$ to S?
   (c) How many mappings are there from S to $P(S)$? Of these, how many are injective?

16. Do Problem 15 with $\{1, 2, 3, 4\}$ replaced by $\{1, 2, 3, 4, 5\}$.

17. Let $S = \{1, 2, 3, 4\}$.
   (a) How many unary operations are there on S?
   (b) How many binary operations are there on S?
   (c) How many of the binary operations are commutative?

18. How many n-ary operations are there on $S = \{1, 2, 3, 4\}$?

19. Let $*$ denote the multiplication operation in $\mathbf{Z}_{1000} = \{\bar{0}, \bar{1}, \ldots, \overline{999}\}$.
   (a) Is the semigroup $(\mathbf{Z}_{1000}, *)$ a monoid? Is it a group? Explain.
   ★ (b) How many elements are there in the principal subsemigroup $\langle \bar{2} \rangle$?
   (c) Is $\langle \bar{2} \rangle$ a monoid? Is it a group? Explain.

20. Do the analogues of (b) and (c) in Problem 19, with $\langle \bar{2} \rangle$ replaced by $\langle \bar{4} \rangle$.

21. Describe a boolean algebra whose carrier has 32 elements.

22. Describe a boolean algebra whose carrier has 64 elements.

23. Show that $x \wedge (y \vee x') = x \wedge y$ for all x and y in the carrier of a boolean algebra.

24. Does $x \vee (y \wedge x') = x \vee y$ for all x and y in the carrier of a boolean algebra?

25. For each of the following boolean functions, tell whether it is a tautology, a contingency, or a contradiction, and explain your answer.
   (a) $\alpha(x, y) = (x \wedge y) \Rightarrow x$.
   (b) $\beta(x, y, z) = [(x \wedge y) \Rightarrow z] \Leftrightarrow [(x \Rightarrow z) \vee (y \Rightarrow z)]$.

26. Do as in Problem 25 for the following boolean functions:
   (a) $\alpha(x, y) = (x \wedge y) \vee (x \wedge y') \vee (x' \wedge y) \vee (x' \wedge y')$.
   (b) $\beta(x, y) = [(x \Rightarrow y) \wedge (x \Rightarrow y')] \Rightarrow x$.

27. Let Y be a subset of X and let H be a subgroup in the group $\mathbf{B}(X)$ of all bijections from X to itself. Let K be the set of all θ in H such that $\theta(y) = y$ for every y in Y. Prove that K is a subgroup in H.

CHAPTER 4

RINGS AND FIELDS

In most sciences one generation tears down what another has built, and what one has established another undoes. In mathematics alone each generation builds a new story to the old structure. —**Hermann Hankel**

In this treatment of abstract algebra, we have chosen to begin with the study of groups. Although the defining properties of a group are few, the structure turns out to be amazingly rich in its complexity, with a number of unexpected properties such as Cayley's Theorem and the whole idea of quotient groups. Furthermore, we find examples of groups in a number of different areas of mathematics and the sciences. But having studied the group structure with various operations (including operations other than the obvious arithmetical ones), we now use this as a foundation for examining algebraic structures that are groups under addition and are partially groups under multiplication. These structures, primarily rings and fields, have as examples number systems that we are accustomed to using all the time. In fact, rings may appear to be almost too familiar. For this reason, one has to be careful to derive their properties from the axioms for rings rather than taking properties for granted merely from familiarity with some special cases of elementary algebra.

We are aware that the integers **Z**, the rational numbers **Q**, the real numbers **R**, and the complex numbers **C** form groups under addition and that the nonzero elements of **Q**, **R**, and **C** form groups under multiplication. But the group theoretic properties are not sufficient to describe all the algebraic properties of these familiar number systems, since to carry out ordinary arithmetical operations we need to have at least one additional property — a distributive law (or in the case of a noncommutative ring, distributive laws) — to tie the two operations together: $a(b + c) = ab + ac$.

The earliest appearance of fields, although they were not at that time formally defined, is in the work of Abel and Galois, not surprisingly, in their study of the solution of polynomial equations by radicals. The Galois Theory, which was used to establish the impossibility of such solution of general higher-degree equations (and which ties together results from group theory and the theory of fields — specifically, extension fields), was first given wide dissemination in some lectures by Richard Dedekind, as well as in his famous supplement to the third edition of P. G. L. Dirichlet's book on number theory, *Vorlesungen über Zahlentheorie.* Leopold Kronecker followed this with an even more general treatment of fields.

4.1 Rings

DEFINITION 1

Ring

A ring R is a set $\hat{R}$ with operations of addition and multiplication satisfying the following three axioms.

(A) R is an abelian group under addition.

(M) R is closed and associative under multiplication.

(D) Multiplication is distributive over addition; that is, for all a, b, and c in $\hat{R}$ one has

$$a(b + c) = ab + ac \quad \text{and} \quad (b + c)a = ba + ca.$$

When not only addition but also multiplication is commutative, the second distributivity formula, $(b + c)a = ba + ca$, follows from the first formula, $a(b + c) = ab + ac$.

As was true in the notation for a group G and its set $\hat{G}$, we shall usually write R for $\hat{R}$ and leave it to the context to indicate whether the ring or its set is involved.

By definition, all rings are commutative under addition; therefore, the phrase ***commutative ring*** is used to refer to a ring commutative under multiplication. The number systems, **Z**, **Q**, **R**, and **C** are examples of infinite commutative rings. Below we construct an infinite number of finite commutative rings. An infinite number of other (finite and infinite) commutative rings will be characterized in Chapter 5.

A finite noncommutative ring is given in Example 2 of this section. Many finite and infinite noncommutative rings are characterized in Section 4.7 below.

Additively, a ring R is a group; hence, R always has an additive identity, called the ***zero element*** and written as 0. Also, every element a of R has an additive inverse, called the ***negative*** of a and denoted by $-a$.

The ring axioms are such that a ring R may or may not have a multiplicative identity. If there is a multiplicative identity in R, it is unique (see the proof of Theorem 2 of Section 2.3) and is called the ***unity*** of R. The unity is represented by 1 (and of course has the property $a \cdot 1 = a = 1 \cdot a$ for all a in R).

The number systems **Z**, **Q**, **R**, and **C** are examples of rings with unity; an example of an infinite commutative ring that does not have a unity is the ring of even integers.

EXAMPLE 1 **The Modular Rings $\mathbf{Z_m}$**

Let m be a positive integer. As in Definition 2 of Section 2.6, let

$$\mathbf{Z_m} = \{\bar{0}, \bar{1}, \bar{2}, \ldots, \overline{m-1}\}$$

and let addition and multiplication in $\mathbf{Z_m}$ be given by $\bar{a} + \bar{b} = \bar{r}$ and $\bar{a} \cdot \bar{b} = \bar{s}$, where r and s are the remainders in the division of $a + b$ and ab, respectively, by m. Since multiplication is commutative, we can show that multiplication is distributive over addition by showing that $\bar{a}(\bar{b} + \bar{c}) = \bar{a} \cdot \bar{b} + \bar{a} \cdot \bar{c}$ for all $\bar{a}$, $\bar{b}$, and $\bar{c}$ in $\mathbf{Z_m}$. We know that $a(b + c) = ab + ac$ in the ring **Z** of the integers. Let r be the remainder when $a(b + c)$ is divided by m. It is easily shown that each of $\bar{a}(\bar{b} + \bar{c})$ and $\bar{a} \cdot \bar{b} + \bar{a} \cdot \bar{c}$ is equal to $\bar{r}$. This establishes distributivity. It now follows from Theorem 1 of Section 2.6 and the definition of a commutative ring that $\mathbf{Z_m}$ is a commutative ring. The rings $\mathbf{Z_1}$, $\mathbf{Z_2}$, ... are called the ***modular rings***. For a fixed m, the ring $\mathbf{Z_m}$ is also called the ***ring of the integers modulo*** m; this is especially true when the bars of the elements $\bar{0}, \bar{1}, \ldots, \overline{m-1}$ are dropped and understood from the context.

DEFINITION 2

Invertible Element of a Ring with Unity

If U is a ring with unity, an element v of U, for which there is a v^{-1} in U with $vv^{-1} = 1 = v^{-1}v$, is called an **invertible**, or a **unit**, of U.

Although the word *unit* is used widely, we prefer *invertible*, since it seems more descriptive and less likely to be confused with *unity*. Also, *invertible* is customary in linear algebra and matrix theory courses.

In the number systems **Z**, **Q**, **R**, and **C**, the multiplicative inverse of v (when v^{-1} exists) is frequently called the ***reciprocal*** of v.

The invertibles of the ring **Z** of the integers are 1 and -1, since these are the only integers with reciprocals in **Z**. In **Q**, **R**, or **C** every nonzero number is an invertible.

THEOREM 1 **Multiplicative Group of Invertibles**

Let U be a ring with unity and let V consist of all the invertibles of U. Then V is a multiplicative group.

The proof of Theorem 1 is left to the reader as Problem 7 of this section.

Distributivity of multiplication over addition enables us to obtain the following multiplicative property of the additive identity 0 of a ring.

THEOREM 2 Multiplication by 0

For all r in a ring R, $r \cdot 0 = 0 = 0 \cdot r$.

Proof Since 0 is the additive identity, $0 + 0 = 0$ and so $r(0 + 0) = r \cdot 0$. By distributivity, this becomes

$$r \cdot 0 + r \cdot 0 = r \cdot 0. \tag{1}$$

Additively, R is a group. Thus we may apply Theorem 2 of Section 2.3 (after translating that result into additive notation) to equation (1) and so find that $r \cdot 0 = 0$. We leave it to the reader (in Problem 10 of this section) to prove that $0 \cdot r = 0$. ∎

The following result is a generalization of one of the "rules of signs" of elementary algebra.

THEOREM 3 A Rule of Signs

If r and s are elements of a ring R,

$$r(-s) = (-r)s = -(rs).$$

Proof By definition of the negative of an element, $s + (-s) = 0$, and so

$$r[s + (-s)] = r \cdot 0.$$

Using distributivity and Theorem 2, we have

$$rs + r(-s) = 0.$$

Hence each of rs and $r(-s)$ is the negative of the other. Similarly, one can show that $(-r)s = -(rs)$. ∎

COROLLARY 1 If r and s are in R, then $(-r)(-s) = rs$.

The proof of this corollary is left to the reader as Problem 12(a) of this section.

We next define the analogue of subgroup in a group.

DEFINITION 3

Subring

A subset S of a ring R such that S is a ring under the operations of R is a **subring** in R.

We note that a subring S in a ring R additively is a normal subgroup of the additive group of R. We consider the integers $\mathbf{Z}$ to be a subring in the rational numbers $\mathbf{Q}$, $\mathbf{Q}$ to be a subring in the real numbers $\mathbf{R}$, and $\mathbf{R}$ to be a subring in the complex numbers $\mathbf{C}$.

THEOREM 4 Subring Conditions

A nonempty subset S of a ring R is a subring in R if and only if S is closed under both subtraction and multiplication — that is, if and only if S contains $a - b$ and ab whenever it contains a and b.

The proof of Theorem 4 is left to the reader as Problem 12(b) of this section.

Using distributivity and mathematical induction, we can establish the following result.

THEOREM 5 General Distributivity

In a ring R, $(a_1 + a_2 + \cdots + a_m)(b_1 + b_2 + \cdots + b_n) = a_1 b_1 + a_1 b_2 + \cdots + a_1 b_n + a_2 b_1 + a_2 b_2 + \cdots + a_2 b_n + \cdots + a_m b_1 + a_m b_2 + \cdots + a_m b_n$.

COROLLARY 2 If a and b are in a ring R and m and n are in $\mathbf{Z}$, $(m \cdot a)(n \cdot b) = (mn) \cdot (ab)$.

The proof is left to the reader.

EXAMPLE 2 Here we show that $R = \{0, b, c, d\}$ is a noncommutative ring under the operations given by the following tables:

+	0	b	c	d
0	0	b	c	d
b	b	0	d	c
c	c	d	0	b
d	d	c	b	0

·	0	b	c	d
0	0	0	0	0
b	0	0	b	b
c	0	0	c	c
d	0	0	d	d

Additively, R is a noncyclic group of order 4, that is, a Klein 4-group. (See Problem 9 of Section 2.3 and Problem 12 of Section 2.5.) It is also clear from the multiplication table that R is closed under multiplication.

Now we consider associativity of multiplication. We let $u = (xy)z$ and $v = x(yz)$, where x, y, and z are any elements of R. If at least one of y and z is in $\{0, b\}$, we see that $u = 0 = v$. In all other cases (when y and z are in $\{c, d\}$), each of u and v is equal to x. Hence, $(xy)z = x(yz)$ for all x, y, and z in R; that is, multiplication in R is associative.

Next we turn to distributivity. Let $s = x(y + z)$ and $t = xy + xz$. If $y = z$, then the addition table shows that $y + z = y + y = 0$ and $t = xy + xz =$

$xy + xy = 0$; in such a case $s = x \cdot 0 = 0$, and so $s = t$. If y or z is 0, it also follows that $s = t$. In the remaining cases, y and z are distinct elements of $S = \{b, c, d\}$ and for such y and z the addition table shows that $y + z$ is the third element of S; all these cases also lead to $s = t$.

Showing the other half of distributivity, $(y + z)x = yx + zx$, is entirely similar. Hence, R is a ring. Since $bc = b$ and $cb = 0$, R is not commutative.

A ring in which the elements are mappings is described in the following example.

EXAMPLE 3 Let $\mathbf{R}$ be the set of all real numbers and S be the subset of all x in $\mathbf{R}$ satisfying $0 < x < 1$. Let M be the set of all mappings from S to $\mathbf{R}$. For α and β in M, we define the sum $\alpha + \beta$ to be the mapping σ from S to $\mathbf{R}$ such that

$$\sigma(x) = \alpha(x) + \beta(x),$$

and we define the product $\alpha\beta$ to be the mapping π from S to $\mathbf{R}$ with

$$\pi(x) = \alpha(x) \cdot \beta(x).$$

It can be shown that these operations make M into a ring. (See Problem 31 below.)

Problems

Problem 31 below is cited in Section 5.4.

1. Let $R = \{0, b\}$ be a ring with two elements.
   (i) Make the table for the additive group of R.
   (ii) Make a multiplication table for R under the assumption that $b^2 = b$.
   (iii) Assume that $b^2 = 0$, and make a multiplication table for R.
   (iv) Does R have a unity in (ii) or (iii)?
2. Let $U = \{0, 1, c\}$ be a ring with three elements (and, of course, with 1 as the unity).
   (i) Explain why $1 + 1 = c$. (See Example 2 of Section 2.3.)
   (ii) Explain why $3 \cdot 1 = 1 + 1 + 1 = 0$ in U.
   (iii) Explain why $c^2 = (1 + 1)(1 + 1) = 1$ in U.
   (iv) Make addition and multiplication tables for U.
3. Let $A = \{u, v, w\}$ have the following addition and multiplication tables.

$+$	u	v	w
u	u	v	w
v	v	w	u
w	w	u	v

$\cdot$	u	v	w
u	u	u	u
v	u	u	u
w	u	u	u

   Is A a ring under these operations? Explain.
4. Let G be an additive group. Describe an operation of multiplication in G that converts G into a ring. (See Problem 3.)

5. Let U be a ring with unity. Show the following.
(i) If $0 = 1$, then U consists of just one element.
(ii) If 0 is an invertible (that is, a unit), $U = \{0\}$.
(iii) If $U = \{0\}$, then 0 is the unity and is an invertible.

6. Prove that, in a ring U with unity, the reciprocal (that is, the multiplicative inverse) of an invertible is unique. (Since U need not be a group under multiplication, one should not use cancellation without proof that is valid under the present hypothesis.)

7. Prove Theorem 1; that is, prove that the set V of invertibles of a ring U with unity is a multiplicative group, by showing the following.
(i) The unity 1 of U is an invertible.
(ii) If v and w are in V, so is vw^{-1}.

8. Make a multiplication table for the group $V = \{1, -1\}$ of invertibles (units) of the integers $\mathbf{Z}$.

9. Let v be an invertible in a ring U. Prove that either $uv = 0$ or $vu = 0$ in U implies $u = 0$.

10. Prove that $0 \cdot r = 0$ for all r in a ring R. (This is the half of Theorem 2 that was left to the reader.)

11. Let S be the subset $\{\bar{0}, \bar{2}, \bar{4}, \bar{6}\}$ of the modular ring
$$\mathbf{Z}_8 = \{\bar{0}, \bar{1}, \bar{2}, \bar{3}, \bar{4}, \bar{5}, \bar{6}, \bar{7}\}.$$
(i) Make the addition and multiplication tables for S.
(ii) Is S a subring in $\mathbf{Z}_8$? Does S have a unity?
(iii) Is the additive group of S cyclic? If so, name a generator.
(iv) Let $U = \{\bar{2} \cdot \bar{a} : \bar{a} \in S\}$, $V = \{\bar{4} \cdot \bar{a} : \bar{a} \in S\}$, and $W = \{\bar{6} \cdot \bar{a} : \bar{a} \in S\}$. [Each of these sets consists of the entries on a row of the multiplication table for S.] Is each of U, V, and W a subring in $\mathbf{Z}_8$?
(v) Let X be U, V, or W. Is $\bar{x} \cdot \bar{a}$ in X for each $\bar{x}$ in X and each $\bar{a}$ in $\mathbf{Z}_8$? (This property of X will be called "closure under right multiplication" in the next section.)

12. (a) Prove the corollary to Theorem 3; that is, prove that $(-r)(-s) = rs$ for all r and s in a ring R.
(b) Prove Theorem 4; that is, prove that a nonempty subset S of a ring R is a subring in R if and only if $a - b$ and ab are in S whenever a and b are in S.

13. Which of the following subsets of the ring $\mathbf{Z}$ of the integers is a subring in $\mathbf{Z}$?
(a) $\{0, 1, -1\}$.
(b) $\{3, 6, 9, \ldots\}$; that is, the positive integral multiples of 3.
(c) $\{0, 5, -5, 10, -10, \ldots\}$; that is, the integral multiples of 5.
(d) $\{\ldots, -4, -2, 0, 2, 4, \ldots\}$; that is, the even integers.

14. Which of the following subsets is a subring in the ring $\mathbf{C}$ of the complex numbers $a + b\mathbf{i}$?
(a) The subset of all $a + b\mathbf{i}$ with a and b rational.
(b) The subset of all $a + b\mathbf{i}$ with $a^2 + b^2 \leq 1$.
(c) The subset S of all the rational numbers of the form $m/2^n$ with m an integer and n a positive integer or zero. (S consists of the rational numbers $0, \pm 1, \pm 2, \ldots$; $\pm 1/2, \pm 3/2, \pm 5/2, \ldots$; $\pm 1/4, \pm 3/4, \pm 5/4, \ldots$; and so on.)

15. Let m be a fixed integer and I consist of all the integral multiples of m. Prove the following.
(i) For every i in I and z in $\mathbf{Z}$, iz is in I.

(ii) I is nonempty and is closed under subtraction.

(iii) I is a subring in **Z** (and hence the additive group of I is a normal subgroup in the additive group of **Z**).

(iv) Let a and b be any elements of **Z**. Let i and i' be elements of I. Then the product $(a + i)(b + i')$ is in the coset $(ab) + I$ of the normal subgroup I in the additive group **Z**.

★ (v) The additive quotient group $\mathbf{Z}/I$ can be made into a ring by using

$$(a + I)(b + I) = (ab) + I$$

as the formula for multipication of cosets.

16. Let I be the subring of all the integral multiples of 3 in the ring **Z** of the integers. Then the cosets of the additive quotient group $\mathbf{Z}/I$ are

$$\bar{0} = 0 + I = \{\ldots, -6, -3, 0, 3, 6, \ldots\},$$
$$\bar{1} = 1 + I = \{\ldots, -5, -2, 1, 4, 7, \ldots\},$$
$$\bar{2} = 2 + I = \{\ldots, -4, -1, 2, 5, 8, \ldots\}.$$

(i) Explain why $\bar{4} = \bar{1}$; that is, $4 + I = \bar{1}$.

(ii) Make the addition table for $\mathbf{Z}/I = \{\bar{0}, \bar{1}, \bar{2}\}$.

(iii) Use the formula of Problem 15(v) to make a multiplication table for $\{\bar{0}, \bar{1}, \bar{2}\}$ and thus convert it into a ring.

17. Among the properties of a subring S in a ring R is that S is nonempty and closed under subtraction. Use this to explain why every subring in the ring **Z** of the integers must be one of the subrings I of Problem 15.

18. Let a be a fixed element of a ring R and let I consist of the products sa for all s in R.

(i) Show that ri is in I for all r in R and all i in I.

(ii) Use Problem 12(b) to prove that I is a subring in R.

19. Let S consist of all the real numbers of the form $a + b\sqrt{2}$ with a and b rational numbers.

(i) Show that S is a subring in the ring **R** of real numbers and hence is a subring in the complex numbers **C**.

(ii) Would S be a subring in **R** if a and b were restricted to integer values?

20. Let T consist of all the real numbers, $a + b\sqrt{3}$ with a and b integers. Show that T is a subring in the ring **R** of real numbers.

21. Let G consist of all the complex numbers $m + ni$ with m and n integers. Show that G is a subring in the ring **C** of complex numbers. (This subring is called the ring of ***gaussian integers***, after Carl Friedrich Gauss.)

22. Find all the invertibles of the ring of gaussian integers. (See Problem 21 above and Example 1 of Section 2.3.)

23. Let $U = \{0, 1, c, d\}$ be a ring (with four elements) and let c and d be invertibles in U. Make the multiplication table for U.

24. Let $U = \{0, 1, c, d\}$ be a ring and let $1 + 1 = c$ in U. Make the addition and multiplication tables for U.

25. Prove that every ring R with 5 elements is commutative.

26. Prove that every ring with 7 elements is commutative.

27. Let two sequences $q_0, q_1, q_2, \ldots$ and $p_0, p_1, p_2, \ldots$ of nonnegative integers be defined by $q_0 = 1$, $p_0 = 0$, $q_{n+1} = q_n + 2p_n$, and $p_{n+1} = q_n + p_n$ for $n = 1, 2, 3, \ldots$.

Emmy Noether
1882–1935

One of the foremost researchers in the theory of rings in the twentieth century was also, possibly, the most important woman in the history of mathematics to date: Emmy Noether. Her father was a well-known mathematician at the University of Erlangen, where she studied and worked until 1916. She then went to work at Göttingen, where both Klein and Hilbert were professors. She remained there until 1933, when she had to leave Germany due to the rise of the Nazis. She came to the United States and became professor of mathematics at Bryn Mawr, but she only lived for a year and a half after arriving in this country.

Although she has been universally acknowledged as one of the great algebraists of history, she was subject during her lifetime to constant difficulties in working in a profession that until recently was an uncommon one for women. Prior to the twentieth century there were few prominent women in mathematics; some of the more famous were Hypatia, a Greek mathematician murdered in Alexandria in A.D. 415; Maria Gaetana Agnesi, the Italian mathematician for whom the famous curve, the "Sail of Agnesi" (mistranslated at some point, and often repeated, as the "Witch of Agnesi"), is named; Sonya Kovalevsky, a Russian mathematician, student of Weierstrass and professor of mathematics at the University of Stockholm; Sophie Germain, a French mathematician who did important work in mathematical physics; and two English mathematicians, Mary Fairfax Somerville and Augusta Ada Lovelace. The latter happened to be the only daughter of the poet, Lord Byron. She worked with Charles Babbage on his Difference Engine and his Analytical Engine, precursors of modern computers.

Professor Noether was never given a professorship at Göttingen, although she was one of the most distinguished mathematicians there. Hilbert once said at a meeting of the faculty held to discuss a possible appointment for her: "I do not see that the sex of the candidate is an argument against her admission as Privatdozent. After all we are a university and not a bathing establishment." In spite of Hilbert's defense, she did not receive the appointment, but in 1922 she finally received the title "nichtbeamteter ausserordentlicher Professor," which she held at Göttingen until she left in 1933. It was a position that involved no duties and, unfortunately, it also involved no salary.

Prove the following for $n = 0, 1, 2, \ldots$.

(i) $(1 + \sqrt{2})^n = q_n + p_n\sqrt{2}$.

(ii) $(q_n + p_n\sqrt{2})(q_n - p_n\sqrt{2}) = (-1)^n$.

(iii) $p_{n+2} = 2p_{n+1} + p_n$ and $q_{n+2} = 2q_{n+1} + q_n$.

★ **28.** Let S be the ring of Problem 19(ii) and let q_n and p_n be as in Problem 27.

(i) Show that every invertible of S is of the form $\pm(q_n \pm p_n\sqrt{2})$ for some non-negative integer n.

(ii) Let V^+ be the set of positive real numbers that are invertibles of S. Show that V^+ is a cyclic multiplicative group.

(iii) List the two elements, each of which is a generator of V^+.

★ **29.** Given that a and b are integers, show that $a + b\sqrt{3}$ is an invertible of the ring T of Problem 20 if and only if $a^2 - 3b^2 = \pm 1$.

30. Let R be the cartesian product $S \times T$ of rings S and T. Let addition and multiplication in R be defined by

$$(s, t) + (s', t') = (s + s', t + t'), \qquad (s, t)(s', t') = (ss', tt').$$

(a) Show that R is a ring (called the ***direct product*** of S and T).

(b) Define addition and multiplication for the direct product $S_1 \times S_2 \times \cdots \times S_n$ of n rings.

31. Let R be a ring, let S be a set, and let F consist of all the mappings from S to R. Define addition and multiplication in F as in Example 3 of this section, and show that these definitions make F into a ring.

★ **32.** Let $\mathbf{R}$ be the set of real numbers and let H be the cartesian product $\mathbf{R} \times \mathbf{R} \times \mathbf{R} \times \mathbf{R}$. Show that H is a noncommutative ring under the operations

$$(a, b, c, d) + (e, f, g, h) = (a + e, b + f, c + g, d + h),$$

$$(a, b, c, d)(e, f, g, h) = (ae - bf - cg - dh, af + be + ch - dg,$$
$$ag - bh + ce + df, ah + bg - cf + de).$$

(H is essentially the ring of quaternions developed by Sir William Rowan Hamilton.)

★ **33.** Find the invertibles of the ring H of Problem 32.

4.2 Ring Homomorphisms and Ideals

Here we introduce the analogues in ring theory of the closely related group theory concepts of homomorphism, normal subgroup, and quotient group.

DEFINITION 1

Ring Homomorphism, Kernel

A mapping θ from a ring R to a ring R' such that, for all r and s in R,

$$\theta(r + s) = \theta(r) + \theta(s), \qquad \theta(rs) = \theta(r)\theta(s)$$

is a **ring homomorphism** from R to R'. The **kernel** of θ is the subset of all k in R for which $\theta(k)$ is the zero of R'.

In other words, a ring homomorphism is a mapping that preserves both additions and multiplications. Clearly, a ring homomorphism from R to R' automatically is a homomorphism from the additive group of R to the additive group of R'.

DEFINITION 2

Isomorphism, Endomorphism, Automorphism

A bijective ring homomorphism from R to R' is a ring **isomorphism.** A ring homomorphism from R to itself is a ring **endomorphism** of R. A ring isomorphism from R to itself is a ring **automorphism** of R.

In our study of groups we found that the kernel of a homomorphism from a group G to a group G' is always a normal subgroup in G. Furthermore, if N is a normal subgroup in a group G, then there exists a homomorphism θ (the natural map) from G to the quotient group G/N such that the kernel of θ is N. Together, these two statements imply that a subset N of a group G is a normal subgroup in G if and only if N is the kernel of a homomorphism from G to a group G'.

This motivates us to investigate the properties possessed by the kernel K of a ring homomorphism θ from R to R'. Since such a θ is a homomorphism from the additive group of R to that of R', K must be a subgroup of the additive group of R. As a subset of R, K has the properties of associativity of multiplication and distributivity. This means that K will have been proved to be a subring of R once we have shown closure of K under multiplication. While demonstrating that K is closed under multiplication, we obtain an important extra property of K, as follows.

Let k be an element of the kernel K and let r be any element of R. Then $\theta(kr) = \theta(k)\theta(r) = 0 \cdot \theta(r) = 0$, and similarly $\theta(rk) = 0$. This means that both kr and rk are in K. Since r is any element of R, it may be any element of K. Thus we have proved that K is closed under multiplication and hence is a subring of R.

The two following definitions help us in stating the properties of a kernel of a ring homomorphism noted above.

DEFINITION 3

Closure Under Left and Right Multiplication

Let J be a subset of a ring R such that both rj and jr are in J for all j in J and all r in R. Then we say that J is **closed under left and right multiplication.**

DEFINITION 4

Ideal in a Ring

Let I be a subring closed under left and right multiplication in a ring R. Then I is an **ideal** in R.

The properties of a kernel found above can now be stated as follows.

LEMMA 1 Kernel of a Ring Homomorphism

The kernel of a ring homomorphism from R to R' is an ideal in R.

We next show how an ideal I in a ring R determines a collection Γ of cosets of I in R and operations of addition and multiplication that make Γ into a ring. This will be the ring theory analogue of the construction of the quotient group G/N from a group G and a normal subgroup N in G.

Under addition, an ideal I in a ring R is a normal subgroup in the additive group of R. Therefore, we use the following approach.

DEFINITION 5 Coset of an Ideal in a Ring

Let I be an ideal in a ring R and let a be any element of R. Then the **ideal coset** for the element a of the ideal I in the ring R is defined to be the same subset of R as the coset $a + I$ considering I to be a subgroup in the additive group of R.

NOTATION 1 The ideal coset for a of I in R will be designated by $a + I$.

The fact that $a + I$ might be either the coset of a subgroup or the coset of an ideal need not cause any difficulty, since $a + I$ has the same elements under each interpretation.

Let Γ be the collection of ideal cosets $a + I$ for all a in R. We can add ideal cosets $a + I$ and $b + I$ in Γ by using the formula

$$(a + I) + (b + I) = (a + b) + I$$

for adding them as normal subgroup cosets. It seems natural to use the formula

$$(a + I)(b + I) = ab + I \tag{M}$$

to introduce an operation of multiplication in Γ. This requires that we show that multiplication of ideal cosets is well defined by formula (M); that is, the product is not changed by using different expressions for the factors.

So let $a + I = a' + I$ and $b + I = b' + I$. This implies that $a' = a + i$ and $b' = b + i_1$ with i and i_1 in I. Now we apply formula (M) to the new expressions for the cosets and have

$$(a' + I)(b' + I) = (a'b') + I.$$

Next we note that $a'b' = (a + i)(b + i_1) = ab + i_2$, where $i_2 = ai_1 + bi + ii_1$. Since I is closed under left multiplication and under addition, i_2 is in I. Hence,

$a'b'$ is in the same coset as ab, and $a'b' + I = ab + I$ by Lemma 2 of Section 2.11 (applied to the subgroup I of the additive group of R).

Therefore, multiplication in Γ is well defined by (M). We leave to the reader the further details of proving the following result.

THEOREM 1 The Ring of Cosets of an Ideal

If I is an ideal in a ring R, the collection Γ of ideal cosets $a + I$ is a ring under the operations given by

$$(a + I) + (b + I) = (a + b) + I,$$

$$(a + I)(b + I) = ab + I.$$

DEFINITION 6 Quotient Ring R/I

The ring Γ described in Theorem 1 is called the **quotient ring** of R by I and is denoted by R/I.

We will see that the quotient rings $\mathbf{Z}/I$ of the ring $\mathbf{Z}$ of the integers by ideals I in $\mathbf{Z}$ are essentially the modular rings $\mathbf{Z}_\mathbf{m}$ described in Example 1 of Section 4.1. But the material developed here, especially Theorem 2 below, will enable us to get new information about the $\mathbf{Z}_\mathbf{m}$ and about the ring $\mathbf{Z}$ from which they were constructed. After studying polynomial rings in Chapter 5, we will use quotient rings to get valuable new rings and additional insight into old rings.

Now we continue the analogy with group theory.

DEFINITION 7 The Natural Map

Let I be an ideal in a ring R. The mapping θ from R to R/I with $\theta(r) = r + I$ is called the **natural map**; that is, the image of an element r of R under the natural map is the ideal coset of I to which r belongs.

THEOREM 2 Properties of the Natural Map

The natural map θ from a ring R to one of its quotient rings R/I is a surjective ring homomorphism with I as kernel.

Proof For all a and b in R,

$$\theta(a + b) = (a + b) + I = (a + I) + (b + I) = \theta(a) + \theta(b),$$

$$\theta(ab) = ab + I = (a + I)(b + I) = \theta(a)\theta(b).$$

This shows that θ is a ring homomorphism. Each coset $r + I$ in R/I is the image under θ of the element r of R; hence, θ is surjective. Also, $\theta(r) = r + I$ is equal to the additive identity I of R/I if and only if r is in I; thus the kernel of θ is the ideal I. ∎

Theorem 2 contains a converse of Lemma 1. We restate Lemma 1 and its converse in the following form.

THEOREM 3 Ideals and Kernels

A subset I of a ring R is an ideal in R if and only if I is the kernel of a ring homomorphism from R to some ring R'.

The next result provides a fairly easy way of showing that a subset of a ring is an ideal.

THEOREM 4 Conditions for an Ideal

Let I be a nonempty subset of a ring R and let I be closed under subtraction and under left and right multiplication. Then I is an ideal in R.

Proof Since closure under left and right multiplication implies closure under multiplication, I is a subring in R. (See Theorem 4 of Section 4.1.) This and closure under left and right multiplication make I an ideal in R. ∎

One way to obtain a subgroup H in a group G is to choose an element a of G and then let H consist of all the elements a^m with m an integer; that is, let H be the cyclic subgroup $[a]$. We next give an analogous procedure for obtaining an ideal in a commutative ring R.

THEOREM 5 Multiples of a Fixed Element

Let a be a fixed element of a commutative ring R and let I be the subset of R consisting of all the multiples ra of the given element a by elements r of R. Then I is an ideal in R.

Proof Let sa and ta be any two elements of I and let u be any element of R. Then $sa - ta = (s - t)a$ and $u(sa) = (us)a$ are elements of I. This and commutativity of R imply that I is closed under subtraction and under left and right multiplication. Also, I is nonempty, since it contains $0 \cdot a = 0$. Theorem 4 now tells us that I is an ideal in R. ∎

DEFINITION 8

Principal Ideal Generated by a

Let a be a fixed element of a commutative ring R and let (a) consist of all products ra with r in R. Then (a) is called the **principal ideal** generated by a in R.

In any ring R, the principal ideal $(0) = \{r \cdot 0 : r \in R\}$ is the singleton subring $\{0\}$. In a ring U with unity, the principal ideal $(1) = \{r \cdot 1 : r \in R\}$ is R itself. In the ring $\mathbf{Z}$ of the integers, the principal ideal $(d) = \{nd : n \in \mathbf{Z}\}$ consists of the same integers as the set $d\mathbf{Z}$ of integral multiples of d (defined in Section 1.2). Some of these principal ideals in $\mathbf{Z}$ are

$$(0) = \{0\},\ (1) = \{n \cdot 1 : n \in \mathbf{Z}\} = \mathbf{Z} = (-1),$$

$$(2) = \{2n : n \in \mathbf{Z}\} = \{\ldots, -4, -2, 0, 2, 4, \ldots\} = (-2),$$

$$(-3) = \{-3n : n \in \mathbf{Z}\} = \{\ldots, 6, 3, 0, -3, -6, \ldots\} = (3).$$

It is easily seen that for $m \neq 0$ the set of elements of the principal ideal (m) is

$$\{\ldots, -3m, -2m, -m, 0, m, 2m, 3m, \ldots\},$$

the same set as that for the cyclic subgroup $[m]$ in the additive group of the integers. If $m = 0$, (m) and $[m]$ each consist only of 0. Hence the sets for (m) and $[m]$ are the same for every integer m.

Now let m be a fixed positive integer. A coset of the principal ideal (m) has the form

$$a + (m) = \{\ldots, a - 2m, a - m, a, a + m, a + 2m, a + 3m, \ldots\}.$$

That is, the coset $a + (m)$ of the principal ideal (m) has the same elements as the coset $a + [m]$ of the cyclic subgroup $[m]$. In Section 3.3, we let $\bar{a}$ be a notation for $a + [m]$, especially when the value of m is known from the context. We now also use $\bar{a}$ to denote the ideal coset $a + (m)$, since $a + [m]$ and $a + (m)$ have the same elements. With this notation, the set of elements of the quotient ring $\mathbf{Z}/(m)$ becomes

$$\{\bar{0}, \bar{1}, \ldots, \overline{m-1}\}.$$

Because addition and multiplication of these cosets $\bar{a}$ is the same as for the elements $\bar{a}$ of the modular ring $\mathbf{Z_m}$, it is also natural to let $\mathbf{Z_m} = \mathbf{Z}/(m)$.

NOTATION 2

The Rings $\mathbf{Z_m}$ and Groups $\mathbf{V_m}$

For m in $\{1, 2, 3, \ldots\}$, the quotient ring $\mathbf{Z}/(m)$ is denoted by $\mathbf{Z_m}$. Also, we use $\mathbf{V_m}$ to designate the multiplicative group of invertibles of $\mathbf{Z_m}$.

With this notation,

$$\mathbf{Z_m} = \{\bar{0}, \bar{1}, \ldots, \overline{m-1}\} = \mathbf{Z}/(m)$$

and the natural map θ from $\mathbf{Z}$ to $\mathbf{Z_m}$ has $\theta(a) = \bar{a} = \bar{r}$, where r is the remainder in division of a by m. For example, when $m = 4$, the natural map from $\mathbf{Z}$ to $\mathbf{Z_4}$ has the table

a	...	-4	-3	-2	-1	0	1	2	3	4	5	...
$\theta(a) = \bar{a}$	...	$\bar{0}$	$\bar{1}$	$\bar{2}$	$\bar{3}$	$\bar{0}$	$\bar{1}$	$\bar{2}$	$\bar{3}$	$\bar{0}$	$\bar{1}$	...

We saw in Problem 5(i) of Section 3.3 that $\bar{a} = \bar{b}$ in $\mathbf{Z_m}$ if and only if $a \equiv b \pmod{m}$. Thus the natural map from $\mathbf{Z}$ to $\mathbf{Z_m}$ converts congruence modulo m into equality in $\mathbf{Z_m}$. We will see in the examples below that this natural map θ enables us to deduce properties of the infinite ring $\mathbf{Z}$ from properties of its finite quotient rings $\mathbf{Z_m}$. But first we emphasize some facts about θ.

LEMMA 2 The Natural Map $\theta : \mathbf{Z} \to \mathbf{Z_m};\ a \mapsto \bar{a} = \bar{r}$

Let m be a positive integer and θ be the natural map from $\mathbf{Z}$ to $\mathbf{Z_m}$. Then:

(i) $\theta(a) = \theta(b)$ [that is, $\bar{a} = \bar{b}$ in $\mathbf{Z_m}$] if and only if $a \equiv b \pmod{m}$.
(ii) $\overline{qm + r} = \bar{r}$ for all integers q and r.
(iii) $\overline{a + b} = \bar{a} + \bar{b}$ and $\overline{ab} = \bar{a} \cdot \bar{b}$ for all integers a and b.

Proof For (i) and (ii), see the answers to Problem 5 of Section 3.3. Since θ is a ring homomorphism (by Theorem 2 above), we have

$$\overline{a+b} = \theta(a+b) = \theta(a) + \theta(b) = \bar{a} + \bar{b},$$
$$\overline{ab} = \theta(ab) = \theta(a)\theta(b) = \bar{a} \cdot \bar{b}.$$

This proves (iii). ∎

EXAMPLE 1 Addition and Multiplication Tables for $\mathbf{Z_6}$

Parts (ii) and (iii) of Lemma 2 help us to add and multiply in $\mathbf{Z_m}$. In particular, we note that in $\mathbf{Z_6}$

$$\bar{4} + \bar{5} = \overline{4+5} = \bar{9} = \overline{6+3} = \bar{3} \quad \text{and} \quad \bar{4} \cdot \bar{5} = \overline{20} = \overline{3 \cdot 6 + 2} = \bar{2}.$$

Similarly, we obtain the information listed in the accompanying tables.

+	$\bar{0}$	$\bar{1}$	$\bar{2}$	$\bar{3}$	$\bar{4}$	$\bar{5}$
$\bar{0}$	$\bar{0}$	$\bar{1}$	$\bar{2}$	$\bar{3}$	$\bar{4}$	$\bar{5}$
$\bar{1}$	$\bar{1}$	$\bar{2}$	$\bar{3}$	$\bar{4}$	$\bar{5}$	$\bar{0}$
$\bar{2}$	$\bar{2}$	$\bar{3}$	$\bar{4}$	$\bar{5}$	$\bar{0}$	$\bar{1}$
$\bar{3}$	$\bar{3}$	$\bar{4}$	$\bar{5}$	$\bar{0}$	$\bar{1}$	$\bar{2}$
$\bar{4}$	$\bar{4}$	$\bar{5}$	$\bar{0}$	$\bar{1}$	$\bar{2}$	$\bar{3}$
$\bar{5}$	$\bar{5}$	$\bar{0}$	$\bar{1}$	$\bar{2}$	$\bar{3}$	$\bar{4}$

·	$\bar{0}$	$\bar{1}$	$\bar{2}$	$\bar{3}$	$\bar{4}$	$\bar{5}$
$\bar{0}$	$\bar{0}$	$\bar{0}$	$\bar{0}$	$\bar{0}$	$\bar{0}$	$\bar{0}$
$\bar{1}$	$\bar{0}$	$\bar{1}$	$\bar{2}$	$\bar{3}$	$\bar{4}$	$\bar{5}$
$\bar{2}$	$\bar{0}$	$\bar{2}$	$\bar{4}$	$\bar{0}$	$\bar{2}$	$\bar{4}$
$\bar{3}$	$\bar{0}$	$\bar{3}$	$\bar{0}$	$\bar{3}$	$\bar{0}$	$\bar{3}$
$\bar{4}$	$\bar{0}$	$\bar{4}$	$\bar{2}$	$\bar{0}$	$\bar{4}$	$\bar{2}$
$\bar{5}$	$\bar{0}$	$\bar{5}$	$\bar{4}$	$\bar{3}$	$\bar{2}$	$\bar{1}$

The two smallest positive primes in the integers $\mathbf{Z}$ are 2 and 3. It can be shown that all the other positive primes p are in one of the two following arithmetic progressions:

$$5, 11, 17, 23, 29, 35, 41, \ldots \tag{A}$$

$$7, 13, 19, 25, 31, 37, 43, \ldots . \tag{B}$$

[See Problem 16(v) below.] The fact that there is a ring homomorphism, the natural map, from the infinite ring $\mathbf{Z}$ to the finite ring $\mathbf{Z}_6$ can be used to obtain fairly efficient techniques for partitioning the set of integers in (A), or in (B), into a subset of composites and a subset of primes. In Example 2, which follows, we show how to do this for the progression (A); we leave the similar problem for progression (B) to the reader as Problem 17 below.

EXAMPLE 2 **Composites in $-1 + 6\mathbf{Z}^+$**

Here we use the multiplication table (in Example 1) for $\mathbf{Z}_6$ to prove that an integer c in the set

$$A = -1 + 6\mathbf{Z}^+ = \{5, 11, 17, \ldots, 6n - 1, \ldots\}$$

is composite if and only if $c = ab$ with a in $-1 + 6\mathbf{Z}^+$ and b in

$$B = 1 + 6\mathbf{Z}^+ = \{7, 13, 19, \ldots, 6n + 1, \ldots\}.$$

Let c be in A. If $c = ab$ with a in A and b in B, then clearly c is composite.

Conversely, let c be a composite integer in A. Then $c = ab$ with a and b integers satisfying $1 < a < c$ and $1 < b < c$. Since c is in A, in the ring $\mathbf{Z}_6$ we have $\bar{c} = \bar{5}$, and so $\bar{a} \cdot \bar{b} = \overline{ab} = \bar{c} = \bar{5}$. But $\bar{5}$ appears only twice in the multiplication table for $\mathbf{Z}_6$ namely, in the products $\bar{1} \cdot \bar{5} = \bar{5}$ and $\bar{5} \cdot \bar{1} = \bar{5}$. Hence, $\bar{a} \cdot \bar{b} = \bar{5}$ in $\mathbf{Z}_6$ implies that $\{\bar{a}, \bar{b}\} = \{\bar{1}, \bar{5}\}$. For definiteness, let $\bar{a} = \bar{5}$ and $\bar{b} = \bar{1}$. Since $a > 1$ and $b > 1$, this means that a is in A and b is in B, as desired.

One can easily deduce from Example 2 that the only composite numbers c with c in the finite arithmetic progression $5, 11, 17, \ldots, 95$ are

$$5 \cdot 7 = 35, \qquad 5 \cdot 13 = 65, \qquad 7 \cdot 11 = 77, \qquad 5 \cdot 19 = 95.$$

Hence the remaining integers

$$5, 11, 17, 23, 29, 41, 47, 53, 59, 71, 83, 89$$

in this finite progression are primes.

We next illustrate the use of the natural map from $\mathbf{Z}$ to a $\mathbf{Z}_m$ to obtain information concerning the squares of integers.

EXAMPLE 3 **No Square in $2 + 3\mathbf{Z}$**

Here we prove that no square of an integer is in the set

$$C = 2 + 3\mathbf{Z} = \{\ldots, -4, -1, 2, 5, 8, \ldots\}.$$

That is, we show that the equation $x^2 = 3y + 2$ has no solution in integers x and y. This is accomplished by assuming that there is such a solution and showing that this assumption leads to a contradiction.

Since $\bar{3} = \bar{0}$ in $\mathbf{Z}_3 = \mathbf{Z}/(3) = \{\bar{0}, \bar{1}, \bar{2}\}$, we can remove the term $3y$ in the equation by using the natural map θ from $\mathbf{Z}$ to $\mathbf{Z}_3$. This map converts $x^2 = 3y + 2$ into $\overline{x^2} = \overline{3y + 2}$. Using Lemma 2(iii), we find that

$$\overline{x^2} = \bar{x}^2 \quad \text{and} \quad \overline{3y + 2} = \bar{3} \cdot \bar{y} + \bar{2} = \bar{0} \cdot \bar{y} + \bar{2} = \bar{2}.$$

Hence the equation becomes $\bar{x}^2 = \bar{2}$. But no square in $\mathbf{Z}_3 = \{\bar{0}, \bar{1}, \bar{2}\}$ equals $\bar{2}$, since

$$\bar{0}^2 = \bar{0}, \quad \bar{1}^2 = \bar{1}, \quad \bar{2}^2 = \bar{4} = \overline{3 + 1} = \bar{1}.$$

This contradiction proves that $2 + 3\mathbf{Z}$ contains no squares.

EXAMPLE 4 **No Integer Solutions to $x^2 - 17y^2 = 855$**

Here we show that the equation

$$x^2 - 17y^2 = 855 \tag{D}$$

has no solution in integers x and y. We remove the influence of the term $-17y^2$ in the equation by using the natural map $\theta : a \mapsto \bar{a}$ from $\mathbf{Z}$ to $\mathbf{Z}_{17}$. Since the ring homomorphism θ preserves additions, subtractions, and multiplications, the existence of integers x and y satisfying (D) would imply that

$$\bar{x}^2 - \overline{17}\bar{y}^2 = \overline{855}. \tag{E}$$

But $\overline{17} = \bar{0}$ and $\overline{855} = \overline{17 \cdot 50 + 5} = \bar{5}$ in $\mathbf{Z}_{17}$. Thus (E) becomes $\bar{x}^2 = \bar{5}$. Now we find all the squares in $\mathbf{Z}_{17}$; for this purpose it is convenient to represent the 17 cosets in $\mathbf{Z}_{17}$ by $\bar{0}, \pm\bar{1}, \pm\bar{2}, \ldots, \pm\bar{8}$. The table

$\bar{x}$	$\bar{0}$	$\pm\bar{1}$	$\pm\bar{2}$	$\pm\bar{3}$	$\pm\bar{4}$	$\pm\bar{5}$	$\pm\bar{6}$	$\pm\bar{7}$	$\pm\bar{8}$
$\bar{x}^2$	$\bar{0}$	$\bar{1}$	$\bar{4}$	$\bar{9}$	$\overline{16}$	$\bar{8}$	$\bar{2}$	$\overline{15}$	$\overline{13}$

shows that $\bar{5}$ is not the square of an element of $\mathbf{Z}_{17}$. Therefore, equation (D) has no solution in the integers.

EXAMPLE 5 **Subrings With or Without a Unity Element**

The ring $\mathbf{Z}$ of all the integers has 1 as its unity element. The subring

$$2\mathbf{Z} = \{\ldots, -4, -2, 0, 2, 4, \ldots\}$$

does not contain the unity of $\mathbf{Z}$, and it is easily seen that $2\mathbf{Z}$ does not have a unity element. That is, there is no u in $2\mathbf{Z}$ such that $un = n$ for all n in $2\mathbf{Z}$.

The ring $\mathbf{Z}_6 = \{\bar{0}, \bar{1}, \bar{2}, \bar{3}, \bar{4}, \bar{5}\}$ has $\bar{1}$ as its unity. This unity is not in the subring $(\bar{2}) = \{\bar{0}, \bar{2}, \bar{4}\}$, but this subring does have $\bar{4}$ as its unity element, since in $\mathbf{Z}_6$ [and hence in $(\bar{2})$], one has

$$\bar{4} \cdot \bar{0} = \bar{0}, \quad \bar{4} \cdot \bar{2} = \overline{6 + 2} = \bar{2}, \quad \bar{4} \cdot \bar{4} = \overline{2 \cdot 6 + 4} = \bar{4}.$$

THEOREM 6 Subrings in Z Are Principal Ideals

Every subring S, and hence every ideal, in $\mathbf{Z}$ is a principal ideal (m) for some m in $\{0, 1, 2, \ldots\}$.

Proof A subring S in $\mathbf{Z}$ is a nonempty subset of $\mathbf{Z}$ and is closed under subtraction. Therefore, it follows from Theorem 3(e) of Section 1.3 that $S = m\mathbf{Z} = (m)$ for some m in $\{0, 1, 2, \ldots\}$. ∎

Problems

Problem 39 below is cited in Section 5.7.

1. Let T be the subring in the real numbers consisting of all numbers of the form $a + b\sqrt{3}$ with a and b integers.

(i) Use the fact that $\sqrt{3}$ is irrational (see Problem 21 of Section 1.4) to show that, if a, b, c, and d are rational numbers with $a + b\sqrt{3} = c + d\sqrt{3}$, then $a = c$ and $b = d$. Thus show that the representation $a + b\sqrt{3}$ for an element of T is unique.

(ii) For every $\alpha = a + b\sqrt{3}$ of T, let $\bar{\alpha} = a - b\sqrt{3}$. Let f be the bijection from T to T given by $f(\alpha) = \bar{\alpha}$. Show that f is a ring automorphism of T.

2. Let $\mathbf{R}$ be the ring of the real numbers and let θ be the mapping from $\mathbf{R}$ to itself with $\theta(x) = |x|$. Show that θ preserves multiplications but does not preserve additions and hence θ is not a ring homomorphism.

3. Let $\mathbf{Z}$ be the ring of the integers and let θ be the mapping from $\mathbf{Z}$ to itself with $\theta(n) = 2n$. Show that θ preserves additions but does not preserve multiplications, and hence θ is not a ring homomorphism.

4. Let θ be the mapping from $\mathbf{Z}_4$ to $\mathbf{Z}_2$ with $\bar{0} \mapsto \bar{0}$, $\bar{1} \mapsto \bar{1}$, $\bar{2} \mapsto \bar{0}$, and $\bar{3} \mapsto \bar{1}$. Explain why the following are true.

(i) $\theta(\bar{a}) = \bar{a}$.

(ii) $\bar{a} + \bar{b} = \bar{c}$ in $\mathbf{Z}_4$ implies that $\bar{a} + \bar{b} = \bar{c}$ in $\mathbf{Z}_2$.

(iii) $\bar{a} \cdot \bar{b} = \bar{d}$ in $\mathbf{Z}_4$ implies that $\bar{a} \cdot \bar{b} = \bar{d}$ in $\mathbf{Z}_2$.

(iv) θ is a ring homomorphism.

5. (a) Show that there is no mapping θ from $\mathbf{Z}_4$ to $\mathbf{Z}_3$ with $\theta(\bar{n}) = \bar{n}$ for all n in $\mathbf{Z}$.

(b) Show that $\theta(\bar{n}) = \bar{n}$ for all n in $\mathbf{Z}$ makes θ a mapping from $\mathbf{Z}_{15}$ to $\mathbf{Z}_3$.

(c) Show that the θ of part (b) is a surjective ring homomorphism, and find its kernel.

6. Let R be a finite commutative ring. Explain why the principal ideal (x) consists of all the entries of the row (or of the column) for x in the multiplication table for R.

7. (a) Show that the mapping θ from $\mathbf{Z}_2$ to $\mathbf{Z}_6$ with $\theta(\bar{0}) = \bar{0}$ and $\theta(\bar{1}) = \bar{3}$ is an injective ring homomorphism.

(b) Show that there is no injective ring homomorphism θ from $\mathbf{Z}_2$ to $\mathbf{Z}_4$.

8. Let $\mathbf{Z}$ and $\mathbf{Q}$ be the rings of the integers and of the rational numbers, respectively. Every rational number r is uniquely expressible in the form s/t with s and t relatively prime integers and t positive. Let the mapping θ from $\mathbf{Q}$ to $\mathbf{Z}$ be given by $\theta(r) = s$. Is θ a ring homomorphism? Explain.

9. For which integer m is the principal ideal (m) in $\mathbf{Z}$ finite?

10. For which integers n does the principal ideal (n) contain all the integers? (There is a negative n as well as a positive one.)

11. Let $m > 1$ so that the quotient ring $\mathbf{Z}/(m) = \mathbf{Z_m} = \{\bar{0}, \bar{1}, \ldots, \overline{m-1}\}$ has at least two elements.

(i) What is the additive identity of $\mathbf{Z_m}$?
(ii) Does $\mathbf{Z_m}$ have a unity? If so, which element is the unity?
(iii) Is the additive group of $\mathbf{Z_m}$ cyclic?
(iv) Is $\bar{0}$ an invertible in $\mathbf{Z_m}$?

12. (i) How many elements are there in $\mathbf{Z_1} = \mathbf{Z}/(1)$?
(ii) Does $\mathbf{Z_1}$ have a unity? If so, which element is the unity?
(iii) Is $\bar{0}$ an invertible in $\mathbf{Z_1}$?

13. The natural map θ from $\mathbf{Z}$ to $\mathbf{Z_{14}}$ is given by

$$\theta(a) = a + (14) = \bar{a}.$$

(i) Which integers are in the kernel K of θ?
(ii) Which integers are in the complete inverse image $\theta^{-1}(\bar{1})$?

14. Answer questions (i) and (ii) of Problem 13 for the natural map θ from $\mathbf{Z}$ to $\mathbf{Z_m}$, where m is an integer greater than 1.

15. The tables of Example 1 may be helpful in the following parts.

(a) Show the negative of each element of $\mathbf{Z_6} = \{\bar{0}, \bar{1}, \bar{2}, \bar{3}, \bar{4}, \bar{5}\}$ in a table.
(b) Show the order of each element of the additive group of $\mathbf{Z_6}$ in a table.
(c) List the elements of the principal ideal $(\bar{a})$ for each $\bar{a}$ in $\mathbf{Z_6}$. [*Hint:* See Problem 6.]
(d) Is the unity $\bar{1}$ in $(\bar{2})$? Is $\bar{2}$ an invertible in $\mathbf{Z_6}$?
(e) Find all $\bar{a}$ and $\bar{b}$ such that $\bar{a} \cdot \bar{b} = \bar{1}$ in $\mathbf{Z_6}$. List the elements of the multiplicative group $\mathbf{V_6}$ of invertibles in $\mathbf{Z_6}$.
(f) Find all $\bar{a}$ and $\bar{b}$ in $\mathbf{Z_6}$ such that $\bar{a} \neq \bar{0}$, $\bar{b} \neq \bar{0}$, but $\overline{ab} = \bar{0}$.
(g) List the squares of the elements of $\mathbf{Z_6}$.
(h) Tabulate $y = x(x + \bar{5})$ and $z = (x + \bar{2})(x + \bar{3})$ for x in $\mathbf{Z_6}$.
(i) Find all solutions of $x(x + \bar{5}) = \bar{0}$ in $\mathbf{Z_6}$.

16. Let p be a prime in the integers $\mathbf{Z}$. Explain why the following are true.

(i) If $\bar{p} = \bar{2}$ in $\mathbf{Z_6}$, then $p = 2$.
(ii) If $\bar{p} = \bar{3}$ in $\mathbf{Z_6}$, then p is 3 or -3.
(iii) If $\bar{p} = \bar{4}$ in $\mathbf{Z_6}$, then $p = -2$.
(iv) $\bar{p}$ cannot be $\bar{0}$ in $\mathbf{Z_6}$.
(v) If $p > 3$, then $\bar{p}$ must be $\bar{1}$ or $\bar{5} = -\bar{1}$ in $\mathbf{Z_6}$.

17. (i) Use Problem 15(e) to explain why a number c in the set $T = \{7, 13, 19, 25, \ldots, 6n + 1, \ldots\}$ is composite if and only if c is the product ab of two smaller integers with either a and b both in T or a and b both in the set

$$S = \{5, 11, 17, 23, \ldots, 6n - 1, \ldots\}.$$

(ii) List all the primes in the finite set $\{7, 13, 19, \ldots, 97\}$.

18. Use Problems 16 and 17 together with Examples 1 and 2 to find all the primes among the first one hundred positive integers.

19. (i) Make the multiplication table for $\mathbf{Z_5} = \{\bar{0}, \bar{1}, \bar{2}, \bar{3}, \bar{4}\}$.
(ii) Give the reciprocal (multiplicative inverse) of each of $\bar{1}, \bar{2}, \bar{3}, \bar{4}$.
(iii) State the order of the group $\mathbf{V_5}$ of invertibles of $\mathbf{Z_5}$.
(iv) Explain why $v^4 = \bar{1}$ for every v in $\mathbf{V_5}$.

(v) Let a be an integer that is not a multiple of 5. Explain why $\bar{a}$ is in $\mathbf{V_5}$ and hence $\bar{a}^4 = \bar{1}$.
(vi) Let a be an integer that is not a multiple of 5. Explain why $5 \mid (a^4 - 1)$ in the integers $\mathbf{Z}$.
(vii) Explain why $\bar{b}^5 = \bar{b}$ for every integer b.
(viii) Explain why $5 \mid (b^5 - b)$ for every integer b.
(ix) List the elements of the principal ideal $(\bar{a})$ for each $\bar{a}$ in $\mathbf{Z_5}$. [*Hint:* See Problem 6.]

20. Let $\mathbf{V_9}$ be the multiplicative group of invertibles of

$$\mathbf{Z_9} = \{\bar{0}, \bar{1}, \bar{2}, \bar{3}, \bar{4}, \bar{5}, \bar{6}, \bar{7}, \bar{8}\}.$$

(i) Make the multiplication table for $\mathbf{Z_9}$.
(ii) Show that $\bar{1}, \bar{2}, \bar{4}, \bar{5}, \bar{7}, \bar{8}$ are invertibles by finding their reciprocals in $\mathbf{Z_9}$.
(iii) Explain why $\bar{0}$, $\bar{3}$, and $\bar{6}$ are not invertibles in $\mathbf{Z_9}$.
(iv) What is the order of $\mathbf{V_9}$?
(v) Explain why $\bar{a}^6 = \bar{1}$ for every $\bar{a}$ in $\mathbf{V_9}$.
(vi) Let a be an integer such that $\gcd(a, 9) = 1$. Explain why $\bar{a}$ is in $\mathbf{V_9}$ and hence $\bar{a}^6 = \bar{1}$.
(vii) Let a be an integer with $\gcd(a, 9) = 1$. Explain why $9 \mid (a^6 - 1)$.
(viii) List the elements of the principal ideal $(\bar{a})$ for each $\bar{a}$ in $\mathbf{Z_9}$. [*Hint:* See Problem 6.]

21. List the elements of each of the following principal ideals in $\mathbf{Z_8}$.
(a) $(\bar{2})$; (b) $(\bar{6})$; (c) $(\bar{3})$; (d) $(\bar{5})$.

22. List the elements of each of the following principal ideals in $\mathbf{Z_7}$.
(a) $(\bar{0})$; (b) $(\bar{1})$; (c) $(\bar{2})$; (d) $(\bar{5})$.

23. (i) Explain why $(\bar{a}) = (\overline{m - a})$ for all $\bar{a}$ in $\mathbf{Z_m}$.
(ii) How many (distinct) principal ideals are there in $\mathbf{Z_{12}}$?

24. (i) Let I be an ideal in a ring U with unity and let I contain an invertible (that is, unit) v of U. Show that $I = U$.
(ii) Explain why $\mathbf{Z_{11}}$ has only two ideals.

25. (a) In $\mathbf{Z_{10}}$, find all x such that $x^2 = x$, and show that the subring $S = (\bar{5}) = \{\bar{0}, \bar{5}\}$ has a unity even though the unity $\bar{1}$ of $\mathbf{Z_{10}}$ is not in S.
(b) In $\mathbf{Z_{16}}$, show that the only subrings S with a unity are $\mathbf{Z_{16}}$ and the single element ring $\{\bar{0}\}$.
(c) In $\mathbf{Z_{30}}$, find all the subrings S with a unity.

26. (i) Explain why a unity u of a subring S in a ring R must satisfy $u^2 = u$.
(ii) Let S be a subring in $\mathbf{Z_m}$. Prove that S has a unity if and only if there exists an element $u^2 = u$ such that S is the principal ideal (u).

27. Let θ be the mapping from $\mathbf{Z_{12}}$ to $\mathbf{Z_6}$ such that the image of the element $\bar{a}$ of $\mathbf{Z_{12}}$ is the element $\overline{4a}$ of $\mathbf{Z_6}$.
(i) Tabulate the mapping θ.
(ii) Prove that θ is a ring homomorphism.
(iii) Show that the image set M of θ is a subring in $\mathbf{Z_6}$.
(iv) Show that $\theta(\bar{1})$ is the unity of the subring M but is not the unity of $\mathbf{Z_6}$.
(v) Explain why there is no homomorphism α from $\mathbf{Z_{12}}$ to $\mathbf{Z_6}$ for which $\alpha(\bar{1}) = \bar{2}$.

28. Let θ be a ring homomorphism from R to R'. Let K and M be the kernel and image set of θ, respectively. Show the following.
(a) M is a subring in R'.
(b) The quotient ring R/K is isomorphic to M. [See Theorem 2(f) of Section 3.3.]

Karl Friedrich Gauss
1777–1855

Gauss was surely one of the giants in the history of mathematics. A precocious youth, he is said to have remarked that he was calculating before he learned to talk. Before reaching the age of twenty, he demonstrated that the 17-sided regular polygon is ... veloped the

Probably his greatest work is his *Disquisitiones Arithmeticae* of 1801, a momumental treatise on the theory of numbers. Gauss once described mathematics as the queen of the sciences and the theory of numbers as the queen of mathematics. In the *Disquisitiones* he developed congruence arithmetic and proved one ... beautiful ... matics: ... ic ... ecture ... nd ... proved ... Gauss ... ear to ... of the ... ad ... ne later ... dditional

Gauss went on to work in many areas of mathematics and physics, astronomy, and geodesics. He showed a reluctance to publish his results, and many discoveries attributed to others were later found in Gauss's diary or in his manuscripts. Probably the most famous instance of this is the simultaneous development of noneuclidean geometry by Bolyai and Lobachevski, a discovery made earlier by Gauss.

He was professor at the University of Göttingen, and never once during his life did he leave Germany. The monument to Gauss in Braunschweig has a base chosen appropriately in the form of a polygon of 17 sides.

Tuesday

check

200 per

/ 400 > June

... he unity of M (but not necessarily the unity of R').
... e group V of invertibles in R, $\theta(v)$ is in the multiplica-... les in M. Hence θ, with its domain restricted to be just ... hism from V to V'.
... e inverse image $\theta^{-1}(m)$ is a coset of the kernel K.
... r of elements in M is an integral divisor of the number of elements ...

29. Explain why every subring, and hence every ideal, in $\mathbf{Z}_m$ is a cyclic subgroup of the additive group of $\mathbf{Z}_m$ and, therefore, is a principal ideal in $\mathbf{Z}_m$.

30. Prove that $(\bar{a}) = (\bar{b})$ in $\mathbf{Z}_m$ if and only if $\gcd(a, m) = \gcd(b, m)$.

31. Use Problems 29 and 30 to find all the subrings in $\mathbf{Z}_{24}$.

32. Find all the subrings in $\mathbf{Z}_{36}$.

33. (i) Prove that there is no surjective ring homomorphism from $\mathbf{Z}_{30}$ to $\mathbf{Z}_7$.
(ii) Describe a ring homomorphism from $\mathbf{Z}_{30}$ to $\mathbf{Z}_7$ that is not surjective.

34. Let m and n be positive integers. Show the following.
(i) If $m \mid n$, then $\bar{a} + \bar{b} = \bar{s}$ and $\bar{a} \cdot \bar{b} = \bar{t}$ in $\mathbf{Z}_n$ imply $\bar{a} + \bar{b} = \bar{s}$ and $\bar{a} \cdot \bar{b} = \bar{t}$ in $\mathbf{Z}_m$.

Richard Dedekind
1831–1916

Dedekind, like Gauss, was born in Braunschweig and became a professor of mathematics at Göttingen, which between the time of Gauss and the rise of the Nazis in the 1930s, was one of the foremost centers of mathematical scholarship in the world. He later taught in Zürich at the Polytechnicum and eventually returned to Braunschweig. It is to Dedekind that we owe the development of the theory of ideals. He was, however, motivated by the work of **E. E. Kummer** (1810–1893) and his "ideal numbers." Kummer's work was part of an unsuccessful attempt to prove the famous Fermat conjecture that, for all integers $n > 2$, there exist no positive integer triples x, y, z such that $x^n + y^n = z^n$. It is one of the many instances in which mathematics developed for very special purposes turns out to be of far wider interest and applicability. The work of Dedekind on rings and ideals was much further developed in the twentieth century at Göttingen by the great mathematician Emmy Noether. (For an interesting account of Dedekind's work on the invention of ideals, see the article by Harold M. Edwards in Phillips, E. R., *Studies in the History of Mathematics*, listed in the Annotated Bibliography.)

(ii) There is a surjective ring homomorphism from $\mathbf{Z}_n$ to $\mathbf{Z}_m$ if and only if $m \mid n$.

35. Let n be a positive integer. Show that there is an injective ring homomorphism from $\mathbf{Z}_2$ to $\mathbf{Z}_{2n}$ if and only if n is odd. (See Problem 7 above.)

36. Let q and q' be the numbers of elements in finite rings R and R', respectively. Let there be an injective ring homomorphism from R to R'. Show that $q \mid q'$.

37. Let n be a positive integer. Explain why there are at most $n + 1$ distinct squares of elements of $\mathbf{Z}_{2n}$.

38. Let n be a positive integer. Explain why there are at most $n + 1$ distinct squares of elements of $\mathbf{Z}_{2n+1}$.

39. Let I_1 and I_2 be ideals in a ring R. Let I be the subset of R consisting of the elements that are in both I_1 and I_2; that is, let

$$I = I_1 \cap I_2.$$

(a) Show that I is an ideal in R.
(b) Explain why I is an ideal in I_1 (and in I_2).

40. Let s and t be integers and let $m = \operatorname{lcm}[s, t]$. Show that $(s) \cap (t) = (m)$ in $\mathbf{Z}$. (See Section 1.6.)

41. Use the natural map from $\mathbf{Z}$ to $\mathbf{Z}_3$ to prove that there are no integers x and y satisfying

$$x^2 - 3y^2 = 992.$$

42. Prove that there are no integers x and y that satisfy

(a) $x^2 + (x + 1)^2 + (x + 2)^2 = y^2$.
(b) $x^2 + (x + 1)^2 + (x + 2)^2 + (x + 3)^2 = y^2$.

43. Prove that there are no squares of integers in the sequence

$$11, 111, 1111, 11111, \ldots$$

of integers with each digit a 1 (and at least two digits).

★ **44.** Prove that there are no integers x and y satisfying

$$x^2 + 3xy - 2y^2 = 122.$$

45. Let $G = \{m + n\mathbf{i} : m, \ n \in \mathbf{Z}\}$ be the ring of gaussian integers. Show that $S = \{m + 2n\mathbf{i} : m, n \in \mathbf{Z}\}$ is a subring in G but is not an ideal in G.

4.3 Congruence in Z, The Euler and Fermat Theorems

Algebra is generous; she often gives more than is asked of her.
—Jean LeRond d'Alembert

In this section we obtain some important results in the Theory of Numbers by applying group theory, especially Lagrange's Theorem, to the multiplicative groups $\mathbf{V_m}$ of the invertibles of $\mathbf{Z_m}$.

We start by characterizing the integers a such that $\bar{a}$ is in $\mathbf{V_m}$.

LEMMA 1 Invertibles of $\mathbf{Z_m}$

An $\bar{a}$ of $\mathbf{Z_m}$ is an invertible if and only if the integers a and m are relatively prime.

Proof First let $\gcd(a, m) = 1$. Then there are integers h and k such that $ah + mk = 1$. (See Corollary (b) to Theorem 2 of Section 1.4.) Using the fact that the natural map from $\mathbf{Z}$ to $\mathbf{Z_m}$ preserves additions and multiplications, we have

$$\bar{1} = \overline{ah + mk} = \bar{a} \cdot \bar{h} + \bar{m} \cdot \bar{k}.$$

Since $\bar{m} = \bar{0}$ in $\mathbf{Z_m}$, this leads to $\bar{1} = \bar{a} \cdot \bar{h}$; thus $\bar{a}$ has a reciprocal $\bar{h}$ in $\mathbf{Z_m}$, and $\bar{a}$ is in $\mathbf{V_m}$.

Conversely, let $\bar{a}$ be an invertible in $\mathbf{Z_m}$; that is, let $\bar{a} \cdot \bar{h} = \bar{1}$ in $\mathbf{Z_m}$. Then $\overline{ah} = \bar{1}$, which implies that $m \mid (1 - ah)$. This in turn means that there is an integer k with $1 - ah = mk$. Then $ah + mk = 1$, and it follows that a and m are relatively prime. ∎

COROLLARY 1 $\mathbf{V_p} = \{\bar{1}, \bar{2}, \ldots, \overline{p - 1}\}$ for all positive primes p.

Proof Since ± 1 and $\pm p$ are the only divisors of p, an integer n is relatively prime to p if and only if n is not a multiple of p. Hence each of $1, 2, \ldots, p - 1$ is relatively prime to p, and it follows from Lemma 1 that $\bar{1}, \bar{2}, \ldots, \overline{p-1}$ are in $\mathbf{V_p}$. Since $\bar{0}$ is the zero of $\mathbf{Z_p}$ and $\mathbf{Z_p}$ has more than one element, $\bar{0}$ is not an invertible. ■

COROLLARY 2 An $\bar{a}$ of $\mathbf{Z_m}$ is an invertible if and only if $\bar{a}$ is a generator of the additive group of $\mathbf{Z_m}$.

The proof of Corollary 2 is left to the reader as Problem 39 of this section.

We recall that $\bar{a} = \bar{b}$ in $\mathbf{Z_m}$ if and only if $a \equiv b \pmod{m}$, and use this fact together with Lagrange's Theorem in the following example. Then we see that the same technique produces important theorems of Euler and Fermat.

EXAMPLE 1 **Congruence Derived from the Order of $\mathbf{V_m}$**

Using Lemma 1, we find that the multiplicative group of invertibles in $\mathbf{Z_{18}} = \{\bar{0}, \bar{1}, \ldots, \overline{17}\}$ is

$$\mathbf{V_{18}} = \{\bar{1}, \bar{5}, \bar{7}, \overline{11}, \overline{13}, \overline{17}\}.$$

Thus $\mathbf{V_{18}}$ has order 6, and it follows from Lagrange's Theorem that $\bar{v}^6 = \bar{1}$ for every $\bar{v}$ in $\mathbf{V_{18}}$. We translate this fact into a statement about the set $\mathbf{Z}$ of the integers as follows.

If 18 and n are relatively prime integers, then $\bar{n}$ is in $\mathbf{V_{18}}$, $\bar{n}^6 = \bar{1}$, and hence $n^6 \equiv 1 \pmod{18}$.

We next give the definition of an important function in the Theory of Numbers.

NOTATION **The Euler ϕ-Function**

Let m be a positive integer. Then $\phi(m)$ denotes the number of integers a in $\{0, 1, 2, \ldots, m-1\}$, such that a and m are relatively prime, that is, such that $\gcd(a, m) = 1$. The function ϕ is known as the **Euler ϕ-function**, or **Euler totient**.

Since an $\bar{a}$ of $\mathbf{Z_m} = \{\bar{0}, \bar{1}, \bar{2}, \ldots, \overline{m-1}\}$ is in $\mathbf{V_m}$ if and only if a and m are relatively prime, it is clear that $\phi(m)$ is the order of $\mathbf{V_m}$. Then it follows from Lagrange's Theorem that

$$\bar{a}^{\phi(m)} = \bar{1}$$

for every $\bar{a}$ in $\mathbf{V_m}$. Translating this into the language of the integers $\mathbf{Z}$, we have the following famous result due to Leonhard Euler. (See the biographical note on p. 244.)

THEOREM 1 Euler's Theorem

Let a and m be relatively prime integers with $m > 1$. Then

$$m \mid (a^{\phi(m)} - 1);$$

that is, $a^{\phi(m)} \equiv 1 \pmod{m}$.

In Section 8.3, we describe an application of Euler's Theorem to modular coding, which has been used to provide secure transfers of funds over telephone wires.

Euler's Theorem is a generalization of the following result due to Pierre Fermat. (See the biographical note on p. 245.)

THEOREM 2 Fermat's Theorem

If p is a positive prime in the integers,

$$a^p \equiv a \pmod{p}$$

for all integers a. If $b \not\equiv 0 \pmod{p}$, then

$$b^{p-1} \equiv 1 \pmod{p}.$$

Proof Corollary 1 to Lemma 1 tells us that $\mathbf{V_p} = \{\bar{1}, \bar{2}, \ldots, \overline{p-1}\}$ and hence that the order of $\mathbf{V_p}$ is $\phi(p) = p - 1$. If $b \not\equiv 0 \pmod{p}$, then $\bar{b} \neq \bar{0}$ in $\mathbf{Z_p}$ and hence $\bar{b}$ is in $\mathbf{V_p}$. Then it follows from Euler's Theorem that $b^{p-1} \equiv 1 \pmod{p}$.

Now let a be any integer. If $\bar{a} \neq \bar{0}$ in $\mathbf{Z_p}$, then $\bar{a}$ is in $\mathbf{V_p}$ and $\bar{a}^{p-1} = \bar{1}$ in $\mathbf{V_p}$, using Lagrange's Theorem. Multiplying both sides by $\bar{a}$ gives us $\bar{a}^p = \bar{a}$, from which it follows that $a^p \equiv a \pmod{p}$. If $\bar{a} = \bar{0}$, then $\bar{a}^p = \bar{0} = \bar{a}$ in $\mathbf{Z_m}$, and it again follows that $a^p \equiv a \pmod{p}$. ■

Next we use the natural map from $\mathbf{Z}$ to $\mathbf{Z_m}$ to show that congruences modulo m can be added, subtracted, or multiplied to obtain new congruences.

Let m be a fixed positive integer and let θ be the natural map from $\mathbf{Z}$ to $\mathbf{Z_m}$. If $\bar{a} = \bar{b}$ and $\bar{c} = \bar{d}$ in $\mathbf{Z_m}$, then, of course,

$$\bar{a} \pm \bar{c} = \bar{b} \pm \bar{d}, \qquad \bar{a} \cdot \bar{c} = \bar{b} \cdot \bar{d}. \tag{F}$$

The equations of (F) and the fact that θ is a ring homomorphism tell us that

$$\overline{a \pm c} = \overline{b \pm d}, \qquad \overline{ac} = \overline{bd}.$$

This may be translated into congruence language as follows.

THEOREM 3 Congruence Preserved Under Addition, Subtraction, Multiplication

If $a \equiv b \pmod{m}$ and $c \equiv d \pmod{m}$, then $a \pm c \equiv b \pm d \pmod{m}$ and $ac \equiv bd \pmod{m}$.

See Problem 21 of this section for some information on congruences and division.

EXAMPLE 2 **Throwing Away Multiples of *m***

Here we use Theorem 3 as an aid in finding the remainder when 1453 is divided by 17. It is easier to divide by 40 than by 17, so we note that

$$1453 = 36 \cdot 40 + 13$$

and hence $1453 \equiv 2 \cdot 6 - 4 = 8 \pmod{17}$. Thus the remainder is 8.

EXAMPLE 3 **No Squares in 2 + 4Z nor in 3 + 4Z**

Modulo 4, an integer n must be congruent to 0, 1, 2, or 3. Then the table

$n \pmod 4$	0	1	2	3
$n^2 \pmod 4$	0	1	0	1

shows that all squares of integers are congruent to 0 or 1 modulo 4, and hence no square is in $2 + 4\mathbf{Z}$ or in $3 + 4\mathbf{Z}$.

EXAMPLE 4 **Order of $\overline{10}$ in $\mathbf{Z}_{17}$**

Here we find the smallest positive integer q such that

$$10^q \equiv 1 \pmod{17}$$

Since 17 is a prime, the order of $\mathbf{V}_{17}$ is $17 - 1 = 16$. This and Lagrange's Theorem imply that the order of $\overline{10}$ in $\mathbf{V}_{17}$ is 2, 4, 8, or 16. Then we note that $\overline{10}^2 = \overline{15}$, $\overline{10}^4 = \overline{15}^2 = (-\bar{2})^2 = \bar{4}$, $\overline{10}^8 = \bar{4}^2 = \overline{16} = -\bar{1}$, and $\overline{10}^{16} = (-\bar{1})^2 = \bar{1}$. These calculations show that $\overline{10}$ has order 16 in $\mathbf{V}_{17}$; that is, 16 is the smallest positive integer q such that $\overline{10}^q = \bar{1}$ in $\mathbf{V}_{17}$. Translating this into the language of congruences, we have 16 as the smallest q with $10^q \equiv 1 \pmod{17}$.

EXAMPLE 5 **Repeating Decimals**

Here we assume as given that $\overline{10}$ has order 5 in $\mathbf{V}_{41}$, and we use this fact to find the unending decimal representation for 1/41 and 6/41. The given fact tells us that $\overline{10}^5 = \bar{1}$ in $\mathbf{V}_{41}$. This implies that

$$41 \mid (10^5 - 1); \qquad \text{that is, } 41 \mid 99999.$$

Division shows that $99999 = 41 \cdot 2439$, and so $1/41 = 2439/99999$. Now the formula

$$1 + r + r^2 + r^3 + \cdots = (1 - r)^{-1}, \quad |r| < 1,$$

for the sum of an infinite geometric progression, gives us

$$\begin{aligned} 1/99999 &= 1/(10^5 - 1) = (10^5 - 1)^{-1} = 10^{-5}(1 - 10^{-5})^{-1} \\ &= 0.00001(1 + 10^{-5} + 10^{-10} + 10^{-15} + \cdots) \\ &= 0.000010000100001 \ldots. \end{aligned}$$

Hence we have the decimal representations

$$1/41 = 2439/99999 = 0.024390243902439\ \ldots,$$

$$6/41 = 14634/99999 = 0.1463414634\ \ldots,$$

in each of which a block of five digits is repeated endlessly.

Problems

Problem 38 below is cited in Section 7.3.

1. (i) List the order of each element of $\mathbf{V}_{18} = \{\bar{1}, \bar{5}, \bar{7}, \overline{11}, \overline{13}, \overline{17}\}$ in a table.
(ii) Is $\mathbf{V}_{18}$ cyclic? If so, which elements are generators?

2. (i) Find the smallest positive integer q such that $v^q = \bar{1}$ for every v in $\mathbf{V}_{12}$.
(ii) Is $\mathbf{V}_{12}$ cyclic? Explain.

3. (i) Show that $4! = 1 \cdot 2 \cdot 3 \cdot 4 \equiv -1 \pmod 5$.
(ii) Show that $1 \cdot 2 \cdot 3 \cdot 4 \cdot 5 \cdot 6 \equiv -1 \pmod 7$.
(iii) Translate the statement in (ii) into an equality in $\mathbf{V}_7$.
(iv) Translate the statement in (ii) into a divisibility statement of the form "$a \mid b$ in **Z**."

4. (i) Show that $3! = 1 \cdot 2\ \ 3 \equiv 2 \pmod 4$.
(ii) Show that $\bar{1} \cdot \bar{2} \cdot \bar{3} \cdot \bar{4} \cdot \bar{5} = \bar{0}$ in $\mathbf{Z}_6$.
(iii) Translate the statement in (ii) into congruence language.
(iv) Translate the statement in (ii) into a divisibility statement of the form "$a \mid b$ in **Z**."

5. Show the following.
(i) $3 \cdot 11 \equiv 1 \pmod{16}$.
(ii) $5 \cdot 13 \equiv 1 \pmod{16}$.
(iii) $7 \cdot 9 \equiv -1 \pmod{16}$.
(iv) $15 \equiv -1 \pmod{16}$.
(v) $1 \cdot 3 \cdot 5 \cdot 7 \cdot 9 \cdot 11 \cdot 13 \cdot 15 \equiv 1 \pmod{16}$.
(vi) The product $v_1 v_2 \cdots v_8$ of the eight elements of the group $\mathbf{V}_{16}$ equals $\bar{1}$.

6. (i) Show that the product of the six elements of $\mathbf{V}_{18}$ is $-\bar{1} = \overline{17}$.
(ii) Translate the statement in (i) into congruence language.

7. Given that $a \equiv 3 \pmod 4$, use Example 3 to prove that there are no integers x and y such that $x^2 + y^2 = a$.

8. Given that $b \equiv 2 \pmod 4$, use Example 3 to prove that there are no integers x and y such that $x^2 - y^2 = b$.

9. Explain why the unity $\bar{1}$ of $\mathbf{Z}_m$ and $-\bar{1} = \overline{m-1}$ are in $\mathbf{V}_m$ for $m \geq 2$.

10. Let $m \geq 2$. Show that $\overline{m-a}$ is in $\mathbf{V}_m$ if and only if $\bar{a}$ is in $\mathbf{V}_m$.

11. Complete the following partial table for the Euler ϕ-function.

m	1	2	3	4	5	6	7	8	9	10
$\phi(m)$	1	1				2	6			

12. Let ϕ be the Euler ϕ-function and find the following.
(i) $\phi(16)$; (ii) $\phi(32)$; (iii) $\phi(2^n)$, where n is a positive integer.

13. Give a formula for $\phi(3^n)$, where n is a positive integer.

14. Let p be a positive prime and n be a positive integer. Give a formula for $\phi(p^n)$.

15. Find the x in $\{1, 2, 3, \ldots, 20\}$ such that $10x \equiv 1 \pmod{21}$.

16. Find the y in $\{1, 2, 3, \ldots, 20\}$ such that $10y \equiv 13 \pmod{21}$.

17. (i) Let a and m be relatively prime integers with $m > 1$. Explain why there is an h in $\{1, 2, \ldots, m-1\}$ such that $ah \equiv 1 \pmod{m}$.
(ii) Give the analogue of (i) for the case $m = 1$.

18. Let a, b, and m be integers with $m > 1$ and $\gcd(a, m) = 1$. Explain why there is a k in $\{0, 1, \ldots, m-1\}$ such that

$$ak \equiv b \pmod{m}.$$

19. Find all elements x of $\mathbf{Z}_{12}$ such that $0, x, 2 \cdot x, 3 \cdot x, \ldots, 11 \cdot x$ are distinct and hence compose all of $\mathbf{Z}_{12}$.

20. Let a and m be integers with $m > 0$. Prove that

$$\bar{0}, \bar{a}, 2 \cdot \bar{a}, \ldots, (m-1) \cdot \bar{a}$$

(are distinct elements of $\mathbf{Z}_m$ and hence compose all of $\mathbf{Z}_m$ if and only if $\gcd(a, m) = 1$.

21. (i) Give an example of integers a, b, c, and m such that $ab \equiv ac \pmod{m}$ but neither $b \equiv c \pmod{m}$ nor $a \equiv 0 \pmod{m}$ is true.
(ii) Prove that $ab \equiv ac \pmod{m}$ and $\gcd(a, m) = 1$ together imply $b \equiv c \pmod{m}$. [*Hint:* "Cancel" a, using Problem 17.]

22. Given that $a \equiv b \pmod{m}$, prove by mathematical induction that $a^n \equiv b^n \pmod{m}$ for all nonnegative integers n.

23. (i) Explain why $10^n \equiv 1 \pmod{9}$ for all nonnegative integers n.
(ii) Let $a_0, a_1, \ldots, a_d$ be integers. Explain why

$$10^d a_d + \cdots + 10a_1 + a_0 \equiv a_d + \cdots + a_1 + a_0 \pmod{9}.$$

(iii) Show that $123456789 \equiv 0 \pmod{9}$.
(iv) Explain why every number obtained by rearranging the digits of 123456789 is a multiple of 9.

24. (i) Explain why $10^n \equiv (-1)^n \pmod{11}$ for all nonnegative integers n.
(ii) Let $a_0, a_1, \ldots, a_d$ be integers. Explain why

$$10^d a_d + \cdots + 10a_1 + a_0 \equiv a_0 - a_1 + a_2 - \cdots + (-1)^d a_d \pmod{11}.$$

(iii) Show that $123456 \equiv 3 \pmod{11}$.
(iv) Explain why no number obtained by rearranging the digits of 123456 can be a multiple of 11.

25. For each of the following powers of 2, find the number in $\{1, 2, 3, 4\}$ to which it is congruent modulo 5.
(i) 2^6; (ii) 2^7; (iii) 2^8; (iv) 2^9; (v) 2^{10}; (vi) 2^{777}.

26. For each of the following values of n, find the number in $\{1, 3, 7, 9\}$ to which 3^n is congruent modulo 10.
(i) 6; (ii) 7; (iii) 8; (iv) 9; (v) 10; (vi) 555.

27. Find x and y in $\{1, 2, 3, \ldots, 12\}$ such that

$$3^{200} \equiv x \pmod{13}, \qquad 94^{200} \equiv y \pmod{13}.$$

28. (i) Find the last digit (units digit) of $a = 7^{123}$. That is, find the remainder in the division of a by 10.
(ii) Find the last digit of 987^{123}.

29. Let a be an integer that is not an integral multiple of 7. Explain why the smallest positive integer q such that $a^q \equiv 1 \pmod 7$ must be 1, 2, 3, or 6.

30. Let a be an integer with $\gcd(a, 19) = 1$. What are the possible values of the least positive integer q such that $a^q \equiv 1 \pmod{19}$?

31. Let m be a positive integer whose last digit (units digit) is 1, 3, 7, or 9. Explain why the following are true.
(i) $\gcd(10, m) = 1$; that is, 10 and m are relatively prime.
(ii) There exist positive integers s such that

$$10^s \equiv 1 \pmod m.$$

(iii) There are integral multiples t of m such that every digit of t is a 9.

32. (a) Find the smallest positive integer r such that $17 | r$ and every digit of r is a 9. (See Example 4 of this section.)
(b) Find the smallest positive integer s such that $13 | s$ and every digit of s is a 9.
(c) Find the smallest positive integer t such that $221 | t$ and every digit of t is a 9.

33. (i) Show that $H = \{\bar{1}, \bar{4}\}$ and $N = \{\bar{1}, \overline{14}\}$ are normal subgroups in $\mathbf{V_{15}}$.
(ii) Show that $\mathbf{V_{15}}/N$ is cyclic and $\mathbf{V_{15}}/H$ is not cyclic.

34. Let N and N' be normal subgroups in G and G', respectively. Show that G/N and G'/N' need not be isomorphic even though G and G' are isomorphic and N and N' are isomorphic. (See Problem 33.)

35. Extend the table of Problem 11 by tabulating $\phi(m)$ for $m = 11, 12, \ldots, 20$.

36. Show the following.
(a) $\phi(1) + \phi(2) + \phi(3) + \phi(6) + \phi(9) + \phi(18) = 18$.
(b) $\phi(1) + \phi(2) + \phi(3) + \phi(5) + \phi(6) + \phi(10) + \phi(15) + \phi(30) = 30$.
[The equations in (a) and (b) are examples of *Gauss's Theorem* that n is the sum of the values of $\phi(d)$ for all positive integral divisors d of n.]

37. Show the following.
(a) $\phi(2)\phi(9) = \phi(18) > \phi(3)\phi(6)$.
(b) $\phi(3)\phi(4) = \phi(12) > \phi(2)\phi(6)$.
(c) $\phi(4)\phi(9) = \phi(36) > \phi(3)\phi(12) > \phi(2)\phi(18) > [\phi(6)]^2$.

38. Explain why one could define $\phi(n)$ to be the number of integers a in $\{1, 2, \ldots, n\}$ such that $\gcd(a, n) = 1$.

39. Show that an $\bar{a}$ of $\mathbf{Z_m}$ is an invertible if and only if $\bar{a}$ is a generator of the additive group of $\mathbf{Z_m}$ (and thus prove Corollary 2 to Lemma 1). [*Hint:* See Problems 39 and 40 of Section 2.7.]

40. Show that $\phi(n)$ is the number of generators in a cyclic group of order n.

41. For which integers a in $\{0, 1, 2, 3, 4\}$ do there exist integers x such that $x^2 \equiv a \pmod 5$?

42. For which integers b in $\{-3, -2, -1, 0, 1, 2, 3\}$ do there exist integers y such that $y^2 \equiv b \pmod 7$?

43. Show that $a^m \equiv a^n \pmod 2$ for integers a, m, and n with $m > 0$ and $n > 0$.

44. (a) Use Fermat's Theorem to show that $a^{2m-1} \equiv a^{2n-1} \pmod 3$ for all integers a and all positive integers m and n.
(b) Show that $a^{2m} \equiv a^{2n} \pmod 3$ for all integers a and all positive integers m and n.

45. Let a, b, c, r, and s be integers. Show the following.
(a) If $a \equiv b \pmod{rs}$, then $a \equiv b \pmod s$.
(b) If $a \equiv b \pmod r$, $a \equiv b \pmod s$, and $m = \operatorname{lcm}[r, s]$, then $a \equiv b \pmod m$.
(c) If $c \equiv a \pmod r$, $c \equiv b \pmod s$, and $d = \gcd(r, s)$, then $a \equiv b \pmod d$.

Leonhard Euler
1707–1783

Euler, the greatest Swiss mathematician and one of the most creative and most prolific mathematicians of all times, spent almost all of his life outside Switzerland. He was a student of Jean Bernoulli in Basel but in 1727 left for St. Petersburg, Russia, to take a position (in medicine and physiology!) at the Imperial Academy there. In 1733, upon the departure of Daniel Bernoulli, the mathematician at the Academy, Euler became the chief mathematician and remained there until he accepted a comparable appointment at the Berlin Academy in 1741. In 1766 he returned to Russia, where he worked until his death in 1783.

Euler's life represents a sharp contrast to those of Abel and Galois, both of whom died young and in rather dramatic or romantic circumstances. Euler had a large family, produced a prodigious number of books and papers, and died at the age of 76 while playing with one of his grandchildren. He was blind during the last 17 years of his life, but this did not keep him from producing further mathematical results. He dictated his results to others. His complete edited works run to over 70 folio volumes.

Euler did work in almost all the branches of mathematics, as well as in areas of physics and astronomy, and even naval architecture and the theory of music. In addition to discovering new mathematical results, he wrote textbooks on algebra and calculus. His writing was a model of clarity. In later years he wrote a three-volume set of letters to a German princess in which he furthered her scientific education by answering questions such as, "Why is the sky blue?"

Euler is responsible for the wide use, if not the first use, of many modern mathematical notations: π, e, $\mathbf{i}$ (all combined in his remarkable formula $e^{\pi \mathbf{i}} = -1$); the abbreviations for the trigonometric functions; $\sum$ for summation; and $f(x)$ for function values. He had a genius for producing beautiful formulas. So many formulas and mathematical ideas are named for Euler, one can easily become confused: Euler's formula $e^{\mathbf{i}x} = \cos x + \mathbf{i} \sin x$ for complex numbers with absolute value 1 (this appears on a Swiss postage stamp commemorating Euler); the Euler formula for polyhedra, $V - E + F = 2$; the Euler constant $\gamma = \lim_{n\to\infty}(1 + \frac{1}{2} + \frac{1}{3} + \cdots + (1/n) - \log n)$; the Euler integrals (the so-called gamma and beta functions); the Euler ϕ-function (see Sections 4.3 and 7.3); and so on.

Euler was the first to publish a proof of the little Fermat theorem (Theorem 2 of Section 4.3 in this text). although Leibniz left a manuscript in which a demonstration appears. Euler also provided the generalization, using his ϕ-function.

Pierre de Fermat
1601–1665

Fermat was a lawyer by profession and served in the Parlement in Toulouse. He is probably the outstanding example of an amateur in mathematics. The word *amateur* is used only to indicate that his great mathematical work was in a field different from his profession; he certainly had the highest professional competence as a mathematician.

He discovered the basic idea of analytic geometry at least a year prior to the publication of Descartes' *La Géométrie*. Furthermore, in the 1630s he developed methods for calculating maxima and minima for certain curves, for determining slopes of tangent lines to curves of the form $y = f(x)$, and for determining the area under the curves $y = x^m$, m rational. Again he did not publish, so the first work on the calculus to appear in print remains the "Nova methodus pro maximis et minimis" of Leibniz of 1684. (Newton, although he had the basic ideas of the calculus earlier, did not publish his results until after that date.)

Much of Fermat's work was motivated by his interest in the classics; the work on analytic geometry and calculus was prompted by his study of Apollonius and Pappus. The contributions for which he is most remembered are those in the Theory of Numbers. He wrote in the margin of the Bachet 1621 edition of Diophantus's *Arithmetica* that he had a proof that for every integer $n > 2$ there exist no positive integer triples x, y, z satisfying $x^n + y^n = z^n$ but that there was too little room in the margin for the proof. This is the famous Fermat Last Theorem (actually a conjecture), still unproved although verified for many values of n.

The "little" Fermat theorem presented in Section 4.3 is a generalization of the special case, known to the Chinese more than 2000 years ago, that $2^p - 2$ is exactly divisible by p, for all positive primes p. The first published proof of the little Fermat theorem was given by Euler.

Fermat published almost nothing, instead making his results known in letters to a French priest, **Marin Mersenne** (1588–1648), who then passed them on to others. Mersenne acted as a clearinghouse for many mathematical results of his era, but he holds a place in the history of the theory of numbers in his own right. Mersenne numbers (integers of the form $2^p - 1$, p a prime) are named for him. Fermat's edition of Diophantus's text was published posthumously in Toulouse in 1670. His own works did not appear until 1679 in his *Varia Opera Mathematica*.

46. Show the following for all integers a.
(a) $a^5 \equiv a \pmod{30}$.
(b) $a^7 \equiv a \pmod{42}$.
(c) $a^{13} \equiv a \pmod{2730}$.
(d) $a^{21} \equiv a \pmod{330}$.

47. Show that $a^p \equiv a \pmod{6p}$ for all integers a and all primes p with $p > 3$.

48. Show that $a^p b \equiv b^p a \pmod{6p}$ for all primes $p > 3$ and all integers a and b.

49. Explain why $a/17$ has a decimal representation in which a block of 16 digits repeats endlessly, for every positive integer a.

50. (i) Find the order of $\bar{2}$, $\bar{5}$, and $\overline{10}$ in $\mathbf{V}_{31}$.
(ii) Find the smallest positive integer m such that in the decimal representation for $1/31$ there is a block of m digits that repeats endlessly.

51. Let d and e be in $\mathbf{Z}^+ = \{1, 2, 3, \ldots\}$, $n = de$, and

$$r \in \{1, 2, 3, \ldots, n\}.$$

Show that $\gcd(r, n) = e$ if and only if there is a k in $\{1, 2, \ldots, d\}$ such that $r = ek$ and $\gcd(k, d) = 1$.

52. Let $[a]$ be a cyclic group of order m, $d \mid m$, and $d > 0$. Prove that there are exactly $\phi(d)$ elements of order d in $[a]$.

53. Let $d_1, d_2, \ldots, d_r$ be all the positive integral divisors of m. Explain why

$$\phi(d_1) + \phi(d_2) + \cdots + \phi(d_r) = m.$$

54. Prove that the number of primitive complex nth roots of unity is $\phi(n)$. (See Problem 55 of Section 2.7.)

55. Find the decimal representations for the following.
(i) $1/101$; (ii) $2/101$; (iii) $3/101$; (iv) $(10a + b)/101$,
with a and b digits in $\{0, 1, \ldots, 9\}$.

56. Find the decimal representations for the following.
(i) $1/1001$; (ii) $2/1001$; (iii) $3/1001$; (iv) $(100a + 10b + c)/1001$,
with a, b, and c digits in $\{0, 1, \ldots, 9\}$.

57. Use $1/7 = 143/1001$ and Problem 56(iv) to find the decimal representation for $1/7$.

58. Find the decimal representation for $1/13$. [*Hint:* $1/13 = 77/1001$.]

4.4 Integral Domains

In Section 4.1 we saw that a product of two elements of a ring is zero if one of the factors is zero. In the familiar number systems — the integers, the rational numbers, the real numbers, and the complex numbers — all of which are rings, the converse is also true; that is, a product is zero only if one of the factors is zero. But in the modular ring $\mathbf{Z}_6$ this converse does not hold, since $\bar{2} \cdot \bar{3} = \bar{0}$ but neither $\bar{2}$ nor $\bar{3}$ is the additive identity $\bar{0}$.

We now introduce terminology that helps us to distinguish between these two types of rings.

DEFINITION 1

0-Divisor in a Ring, Regular Element

If $ab = 0$ with $a \neq 0$ and $b \neq 0$ in a ring R, then a and b are said to be **0-divisors** (zero-divisors) in R. An element of R that is not a 0-divisor is a **regular element**.

If a ring R has 0-divisors, a quadratic equation $ax^2 + bx + c = 0$ can have more than two solutions in R [as one can see from Parts (h) and (i) of Problem 15 in Section 4.2].

DEFINITION 2

Integral Domain

A commutative ring D with a unity $1 \neq 0$ and with no 0-divisors is called an **integral domain**.

As one might guess from the name, the ring $\mathbf{Z}$ of the integers is an integral domain. So are the number systems $\mathbf{Q}$, $\mathbf{R}$, and $\mathbf{C}$. $\mathbf{Z_6}$ is not an integral domain, since some of its elements (namely, $\bar{2}$, $\bar{3}$, and $\bar{4}$) are 0-divisors.

When m is composite, one can readily show that $\mathbf{Z_m}$ has 0-divisors and hence is not an integral domain. (See Problem 5 of this section.) It can also be shown that $\mathbf{Z_p}$ is an integral domain for all positive primes p. (See Problem 7 of this section.)

In $\mathbf{Z_6}$, $\bar{2} \cdot \bar{1} = \bar{2} \cdot \bar{4}$ but $\bar{1} \neq \bar{4}$; this shows the danger in attempting to cancel a 0-divisor. The following result indicates a situation in which we are allowed to cancel.

THEOREM 1 **Multiplicative Cancellation in Rings**

In a ring R, let c be neither 0 nor a 0-divisor. Then either $ca = cb$ or $ac = bc$ implies that $a = b$.

Proof We deal with the case in which $ac = bc$ and leave the other case to the reader. Adding the same element $-bc$ to the given equal products, we have

$$ac + (-bc) = bc + (-bc),$$

which implies that

$$[a + (-b)]c = 0.$$

Since c is neither 0 nor a 0-divisor, the other factor $a + (-b)$ must be zero. Adding b to both sides of $a + (-b) = 0$ then leads to $a = b$, as desired. ∎

COROLLARY Cancellation in an Integral Domain

In an integral domain D, $c \neq 0$ and either $ac = bc$ or $ca = cb$ imply that $a = b$.

THEOREM 2 A Converse of Theorem 1

Let c be a nonzero element of a ring R such that each of $ac = bc$ and $ca = cb$ implies that $a = b$ in R. Then c is not a 0-divisor.

Proof If c were a 0-divisor, there would be an element $d \neq 0$ in R such that either $cd = 0$ or $dc = 0$. We deal with the case $cd = 0$; the other case is similar. Now $cd = 0$ and $c \cdot 0 = 0$ imply that $cd = c \cdot 0$. Then the hypothesis allowing cancellation of c gives us the contradiction $d = 0$, which proves the theorem. ∎

COROLLARY Let D be a commutative ring with unity $1 \neq 0$ in which each of $ac = bc$ and $ca = cb$ implies that either $c = 0$ or $a = b$. Then D is an integral domain.

Next we note a special property of the additive group of an integral domain.

THEOREM 3 Additive Order of Elements of *D*

In the additive group of an integral domain D, either all the elements except 0 have infinite order or all the nonzero elements have the same finite order.

Proof Here we denote the unity element 1 of D by e. Let a be any nonzero element of D. Using distributivity and mathematical induction, one can show that, for all positive integers n,

$$n \cdot a = a + \cdots + a = ea + \cdots + ea = (e + \cdots + e)a = (n \cdot e)a. \qquad \text{(A)}$$

Let e have finite order q (in the additive group of D); then $q \cdot e = 0$, and replacing n by q in (A) gives us $q \cdot a = 0 \cdot a = 0$. This implies that a has finite order q' with $q' \mid q$.

Now let a have finite order q'; then $q' \cdot a = 0$, and replacing n by q' in (A) leads to $0 = q' \cdot a = (q' \cdot e)a$. Since a is neither 0 nor a 0-divisor, this implies that $q' \cdot e = 0$ and we see that e has finite order q with $q \mid q'$.

It follows that if either e or a has finite order, they both have the same finite order. This means that all nonzero elements of D have the same order. ∎

DEFINITION 3

Characteristic of D

If the unity 1 of an integral domain D has finite order q in the additive group of D, one says that D has **characteristic** q. If 1 has infinite order, D has **characteristic** 0.

For example, $\mathbf{Z}$ has characteristic 0, $\mathbf{Z}_2$ has characteristic 2, and $\mathbf{Z}_6$ does not have a characteristic (since it is not an integral domain).

Since $1 \neq 0$ in an integral domain D, the characteristic of D cannot be 1. We will see in Theorem 4 and Problem 11 below that the characteristic of D is either 0 or a prime.

EXAMPLE 1 Now, given that $4 \cdot 1 = 0$ in an integral domain D, we show that $2 \cdot 1 = 0$. Using associativity and distributivity, we have

$$\begin{aligned} 4 \cdot 1 = 1 + 1 + 1 + 1 &= (1 + 1) + (1 + 1) \\ &= 1(1 + 1) + 1(1 + 1) \\ &= (1 + 1)(1 + 1). \end{aligned}$$

Then the hypothesis $4 \cdot 1 = 0$ gives us

$$(1 + 1)(1 + 1) = 0. \qquad \text{(B)}$$

Since D has no 0-divisors, one of the factors on the left side of (B) must be zero. Hence $1 + 1 = 2 \cdot 1 = 0$.

The following is the natural analogue of subgroup and subring.

DEFINITION 4

Subdomain

A **subdomain** in an integral domain D is a subset S of D such that S is an integral domain under the operations in D.

It is left to the reader, in Problem 19 below, to characterize the subrings in an integral domain D that are subdomains.

THEOREM 4 **The Characteristic Is Zero or a Prime**

The characteristic q of an integral domain D is either 0 or a positive prime. If D is finite with n elements, then $q \neq 0$ and $q \mid n$.

Proof The overall proof can be broken down into proofs of the following statements:

(i) $q \neq 1$.

(ii) q is not composite, and hence either $q = 0$ or q is a positive prime.

(iii) If D has n elements, then $q \neq 0$ and $q \mid n$.

For (i), we note that q is the order of 1 in the additive group of D, only the identity 0 in this group has order 1, and $1 \neq 0$. Hence, $q \neq 1$. The proof of (ii) is left to the reader as Problem 11 below. For (iii), we use the facts that q is the order of an element of the additive group of D and that n is the order of this group. These facts and Lagrange's Theorem (Theorem 1 of Section 2.11) imply that $q \mid n$. ■

Problems

In this problem set, D is always an integral domain.

1. Let $D = \{0, 1, c, d\}$. Explain why the following are true, and then make the addition and multiplication tables for D.

(i) $4 \cdot 1 = 0$.

(ii) $2 \cdot 1 = 0$. [*Hint:* See Example 1.]

(iii) D has characteristic 2. [Do not use Theorem 4.]

(iv) $x + x = 0$ for each x in D, and hence additively D is a Klein 4-group.

(v) $cd \neq 0$.

(vi) $cd \neq d$ and $cd \neq c$.

(vii) $cd = 1$.

(viii) $\{1, c, d\}$ is a multiplicative group with $d = c^2$.

2. Make the addition and multiplication tables for an integral domain $D = \{0, 1, c\}$ with three elements. [*Hint:* See Problem 2 of Section 4.1.]

3. Show that $\bar{2}$ is the only 0-divisor in $\mathbf{Z}_4$.

4. Find all the 0-divisors in each of the following.

(a) $\mathbf{Z}_8$; (b) $\mathbf{Z}_9$; (c) $\mathbf{Z}_{10}$.

5. Prove that $\mathbf{Z}_m$ is not an integral domain if m is composite.

6. Let D have m elements and let q be the characteristic of D.

(a) Explain why $q \mid m$. Show that the additive group of D is not cyclic unless m is prime. [*Hint:* See Theorem 3.]

★ (b) Use Cauchy's Theorem or a Sylow Theorem (see Section 2.15) to prove that m is a power of a prime. (It can be proved that, if p and d are positive integers with p prime, then there exists an integral domain with p^d elements. Example 2 of Section 5.10 will illustrate the construction of such integral domains.)

7. (i) Show that an invertible v in a ring U with unity is never a 0-divisor.

(ii) Explain why $\mathbf{Z}_p$ is an integral domain for every positive prime p.

(iii) What is the characteristic of $\mathbf{Z}_p$?

8. Let a, b, and c be elements of a ring R with c neither zero nor a 0-divisor. Show that $ca = cb$ implies that $a = b$. (This is the part of the proof of Theorem 1 that was left to the reader.)

9. Prove the corollary to Theorem 2; that is, prove that, if U is a commutative ring with unity $1 \neq 0$ and U has the nonzero multiplicative cancellation property, then U is an integral domain.

10. Let $6 \cdot 1 = 0$ in D. Use the fact that D has no 0-divisors to show that either $2 \cdot 1 = 0$ or $3 \cdot 1 = 0$ and hence that the characteristic of D is 2 or 3.

11. (i) Explain why $(st) \cdot 1 = (s \cdot 1)(t \cdot 1)$ in an integral domain D, for all s and t in $\mathbf{Z}^+$.
(ii) Show that an integral domain D cannot have a composite characteristic $q = st$ (with s and t in $\{2, 3, \ldots, q - 1\}$), and thus complete the proof of Theorem 4.

12. Let there be n distinct elements in $D = \{0, d_2, d_3, \ldots, d_n\}$. Show the following.
(i) For every k with $2 \leq k \leq n$, the products $d_2 d_k, d_3 d_k, \ldots, d_n d_k$ are all different from one another.
(ii) Each d_k is an invertible in D.
(iii) $\{d_2, d_3, \ldots, d_n\}$ is a multiplicative group.

13. Show that $x^2 - (r + s)x + rs = 0$ in D if and only if $x = r$ or $x = s$.

14. (i) Show that $\mathbf{Z}_6$ has 4 elements x such that $x(x + 1) = 0$.
(ii) How many elements x of an integral domain satisfy $x(x + 1) = 0$?

15. Let D have characteristic 2. Show the following in D.
(i) $(a + b)^2 = a^2 + b^2$.
(ii) $(a + b)^4 = a^4 + b^4$.
(iii) $(a + 1)^{10} = (a + 1)^8(a + 1)^2 = (a^8 + 1)(a^2 + 1)$
$= a^{10} + a^8 + a^2 + 1$.
(iv) The binomial coefficient $\binom{10}{k}$ is even unless k is in $\{0, 2, 8, 10\}$.

16. Let D have characteristic 3. Show the following in D.
(i) $(a + b)^3 = a^3 + b^3$.
(ii) $(a + b)^9 = a^9 + b^9$.
(iii) $(a + 1)^{15} = (a^9 + 1)(a^6 + 2 \cdot a^3 + 1) = a^{15} + 2 \cdot a^{12} + a^9 + a^6 + 2 \cdot a^3 + 1$.
(iv) $3 \mid \binom{15}{k}$ unless k is in $\{0, 3, 6, 9, 12, 15\}$.

17. Prove that a subring S in D has a unity if and only if the unity 1 of D is in S or $S = \{0\}$.

18. Is the ring G of gaussian integers an integral domain? (G consists of the complex numbers $m + ni$, with m and n in $\mathbf{Z}$.)

19. What conditions suffice to show that a subring S in D is also an integral domain, that is, for S to be a subdomain?

20. Let a, b, and p be integers with p a positive prime. Use Problem 7(ii) to give another proof that $p \mid (ab)$ implies either $p \mid a$ or $p \mid b$.

21. Prove that there is no integral domain with 6 elements.

22. Let $a \neq 0$ in D. Show that $a = -a$ if and only if D has characteristic 2.

23. Let D have characteristic q. Let V be the multiplicative group of invertibles in D. Let θ be the mapping from V to V with $\theta(v) = v^2$. Show the following.
(i) θ is a group homomorphism.
(ii) The only solutions of $x^2 = b^2$ in D are $x = b$ and $x = -b$. (These are equal when $q = 2$.)
(iii) The kernel of θ is $\{1, -1\}$ if $q \neq 2$ and is $\{1\}$ if $q = 2$.
(iv) If m is in the image set M of θ, $\theta^{-1}(m)$ has two elements if $q \neq 2$ and one element if $q = 2$.

(v) If V is finite, M is a subgroup in V with index 2 when $q \neq 2$ and with index 1 when $q = 2$.

24. Let V be the multiplicative group of the invertibles in D. Show the following.
(i) If x is in V and $x^{-1} = x$, then $x = 1$ or $x = -1$. [See Problem 23(iii).]
(ii) If V is finite, it has an even number of elements v such that $v^{-1} \neq v$, and the product of all these elements v is 1.
(iii) If V is finite, V has odd order if and only if D has characteristic 2.
(iv) If V has finite order s, the product $v_1 v_2 \cdots v_s$ of all the elements of V equals -1. (If the characteristic of D is 2, then $-1 = 1$.)

25. Let p be a positive prime in the integers $\mathbf{Z}$ and let $\mathbf{V_p} = \{\bar{1}, \bar{2}, \ldots, \overline{p-1}\}$ be the multiplicative group of the invertibles of $\mathbf{Z_p} = \mathbf{Z}/(p)$. Show the following.
(i) $\bar{1} \cdot \bar{2} \cdot \bar{3} \cdots \overline{p-2} = \bar{1}$ in $\mathbf{V_p}$. [See Problem 24(ii).]
(ii) $\bar{1} \cdot \bar{2} \cdot \bar{3} \cdots \overline{p-1} = \overline{p-1} = -\bar{1}$ in $\mathbf{V_p}$.
(iii) $(p-1)! + 1$ is a multiple of p in $\mathbf{Z}$.
(iv) $(p-1)! \equiv -1 \pmod{p}$.
[This is named Wilson's Theorem after John Wilson (1741–1793). See the biographical note on Lagrange on p. 110.]

26. Let p be a positive prime in $\mathbf{Z}$. Show that:
(i) $(p-2)! - 1$ is a multiple of p in $\mathbf{Z}$.
(ii) $(p-2)! \equiv 1 \pmod{p}$.

27. Find the remainder when each of the following is divided by the prime 101.
(a) 100!; (b) 99!; ★ (c) 98!; ★ (d) 97!; ★ (e) 96!.

28. Find the remainder when each of the following is divided by 103:
(a) 102!; (b) 101!; ★ (c) 100!; ★ (d) 99!; ★ (e) 98!.

29. Explain why $(29!)^2 \equiv 1 \pmod{59}$ and $(30!)^2 \equiv -1 \pmod{61}$.

30. Let U be a (not necessarily commutative) ring with unity and no 0-divisors. Do all nonzero elements of U have the same order in the additive group of U? Explain.

31. Let c be the characteristic of an integral domain D and m be a positive integer such that $m \cdot x = 0$ for all x in D. Prove that $c \mid m$.

32. Find a 0-divisor in the direct product $\mathbf{Z_2} \times \mathbf{Z_2}$. (See Problem 30 of Section 4.1 for the definition of direct product.)

33. Is the direct product $D \times D'$ of integral domains ever an integral domain? Explain.

★ **34.** Let n, p, and q be positive integers with p and q distinct primes. Also, let $p \mid n$ and $q \mid n$. Prove that there is no integral domain D with n elements. [*Hint:* Use Theorem 3 of this section.]

4.5 Fields

Let U be a ring with unity in which $0 \neq 1$. By Theorem 2 of Section 4.1, we then have $0 \cdot a = 0 \neq 1$ for every a in U. Hence the additive identity 0 has no multiplicative inverse; that is, 0 is not an invertible. This means that U is not a group under multiplication.

However, the set obtained by deleting 0 from U may be a multiplicative group, as is true in the number systems $\mathbf{Q}$, $\mathbf{R}$, and $\mathbf{C}$ and in the $\mathbf{Z_p}$ with p prime.

DEFINITION 1

Division Ring, Field, Skew-Field

A **division ring** is a ring with unity $1 \neq 0$ in which every nonzero element is an invertible. A **field** is a commutative division ring. A **skew-field** is a noncommutative division ring.

In a field there are no 0-divisors, since 0 is not a 0-divisor and every nonzero element is an invertible and thus is not a 0-divisor. [See Problem 7(i) of Section 4.4.] It follows that every field is an integral domain. The integers $\mathbf{Z}$ furnish an example of an integral domain that is not a field. Thus the fields F form a proper subcollection of the integral domains D, the integral domains form a proper subcollection of the rings U with unity, and the rings with unity in turn form a proper subcollection of the rings R. These collections are represented by the nest of circles in Figure 4.1.

FIGURE 4.1

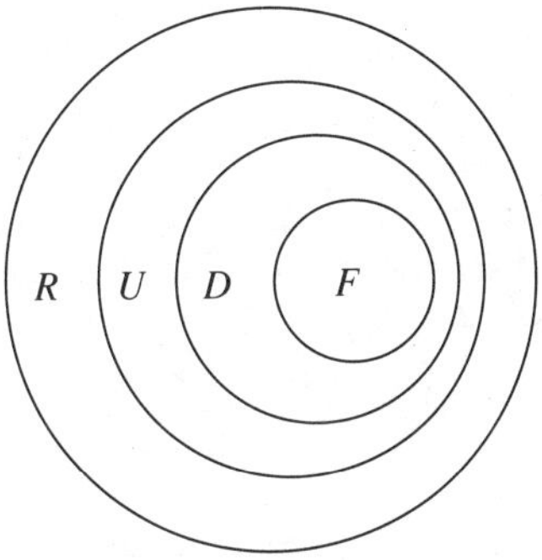

We have defined a field to be a commutative division ring. Alternatively, a field could be defined to be an integral domain in which every nonzero element is an invertible. Because of the importance of fields in abstract algebra, it is also worth giving the following direct characterization of fields.

THEOREM 1

Alternate Characterization of a Field

A set X closed under operations of addition and multiplication is a field if and only if it has all three of the following properties:

(A) X is an abelian group under addition, with 0 as the identity.
(M) X with 0 deleted is an abelian group under multiplication. (This is the **multiplicative group** of X.)
(D) Multiplication is distributive over addition in X.

Proof If X is a field, its nonzero elements are all invertibles, and hence they form a multiplicative group by Theorem 1 of Section 4.1; then one sees readily that X has properties (A), (M), and (D). We have to establish the converse. So let X have these properties. Using properties (D) and (A), we can show, as in Theorem 2 of Section 4.1, that

$$x \cdot 0 = 0 = 0 \cdot x \quad \text{for all } x \text{ in } X. \tag{1}$$

It then follows from property (M) that

$$xy = yx \quad \text{for any nonzero elements } x \text{ and } y \text{ in } X. \tag{2}$$

$$x(yz) = (xy)z \quad \text{for any nonzero } x, y, \text{ and } z \text{ in } X. \tag{3}$$

With the help of (1), (2), and (3), one readily sees that multiplication is a commutative and associative operation on all of X. This shows that X is a commutative ring. It then follows from (M) that X is a commutative division ring; so X is a field. ■

Some infinite fields are the rational numbers $\mathbf{Q}$, the real numbers $\mathbf{R}$, and the complex numbers $\mathbf{C}$. In Section 4.7 we describe the infinite skew-field of the quaternions.

The $\mathbf{Z_p}$ with p prime are examples of finite fields. A theorem of J. H. M. Wedderburn states that there are no finite skew-fields; that is, every finite division ring is a field. [For a comparatively elementary proof, see I. N. Herstein, "Wedderburn's Theorem and a Theorem of Jacobson," *American Mathematical Monthly* 68 (1961): 249–51.]

We next define an analogue of subgroup and subring.

DEFINITION 2

Subfield, Extension Field

A subset F of a field E is a **subfield** in E if F is a field under the operations of E. If F is a subfield in E, one also says that E is an **extension field**, or **superfield**, over F.

THEOREM 2 **Subfield Conditions**

Let F be a subset with at least two elements in a field E. Then F is a subfield in E if and only if F is closed under subtraction and under division by nonzero elements.

The proof is left to the reader as Problem 25 of this section.

TABLE 4.1 **Summary of Axioms for Groups, Rings, and Fields**

Properties Under Addition	Properties Under Multiplication
1. Closure	6. Closure
2. Associativity	7. Associativity
3. Existence of an identity	8. Existence of an identity
4. Existence of inverses	9. Existence of inverses (for nonzero elements)
5. Commutativity	9′. Cancellation law (or no 0-divisors)
	10. Commutativity

Properties Under Addition and Multiplication
11. Distributivity
12. $0 \neq 1$

Algebraic Structure	Required Properties
Additive group	1–5
Multiplicative group	6–9
Ring	1–7, 11
Ring with unity	1–8, 11
Commutative ring	1–7, 10, 11
Integral domain	1–8, 9′, 10–12
Division ring	1–9, 11, 12
Skew-field	1–9, 11, 12, and negation of 10
Field	1–12

In Table 4.1 we summarize the axioms for groups and for the types of rings (including fields) discussed in this text.

Since a field or an integral domain is a ring, the definitions of ring homomorphism, isomorphism, automorphism, and endomorphism apply to fields and integral domains. Also, the ***characteristic of a field*** is its characteristic as an integral domain.

It can be shown (see Problem 26 below) that the intersection P of the collection of all the subfields in a field F is also a subfield in F; this smallest subfield P is called the ***prime subfield*** in F. It is also left to the reader to show that the prime subfield P is isomorphic to the rational numbers $\mathbf{Q}$ when F has characteristic 0 (see Problem 27 below) and that P is isomorphic to $\mathbf{Z_p}$ when F has characteristic p (see Problem 28 below).

Problems

1. In $\mathbf{Z_5} = \{\bar{0}, \bar{1}, \bar{2}, \bar{3}, \bar{4}\}$, find $\bar{3}^{-1}$ and the x such that $\bar{3}x = \bar{2}$.

2. In $\mathbf{Z_7}$, find $\bar{2}^{-1}$ and the x such that $\bar{2}x = \bar{5}$.

3. Tabulate the mapping θ from $\mathbf{Z}_5$ to itself with $\theta(x) = x^2$, and thus find all solutions x (if any) of the following equations.
(a) $x^2 = \bar{0}$; (b) $x^2 = \bar{1}$; (c) $x^2 = \bar{2}$; (d) $x^2 = \bar{3}$; (e) $x^2 = \bar{4}$.

4. In $\mathbf{Z}_7$, find all solutions x of the following equations.
(a) $x^2 = \bar{0}$; (b) $x^2 = \bar{1}$; (c) $x^2 = \bar{2}$; (d) $x^2 = \bar{3}$; (e) $x^2 = \bar{4}$;
(f) $x^2 = \bar{5}$; (g) $x^2 = \bar{6}$.

5. Let a and b be elements of a field F with $a \neq 0$. Explain why there is a unique x satisfying $ax + b = 0$.

6. Let a, b, and c be elements of a field F with $a \neq 0$. Must the equation $ax^2 + bx + c = 0$ have a solution in F? Explain.

7. Let $\mathbf{C}$ be the field of complex numbers. Every α in $\mathbf{C}$ is uniquely expressible as $\alpha = a + b\mathbf{i}$, with a and b real and $\mathbf{i}^2 = -1$. Let $\bar{\alpha} = a - b\mathbf{i}$ be the ***conjugate*** of α, and let $|\alpha| = \sqrt{a^2 + b^2}$ be the absolute value of α. Let f be the mapping from $\mathbf{C}$ to itself with $f(\alpha) = \bar{\alpha}$. Show the following.
(i) f is a bijection.
(ii) $\overline{\alpha + \beta} = \bar{\alpha} + \bar{\beta}$, $\overline{\alpha\beta} = \bar{\alpha} \cdot \bar{\beta}$; that is, the bijection f is a field automorphism of $\mathbf{C}$.
(iii) $|\alpha|^2 = \alpha\bar{\alpha}$.
(iv) $|\alpha\beta| = |\alpha| \cdot |\beta|$. [It may be helpful to use parts (iii) and (ii). This formula implies that the mapping g with $g(\alpha) = |\alpha|$ is a homomorphism from the multiplicative group of $\mathbf{C}$ to the multiplicative group of the positive real numbers.]

8. Let T be the set of all real numbers of the form $a + b\sqrt{2}$ with a and b rational. The ***conjugate*** $\bar{\alpha}$ and ***norm*** $\|\alpha\|$ of $\alpha = a + b\sqrt{2}$ are defined by $\bar{\alpha} = a - b\sqrt{2}$ and $\|\alpha\| = \alpha\bar{\alpha}$.
(i) Use the fact that $\sqrt{2}$ is irrational to show that, if $a + b\sqrt{2} = c + d\sqrt{2}$, then $a = c$ and $b = d$.
(ii) Show that T is a field.
(iii) Show that the mapping f with $f(\alpha) = \bar{\alpha}$ is a field automorphism of T.
(iv) Show that $\|\alpha\beta\| = \|\alpha\| \cdot \|\beta\|$ for all α and β in T.

9. Explain why a finite integral domain must be a field. (See Problem 12 of Section 4.4.)

10. Let F be a finite field with m elements. Explain why the following are true.
(i) $x^{m-1} = 1$ for all nonzero x in F. [*Hint:* Use Lagrange's Theorem.]
(ii) $x^m - x = 0$ for all x in F.

11. Let $F = \{0, 1, c\}$ be a field with 3 elements.
(i) Explain why both the additive group and the multiplicative group of F are cyclic.
(ii) Explain why F must have characteristic 3.
(iii) Make the addition and multiplication tables for F.
(iv) Explain why any two fields with 3 elements are isomorphic.
(v) List the elements of each principal ideal (f) in F.

12. Let F be a field with a prime number p of elements. Explain why the following are true.
(i) The additive group of F is cyclic, with 1 as a generator.
(ii) $F = \{0, 1, 2 \cdot 1, 3 \cdot 1, \ldots, (p-1) \cdot 1\}$.
(iii) $s \cdot 1 = t \cdot 1$ in F with s and t in $\mathbf{Z}$ if and only if $s \equiv t \pmod{p}$.
(iv) $(a \cdot 1)(b \cdot 1)$ in F is the $c \cdot 1$ with $ab \equiv c \pmod{p}$ and c in $\{0, 1, \ldots, p-1\}$.
(v) F is isomorphic to $\mathbf{Z}_p$.

(vi) The product of the nonzero elements of F equals -1. [*Hint:* See Problem 24(iv) of Section 4.4.]

13. Let F be a field with characteristic 3 and let $1 + 1$ be denoted by 2 in F. Show that $P = \{0, 1, 2\}$ is a subfield in F.

14. Let F be a field with characteristic p, a prime. In F, let $2 \cdot 1, 3 \cdot 1, \ldots, (p-1) \cdot 1$ be denoted by $2, 3, \ldots, p-1$. Show that $P = \{0, 1, 2, \ldots, p-1\}$ is a subfield in F. (The field P can be shown to be the prime subfield in F. See Problem 28 below.)

15. Show that the only automorphism of $\mathbf{Z}_5$ is the identity automorphism θ with $\theta(\bar{a}) = \bar{a}$ for each $\bar{a}$ in $\mathbf{Z}_5$).

16. (i) Repeat Problem 15, with $\mathbf{Z}_5$ replaced by $\mathbf{Z}_p$ (where p is a positive prime).
(ii) How many automorphisms are there of a field with a prime number of elements?

17. In $\mathbf{Z}_{19}$, find x and y satisfying the simultaneous equations

$$\bar{2}x + \bar{7}y = \bar{4}$$
$$\bar{9}x + \bar{4}y = \bar{1}.$$

18. Let a, b, c, d, h, and k be elements of a field F, with $ad - bc \neq 0$. Show that there exist unique elements x and y in F satisfying the simultaneous equations

$$ax + by = h$$
$$cx + dy = k.$$

19. Let F be a field with 8 elements. Explain why the following hold.
(i) The characteristic of F must be 2.
(ii) Every nonzero a of F satisfies $a^7 - 1 = 0$.
(iii) Every x of F satisfies

$$x(x-1)(x^3 + x^2 + 1)(x^3 + x + 1) = x^8 - x = 0.$$

20. Let F be a field with 9 elements. Explain why the following hold.
(i) The characteristic of F must be 3.
(ii) Every x of F satisfies

$$x(x-1)(x+1)(x^2+1)(x^2+x-1)(x^2-x-1) = x^9 - x = 0.$$

21. Explain why the only ideals in a field F are $\{0\}$ and F itself.

22. Let M be the image set of a homomorphism θ from a field F to a field F'. Explain why either M is isomorphic to F or $M = \{0\}$. [*Hint:* See the preceding problem.]

23. Is $\theta(1) = 1$ necessarily true for every field homomorphism θ from F to F'? Explain.

24. Let θ be a field isomorphism from F to F'. Explain why the following are true.
(i) $\theta(a) \neq 0$ for all nonzero a of F.
(ii) The multiplicative groups of F and F' are isomorphic.
(iii) $\theta(1) = 1$.
(iv) $\theta(a^{-1}) = [\theta(a)]^{-1}$ for every nonzero a of F.

25. Let F be a subset, with more than one element, of a field E. Prove that F is a subfield in E if and only if F is closed under subtraction and under division by nonzero elements. (This is the proof of Theorem 2 that was left to the reader.)

26. Let P be the intersection of all the subfields in a field F. Prove that P is a subfield in F and hence is the smallest subfield in F. (This smallest subfield P is the prime subfield in F.)

27. Explain why the following are true.
(a) In a ring U with unity, $\{n \cdot 1 : n \in \mathbf{Z}\}$ is a subring.

(b) The prime subfield P in a field F consists of all elements of the form $(m \cdot 1)(n \cdot 1)^{-1}$, with m and n integers and $n \cdot 1 \neq 0$.

(c) The prime subfield P in a field with characteristic 0 is isomorphic to the field $\mathbf{Q}$ of the rational numbers.

28. Prove that the prime subfield P in a field F with characteristic p is isomorphic to $\mathbf{Z}_p$. (See Problems 26 and 14.)

29. Let $F = \{0, f_2, f_3, \ldots, f_t\}$ be a finite field. Explain why $f_2 f_3 \cdots f_t = -1$.

30. Let F be as in Problem 29. Explain why the following are true.

(i) The mapping β from F to itself with $\beta(x) = x + 1$ is a bijection.

(ii) $(f_2 + 1)(f_3 + 1) \cdots (f_t + 1) = 0$.

31. Let F be a subfield in E and let H consist of all the field automorphisms θ of E such that $\theta(f) = f$ for all f in F. Show that H is a subgroup in the group $\mathbf{B}(E)$ of all bijections from E to itself.

32. Let R be a commutative ring with 0-divisors; that is, let there be nonzero elements r and s in R with $rs = 0$. Explain why R cannot be a subring in a field F.

33. Let θ be a ring homomorphism from a ring U with unity to a field F. Prove that either $\theta(1) = 1$ or $\theta(u) = 0$ for all u in U.

4.6 Ordered Integral Domains and Fields

The integral domain $\mathbf{Z}$ has a subset $\mathbf{Z}^+ = \{1, 2, 3, \ldots\}$ that is closed under addition and under multiplication and has the property that $\mathbf{Z}$ is partitioned into the three subsets: $\{0\}$, $\mathbf{Z}^+$, and the set $\mathbf{Z}^- = \{-1, -2, -3, \ldots\}$ of the negatives of the elements of $\mathbf{Z}^+$. The order relation $x < y$ in $\mathbf{Z}$ can be defined in terms of this partitioning. We indicate below how this may be done in the process of considering the generalization to integral domains.

DEFINITION 1 **Ordered Integral Domain**

Let D^+ be a subset of an integral domain D. Then D is said to be **ordered** by D^+ if D^+ has the following two properties:

O_1 ***Closure*** D^+ is closed under addition and multiplication.

O_2 ***Trichotomy*** If a is an element of D, then one and only one of the following holds:

(i) $a = 0$; (ii) a is in D^+; (iii) the negative $-a$ of a is in D^+.

For example, the integral domain $\mathbf{Z} = \{\ldots, -1, 0, 1, \ldots\}$ of the integers is ordered by its subset $\mathbf{Z}^+ = \{1, 2, \ldots\}$, since properties O_1 and O_2 hold when

D and D^+ are replaced by $\mathbf{Z}$ and $\mathbf{Z}^+$, respectively. (It can be shown that this $\mathbf{Z}^+$ is the only subset ordering $\mathbf{Z}$. See Problem 17 below.)

DEFINITION 2 **Positive Elements, Negative Elements**

Let an integral domain D be ordered by a given subset D^+. For this ordering, the elements of D^+ are called the **positive** elements of D. If $-a$ is positive, one says that a is **negative**. The notation $x < y$ means that $y - x$ is positive, that is, $y - x$ is in D^+. Also, $y > x$ is equivalent to $x < y$, and $a \leq b$ means that $a < b$ or $a = b$.

LEMMA 1 In an ordered integral domain D, an element x is positive if and only if $x > 0$, and y is negative if and only if $y < 0$.

Proof From Definition 2, $x > 0$ if and only if $x - 0 = x$ is positive, and $y < 0$ if and only if $0 - y = -y$ is positive. But Definition 2 also tells us that $-y$ is positive if and only if y is negative. ■

We next illustrate the derivation of order properties from the order axioms O_1 and O_2.

THEOREM 1 **Positive Multiplication Preserves Order**

If $a < b$ and $c > 0$ in an ordered integral domain D, then $ac < bc$.

Proof The hypotheses $a < b$ and $c > 0$ tell us that $b - a$ and c are positive. Closure of D^+ under multiplication then implies that $(b - a)c = bc - ac$ is positive. Finally, $ac < bc$, by definition of $x < y$. ■

THEOREM 2 **Nonzero Squares Are Positive**

If $x \neq 0$ in an ordered integral domain, then $x^2 > 0$.

Proof Let $x \neq 0$. From the trichotomy axiom O_2 it follows that either x is positive or $-x$ is positive. If x is positive, $x^2 = x \cdot x$ is positive, by closure of D^+ under multiplication. If $-x$ is positive, we similarly see that $x^2 = (-x)(-x)$ is positive. Hence, $x^2 > 0$ whenever $x \neq 0$, and the proof is complete. ■

The result in Theorem 2 shows that $y^2 \geq 0$ for every element y of an ordered integral domain D. Also, it follows from Theorem 2 and $1 \neq 0$ that $1^2 = 1 > 0$.

DEFINITION 3 **Absolute Value**

Let a be an element of an ordered integral domain D. The **absolute value** of a, denoted by $|a|$, is defined as follows.
(i) If a is positive or $a = 0$, then $|a| = a$.
(ii) If a is negative, then $|a|$ is the positive element $-a$.

For all x in an ordered integral domain, this definition implies that $|x| \geq 0$, that $|x| = 0$ if and only if $x = 0$, and that $|x| > 0$ if $x \neq 0$.

Examples of ordered integral domains are the integers **Z**, the rational numbers **Q**, and the real numbers **R**. It is not possible to order the complex numbers **C**, that is, to find a subset $\mathbf{C}^+$ of **C** with the closure and trichotomy properties O_1 and O_2. (The proof of this is left to the reader as Problem 21 below.)

THEOREM 3 **Rules of Signs**

In an ordered integral domain D, let $ab = c$.
(a) If $a > 0$ and $b > 0$, then $c > 0$.
(b) If $a < 0$ and $b < 0$, then $c > 0$.
(c) If $a > 0$ and $b < 0$, then $c < 0$.

Proof Part (a) follows from the closure axiom O_1 and the fact that $x > 0$ is equivalent to $x \in D^+$. For part (b), let $a < 0$ and $b < 0$. Then $-a = u$ and $-b = v$ with u and v in D^+. It follows that $c = ab = (-u)(-v) = uv$ is in D^+ by closure of D^+ under multiplication. Hence, $c > 0$. For part (c), let $a > 0$ and $b < 0$. Then $-b = d$ with d in D^+ and $c = ab = a(-d) = -(ad) = -f$ with f in D^+. Finally, $c = -f$ with f in D^+ implies that $c < 0$. ■

THEOREM 4 **Two out of Three Rule**

If $ab = c$, $a > 0$, and $c > 0$ in an ordered integral domain D, then $b > 0$.

Proof If $b = 0$, then $c = ab = a \cdot 0 = 0$, which contradicts $c > 0$. If $b < 0$, then $c = ab < 0$ by Theorem 3(c); this also contradicts $c > 0$. The only remaining possibility is that $b > 0$. ■

An ***ordered field*** is a field that is ordered as an integral domain.

THEOREM 5 Reciprocals of Positive and of Negative Elements

In an ordered field F, if $a > 0$ then $a^{-1} > 0$, and if $b < 0$ then $b^{-1} < 0$.

Proof Let $a > 0$. Then it follows from $aa^{-1} = 1$, $1 > 0$, and Theorem 4 that $a^{-1} > 0$. Now let $b < 0$. Then $b = -a$ with $a > 0$, and hence $a^{-1} > 0$. Since $b^{-1} = -a^{-1}$, we have $b^{-1} < 0$. ∎

Problems

In the problems below, D always denotes an ordered integral domain, and a, b, c, and d always are elements of D.

1. Prove the following.
(a) $a < b$ and $b < c$ imply that $a < c$.
(b) $a < b$ and $c < 0$ imply that $ca > cb$.
(c) $a < b$ if and only if $a + c < b + c$.
(d) $a < b$ and $c < d$ imply that $a + c < b + d$.

2. By a consideration of cases, prove that $a \leq b$ and $c \leq d$ imply that $a + c \leq b + d$.

3. Let $a > 0$, $b > 0$, $c < 0$, and $d < 0$. Prove the following.
(a) $ab > 0$. (b) $ac < 0$.
(c) $cd > 0$. (d) $2 \cdot a > 0$ (that is, $a + a > 0$).
(e) $2 \cdot c < 0$. (f) $c < a$.

4. Show that $2 \cdot a < 2 \cdot b$ if and only if $a < b$.

5. Show the following.
(a) $|-a| = |a|$. (b) $a \leq |a|$.
(c) $-b \leq |b|$. (d) $-|a| \leq a$.
(e) $a + b \leq |a| + |b|$. (f) $-(a + b) \leq |a| + |b|$.
(g) $|a + b| \leq |a| + |b|$. (h) $|c| - |a| \leq |c - a|$.
(i) $|c| - |b| \leq |c + b|$. (j) $|(|a| - |b|)| \leq |a \pm b|$.

6. Show the following.
(a) $a^2 - 2 \cdot ab + b^2 \geq 0$. (b) $a^2 + b^2 \geq 2 \cdot ab$.
(c) $a^2 + b^2 + c^2 \geq bc + ac + ab$. (d) $a^2 + b^2 \geq ab$.
(e) $a^2 + b^2 \geq -ab$.

7. Given that $a \neq b$, show the following.
(i) $a^2 - 2 \cdot ab + b^2 > 0$.
(ii) $a^2 + b^2 > 2 \cdot ab$.
(iii) $a^2 + b^2 + c^2 > bc + ac + ab$.
(iv) $a^2 + b^2 > ab$.

8. Use trichotomy to show that exactly one of the following is true for fixed b and c.
(i) $b = c$; (ii) $b > c$; (iii) $b < c$.

9. Given that $0 < a < b$ in an ordered field F, show that $0 < 1/b < 1/a$.

10. Given that $a < b < 0$ in an ordered field F, show that $1/b < 1/a < 0$.

11. Given that $a \geq 1$ and $b \geq 1$, show that $ab + 1 \geq a + b$.

12. Let $a > 1$, $b > 1$, and $c > 1$. Show that

$$abc + a + b + c > bc + ac + ab + 1.$$

13. Given that $a > 0$, prove that $n \cdot a > 0$ for all n in $\mathbf{Z}^+$.

14. Explain why an ordered integral domain has characteristic 0.

15. Explain why an ordered integral domain must be infinite.

16. Can $\mathbf{Z_p}$ be an ordered integral domain for any prime p? Explain.

17. We know that the subset $\mathbf{Z}^+ = \{1, 2, 3, \ldots\}$ of the integers $\mathbf{Z}$ satisfies the order axioms O_1 and O_2. Assume that P is a subset of $\mathbf{Z}$ that also has properties O_1 and O_2, and prove that P must be $\mathbf{Z}^+$. (This shows that there is only one way to order $\mathbf{Z}$.)

18. Let S be a subdomain in an ordered domain D. Describe a method of ordering S.

19. Prove that there is only one way to order the rational numbers $\mathbf{Q}$.

20. Prove that every ordered integral domain D has a subdomain D' that is isomorphic to the integers $\mathbf{Z}$.

21. Prove as follows that the complex numbers $\mathbf{C}$ cannot be ordered. Assume that a subset $\mathbf{C}^+$ has properties O_1 and O_2, show that both 1 and -1 are in $\mathbf{C}^+$, and note the contradiction.

★ **22.** Prove that there is only one way to order the real numbers. (You may use the facts that every positive real number is the square of a real number and that there is a rational number between any two real numbers.)

★ **23.** Prove that the field $\mathbf{Q}$ of rational numbers can be ordered in only one way; that is, there is only one way to select a subset $\mathbf{Q}^+$ satisfying the order axioms O_1 and O_2.

24. Prove that every ordered field has a subfield isomorphic to $\mathbf{Q}$.

25. Find all the real numbers x such that

$$|x - 4| + |x - 7| = 6.$$

(It may be helpful to graph $y = |x - 4| + |x - 7|$.)

26. For each of the following, find all the rational numbers x that satisfy the equation.
(a) $|x - 1| + |x - 2| + |x - 3| = 5/2$.
(b) $|x - 1| + |x - 2| + |x - 3| = 4$.
(c) $|x - 1| + |x - 2| + |x - 3| = 1$.

27. Let c be a positive composite integer. Prove that there exists a prime p in $\mathbf{Z}$ with $1 < p^2 \le c$ and $p \mid c$.

28. Use the fact that no one of the primes

$$2, 3, 5, 7, 11$$

is an integral divisor of 131 to prove that 131 is a prime.

29. Prove that there do not exist integers m and n with

$$4 < m^2 \le n \quad \text{and} \quad 2n < (m + 1)^2.$$

★ **30.** Find the largest positive integer n such that the sum of the squares of the digits of n exceeds n.

31. (i) Find all triples of positive integers $\{x, y, z\}$ such that

$$yz + xz + xy = xyz + 1.$$

[*Hint:* Assume that $0 < x < y < z$.]
(ii) Find all triples of positive integers $\{x, y, z\}$ for which there exists a positive integer k with $yz + xz + xy = kxyz + 1$.

32. Find all triples of positive integers $\{r, s, t\}$ such that simultaneously $\bar{s} \cdot \bar{t} = \bar{1}$ in $\mathbf{Z_r}$, $\bar{r} \cdot \bar{t} = \bar{1}$ in $\mathbf{Z_s}$, and $\bar{r} \cdot \bar{s} = \bar{1}$ in $\mathbf{Z_t}$.

33. Let $m = 10^{150} + 3 \cdot 10^{100} + 1$. Use the fact that $(10^{50})^3 < m < (10^{50} + 1)^3$ to explain why m is not the cube of an integer.

34. (i) Show that $(a + 3)^3 < a(a + 4)(a + 6) < (a + 4)^3$ for all real $a \geq 7$.
★ (ii) Use (i) to prove that the product of six consecutive positive integers is never the cube of an integer.

35. Find all solutions of $y^2 = x^4 + 2x^3 + 2x^2 + 2x + 5$ in integers x and y.

★ **36.** Find all solutions in integers x and y of the equation

$$y^2 = x^4 + 6x^3 + 10x^2 + 15x - 13.$$

4.7 Matrices, Quaternions

There is an astonishing imagination, even in the science of mathematics .... We repeat, there is far more imagination in the head of Archimedes than in that of Homer.
—François Marie Arouet de Voltaire

Multiplication is commutative, except before breakfast.
—William Feller

Here we sample the branch of algebra called "matrix theory." We present material on m by n matrices to support our work on coding theory in Chapter 8, and then we specialize to 2 by 2 matrices.

Let R be a commutative ring. An ***m* × *n* matrix** (read as "m by n matrix") over R is a rectangular array

$$A = \begin{pmatrix} a_{11} & a_{12} & a_{13} & \cdots & a_{1n} \\ a_{21} & a_{22} & a_{23} & \cdots & a_{2n} \\ \vdots & & & & \\ a_{m1} & a_{m2} & a_{m3} & \cdots & a_{mn} \end{pmatrix} \tag{1}$$

with m rows and n columns and with each entry a_{ij} in R. For short, we write this matrix A as (a_{ij}). If we wish to indicate its size (that is, the number of rows and number of columns), we write it as $(a_{ij})_{m \times n}$. If $m = n$, A is a ***square matrix***.

NOTATION **Set of $m \times n$ Matrices over R**

Let $M(m, n, R)$ denote the set of $m \times n$ matrices over the ring R.

DEFINITION 1 **Matrix Addition**

Let $A = (a_{ij})$ and $B = (b_{ij})$ be in $M(m, n, R)$. Then $A + B$ is the matrix $C = (c_{ij})$ in $M(m, n, R)$ with $c_{ij} = a_{ij} + b_{ij}$ for $1 \leq i \leq m$ and $1 \leq j \leq n$.

This definition makes $M(m, n, R)$ into an additive group isomorphic to the direct product $R \times R \times \cdots \times R$ of mn copies of the additive group of R.

The product AB is defined only when the number of columns in A is the same as the number of rows in B; before giving that definition, we need some preliminary concepts.

DEFINITION 2

Transpose

The **transpose** A^T of an $m \times n$ matrix $A = (a_{ij})$ is the $n \times m$ matrix $B = (b_{ij})$ with $b_{ij} = a_{ji}$ for $1 \leq i \leq n$ and $1 \leq j \leq m$.

Thus the rows of A are the columns of A^T (and the columns of A are the rows of A^T). As an example, we note that the transpose of

$$A = \begin{pmatrix} 1 & 2 & 3 \\ 4 & 5 & 6 \end{pmatrix} \quad \text{is} \quad A^T = \begin{pmatrix} 1 & 4 \\ 2 & 5 \\ 3 & 6 \end{pmatrix}.$$

Clearly, $(A^T)^T = A$ for all matrices A; that is, the transpose of the transpose of A is the original matrix A.

DEFINITION 3

n-Vector

A $1 \times n$ matrix over R is also called an **n-vector** over R. If one separates the entries of an n-vector by commas, it becomes an element of the cartesian product $R \times R \times \cdots \times R$ of n copies of R.

DEFINITION 4

Rows and Columns

Let $A = (a_{ij})$ be the matrix in $M(m, n, R)$ of display (1). For $1 \leq i \leq m$, the n-vector $V_i = (a_{i1} \ a_{i2} \ \ldots \ a_{in})$ is the ith row of $A = (a_{ij})$. For $1 \leq j \leq n$, the transpose W_j^T of the m-vector $W_j = (a_{1j} \ a_{2j} \ \ldots \ a_{mj})$ is the jth column of A.

For example, let

$$A = \begin{pmatrix} 1 & 2 & 3 \\ 6 & 5 & 4 \end{pmatrix}. \tag{2}$$

Then the first row in $V_1 = (1 \quad 2 \quad 3)$, and the second row is $V_2 = (6 \quad 5 \quad 4)$. The columns of A, in order, are

$$\begin{pmatrix} 1 \\ 6 \end{pmatrix} = (1 \quad 6)^T, \quad \begin{pmatrix} 2 \\ 5 \end{pmatrix} = (2 \quad 5)^T, \quad \text{and} \quad \begin{pmatrix} 3 \\ 4 \end{pmatrix} = (3 \quad 4)^T.$$

DEFINITION 5

Dot Product

Let $U = (u_1 \ u_2 \ \ldots \ u_n)$ and $V = (v_1 \ v_2 \ \ldots \ v_n)$. Then the **dot product** $U \cdot V$ of these two n-vectors is the sum of products

$$u_1 v_1 + u_2 v_2 + \cdots + u_n v_n.$$

The dot product is also called the **scalar product.**

For example, the dot product $V_1 \cdot V_2$ of the row vectors of the matrix A in display (2) is

$$(1 \quad 2 \quad 3) \cdot (6 \quad 5 \quad 4) = 1 \cdot 6 + 2 \cdot 5 + 3 \cdot 4 = 28.$$

DEFINITION 6

Product of Matrices of Compatible Sizes

Let $A = (a_{ij})_{r \times s}$ and $B = (b_{ij})_{s \times t}$ be matrices over R with the number of columns in A the same as the number of rows in B. Let V_i be the ith row of A and W_j^T be the jth column of B. Then the **product** $C = AB$ is the $r \times t$ matrix (c_{ij}) with the dot product $V_i \cdot W_j$ as the entry c_{ij}.

For example, one can see that

$$\begin{pmatrix} 1 & 2 & 3 & 0 \\ 6 & 5 & 4 & 0 \\ 7 & 8 & 9 & 1 \end{pmatrix} \begin{pmatrix} 1 & -1 \\ 0 & 7 \\ 0 & 4 \\ 2 & 3 \end{pmatrix} = \begin{pmatrix} 1 & 25 \\ 6 & 45 \\ 9 & 88 \end{pmatrix}.$$

The entry c_{32} in the product matrix is the dot product

$$\begin{aligned} V_3 \cdot W_2 &= (7 \quad 8 \quad 9 \quad 1) \cdot (-1 \quad 7 \quad 4 \quad 3) \\ &= 7(-1) + 8 \cdot 7 + 9 \cdot 4 + 1 \cdot 3 \\ &= -7 + 56 + 36 + 3 = 88, \end{aligned}$$

and the other entries are calculated similarly.

Multiplication of matrices is not commutative; that is, AB and BA need not be equal. Moreover, AB may exist without BA being defined.

THEOREM 1 **Transpose of a Product**

$(AB)^T = B^T A^T$ (whenever the product AB is defined).

The proof is left to the reader.

Theorem 1 states that the transpose of a product is the product of the transposes in reverse order.

Square matrices of fixed size form a ring; that is, $M(n, n, R)$ is a ring for each positive integer n. The case $n = 2$ is especially useful for abstract algebra, and we devote the rest of this section to that case. We abbreviate $M(2, 2, R)$ as $M(R)$. Addition and multiplication in $M(R)$ are given by

$$\begin{pmatrix} a & b \\ c & d \end{pmatrix} + \begin{pmatrix} a' & b' \\ c' & d' \end{pmatrix} = \begin{pmatrix} a + a' & b + b' \\ c + c' & d + d' \end{pmatrix},$$

$$\begin{pmatrix} a & b \\ c & d \end{pmatrix} \begin{pmatrix} a' & b' \\ c' & d' \end{pmatrix} = \begin{pmatrix} aa' + bc' & ab' + bd' \\ ca' + dc' & cb' + dd' \end{pmatrix}. \tag{3}$$

Additively $M(R)$ is an abelian group, since it is the direct product $R \times R \times R \times R$ of four copies of the additive group of R. Since a ring R is closed under addition and multiplication, formula (3) shows that $M(R)$ is closed under multiplication. We omit the straightforward (but tedious) details of verifying associativity of multiplication and distributivity for $M(R)$ and thus completing the proof that $M(R)$ is a ring.

Let

$$\alpha = \begin{pmatrix} a & b \\ c & d \end{pmatrix}, \quad O = \begin{pmatrix} 0 & 0 \\ 0 & 0 \end{pmatrix}, \quad I = \begin{pmatrix} 1 & 0 \\ 0 & 1 \end{pmatrix}. \tag{4}$$

For any ring R, the matrix O is the ***zero*** 0 of $M(R)$. If R has a unity 1, the matrix I is the ***unity*** 1 of $M(R)$.

EXAMPLE 1 **Multiplication Not Commutative**

The following products show that $M(R)$ is not commutative when $rs \neq 0$ for some r and s in R:

$$\begin{pmatrix} 0 & r \\ 0 & 0 \end{pmatrix} \begin{pmatrix} 0 & 0 \\ s & 0 \end{pmatrix} = \begin{pmatrix} rs & 0 \\ 0 & 0 \end{pmatrix}, \quad \begin{pmatrix} 0 & 0 \\ s & 0 \end{pmatrix} \begin{pmatrix} 0 & r \\ 0 & 0 \end{pmatrix} = \begin{pmatrix} 0 & 0 \\ 0 & sr \end{pmatrix}.$$

DEFINITION 7 **Determinant of a 2 × 2 Matrix**

The **determinant** of the matrix α of (4) is the element $ad - bc$ of R and is denoted by $\det \alpha$.

We note that the determinant of the unity matrix of display (4) is $1 \cdot 1 - 0 \cdot 0 = 1$. The mapping with $\alpha \mapsto \det \alpha$ plays an important role in matrix theory. One of its key properties is given in the following result.

THEOREM 2 Determinant of a Product

Let R be a commutative ring and let α and β be in $M(R)$. Then $\det(\alpha\beta) = (\det \alpha)(\det \beta)$.

The straightforward proof of this theorem is left to the reader as Problem 27 of this section.

THEOREM 3 Invertible Matrices

Let U be a commutative ring with unity. Then a matrix α is an invertible in $M(U)$ if and only if $\det \alpha$ is an invertible in U.

Proof Let $\alpha\beta = 1$ in $M(U)$. Then

$$(\det \alpha)(\det \beta) = \det(\alpha\beta) = \det 1 = 1,$$

and hence $\det \alpha$ is an invertible in U.

Conversely, let α be as in display (4) and let $g = \det \alpha = ad - bc$ be an invertible in U. Then we can let $a' = dg^{-1}$, $b' = -bg^{-1}$, $c' = -cg^{-1}$, $d' = ag^{-1}$, and

$$\alpha' = \begin{pmatrix} a' & b' \\ c' & d' \end{pmatrix}.$$

Using multiplication formula (3), we then see that $\alpha\alpha'$ and $\alpha'\alpha$ are both the unity of $M(R)$; that is, α' is the multiplicative inverse of α. ■

We next consider some important subrings in rings of matrices.

THEOREM 4 Matric Representation of the Complex Numbers

Let $\mathbf{R}$ be the field of real numbers and let K consist of the matrices in $M(\mathbf{R})$ of the form

$$\alpha = \begin{pmatrix} a & b \\ -b & a \end{pmatrix}. \tag{5}$$

Then K is a subring in $M(\mathbf{R})$, and K is a field isomorphic to the complex numbers $\mathbf{C}$.

The proof of Theorem 4 is left to the reader as Problem 28 below.

THEOREM 5 The Quaternions as Matrices

Let $\mathbf{C}$ be the field of complex numbers and let H be the subset of $M(\mathbf{C})$ consisting of the matrices

$$\alpha = \begin{pmatrix} a + b\mathbf{i} & c + d\mathbf{i} \\ -c + d\mathbf{i} & a - b\mathbf{i} \end{pmatrix}. \tag{6}$$

Then H is an infinite skew-field.

Partial Proof We leave the proof that H is a noncommutative subring in $M(\mathbf{C})$ to the reader as Problem 29 below. Assuming that H is a noncommutative subring, we now show it to be an infinite skew-field. Let α be as in (6). Then

$$\det \alpha = (a + b\mathbf{i})(a - b\mathbf{i}) - (c + d\mathbf{i})(-c + d\mathbf{i}) = a^2 + b^2 + c^2 + d^2.$$

Since squares of real numbers are never negative, this means that $\det \alpha = 0$ only when $a = b = c = d = 0$, that is, only if α is the zero matrix of $M(\mathbf{C})$. With the help of Theorem 3 and the fact that conjugation (that is, the mapping $\theta : a + b\mathbf{i} \mapsto a - b\mathbf{i}$) is a field automorphism of $\mathbf{C}$, we can show that every nonzero α in H has an inverse in H and hence that H is a skew-field. Clearly H is infinite. ∎

We next present some other notations for the elements of the field H of Theorem 5. Temporarily, let the matrix α of (6) also be denoted by the ordered quadruple $[a, b, c, d]$. Let

$$i^* = [0, 1, 0, 0], \qquad j^* = [0, 0, 1, 0], \quad k^* = [0, 0, 0, 1]$$

and let $a^* = [a, 0, 0, 0]$ for every real number a. Then it is straightforward to verify that

$$a^* + b^*i^* + c^*j^* + d^*k^* = [a, b, c, d], \tag{7}$$

and that the subset C^* of H consisting of all elements of the form $a^* + b^*i^*$ is isomorphic to $\mathbf{C}$. This isomorphism makes it possible to delete the asterisks in (7) without causing any confusion. Thus we obtain the expressions for the quaternions,

$$\alpha = a + b\mathbf{i} + c\mathbf{j} + d\mathbf{k}, \tag{8}$$

used by their originator, Sir William Rowan Hamilton. (See the biographical note on p. 271.) Generally we use the notation (8) for the elements of the skew-field H of quaternions, and we consider the complex numbers $\mathbf{C}$ and the real numbers $\mathbf{R}$ to be subrings in H.

DEFINITION 8

Conjugate and Norm of a Quaternion

The **conjugate** of $\alpha = a + b\mathbf{i} + c\mathbf{j} + d\mathbf{k}$ is $\bar{\alpha} = a - b\mathbf{i} - c\mathbf{j} - d\mathbf{k}$. The **norm** of α is $\|\alpha\| = \alpha\bar{\alpha}$.

Calculation shows that $\|a + b\mathbf{i} + c\mathbf{j} + d\mathbf{k}\| = a^2 + b^2 + c^2 + d^2$; thus $\|\alpha\|$ for the α of (8) is the same as $\det \alpha$ for the α of (6). Then it follows from Theorem 2 that $\|\alpha\beta\| = \|\alpha\| \cdot \|\beta\|$.

Addition or subtraction of quaternions is easy in the form (6) and hence in the form (8). Multiplication and division of quaternions expressed as in (8) is facilitated by verifying the following:

(A) $\mathbf{i}^2 = \mathbf{j}^2 = \mathbf{k}^2 = -1$.

(B) $\mathbf{ij} = -\mathbf{ji} = \mathbf{k}, \mathbf{jk} = -\mathbf{kj} = \mathbf{i}, \mathbf{ki} = -\mathbf{ik} = \mathbf{j}$.

(C) The center of the multiplicative group (of nonzero elements) of H consists of the nonzero real numbers.

(D) If $\alpha \neq 0$, then $\alpha^{-1} = \|\alpha\|^{-1}\bar{\alpha}$.

(E) $\overline{\alpha + \beta} = \bar{\alpha} + \bar{\beta}, \overline{\alpha\beta} = \bar{\beta} \cdot \bar{\alpha}$.

The proof of these results is left to the reader in Problem 13 through 20 for this section.

The mapping f from H to H with $f(\alpha) = \bar{\alpha}$ is a bijection with $f(\alpha + \beta) = f(\alpha) + f(\beta)$ and $f(\alpha\beta) = f(\beta)f(\alpha)$; such a mapping is an ***anti-automorphism*** of the skew-field H.

THEOREM 6 Ring of Hamiltonian Integers

Let W be the subset of H consisting of the quaternions $\alpha = a + b\mathbf{i} + c\mathbf{j} + d\mathbf{k}$ with a, b, c, and d integers. Then W is a subring in H.

Proof It is easily seen that W is nonempty and is closed under subtraction and multiplication. Hence W is a subring in H. ∎

THEOREM 7 The Quaternion Group

The multiplicative group of invertibles of the ring W of hamiltonian integers is

$$\{1, -1, \mathbf{i}, -\mathbf{i}, \mathbf{j}, -\mathbf{j}, \mathbf{k}, -\mathbf{k}\}. \tag{9}$$

Proof Let $\alpha = a + b\mathbf{i} + c\mathbf{j} + d\mathbf{k}$ with a, b, c, and d integers. Then $\|\alpha\| = a^2 + b^2 + c^2 + d^2$ is a nonnegative integer. If $\alpha\beta = 1$ in W, then $\|\alpha\| \cdot \|\beta\| = \|\alpha\beta\| = \|1\| = 1$ in $\mathbf{Z}$. It follows from the last two statements that $\|\alpha\| = 1$. This implies that α is an invertible only if one of a, b, c, d is chosen as ± 1 and the others as 0, that is, only if α is in the set of display (9). Conversely, one sees easily that each of the eight elements of that set is an invertible in W. ∎

Problems

In this problem set, a, b, c, and d always designate real numbers.

1. Describe a noncommutative ring with 16 elements.

2. Show that there exists a noncommutative ring with n^4 elements for every integer $n > 1$. (The case $n = 2$ is taken up in Problem 1.)

3. (i) In $M(\mathbf{Z}_2)$, find the product
$$\begin{pmatrix} 1 & 1 \\ 1 & 1 \end{pmatrix}\begin{pmatrix} 1 & 1 \\ 1 & 1 \end{pmatrix}.$$
(ii) Does the matrix ring $M(\mathbf{Z}_2)$ have 0-divisors? Explain.

4. (i) In $M(\mathbf{Q})$, find
$$\begin{pmatrix} 1 & 0 \\ 1 & 0 \end{pmatrix}\begin{pmatrix} 0 & 0 \\ 1 & 1 \end{pmatrix}.$$
(ii) Does $M(\mathbf{Q})$ have 0-divisors? Explain.

5. What are the entries of the zero matrix (additive identity) in $M(2, 3, R)$?

6. What are the entries of the additive identity $0_{m \times n}$ in $M(m, n, R)$?

7. Describe a matrix $A = (a_{ij})$ in $M(10, 10, \mathbf{Z}_{10})$ such that A is not the additive identity $0_{10 \times 10}$ but $A^2 = 0_{10 \times 10}$.

8. Describe a matrix $B = (b_{ij})$ in $M(m, m, \mathbf{Z}_m)$ such that $B \neq 0_{m \times m}$ and $B^2 = 0_{m \times m}$.

9. (i) Give an example of matrices A, B, C in $M(\mathbf{Q})$ with $AC = BC$, $A \neq B$, and C not the zero matrix.
(ii) Let $F = \begin{pmatrix} 5 & 0 \\ 0 & 5 \end{pmatrix}$. If $A^2 = FA$ in $M(\mathbf{Q})$, must $A = F$?

10. (i) Do Problem 9(i) with $M(\mathbf{Q})$ replaced by $M(m, m, \mathbf{Z}_m)$.
(ii) If $AB = AC$ in $M(R)$ and A is invertible, must $B = C$? Explain.

11. In $M(\mathbf{Z}_2)$, find the product
$$(1 \quad 1 \quad 0)\begin{pmatrix} 1 & 0 & 0 & 1 \\ 0 & 1 & 0 & 1 \\ 0 & 0 & 1 & 1 \end{pmatrix}.$$

12. Do as in Problem 11, with the matrix $(1 \quad 1 \quad 0)$ replaced by the following.
(a) $(0 \quad 0 \quad 1)$; (b) $(1 \quad 1 \quad 1)$; (c) $(0 \quad 0 \quad 0)$.

13. Verify that $\mathbf{i}^2 = \mathbf{j}^2 = \mathbf{k}^2 = -1$ in the quaternions.

14. Verify that $\mathbf{ij} = -\mathbf{ji} = \mathbf{k}$. Then use this result, Problem 13, and associativity to show that
$$\mathbf{jk} = -\mathbf{kj} = \mathbf{i}, \qquad \mathbf{ki} = -\mathbf{ik} = \mathbf{j}.$$

15. Let $\alpha = a + b\mathbf{i} + c\mathbf{j} + d\mathbf{k}$. Show the following.
(a) $\alpha\mathbf{i} = \mathbf{i}\alpha$ if and only if $c = 0 = d$.
(b) $\alpha\mathbf{i} = \mathbf{i}\bar{\alpha}$ if and only if $b = 0$.

16. Let $\alpha = a + b\mathbf{i} + c\mathbf{j} + d\mathbf{k}$. Give the conditions on a, b, c, and d for each of the following to hold.
(a) $\alpha\mathbf{j} = \mathbf{j}\alpha$; (b) $\alpha\mathbf{k} = \mathbf{k}\alpha$; (c) $\alpha\mathbf{j} = \mathbf{j}\bar{\alpha}$.

17. Show that under multiplication the quaternion α of Problem 15 commutes with all quaternions if and only if α is real — that is, if and only if $b = c = d = 0$.

18. In the skew-field H of quaternions, show the following.
(a) $\alpha^{-1} = \|\alpha\|^{-1}\bar{\alpha}$ if $\alpha \neq 0$.
(b) $\|\bar{\alpha}\| = \|\alpha\|$.
(c) $\|\alpha^{-1}\| = \|\alpha\|^{-1}$ if $\alpha \neq 0$.

19. Show that $\overline{\alpha + \beta} = \bar{\alpha} + \bar{\beta}$ in H.

Sir William Rowan Hamilton
1805–1865

Hamilton (not to be confused with the roughly contemporaneous Scottish philosopher, Sir William Hamilton) started out as a prodigious linguist. By the age of five he could read Latin, Greek, and Hebrew, and he knew several oriental languages by the age of ten. His family was originally Scottish, but he was born in Ireland and attended Trinity College, Dublin. At the age of twenty-two he was appointed Royal Astronomer of Ireland.

Although he made significant contributions in physics and astronomy, in mathematics his name is usually associated with his attempt to generalize complex numbers to a higher-dimensional system, algebraic entities he called *quaternions*. For ten years Hamilton worked on the problem of the definition of multiplication for such a system and finally discovered that such a multiplication would have to be noncommutative. The discovery of the necessary relationships was made while he was walking along the Royal Canal in Dublin, and he inscribed the basic formulas on the stone of Brougham Bridge. Today one can still locate a tablet on the bridge which reads: "Here as he walked by on the 16th of October 1843 Sir William Rowan Hamilton in a flash of genius discovered the fundamental formula for quaternion multiplication $i^2 = j^2 = k^2 = ijk = -1$ and cut it in a stone of this bridge." He published his full results in 1853 in his *Lectures on Quaternions*.

The development of a noncommutative system was a remarkable breakthrough in algebra — a step comparable to the abandonment of the parallel postulate in the development of noneuclidean geometry. Present-day applications make great use of a related system with noncommutative multiplication, the 3-dimensional vectors of **Josiah Willard Gibbs** (1839–1903), the American mathematician-physicist at Yale. With the introduction of Gibbs's vectors, it was thought for a number of years that quaternions were destined to remain a footnote in history. In very recent times, quaternions have found a new role in discussions of knot theory in topology and in string theory in mathematical physics, reasserting the fact that some results in mathematics, though unfashionable for certain periods, often come back as new applications or relationships are discovered.

20. Show that $\overline{\alpha\beta} = \bar{\beta} \cdot \bar{\alpha}$ in H.

21. Show that $a + b\mathbf{i} + c\mathbf{j} + d\mathbf{k} = (a + b\mathbf{i}) + (c + d\mathbf{i})\mathbf{j}$.

22. Let z_1, z_2, w_1, and w_2 be complex numbers. Show that, in H,

$$(z_1 + z_2\mathbf{j})(w_1 + w_2\mathbf{j}) = (z_1w_1 - z_2\bar{w}_2) + (z_1w_2 + z_2\bar{w}_1)\mathbf{j}.$$

23. In the skew-field H of quaternions, show the following.

(i) $(a + b\mathbf{i} + c\mathbf{j} + d\mathbf{k})^2 = (a^2 - b^2 - c^2 - d^2) + 2ab\mathbf{i} + 2ac\mathbf{j} + 2ad\mathbf{k}$.
(ii) $[(1 + \mathbf{i} + \mathbf{j} + \mathbf{k})/2]^2 = (-1 + \mathbf{i} + \mathbf{j} + \mathbf{k})/2$.

24. In W, show that $\alpha^2 = -1$ if and only if α is in

$$\{\mathbf{i}, -\mathbf{i}, \mathbf{j}, -\mathbf{j}, \mathbf{k}, -\mathbf{k}\}.$$

25. Let f be an ***antiautomorphism*** of the skew-field H over the real numbers $\mathbf{R}$; that is, let f be a bijection from H to itself with

$$f(a) = a, \qquad f(\alpha + \beta) = f(\alpha) + f(\beta), \qquad f(\alpha\beta) = f(\beta)f(\alpha),$$

for all a in $\mathbf{R}$ and all α and β in H. Show the following.
(i) $[f(\alpha)]^2 = -1$ for α in $\{\mathbf{i}, -\mathbf{i}, \mathbf{j}, -\mathbf{j}, \mathbf{k}, -\mathbf{k}\}$.
★ (ii) There are at least 24 such antiautomorphisms f.

★ **26.** Show that there are at least 24 automorphisms of H over the real numbers $\mathbf{R}$, that is, bijections g from H to H with

$$g(a) = a, \qquad g(\alpha + \beta) = g(\alpha) + g(\beta), \qquad g(\alpha\beta) = g(\alpha)g(\beta),$$

for all a in $\mathbf{R}$ and all α and β in H.

27. Prove that $\det(\alpha\beta) = (\det \alpha)(\det \beta)$ in $M(R)$ (and thus prove Theorem 2).

28. Let K consist of all the matrices in $M(\mathbf{R})$ of the form

$$\alpha = \begin{pmatrix} a & b \\ -b & a \end{pmatrix}.$$

Prove that K is isomorphic to the field $\mathbf{C}$ of complex numbers (and thus prove Theorem 4).

29. Let H be the subset of $M(\mathbf{C})$ consisting of the (quaternion) matrices

$$\alpha = \begin{pmatrix} a + b\mathbf{i} & c + d\mathbf{i} \\ -c + d\mathbf{i} & a - b\mathbf{i} \end{pmatrix}.$$

Prove that H is a noncommutative subring in $M(\mathbf{C})$. (This is the part of Theorem 5 that was left to the reader.)

30. Show that $\|\alpha^{-1}\beta^{-1}\alpha\beta\| = 1$ for all quaternions α and β.

31. Show that there are six elements with order 4 in the quaternion group

$$\{1, -1, \mathbf{i}, -\mathbf{i}, \mathbf{j}, -\mathbf{j}, \mathbf{k}, -\mathbf{k}\}$$

of invertibles of the ring of hamiltonian integers and that this group is isomorphic to the group of Example 1 of Section 3.4.

32. Let V be the multiplicative group of nonzero quaternions and let S consist of all quaternions α with $\|\alpha\| = 1$. Show that S is a subgroup in V.

4.8 Embedding a Ring in a Field

The elements of the field $\mathbf{Q}$ of rational numbers are of the form ab^{-1}, with a and $b \neq 0$ in the integral domain $\mathbf{Z}$ of the integers. In this section, we generalize on this relationship between $\mathbf{Z}$ and $\mathbf{Q}$. More specifically, for certain commutative rings R we construct a field F related to R very much as $\mathbf{Q}$ is to $\mathbf{Z}$.

Our first step is to note that, if R is a subring in a field F, then R has no 0-divisors, since $ab = 0$ and $b \neq 0$ in F imply that

$$a = abb^{-1} = 0 \cdot b^{-1} = 0.$$

Hence we restrict ourselves here to rings R with no 0-divisors. Also, we ignore the trivial ring $\{0\}$.

THEOREM 1 Field of Fractions

Let R be a commutative ring with at least one nonzero element and no 0-divisors. Then there is a field F such that the following hold:
(a) F has a subring R_1 isomorphic to R.
(b) Every element of F is the quotient of elements in R_1.

Proof Let P be the subset of the cartesian product $R \times R$ consisting of all the ordered pairs

$$(r, s), \text{ with } r \text{ and } s \text{ in } R \text{ and } s \neq 0.$$

Let E be the relation on P such that

$$(a, b)E(c, d) \text{ if and only if } ad = bc.$$

This relation is easily seen to be reflexive, symmetric, and transitive; thus E is an equivalence relation.

Let $C(a, b)$ denote the subset of all (r, s) in P with

$$(a, b)E(r, s).$$

Then the family F of all these equivalence classes $C(a, b)$ is a partition of P. We next define operations in F that make F into a field.

It is natural to try to define addition and multiplication in F using the formulas suggested by the operations in the rational numbers — that is, to let

$$C(a, b) + C(c, d) = C(ad + bc, bd) \qquad \text{(A)}$$

$$C(a, b)C(c, d) = C(ac, bd). \qquad \text{(M)}$$

But this requires that the operations in F be well defined by formulas (A) and (M); that is, the use of any other representation for an equivalence class $C(c, d)$ as a term in (A) or a factor in (M) should not cause a change in the sum or product. Hence we present the following preliminary result. ■

LEMMA 1 Addition and Multiplication Are Well Defined

If $(c, d)E(c', d')$, then $(ad + bc, bd)E(ad' + bc', bd')$ and $(ac, bd)E(ac', bd')$.

The proof is left to the reader as Problems 1 and 2 of this section.

Continuation of the Proof of the Theorem

We note that (a, b) and (c, d) play the same role in (A) and (M), since R is commutative. Therefore, it is not necessary to check that the replacement of (a, b) by an equivalent (a', b') also would not alter the results in (A) and (M).

It is left to the reader, as Problems 1 through 7 of this section, to prove that F is a commutative field.

Let b be a fixed nonzero element of R, and let θ be the mapping from R to F with $\theta(r) = C(rb, b)$. It can be shown [see Problem 7(d) of this section] that θ is an injective ring homomorphism. This implies that the image set R_1 of θ is isomorphic to R. It is also clear that the only subfield of F that contains R_1 is F itself; hence, F is the smallest field containing R_1. One easily verifies that every $C(r, s)$ in F is the quotient $C(rb, b) \div C(sb, b)$ of elements in R_1. ■

The notation $C(r, s)$ has helped us to keep in mind that each member of F is an equivalence class; now we would like to replace $C(r, s)$ with symbolism carrying the connotation of a fraction or quotient. If the technique of Theorem 1 is applied to the case in which R is the integral domain $\mathbf{Z}$ of the integers, the resulting field F is isomorphic to the rational numbers $\mathbf{Q}$. Since each element of $\mathbf{Q}$ may be expressed as r/s or rs^{-1}, with r and s in $\mathbf{Z}$ and $s \neq 0$, this motivates us to identify the original ring R of Theorem 1 with its isomorphic subring R_1 in F and to use r/s or rs^{-1} as a notation for $C(r, s)$. There are other good reasons for this procedure. One is that ring theory in abstract algebra is concerned with the structure of rings and not with the names for their elements. Another is that this is the best way to avoid giving up an old familiar ring R or having two isomorphic rings to deal with extensively. For example, we prefer not to give up the integral domain $\mathbf{Z}$ of the integers when we "embed" it in the field $\mathbf{Q}$.

Problems

In Problems 1–7 below, R is a commutative ring with at least two elements and no 0-divisors; also P, E, and F are as described in the text above.

1. Given $(c, d)E(c', d')$, show that $(ad + bc, bd)E(ad' + bc', bd')$.
2. Given $(c, d)E(c', d')$, show that $(ac, bd)E(ac', bd')$.
3. Prove that F is an abelian group under addition.
4. Prove that multiplication is associative in F.
5. Prove that multiplication is distributive over addition in F.
6. Prove that multiplication is commutative in F.
7. Let b be a nonzero element of R. Show the following.
   (a) $C(0, b)$ is the zero of F.
   (b) $C(b, b)$ is the unity of F.
   (c) $[C(r, b)]^{-1} = C(b, r)$ for all nonzero r in R.
   (d) The mapping θ from R to F with $\theta(r) = C(rb, b)$ is an injective ring homomorphism.
8. Here let R be the ring consisting of all the even integers and F be its corresponding field of fractions (as defined in Theorem 1). Prove that F is isomorphic to the field $\mathbf{Q}$ of rational numbers.
9. Show that a field R is isomorphic with its field of fractions F.

★

4.9 Characterizations of Z, Q, R, and C

This therefore is Mathematics, she reminds you of the invisible forms of the soul; she gives life to her own discoveries; she awakens the mind and purifies the intellect; she brings light to our intrinsic ideas; she abolishes oblivion and ignorance which are ours by birth .... —**Proclus Diadochus**

In this section we look at the familiar number systems **Z**, **Q**, **R**, and **C** from the point of view of abstract ring and field theory.

The integers **Z** may be characterized as the smallest ordered integral domain, since every ordered integral domain D has a subdomain D' that is isomorphic to **Z**. (See Problem 20 of Section 4.6.)

Similarly, the field **Q** of the rational numbers is the smallest field of characteristic 0, or the smallest ordered field. (See Problem 27 of Section 4.5 and Problem 24 of Section 4.6.)

We shall characterize the field **R** of real numbers after introducing some new terminology.

DEFINITION 1

Dedekind Cut

Let F be an ordered field; then a **Dedekind Cut** in F is a partitioning of F into a pair $\{A, B\}$ of subsets with the two following properties:

(i) $A \neq \varnothing, B \neq \varnothing, A \cup B = F, A \cap B = \varnothing$.

(ii) For every a in A and b in B, $a < b$.

DEFINITION 2

Completely Ordered Field

If for every Dedekind Cut $\{A, B\}$ in a field F there is an element c in F that is either the greatest element in A or the least element in B, then F is said to be **completely ordered** or **complete**.

It can be shown that the field **Q** of the rational numbers is not complete.

Dedekind (see the biographical note on p. 236) gave a constructive definition of the real numbers, using cuts in the rational numbers; from this definition it follows that the field **R** of the real numbers is completely ordered. Then it can be shown that in fact **R** is essentially the only completely ordered field; that is, every completely ordered field is isomorphic to **R**.

Before giving a characterization of this type for the complex numbers **C**, we indicate some relations between the additive and multiplicative groups of **C** and groups that can be constructed from the real numbers **R**.

The mapping θ with $\theta(a + b\mathbf{i}) = (a, b)$ is an isomorphism from the additive group of the complex numbers to the direct product of the additive group of the real numbers with itself. (See Problem 13 of Section 3.6.)

Let $\mathbf{R}^+$ denote the multiplicative group of the positive real numbers and let T be the quotient group $\mathbf{R}/[2\pi]$ of the additive group **R** of the real numbers by its cyclic subgroup $[2\pi]$. Since every nonzero complex number has a unique expression of the form

$$r(\cos x + \mathbf{i} \sin x), \quad r > 0,\ 0 \le x < 2\pi$$

and multiplication of complex numbers satisfies

$$[r_1(\cos x_1 + \mathbf{i} \sin x_1)][r_2(\cos x_2 + \mathbf{i} \sin x_2)]$$
$$= r_1 r_2[\cos(x_1 + x_2) + \mathbf{i} \sin(x_1 + x_2)],$$

the mapping f with

$$r(\cos x + \mathbf{i} \sin x) \overset{f}{\mapsto} (r, x)$$

is easily shown to be an isomorphism from the multiplicative group of the nonzero complex numbers to the direct product $\mathbf{R}^+ \times T$. (See Problem 16 of Section 3.6.)

We now define a phrase that allows us to characterize **C** as the smallest field with certain properties.

DEFINITION 3

Algebraically Closed Field

A field F is **algebraically closed** if for every positive integer d and any $d + 1$ elements $a_0, a_1, \ldots, a_d$ of F with $a_d \neq 0$ there exists an element s of F such that

$$a_d s^d + \cdots + a_1 s + a_0 = 0.$$

With terminology from the next chapter, this can be restated as: F is algebraically closed if every polynomial equation of positive degree with coefficients in F has a root in F.

The field **R** of the real numbers is not algebraically closed, since the second-degree polynomial equation

$$x^2 + 1 = 0$$

has no real root. (Since **R** is an ordered field, the square of a real number r can never be negative, and hence $r^2 \neq -1$ for all r in **R**.)

In 1799, Gauss proved that the field **C** of the complex numbers is algebraically closed. (See the biographical note on p. 235.) This means that every polynomial equation of positive degree with complex coefficients has at least one complex root.

Hence **C** is an algebraically closed extension field (superfield) over the completely ordered field **R**. In fact, **C** may be characterized as the smallest algebraically closed extension field over a complete field.

Review Problems

1. In a ring U with unity 1, use both left and right distributivity to expand

$$(a + b)(1 + 1),$$

and then use the results to show that commutativity of addition is a consequence of the other axioms for a ring with unity.

2. Explain why $r^2 \cdot r = r \cdot r^2$ for every r of a ring R, even when R is noncommutative.

3. Which elements of $\mathbf{Z}_{42}$ are invertibles?

4. Which elements of $\mathbf{Z}_{43}$ are invertibles?

5. Find all the primes in the finite arithmetic progression

$$101, 107, 113, 119, \ldots, 191, 197.$$

6. Find all the primes in the finite arithmetic progression

$$103, 109, 115, 121, \ldots, 193, 199.$$

7. Find the following in $\mathbf{Z}_{20}$.
(i) All the elements x such that $x^2 = x$.
(ii) A subring S with a unity u different from $\bar{1}$.

8. Find all the subrings in each of the following.
(a) $\mathbf{Z}_{29}$; (b) $\mathbf{Z}_{45}$.

9. Show that there is no injective homomorphism from $\mathbf{Z}_2$ to $\mathbf{Z}_8$.

10. Tabulate an injective homomorphism from $\mathbf{Z}_2$ to $\mathbf{Z}_{10}$.

11. Find the integer x with both $0 < x < 41$ and $19x \equiv 22 \pmod{41}$.

12. Find the integer y with both $0 < y < 144$ and $35y \equiv 42 \pmod{144}$.

13. (i) In $\mathbf{Z}_{30}$, list each invertible v and its inverse v^{-1} in a table.
(ii) Show that

$$1 \cdot 7 \cdot 11 \cdot 13 \cdot 17 \cdot 19 \cdot 23 \cdot 29 \equiv 11 \cdot 19 \cdot 29 \equiv -11 \cdot 19 \equiv 1 \pmod{30}.$$

14. Let α be the mapping from $\mathbf{V}_{30}$ to itself with $\alpha(x) = x^3$.
(i) Tabulate α.
(ii) List the elements with order 3 in $\mathbf{V}_{30}$.
(iii) Is α a group homomorphism? Explain.

15. Let ϕ be the Euler ϕ-function. Show the following.
(a) $\phi(3)\phi(8) = \phi(24) = 2\phi(4)\phi(6)$.
(b) $\phi(1) + \phi(2) + \phi(4) + \phi(5) + \phi(10) + \phi(20) = 20$.

16. (a) Show that $\phi(2)\phi(9) = \phi(18) = (3/2)\phi(3)\phi(6)$.
(b) Show that $\phi(1) + \phi(2) + \phi(11) + \phi(22) = 22$.

17. Using gcd(3, 10) = 1, $\phi(10) = 4$, and Euler's Theorem to find the following.
(i) The last digit (units digit) of 3^{597}.
(ii) The last digit of 943^{597}.

18. Find the last digit of 689^{1287}.

19. Use Fermat's Theorem to show that $a^n \equiv a \pmod 7$ for all integers a and for $n = 7$, 13, 19, 25, ..., that is, for $n = 6m + 1$ with m a positive integer.

20. Characterize the positive integers n such that $a^n \equiv a \pmod{11}$ for all integers a.

21. Show that $a^{19} \equiv a \pmod{798}$ for all integers a.

22. Show that $a^{17} \equiv a \pmod{510}$ for all integers a.

23. Let p and n be integers with p a prime. Given that $p \mid n^2$, explain why both $p \mid n$ and $p^2 \mid n^2$.

24. Let p and n be integers with p a prime. Given that $p \mid n^3$, explain why $p^3 \mid n^3$.

25. Use Problem 23 to prove that there do not exist integers x and y such that $25x + 10 = y^2$.

26. In each of the following, prove that there are no integers x and y satisfying the equation.
(a) $9x^2 + 9x + 3 = y^2$.
(b) $54x + 15 = y^3$.

27. Let a and m be integers. Show the following.
(a) $(m - a)^2 \equiv a^2 \pmod m$.
(b) $(2m - a)^2 \equiv a^2 \pmod{4m}$.
(c) If $a^2 \equiv 6 \pmod{10}$, then $a^2 - 6 \equiv 10 \pmod{20}$.

28. (i) Show that for all integers a,

$$a^2 \equiv (50 - a)^2 \equiv (50 + a)^2 \equiv (100 - a)^2 \pmod{100}.$$

(ii) Find all the numbers b in $\{1, 2, 3, \ldots, 99\}$ such that $b^2 \equiv 44 \pmod{100}$.
(iii) Find all c in $\{1, 2, 3, \ldots, 99\}$ such that $c^2 \equiv 444 \pmod{1000}$.

29. Let a and m be integers. Prove the following.
(i) If $a \equiv b \pmod{2m}$, then $a^2 \equiv b^2 \pmod{4m}$.
(ii) If $a \equiv b \pmod{3m}$, then $a^3 \equiv b^3 \pmod{9m}$.
(iii) $(2a - 1)^8 \equiv 1 \pmod{32}$.
(iv) $(3a + 1)^{81} \equiv 1 \pmod{243}$.

30. Let n be an integer. Show the following.
(a) Either $n \equiv 0 \pmod 5$ or $n^8 + 3n^4 \equiv 4 \pmod{10}$.
★ (b) If $b \not\equiv 1 \pmod 2$ and $b \not\equiv 0 \pmod{10}$, then $b^4 \equiv 6 \pmod{10}$, $b^{20} \equiv 76 \pmod{100}$, and $b^{200} \equiv 376 \pmod{1000}$.

31. (i) Make the multiplication table for $\mathbf{Z_8}$.
(ii) Show that $v^2 = \bar{1}$ for every invertible v of $\mathbf{Z_8}$.
(iii) Given that $\bar{b} \cdot \bar{c} = -\bar{1} = \bar{7}$ in $\mathbf{Z_8}$, show that $\bar{b}$ and $\bar{c}$ are invertibles in $\mathbf{Z_8}$, $\bar{b} \neq \bar{c}$, and $8 \mid (b + c)$ in $\mathbf{Z}$.

32. Do analogues of Problem 31(iii) in which $\mathbf{Z_8}$ is replaced by $\mathbf{Z_m}$ for each of the following values of m.
(a) 2; (b) 3; (c) 4; (d) 6; (e) 12; (f) 24.

33. Let k be a positive integer and let $d_1, d_2, \ldots, d_s$ be all the positive integral divisors of $24k - 1$. Use Problem 32(f) to explain why $24 \mid (d_1 + d_2 + \cdots + d_s)$.

34. Let W_m be the set of all v^2 in $\mathbf{V_m}$.
(a) Explain why W_m is a subgroup in $\mathbf{V_m}$.

(b) Use the table in Example 4 of Section 4.2 to show that W_{17} has index 2 in $\mathbf{V}_{17}$.
(c) Find the smallest m such that W_m has index 4 in $\mathbf{V}_m$.

35. (i) Show that $\mathbf{V}_5$, $\mathbf{V}_8$, $\mathbf{V}_{10}$, and $\mathbf{V}_{12}$ all have the same order.
(ii) Split the groups of (i) into two classes such that in each class any two groups are isomorphic.

36. Do as in the preceding problem with $\mathbf{V}_{15}$, $\mathbf{V}_{16}$, $\mathbf{V}_{20}$, $\mathbf{V}_{24}$, and $\mathbf{V}_{30}$.

37. Show that the group $\mathbf{V}_{40}$ is solvable by producing an appropriate chain of subgroups and indicating their relevant properties.

38. Do as in the preceding problem with $\mathbf{V}_{44}$ instead of $\mathbf{V}_{40}$.

39. Show that the multiplicative group $\mathbf{V}_m$ of invertibles in $\mathbf{Z}_m$ is cyclic for each of the following values of m.
(a) 2; (b) 3; (c) 4; (d) 5; (e) 6; (f) 7; (g) 9; (h) 10;
(i) 11; (j) 13; (k) 14; (l) 17.

40. Show that $\mathbf{V}_m$ is not cyclic for the following values of m.
(a) 8; (b) 12; (c) 15; (d) 16; (e) 20.

41. (i) Tabulate the homomorphism θ from $\mathbf{V}_{28}$ to $\mathbf{V}_{28}$ given by $\theta(x) = x^2$.
(ii) List the elements of the kernel K of θ.
(iii) Tabulate the natural map α from $\mathbf{V}_{28}$ to $\mathbf{V}_{28}/K$. (Use letters $K, L, \ldots$ for the cosets of K in $\mathbf{V}_{28}$.)
(iv) Make a multiplication table for the quotient group $\mathbf{V}_{28}/K$.

42. (i) List the elements of $\mathbf{V}_{35}$, and thus find its order q.
(ii) Find the orders of $\bar{1}$, $\bar{6}$, $\overline{29}$, and $\overline{34}$ in $\mathbf{V}_{35}$.
(iii) Show that $\mathbf{V}_{35}$ has too many elements of order 2 to be cyclic. (How many elements of order 2 are there in a cyclic group of order q?)
(iv) Show that $N = \{\bar{1}, \bar{6}, \overline{29}, \overline{34}\}$ is a normal subgroup in $\mathbf{V}_{35}$.
(v) Show that the quotient group $\mathbf{V}_{35}/N$ is cyclic.
(vi) Tabulate the natural map α from $\mathbf{V}_{35}$ to $\mathbf{V}_{35}/N$. (Use symbols $N, B, C, D, \ldots$ for the cosets of N in $\mathbf{V}_{35}$.)
(vii) Show that $\mathbf{V}_{35}$ is solvable.

43. Tabulate one of the isomorphisms from $\mathbf{V}_{14}$ to $\mathbf{V}_{18}$.

44. Tabulate one of the isomorphisms from $\mathbf{V}_9$ to $\mathbf{V}_{18}$.

45. (a) Show that $\mathbf{V}_8$ is isomorphic to the direct product of two groups of order 2.
(b) Is the same true of $\mathbf{V}_{12}$? Explain.
(c) Is the same true of $\mathbf{V}_5$ or of $\mathbf{V}_{10}$? Explain.

46. Show that $\mathbf{V}_{16}$ is isomorphic to the direct product of two cyclic groups, with one of order 4 and the other of order 2.

47. Explain why the characteristic of an integral domain cannot be 4, 6, 8, 9, 10, 12, 14, or 15.

48. Use the fact that every abelian group of order 38 is cyclic to prove that there is no integral domain with 38 elements.

49. Let r, s, and t be in an integral domain D. Show that

$$x^3 - (r + s + t)x^2 + (st + rt + rs)x - rst = 0$$

with $x \in D$ if and only if $x \in \{r, s, t\}$.

50. Let D be an integral domain with characteristic 5. Show that in D the following are true.
(i) $(a + b)^5 = a^5 + b^5$.

(ii) $(a+b)^{25} = a^{25} + b^{25}$.

51. Show that $x^2 + 1 \geq 2 \cdot x$ for all x in an ordered integral domain D.

52. Show that $x^4 + 6 \cdot x^2y^2 + y^4 \geq 4 \cdot x^3y + 4 \cdot xy^3$ for all x and y in an ordered integral domain D.

53. Let u and v be in a ring U with unity and let v be an invertible. Show that there is a unique x in U such that $vx = u$.

54. Let a, b, c, d, h, and k be in a commutative ring U with unity and let $ad - bc$ be an invertible v. Show that there is a unique solution in U for the following system of two simultaneous equations:

$$\begin{cases} ax + by = h \\ cx + dy = k. \end{cases}$$

55. For which prime p is $\mathbf{Z_p}$ isomorphic to a subfield in a field F with 128 elements?

56. Let F be a field with 625 elements. How many elements are there in the intersection of all the subfields in F?

57. Show that $x^2 + x^{-2} \geq 1 + 1$ for all nonzero x in an ordered field F.

58. Use the fact that every positive number y in the real numbers $\mathbf{R}$ has square roots in $\mathbf{R}$ to show that, in $\mathbf{R}$, $y > 0$ implies that

$$y + y^{-1} \geq 2.$$

59. Find the remainder when each of the following factorials is divided by the prime 139.
(a) 138!; (b) 137!; ★ (c) 136!; ★ (d) 135!.

60. Use Wilson's Theorem to show the following.
(a) $(50!)^2 \equiv -1 \pmod{101}$.
(b) $(51!)^2 \equiv 1 \pmod{103}$.

61. Let a and b be in a commutative ring R. Given that ab is neither 0 nor a 0-divisor, prove the following.
(a) $a \neq 0$ and $b \neq 0$.
(b) Neither a nor b is a 0-divisor.

62. Describe two ideals in the field $\mathbf{Q}$ of the rational numbers, and explain why $\mathbf{Q}$ has only two ideals.

63. How many ideals are there in $\mathbf{Z_{39}}$? Explain.

64. How many ideals are there in $\mathbf{Z_{101}}$? Explain.

65. How many ring homomorphisms are there from $\mathbf{Z_{17}}$ to $\mathbf{Z_{17}}$? Explain.

66. Let p be a positive prime in $\mathbf{Z}$. How many ring homomorphisms are there from $\mathbf{Z_p}$ to $\mathbf{Z_p}$? Explain.

67. Describe a homomorphism θ from $\mathbf{Z_{54}}$ to some ring R such that the kernel of θ is $(\bar{3})$.

68. Let J be the intersection of a collection Γ of subrings in a ring R. Show the following.
(a) J is a subring in R.
(b) If each subring of Γ is an ideal, so is J.
(c) If each subring of Γ is an integral domain, so is J.
(d) If each subring of Γ is a field, so is J.

69. Let U be a commutative ring with unity and let $a \in U$. Explain the relationships among the following.

(i) The principal ideal (a) in U.
(ii) The subring generated by $\{a\}$ in U; that is, the intersection D of all the subrings S in U with $a \in S$.
(iii) The smallest subring J in U with $a \in J$.

70. Let a and b be integers, not both zero. Let I be the smallest ideal, containing both a and b, in the ring of the integers $\mathbf{Z}$. (I is the intersection of all the ideals containing a and b.) Explain why $\gcd(a, b)$ is the unique positive generator of I.

CHAPTER 5

POLYNOMIALS

In this chapter we continue the work begun in Chapter 4, concentrating on the algebra of polynomials, which provided motivation for the development of the theory of rings. The concept of a ring was introduced by Richard Dedekind, and much additional work in the area was done by Leopold Kronecker, but the abstract theory of rings has taken place in the twentieth century. The word *ring* was introduced by the eminent Göttingen mathematician, David Hilbert.

The whole theory of rings and ideals was made more systematic by Emmy Noether at Göttingen. She provided the axiomatic underpinnings that made possible the enormous growth in this branch of mathematics since her work in the 1920s and 1930s.

Here we will take up not only the algebraic structure of these polynomials but some of their more prosaic properties as well — for example, their use to approximate more complicated functions (since polynomials are generally easier to deal with), and problems of factoring polynomials usually encountered in an earlier course in college algebra. In addition we will justify the technique of partial fractions expansion, known to every calculus student but rarely justified in a calculus course.

In this chapter we restrict ourselves to commutative rings. We always use R to denote a commutative ring; U, a commutative ring with unity; D, an integral domain; and F, a field.

5.1 Polynomial Extensions of Rings

The ring $\mathbf{Z}$ of the integers is a proper subring in the complex numbers $\mathbf{C}$. One element of $\mathbf{C}$ that is not in $\mathbf{Z}$ is the real number $\pi = 3.14159\ldots$, the ratio of the circumference of a circle to its diameter. We ask the question: "If we adjoin π to $\mathbf{Z}$, what other numbers must we also include in order to have a ring?" This question is answered in the following example.

EXAMPLE 1 **The Ring $\mathbf{Z}[\pi]$**

Let S designate a subring in $\mathbf{C}$ such that S contains π and all the integers. (The ring $\mathbf{R}$ of the real numbers and $\mathbf{C}$ itself are such subrings.) Since S is closed under multiplication, it must contain the powers $\pi, \pi^2, \pi^3, \ldots$ of π and also must contain all the numbers of the form $a\pi^j$, where j is a nonnegative integer and a is any integer. Closure of S under addition then implies that S contains all the numbers of the form

$$\alpha = a_d\pi^d + a_{d-1}\pi^{d-1} + \cdots + a_1\pi + a_0, \qquad a_j \text{ in } \mathbf{Z}. \tag{1}$$

An α of the form (1) is called a polynomial in π with integer coefficients. The set of all such polynomials is denoted by $\mathbf{Z}[\pi]$.

We see that π is in $\mathbf{Z}[\pi]$ by letting $d = 1$, $a_1 = 1$, and $a_0 = 0$ in (1). If we let $d = 0$ and let a_0 be any integer n, the α of (1) becomes n. Hence, $\mathbf{Z}[\pi]$ contains π and all the integers. It can also be shown that $\mathbf{Z}[\pi]$ is closed under subtraction and multiplication and hence is a subring in $\mathbf{C}$. We noted above that each α of $\mathbf{Z}[\pi]$ is in every subring S in $\mathbf{C}$ that contains π and all the integers. Therefore, $\mathbf{Z}[\pi]$ is the desired smallest of these subrings S. (See the definition in Section 3.5 of the smallest set in a collection.)

EXAMPLE 2 **The Ring $\mathbf{Z}[\mathbf{i}]$**

Now we find the smallest subring in $\mathbf{C}$ that contains the complex number $\mathbf{i}$ and all the integers. In a manner similar to that of Example 1, we see that the subring sought here consists of all the complex numbers of the form

$$\beta = b_d\mathbf{i}^d + b_{d-1}\mathbf{i}^{d-1} + \cdots + b_1\mathbf{i} + b_0, \qquad b_j \text{ in } \mathbf{Z}. \tag{2}$$

Such a number β is called a polynomial in $\mathbf{i}$ with integer coefficients, and the ring of all these polynomials is denoted by $\mathbf{Z}[\mathbf{i}]$.

Since $\mathbf{i}^2 = -1$, $\mathbf{i}^3 = -\mathbf{i}$, $\mathbf{i}^4 = 1$, $\mathbf{i}^5 = \mathbf{i}, \ldots$, the polynomial β of (2) can be rewritten in the simpler form

$$\beta = a_0 + a_1\mathbf{i}, \qquad a_0 \text{ and } a_1 \text{ in } \mathbf{Z}. \tag{3}$$

Hence, $\mathbf{Z}[\mathbf{i}]$ is the ring of gaussian integers discussed in Problem 21 of Section 4.1.

The simplification from a form such as (1) or (2), in which the number of terms is arbitrary, to a form such as (3), in which the number of terms is limited, is not possible for $\mathbf{Z}[\pi]$. The following terminology will help us discuss this difference between the nature of $\mathbf{Z}[\pi]$ and that of $\mathbf{Z}[\mathbf{i}]$.

DEFINITION 1 **Algebraic over R**

Let R be a subring in a commutative ring T. An element t of T is said to be **algebraic over R** if there is a nonnegative integer d and elements $a_d, a_{d-1}, \ldots, a_0$ of R such that $a_d \neq 0$ and, in T,

$$a_dt^d + a_{d-1}t^{d-1} + \cdots + a_1t + a_0 = 0.$$

DEFINITION 2

Transcendental over R, Indeterminate over R

Let R be a subring in a commutative ring T. An element t of T is said to be **transcendental over R**, or an **indeterminate over R**, if t is not algebraic over R.

Clearly, if t is transcendental over R and satisfies an equation

$$a_d t^d + a_{d-1}t^{d-1} + \cdots + a_1 t + a_0 = 0$$

with the a_i in R, then all of the coefficients a_i must be 0.

Let c be a complex number. Then c is said to be ***algebraic***, or an ***algebraic number***, if c is algebraic over the integers **Z**. Similarly, c is ***transcendental*** or a ***transcendental number***, if c is transcendental over **Z**.

Clearly every complex number is either algebraic or transcendental. One way to show that **i** is algebraic is to note that

$$a_2 \mathbf{i}^2 + a_1 \mathbf{i} + a_0 = 0$$

when we let $a_2 = 1$, $a_1 = 0$, and $a_0 = 1$ and that some of these integer coefficients are not zero.

In 1882 C. L. F. Lindemann established the far from trivial result that π is transcendental. (See the biographical note near the close of this section.) We now show that this implies that a given number α in $\mathbf{Z}[\pi]$ has a unique expression as a polynomial of the form (1) above. To do this, we assume that α is expressible both as

$$\alpha = a_d \pi^d + a_{d-1}\pi^{d-1} + \cdots + a_1\pi + a_0$$

and as

$$\alpha = b_e \pi^e + b_{e-1}\pi^{e-1} + \cdots + b_1\pi + b_0,$$

with the coefficients $a_d, \ldots, a_0, b_e, \ldots, b_0$ all integers. It is convenient to make $e = d$; we do this by inserting terms with zero coefficients if necessary. Then

$$0 = \alpha - \alpha = (a_d - b_d)\pi^d + \cdots + (a_1 - b_1)\pi + (a_0 - b_0). \tag{4}$$

Since π is transcendental, it follows from (4) that $a_i - b_i = 0$, and hence $a_i = b_i$ for $0 \le i \le d$. This means that the expression (1) for a given α in $\mathbf{Z}[\pi]$ is unique.

The polynomial form (2) for a number in $\mathbf{Z}[\mathbf{i}]$ is not unique; for example,

$$2\mathbf{i}^3 + 3\mathbf{i}^2 - 8\mathbf{i} + 1 = -10\mathbf{i} - 2.$$

However, a number in $\mathbf{Z}[\mathbf{i}]$ does have a unique expression of the form $m + n\mathbf{i}$, with m and n integers.

Next we generalize on these two examples. Let R be a subring in a commutative ring T and let t be an element of T. The notation $R[t]$ may be used to designate the smallest subring S in T such that t and all the elements of R are in S; one easily sees, as in the examples above, that $R[t]$ consists of all the polynomials

$$\alpha = a_d t^d + a_{d-1} t^{d-1} + \cdots + a_0 \qquad (a_j \text{ in } R)$$

in t with coefficients in R.

If both t_1 and t_2 are transcendental over R, it can be shown that there is an isomorphism f from $R[t_1]$ to $R[t_2]$ with $f(a) = a$ for all a in R and $f(t_1) = t_2$. That is, the structure of $R[t]$ does not depend on the choice of t, as long as t is transcendental over R. Yet another way to say this is that t is a "place holder" in the expression for a polynomial.

This suggests that we might be able to construct a substitute for the ring of polynomials $R[t]$ without the hypothesis that there is an element t transcendental over R. Such a construction follows for the case in which R is a commutative ring U with unity. (Theorem 1 of Section 4.8 tells us that a commutative ring R, with at least two elements and no 0-divisors, can be embedded in a field F; and every field is a commutative ring with unity.)

THEOREM 1 **Extension of a Commutative Ring**

Let U be a commutative ring with unity. Then there exists a ring S^* in which there is a subring U^* isomorphic to U and an element x^* that is transcendental over U^*.

Proof Let S^* consist of the infinite sequences

$$\alpha^* = (a_0, a_1, a_2, \ldots), \qquad a_j \text{ in } U,$$

in which only a finite number of the a_j may be nonzero. That is, for each α^* in S^* there is a fixed d such that $a_j = 0$ for $j > d$.

Using the operations in $U[t]$ as a guide, we define addition and multiplication of sequences

$$\alpha^* = (a_0, a_1, \ldots), \qquad \beta^* = (b_0, b_1, \ldots)$$

in S^* by

$$\alpha^* + \beta^* = (a_0 + b_0, a_1 + b_1, a_2 + b_2, \ldots), \tag{A}$$

$$\alpha^* \beta^* = (c_0, c_1, c_2, \ldots) \quad \text{with } c_k = \sum_{i+j=k} a_i b_j. \tag{M}$$

(This "sigma" notation states that c_k is the sum of all the products $a_i b_j$ in which $i + j = k$.) We leave it to the reader to show that the operations (A) and (M) convert S^* into a commutative ring.

Let U^* be the subset of S^* consisting of all the sequences $(a_0, 0, 0, \ldots)$, that is, of all

$$(a_0, a_1, a_2, \ldots), \qquad a_j = 0 \text{ for } j > 0.$$

It is easily shown that U^* is a subring in S^*, U and U^* are isomorphic, $x = (0, 1, 0, 0, \ldots)$ is transcendental over U^*, and $S^* = U^*[x]$. ∎

The construction outlined in Theorem 1 provides partial support for the following statement.

Carl Louis Ferdinand Lindemann
1852–1939

Lindemann was a student of Karl Weierstrass and later became a professor of mathematics in Königsberg and Munich. In 1882 he published a paper in the *Mathematische Annalen* showing that $e^{ix} + 1 = 0$ cannot be satisfied by an algebraic x; and since Euler had proved that it was satisfied by π, it follows that π is not algebraic. That is, π must be a transcendental number. This result settled the classical problem of whether the "circle can be squared"— that is, of whether one can construct with straightedge and compass a square with the same area as a given circle. (Constructions of this type are discussed in Chapter 6.) It had been shown earlier by Johann Heinrich Lambert in 1770 and Adrien-Marie Legendre in 1794 that π and π^2 were irrational, but this did not settle the classical construction problem, since they might be irrational and still constructible by straightedge and compass.

The Lindemann result on π followed the proof of 1873 by **Charles Hermite** (1822–1901) that e is transcendental. This result had appeared in the *Comptes Rendus de l'Académie des Sciences, Paris.*

ASSUMPTION Existence of Transcendental Elements

Let R be a commutative ring. Then there exist a commutative ring T in which R is a subring and an element x in T that is transcendental over R.

Theorem 1 supports the above assumption in the following sense. If we have a commutative ring U with unity and desire an indeterminate x over U, we can replace U with its isomorphic U^* and then drop the awkward asterisks as a means of simplifying the notation. Some mathematicians refer to this process as justifiable "abuse of the language."

Problems

1. Show that $\sqrt{2}$ is algebraic.

2. Show that $\sqrt[3]{5}$ is algebraic.

3. Let α be the complex number $2 + 3\mathbf{i}$. Find integers a_1 and a_0 such that $\alpha^2 + a_1\alpha + a_0 = 0$, and thus show that α is algebraic.

4. Show that the complex number $\beta = 7 - 5\mathbf{i}$ is algebraic.

5. Use Lindemann's result that π is transcendental to show that π^2 is transcendental.

6. Show that $\pi + 1$ is transcendental.

7. Show that each of the following is algebraic.
(a) $\sqrt{2} + \sqrt{3}$.
(b) $\sqrt[3]{5} + \sqrt[3]{25}$.
★ (c) $\sqrt{3} + \sqrt[3]{2}$.

David Hilbert
1862–1943

It is often said that Hilbert was the last mathematician who was able to know something about the whole range of mathematical knowledge (although the same thing is said of Henri Poincaré as well). As mathematics has grown, both in the amount known about existing branches and in the emergence of new branches, it has become very difficult for almost all mathematicians to know much outside their specialized areas of interest. But Hilbert made significant contributions to a startlingly broad range of mathematical fields: foundations of geometry, integral equations, the theory of algebraic invariants, algebraic numbers, the calculus of variations, the foundations of mathematics, and theoretical physics, to name a few. Through his study of invariants and his approach to problems in this area, he paved the way for the modern treatment of abstract algebra.

Educated at the University of Königsberg, he became professor at Göttingen in 1895, where he remained until his retirement in 1930. A famous quote of Hilbert's was made during a talk he gave in Königsberg (one that is preserved on a sound recording); it also appears on the marker of his grave in Göttingen: "Wir mussen wissen, Wir werden wissen" (We must know, we will know).

Hilbert is often remembered for the talk he gave at the Second International Mathematical Congress in Paris in 1900. He presented 23 problems he suggested for research in the twentieth century. This set of problems proved to be remarkably challenging and has stimulated some of the greatest mathematical achievements of this century to date. Solving any one of Hilbert's 23 problems has practically assured the solver or solvers of a permanent place in the pantheon of mathematicians. It is not entirely clear how many of the problems have been solved, because in the light of subsequent developments some have been recast or written in more general form and some have partial solutions. But roughly four or five remain to be solved as we approach the end of the century.

8. Let **Q** denote the rational numbers. Explain why every element of $\mathbf{Q}[\sqrt[3]{7}]$ can be written as

$$a_0 + a_1\sqrt[3]{7} + a_2(\sqrt[3]{7})^2,$$

with the a_j rational.

9. Let **R** and **C** denote the real and the complex numbers, respectively. Let $a + b\mathbf{i}$ be a fixed complex number with $b \neq 0$. Explain why

$$\mathbf{R}[a + b\mathbf{i}] = \mathbf{R}[\mathbf{i}] = \mathbf{C}.$$

10. Let m be a fixed integer. Explain why

$$\mathbf{Z}[m + \mathbf{i}] = \mathbf{Z}[\mathbf{i}].$$

11. What is the smallest subring S in **C** such that 1 and $\sqrt{2}$ are in S?

12. What is the smallest subring T in **C** such that $\sqrt{2}$ is in T?

13. Let **Q** be the field of rational numbers.

(a) In $\mathbf{Q}[\sqrt{2}]$, express the reciprocal of $3 - \sqrt{2}$ in the form $h + k\sqrt{2}$, with h and k in $\mathbf{Q}$.

(b) Explain why $\mathbf{Q}[\sqrt{2}]$ is the field T of Problem 8, Section 4.5.

14. (a) In $\mathbf{Q}[\sqrt{5}]$, express the reciprocal of $4 + 7\sqrt{5}$ in the form $h + k\sqrt{5}$, with h and k in $\mathbf{Q}$.

(b) Show that $\mathbf{Q}[\sqrt{5}]$ is a subfield in the real numbers $\mathbf{R}$.

★ **15.** Prove that $\mathbf{Z}[\sqrt{2}]$ and $\mathbf{Z}[\sqrt{3}]$ are not isomorphic rings.

5.2 Polynomials over a Commutative Ring

Next we give a formal definition that is consistent with the discussion in the previous section.

DEFINITION 1

Polynomial over *R*

Let x be an indeterminate over a commutative ring R. An expression of the form

$$\alpha = a_d x^d + a_{d-1}x^{d-1} + \cdots + a_1 x + a_0, \qquad \text{(P)}$$

with the coefficients $a_d, \ldots, a_0$ in R, is a **polynomial in x over R**.

DEFINITION 2

Degree of a Polynomial

If $a_d \neq 0$ in the expression (P), the polynomial α has **degree** d, and we write $\deg \alpha = d$.

In (P), if a coefficient $a_j = 0$, the term $a_j x^j$ may be omitted; if $a_j = 1$, the term $a_j x^j$ may be written as x^j. Each element r of R is in the form (P) with $d = 0$ and $a_0 = r$. Hence every nonzero r of R is a polynomial of degree 0 over R. Moreover, $\deg \alpha = 0$ if and only if $\alpha = r$ with r a nonzero element of R.

The 0 of R is also a polynomial over R; it is the only polynomial without degree.

NOTATION 1 ***R*[*x*]**

The set of all polynomials in x over R is denoted by $R[x]$. (One may read "$R[x]$" as "R bracket x.")

In characterizing addition and multiplication in $R[x]$, it is convenient to express any α and β in $R[x]$ as

$$\alpha = a_d x^d + a_{d-1} x^{d-1} + \cdots + a_1 x + a_0, \qquad a_j \text{ in } R,$$
$$\beta = b_d x^d + b_{d-1} x^{d-1} + \cdots + b_1 x + b_0, \qquad b_j \text{ in } R.$$

(We do not assume that deg α = deg β, but merely insert terms with zero coefficients when that is needed to achieve this uniformity of appearance.)

Then the operations of addition and multiplication in $R[x]$ are given by the following

$$\alpha + \beta = c_d x^d + c_{d-1} x^{d-1} + \cdots + c_1 x + c_0, \tag{A}$$
$$\alpha\beta = e_{2d} x^{2d} + e_{2d-1} x^{2d-1} + \cdots + e_0, \tag{M}$$

where $c_j = a_j + b_j$ (for $0 \le j \le d$) and e_k is the sum of all the products $a_i b_j$ in which $i + j = k$ (for $0 \le k \le 2d$).

We omit the details of showing that with these operations $R[x]$ is a commutative ring with R as a subring. It is clear that the 0 of R is also the zero of $R[x]$ and that the unity of R (if R has a unity) is the unity of $R[x]$.

Since x is transcendental over R, a polynomial

$$\alpha = a_d x^d + \cdots + a_1 x + a_0, \qquad a_j \text{ in } R, \tag{P}$$

is equal to the zero polynomial if and only if all the coefficients a_j are zero. As is true for polynomials in π over **Z** (see Section 5.1), it then follows that the expression (P) for a fixed α in $R[x]$ is unique, except for the possible omission of terms with zeros as coefficients.

If R has at least one nonzero element r (that is, if R has at least two elements), there are an infinite number of polynomials in $R[x]$, since, for example,

$$r, rx, rx^2, \ldots$$

are distinct polynomials in $R[x]$.

We next introduce some notation that helps us to state properties of multiplication in $R[x]$.

DEFINITION 3

Leading Coefficient

The **leading coefficient** $L(\alpha)$ of a nonzero polynomial α in $R[x]$ is the coefficient a_d in the expression

$$\alpha = a_d x^d + \cdots + a_0, \qquad a_j \in R, \qquad a_d \neq 0;$$

if α is the zero polynomial, its leading coefficient $L(\alpha)$ is the zero element of R.

For example, if α is the polynomial $5x^3 - 7x + 8$ in $\mathbf{Q}[x]$, then $L(\alpha) = 5$.

DEFINITION 4

Monic Polynomial

Let U be a ring with unity. An α in $U[x]$ is **monic** if its leading coefficient is 1.

In $\mathbf{Q}[x]$, the polynomial $\alpha = 2x + 1$ is not monic, because $L(\alpha) = 2$; but the polynomial $\beta = x^6 + 13x^3 - 7$ is monic, since $L(\beta) = 1$. If R is the ring $2\mathbf{Z}$ of the even integers, there are no monic polynomials in $R[x]$, because R has no unity.

When studying polynomials in elementary algebra, one usually has the coefficients in the integral domain $\mathbf{Z}$ or one of the fields $\mathbf{Q}$, $\mathbf{R}$, and $\mathbf{C}$. Such experience leads us to expect that $\deg(\alpha\beta) = \deg \alpha + \deg \beta$ for all nonzero α and β, that $\alpha\beta = 0$ if and only if α or β is 0, and that there is a form of unique factorization. Below, we shall see that these properties do go over to polynomials with coefficients in any field. However, they need not apply to polynomials with coefficients in a ring with 0-divisors, as we see in the following example.

EXAMPLE 1 In $\mathbf{Z}_6[x]$, we have

(a) $(\bar{3}x + \bar{2})(\bar{2}x + \bar{5}) = \bar{6}x^2 + (\overline{15} + \bar{4})x + \overline{10}$
$= \overline{19}x + \overline{10} = x + \bar{4}$,

(b) $(\bar{3}x + \bar{3})(\bar{4}x^2 + \bar{2}) = \overline{12}x^3 + \overline{12}x^2 + \bar{6}x + \bar{6} = \bar{0}$,

and

(c) $(x + \bar{1})(x + \bar{2}) = x^2 + \bar{3}x + \bar{2} = (x + \bar{4})(x + \bar{5})$.

The multiplications in Example 1 above suggest the following result.

LEMMA 1 **Multiplication in $R[x]$**

Let $\alpha \neq 0$ and $\beta \neq 0$ in $R[x]$.

(a) If $\alpha\beta \neq 0$, then $\deg(\alpha\beta) \leq \deg \alpha + \deg \beta$.

(b) If $L(\beta)$ is not a 0-divisor in R, then $L(\alpha\beta) = L(\alpha)L(\beta)$, $\alpha\beta \neq 0$, and

$$\deg(\alpha\beta) = \deg \alpha + \deg \beta.$$

COROLLARY In $U[x]$, a monic polynomial is never a 0-divisor.

Lemma 1 and its corollary follow immediately from formula (M) for multiplication in $R[x]$.

Since an integral domain D or a field F is a ring, the results in Lemma 1 apply to $D[x]$ and $F[x]$. We also have the following improvements and additions.

LEMMA 2 Polynomials over an Integral Domain D

(a) $L(\alpha\beta) = L(\alpha)L(\beta)$ for all α and β in $D[x]$.
(b) If $\alpha \neq 0$ and $\beta \neq 0$ in $D[x]$, then $\alpha\beta \neq 0$ and

$$\deg(\alpha\beta) = \deg \alpha + \deg \beta.$$

(c) $D[x]$ is an integral domain.
(d) The invertibles of $D[x]$ are just the invertibles of D.

Proof Parts (a) and (b) follow from formula (M) for polynomial multiplication and the fact that an integral domain D has no 0-divisors. Part (b) then implies that $D[x]$ has no 0-divisors. Since the unity 1 of D is also the unity of $D[x]$, this means that $D[x]$ is an integral domain.

Let α be an invertible of $D[x]$. Then there is a β in $D[x]$ such that $\alpha\beta = 1$. It follows that $\alpha \neq 0$ and $\beta \neq 0$. Using part (b), we have

$$\deg \alpha + \deg \beta = \deg(\alpha\beta) = \deg 1 = 0.$$

Since $\deg \alpha$ and $\deg \beta$ are nonnegative, we see that

$$\deg \alpha = \deg \beta = 0.$$

Hence we have shown that the invertible α of $D[x]$ is in D. ∎

LEMMA 3 The Invertibles of $F[x]$

An α of $F[x]$ is an invertible if and only if α is a nonzero element of the field F.

This is an immediate consequence of Lemma 2(d) and the fact that every nonzero element of a field F is an invertible.

EXAMPLE 2 All the First-Degree Polynomials in $\mathbf{Z}_2[x]$

An α in $\mathbf{Z}_2[x]$ with $\deg \alpha = 1$ has the form $\alpha = a_1 x + a_0$ with a_1 and a_0 in $\mathbf{Z}_2 = \{\bar{0}, \bar{1}\}$ and $a_1 \neq \bar{0}$. Hence, a_1 must be $\bar{1}$ and α must be x or $x + \bar{1}$.

If one is dealing extensively with $\mathbf{Z}_p$ or $\mathbf{Z}_p[x]$, it may be convenient to drop the bars on the symbols $\bar{0}, \bar{1}, \ldots, \overline{p-1}$ for the elements of $\mathbf{Z}_p$. Also the following compact notation may be helpful.

NOTATION 2 $[a_d a_{d-1} \cdots a_0]_p$

If each a_i is in $\{0, 1, \ldots, p-1\}$, the polynomial

$$\bar{a}_d x^d + \bar{a}_{d-1} x^{d-1} + \cdots + \bar{a}_1 x + \bar{a}_0$$

in $\mathbf{Z_p}[x]$ may be represented by $[a_d a_{d-1} \ldots a_0]_p$ or just by $a_d a_{d-1} \ldots a_0$ when the value of p is known from the context.

For example, in $\mathbf{Z_2}[x]$ one can write $x = [10]_2 = 10$, $x + \bar{1} = [11]_2 = 11$, and $x^2 + \bar{1} = [101]_2 = 101$. In $\mathbf{Z_3}[x]$ one has $x^2 + 2 = [102]_3 = 102$.

DEFINITION 5

Reducible Polynomial, Irreducible Polynomial

A nonzero α is **reducible** in $R[x]$ if there exist β and γ in $R[x]$ such that $\alpha = \beta\gamma$, $\deg \beta < \deg \alpha$, and $\deg \gamma < \deg \alpha$. A polynomial is **irreducible** if it has positive degree and is not reducible.

Briefly, a polynomial is reducible if it can be factored into polynomials with "reduced" degrees. A "reducible polynomial" is analogous to a "composite integer," since an integer is composite if and only if it is the product of two integers with smaller absolute values. We discuss another analogue in the next section.

EXAMPLE 3 **Reducible Polynomials in $\mathbf{Z_2}[x]$**

Now we find all the reducible polynomials of degree 2 in $\mathbf{Z_2}[x]$. We saw in Example 2 that x and $x + \bar{1}$ are the only first-degree polynomials in $\mathbf{Z_2}[x]$. Since $\mathbf{Z_2}$ is an integral domain, in $\mathbf{Z_2}[x]$ a second-degree polynomial is reducible if and only if it is the product of two first-degree polynomials. (See the definition of reducible polynomials and Lemma 2.) Hence the reducible polynomials of degree 2 in $\mathbf{Z_2}[x]$ are

$$x^2, \qquad (x + \bar{1})^2 = x^2 + \bar{1}, \qquad x(x + \bar{1}) = x^2 + x.$$

EXAMPLE 4 **Irreducible Polynomials in $\mathbf{Z_2}[x]$**

Here let us find all the irreducible polynomials of degree 2 in $\mathbf{Z_2}[x]$. We seek polynomials of the form

$$x^2 + a_1 x + a_0,$$

with a_1 and a_0 in $\{\bar{0}, \bar{1}\}$, that are not included among the answers in Example 3. The only such polynomial is

$$x^2 + x + \bar{1}.$$

EXAMPLE 5 Here we show that a first-degree α in $R[x]$ is irreducible. Let $\alpha = \beta\gamma$ in $R[x]$. We see that $\deg \beta < \deg \alpha = 1$ and $\deg \gamma < \deg \alpha = 1$ would imply that

$\deg \beta = 0 = \deg \gamma$. Then it would follow from Lemma 1(a) that

$$\deg \alpha = \deg(\beta\gamma) \leq \deg \beta + \deg \gamma = 0 + 0 = 0.$$

This would contradict the hypothesis $\deg \alpha = 1$; hence either $\deg \beta \geq \deg \alpha$ or $\deg \gamma \geq \deg \alpha$, and α is irreducible.

EXAMPLE 6 Let $\alpha = \bar{2}x + \bar{1}$ in $\mathbf{Z}_6[x]$. Is α an invertible? Let us see if there is a β in $\mathbf{Z}_6[x]$ such that $\alpha\beta = 1$, that is, if there are $b_0, \ldots, b_d$ in $\mathbf{Z}_6$ such that

$$(\bar{2}x + \bar{1})(b_d x^d + b_{d-1}x^{d-1} + \cdots + b_0) = 1. \tag{1}$$

Expanding the left side of (1) gives us

$$\bar{2}b_d x^{d+1} + (b_d + \bar{2}b_{d-1})x^d + \cdots + (b_1 + \bar{2}b_0)x + b_0 = 1. \tag{2}$$

Equating coefficients on the two sides of (2), we find that

$$\bar{2}b_d = b_d + \bar{2}b_{d-1} = \cdots = b_1 + \bar{2}b_0 = \bar{0}, \quad b_0 = \bar{1}.$$

But $\bar{2}b_d = \bar{0}$ and $b_d \neq \bar{0}$ in $\mathbf{Z}_6$ requires that $b_d = \bar{3}$, and then

$$b_d + \bar{2}b_{d-1} = \bar{3} + \bar{2}b_{d-1} = 0$$

is not true for any b_{d-1} in $\mathbf{Z}_6$. Hence α is not an invertible. (In this proof we tacitly assumed that $d > 0$, but it is easily seen that no β with $\deg \beta = 0$ satisfies $\alpha\beta = 1$.)

DEFINITION 6

Polynomials in $x_1, x_2, \ldots, x_n$

Let R be a subring in a commutative ring T, and $x_1, x_2, \ldots, x_n$ be elements of T. Let $T_0 = R$ and $T_i = T_{i-1}[x_i]$ for $i = 1, 2, \ldots, n$. Then T_n is the **ring of polynomials** in $x_1, x_2, \ldots, x_n$ over R, and T_n is denoted by $R[x_1, x_2, \ldots, x_n]$.

For example, $x^4 - 3xy + y^2$ is a polynomial in x and y over $\mathbf{Z}$.

DEFINITION 7

Independent Indeterminates

Let $R[x_1, x_2, \ldots, x_n]$ be as in Definition 6. Then $x_1, x_2, \ldots, x_n$ are **independent indeterminates** over R if x_i is an indeterminate over $R[x_1, \ldots, x_{i-1}, x_{i+1}, \ldots, x_n]$ for $i = 1, 2, \ldots, n$.

DEFINITION 8

Derivative of a Polynomial in x

The **derivative** of a polynomial $\alpha = a_d x^d + a_{d-1}x^{d-1} + \cdots + a_1 x + a_0$ in $R[x]$ is the polynomial

$$\alpha' = [d \cdot a_d]x^{d-1} + [(d-1) \cdot a_{d-1}]x^{d-2} + \cdots + 2 \cdot a_2 x + a_1.$$

[Recall that $n \cdot a$ means the sum $a + a + \cdots + a$ of n terms, each of which is a, when n is a positive integer.]

Problems

1. (i) Explain why every nonzero polynomial in $\mathbf{Z}_2[x]$ is monic.
   (ii) List all the reducible polynomials with degree 3 in $\mathbf{Z}_2[x]$.
   (iii) List all the irreducible polynomials with degree 3 in $\mathbf{Z}_2[x]$.
2. Given that α and β are monic polynomials in $U[x]$, explain why $\alpha\beta$ is monic and $\deg(\alpha\beta) = \deg \alpha + \deg \beta$.
3. (i) Use the $a_d a_{d-1} \ldots a_0$ notation introduced after Example 2 to list six monic reducible polynomials of degree 2 in $\mathbf{Z}_3[x]$.
   (ii) Use the $a_d a_{d-1} \ldots a_0$ notation to list three monic irreducible polynomials of degree 2 in $\mathbf{Z}_3[x]$.
4. Is every nonzero polynomial in $\mathbf{Z}_2[x]$ monic? Explain.
5. How many monic polynomials with degree d are there in $\mathbf{Z}_\mathbf{m}[x]$?
6. How many polynomials with degree d are there in $\mathbf{Z}_\mathbf{m}[x]$?
7. Explain why the following are true.
   (i) A 0-divisor in R is also a 0-divisor in $R[x]$.
   (ii) If $R[x]$ has 0-divisors, so does R.
   (iii) $R[x]$ is an integral domain if and only if R is an integral domain.
   (iv) $\mathbf{Z}_\mathbf{m}[x]$ is an integral domain if and only if the positive integer m is a prime in $\mathbf{Z}$.
8. Let α, β, and γ be in $R[x]$ with $\alpha \neq 0$ and the leading coefficient $L(\alpha)$ not a 0-divisor in R. Explain why $\beta = \gamma$ follows from $\alpha\beta = \alpha\gamma$ or $\beta\alpha = \gamma\alpha$.
9. Let β be in $U[x]$, and let the leading coefficient $L(\beta)$ be an invertible v of U. Explain why $v^{-1}\beta$ is monic and has the same degree as β.
10. What is the only monic α in $U[x]$ with $\deg \alpha = 0$?
11. Can $\deg(\alpha + \beta)$ be larger than both $\deg \alpha$ and $\deg \beta$? Explain.
12. Let α and β be nonzero polynomials in $R[x]$ with $\deg(\alpha + \beta)$ smaller than both $\deg \alpha$ and $\deg \beta$.
    (i) What must be the relation between $\deg \alpha$ and $\deg \beta$?
    (ii) What must be the relation between $L(\alpha)$ and $L(\beta)$?
13. Show the following.
    (a) $(x + \bar{2})(x + \bar{3}) = x(x + \bar{5})$ in $\mathbf{Z}_6[x]$.

(b) $(x + \bar{2})(x + \bar{5}) = x(x + \bar{7})$ in $\mathbf{Z}_{10}[x]$.
(c) $(x + \bar{3})(x + \bar{5}) = x(x + \bar{8})$ in $\mathbf{Z}_{15}[x]$.

14. Show the following.
(a) $(x + \bar{2})^2 = x^2$ in $\mathbf{Z}_4[x]$.
(b) $(x + \bar{4})^2 = x^2$ in $\mathbf{Z}_8[x]$.
(c) $(x + \bar{6})^2 = x^2$ in $\mathbf{Z}_{12}[x]$.

15. Find all the factorizations of x^2 as a product $\alpha\beta$ of monic first-degree polynomials in $\mathbf{Z}_9[x]$.

16. Show that in $\mathbf{Z}_5[x]$ there is one and only one way to factor $x^2 + x - \bar{1}$ as a product of two monic first-degree polynomials.

17. In $\mathbf{Z}_2[x]$, show that

$$x(x + \bar{1})(x^3 + x + \bar{1})(x^3 + x^2 + \bar{1}) = x^8 - x.$$

18. In $\mathbf{Z}_3[x]$, show that

$$x(x - 1)(x + 1)(x^2 + 1)(x^2 + x - \bar{1})(x^2 - x - \bar{1}) = x^9 - x.$$

19. Let D be an ordered integral domain. Show that $D[x]$ may be ordered by letting $D[x]^+$ consist of all $\alpha = a_d x^d + \cdots + a_0$ of $D[x]$ such that a_d is in D^+.

20. (i) Let D be an integral domain and let $D[x]$ be ordered. Explain a method of ordering D.
(ii) Let $\mathbf{C}$ be the field of complex numbers. Can $\mathbf{C}[x]$ be ordered? Explain.

21. In the polynomial ring $R[x]$, prove that

$$(\alpha + \beta)' = \alpha' + \beta', \quad (\alpha\beta)' = \alpha\beta' + \alpha'\beta,$$

where the derivatives α', β', $(\alpha + \beta)'$, and $(\alpha\beta)'$ are as specified in Definition 8.

22. (a) Use Problem 21 to prove that $(\alpha^n)' = n \cdot \alpha^{n-1}\alpha'$ for all α in $R[x]$ and all nonnegative integers n.
(b) Given that R has a nonzero element r, show that the mapping D from $R[x]$ to itself with $D(\alpha) = \alpha'$ is not injective.

23. In $U[x]$, expand $(x - 1)(a_2 x^2 + a_1 x + a_0)$ in the form

$$b_3 x^3 + b_2 x^2 + b_1 x + b_0,$$

and show that $b_3 + b_2 + b_1 + b_0 = 0$.

24. In $U[x]$, expand $(x + 1)(a_3 x^3 + a_2 x^2 + a_1 x + a_0)$ in the form

$$b_4 x^4 + b_3 x^3 + b_2 x^2 + b_1 x + b_0$$

and show that $b_4 - b_3 + b_2 - b_1 + b_0 = 0$.

25. Generalize on Problem 23.

26. Generalize on Problem 24.

27. In $\mathbf{Z}_3[x]$, show the following.
(i) $(x + \bar{1})^3 = x^3 + \bar{1}$.
(ii) $(x + \bar{1})^9 = [(x + \bar{1})^3]^3 = x^9 + \bar{1}$.
(iii) $(x + \bar{1})^{15} = (x + \bar{1})^9[(x + \bar{1})^3]^2$
$= (x^9 + \bar{1})(x^3 + \bar{1})^2$
$= (x^9 + \bar{1})(x^6 + \bar{2}x^3 + \bar{1})$
$= x^{15} + \bar{2}x^{12} + x^9 + x^6 + \bar{2}x^3 + \bar{1}$.

28. In $\mathbf{Z}_5[x]$, show the following.
(i) $(x + \bar{1})^5 = x^5 + \bar{1}$.

(ii) $(x + \bar{1})^{25} = x^{25} + \bar{1}$.
(iii) $(x + \bar{1})^{96} = (x^{75} + \bar{3}x^{50} + \bar{3}x^{25} + \bar{1})(x^{20} - x^{15} + x^{10} - x^5 + \bar{1})(x + \bar{1})$.

29. Let f be the mapping from $R[x]$ to R such that the image of $a_d x^d + \cdots + a_1 x + a_0$ is a_0.
(i) Show that f is a surjective ring homomorphism.
(ii) Give a generator for the kernel K of f.
(iii) Let I be the principal ideal (x) in $R[x]$. Explain why the quotient ring $R[x]/I$ is isomorphic to R.

30. Let f be the mapping of Problem 29. Let S be a subset of $R[x]$ and T consist of the images $f(\alpha)$ for all the α in S.
(i) Prove that T is a subring in R if S is a subring in $R[x]$.
(ii) Prove that T is an ideal in R if S is an ideal in $R[x]$.

31. Let m be a fixed positive integer and let $a \mapsto \bar{a}$ be the natural map from $\mathbf{Z}$ to $\mathbf{Z_m}$. Let g be the mapping from $\mathbf{Z}[x]$ to $\mathbf{Z_m}$ such that

$$g(a_d x^d + \cdots + a_1 x + a_0) = \bar{a}_0 = a_0 + (m).$$

Show that g is a surjective ring homomorphism.

32. Let m be a fixed positive integer and let $a \mapsto \bar{a}$ give the natural map from $\mathbf{Z}$ to $\mathbf{Z_m}$. Let h be the mapping from $\mathbf{Z}[x]$ to $\mathbf{Z_m}[x]$ with

$$a_d x^d + \cdots + a_1 x + a_0 \overset{h}{\mapsto} \bar{a}_d x^d + \cdots + \bar{a}_1 x + \bar{a}_0.$$

(i) Is h injective?
(ii) Is h surjective?
(iii) Is h a ring homomorphism?

33. Let p be a positive prime in $\mathbf{Z}$ and let $a \mapsto \bar{a}$ give the natural map from $\mathbf{Z}$ to $\mathbf{Z_p}$. Let f be the mapping from $\mathbf{Z}[x]$ to $\mathbf{Z_p}[x]$ with

$$f(a_d x^d + \cdots + a_1 x + a_0) = \bar{a}_d x^d + \cdots + \bar{a}_1 x + \bar{a}_0.$$

(i) Show that $f(\alpha) = 0$ if and only if $p \mid a_i$ for $0 \le i \le d$.
(ii) Given that $f(\beta\gamma) = 0$, show that either $f(\beta) = 0$ or $f(\gamma) = 0$.
(iii) If $\beta\gamma$ is monic, show that $f(\beta) \neq 0$ and $f(\gamma) \neq 0$.

34. Let $r \mapsto r'$ give a ring homomorphism f from R to R'. Let g be the mapping from $R[x]$ to $R'[x]$ with

$$g(a_d x^d + \cdots + a_1 x + a_0) = a'_d x^d + \cdots + a'_1 x + a'_0.$$

★ (i) Prove that g is a ring homomorphism.
(ii) Prove that g is injective if and only if f is injective.
(iii) Prove that g is surjective if and only if f is surjective.

35. Let f be the mapping from $R[x]$ to itself with

$$a_d x^d + a_{d-1} x^{d-1} + \cdots + a_1 x + a_0 \mapsto a_d x^{2d} + a_{d-1} x^{2d-2} + \cdots + a_1 x^2 + a_0.$$

(i) Is f injective?
(ii) Is f surjective?
(iii) Is f a ring homomorphism?

36. Let α be in $R[x]$. Prove the following.
(i) There is exactly one ring homomorphism f from $R[x]$ to itself with $r \mapsto r$ for all r in R and $x \mapsto \alpha$.
(ii) The f of part (i) is surjective if and only if $\alpha = ax + b$ with a and b in R and a an invertible.

37. Explain why $R[x, y] = R[y, x]$. (See Definition 6.)

38. Explain why $R[x, y, z] = R[y, z, x] = R[y, x, z]$.

5.3 Divisibility in Commutative Rings

The main result of this section is the division algorithm for polynomials. First we present generalizations to an arbitrary commutative ring R of some divisibility notation previously given for the ring $\mathbf{Z}$ of the integers. This material is given here rather than in Chapter 4 because the polynomial rings provide a more meaningful setting.

Let r and s be elements of R. Then $s \mid r$ is defined to mean that there exists an element t of R such that $r = st$. When $s \mid r$, one also says that s is a ***divisor*** of r in R and that r is a ***multiple*** of s in R.

EXAMPLE 1 **$\beta \mid \alpha$ but Not $\alpha \mid \beta$**

Here our work is in $\mathbf{Z}_6[x]$. Let $\alpha = x + \bar{5}$ and $\beta = \bar{2}x + \bar{1}$. The equation

$$(\bar{2}x + \bar{1})(\bar{3}x + \bar{5}) = x + \bar{5}$$

shows that $\beta \mid \alpha$. We prove that α is not a divisor of β by assuming that $\beta = \alpha\gamma$ with γ in $\mathbf{Z}_6[x]$, and obtaining a contradiction. Since α is monic, it follows from Lemma 1(b) of Section 5.2 that $\beta = \alpha\gamma$ would imply that

$$1 = \deg \beta = \deg \alpha + \deg \gamma = 1 + \deg \gamma$$

and hence that $\deg \gamma = 0$. However, the only polynomials of degree zero in $\mathbf{Z}_6[x]$ are $\bar{1}, \bar{2}, \bar{3}, \bar{4}$, and $\bar{5}$, and it is easily seen that none of these can be chosen as γ so as to have $\alpha\gamma = \beta$.

EXAMPLE 2 **Both $\alpha \mid \beta$ and $\beta \mid \alpha$**

In $\mathbf{Z}_6[x]$, let $\alpha = x + \bar{5}$ and $\beta = \bar{5}x + \bar{1}$. The equations

$$\beta = \bar{5}\alpha, \qquad \alpha = \bar{5}\beta,$$

show that $\alpha \mid \beta$ and $\beta \mid \alpha$.

DEFINITION 1

Associates in a Commutative Ring *R*

If both $a \mid b$ and $b \mid a$ in R, one says that a and b are **associates** of each other in R.

For example, the associates of a nonzero n in $\mathbf{Z}$ are n and $-n$; the only associate of 0 in $\mathbf{Z}$ is 0 itself.

Example 2 above shows that $x + \bar{5}$ and $\bar{5}x + \bar{1}$ are associates in $\mathbf{Z}_6[x]$, while Example 1 shows that $x + \bar{5}$ and $\bar{2}x + \bar{1}$ are not associates in $\mathbf{Z}_6[x]$.

LEMMA 1 **Associates in an Integral Domain *D***

The following are all equivalent in D:

(i) a and b are associates; that is, $a \mid b$ and $b \mid a$.

(ii) $a = vb$ (and $b = v^{-1}a$), with v an invertible in D.
(iii) $(a) = (b)$; that is, a and b are generators of the same principal ideal in D.

Proof Let $a \mid b$ and $b \mid a$. Then there are u and v in D such that $b = ua$ and $a = vb$. Since D is an integral domain, one may cancel a nonzero a in $1 \cdot a = vua$ and so obtain $1 = vu$, which means that v and u are invertibles. On the other hand, $a = 0$ implies that $b = u \cdot 0 = 0$, and so $a = 1 \cdot b$ and $b = 1 \cdot a$ (with 1 an invertible).

Thus we see that, whether $a = 0$ or $a \neq 0$, each of a and b is the other multiplied by an invertible. This shows that statement (ii) follows from statement (i). The converse follows immediately, as does the equivalence of statements (i) and (iii). ∎

COROLLARY Associates in $D[x]$

Polynomials α and β are associates in $D[x]$ if and only if $\alpha = v\beta$ with v an invertible in D. If α and β are associates, $\deg \alpha = \deg \beta$.

The proof is left to the reader as Problem 13(a) of this section. Note that the $D[x]$ of the corollary may be taken as the D of Lemma 1.

In the rest of this section, we generalize some concepts previously introduced for the integers $\mathbf{Z}$ by applying them to general commutative rings U with unity or to polynomial rings $U[x]$.

DEFINITION 2 Prime in U

In a commutative ring U with unity, let p be an element that is not 0 and is not an invertible. Also, let every divisor of p be either an invertible or an associate of p. Then p is a **prime** in U.

DEFINITION 3 Composite in U

An element a in U is **composite** if a is not 0, not an invertible, and not a prime.

The following examples show that our concepts of "prime polynomial" and "irreducible polynomial" are distinct analogues of "prime integer." However, in Theorem 1 below we show that they are equivalent for $F[x]$, that is, when the coefficients are in a field.

EXAMPLE 3 **Irreducible but Not Prime**

Here we find a polynomial in $\mathbf{Z}[x]$ that is irreducible but is not a prime. The polynomial $\alpha = 5x + 10$ is irreducible, since $\deg \alpha = 1$. (See Example 5 of Section 5.2.) The divisor 5 of α is not an invertible in $\mathbf{Z}[x]$, since it is not an invertible in the integral domain $\mathbf{Z}$. Also, $\deg 5 \neq \deg \alpha$ implies that 5 is not an associate of α. Hence it follows from the definition of a prime that α is not a prime.

EXAMPLE 4 **This Time in $\mathbf{Z}_6[x]$**

Now we show that the monic polynomial $\alpha = x + \bar{5}$ is irreducible but is not a prime in $\mathbf{Z}_6[x]$. Since $\deg \alpha = 1$, α is irreducible. (See Example 5 of Section 5.2.) In Example 1 above, it was shown that α has a divisor $\beta = \bar{2}x + 1$, which is not an associate of α. As β is also not an invertible (see Example 6 of Section 5.2), it follows that α is not a prime.

THEOREM 1 **Primes and Irreducibles in $F[x]$**

Let F be a field. Then a polynomial α is prime in $F[x]$ if and only if α is irreducible in $F[x]$.

Proof Let α be an irreducible polynomial and let $\alpha = \beta\gamma$ in $F[x]$. Since α is irreducible, either $\deg \beta \geq \deg \alpha$ or $\deg \gamma \geq \deg \alpha$; for definiteness, let $\deg \beta \geq \deg \alpha$. Then

$$\deg \alpha = \deg(\beta\gamma) = \deg \beta + \deg \gamma$$

implies that $\deg \gamma = 0$. Hence γ is a nonzero element of F and so is an invertible in F and in $F[x]$. That makes β an associate of α and shows that every divisor of α is either an invertible or an associate of α; that is, α is a prime.

Conversely, let α be a prime in $F[x]$. Then $\alpha \neq 0$ and $\deg \alpha > 0$, since a polynomial of degree 0 is an invertible in $F[x]$ and a prime is not an invertible. Let $\alpha = \beta\gamma$. Then the hypothesis that α is a prime implies that either β is an invertible, and so of degree 0, or β is an associate of α, and hence $\deg \beta = \deg \alpha$. The same is true of γ. If both $\deg \beta < \deg \alpha$ and $\deg \gamma < \deg \alpha$, it would follow that $\deg \beta = 0 = \deg \gamma$ and

$$\deg \alpha = \deg \beta + \deg \gamma = 0 + 0 = 0.$$

But $\deg \alpha > 0$. Hence the assumption that $\deg \beta < \deg \alpha$ and $\deg \gamma < \deg \alpha$ is false; that is, the prime α is irreducible. ∎

A ring is closed under addition, subtraction, and multiplication but may or may not contain the quotient of two given elements. In rings of polynomials, as in the ring $\mathbf{Z}$ of integers, there is a division algorithm — that is, a process of dividing α by β so as to obtain a quotient γ and a remainder ρ of lower degree than β. (There are restrictions on the divisor β.) Since the steps of the algorithm are familiar from elementary algebra, we shall not describe them in detail. However, we present a theorem that has many important applications and also implicitly characterizes the algorithm.

For practical and theoretical reasons, it is convenient to start by considering the case in which β is monic.

LEMMA 2 Division by a Monic Polynomial

Let α and β be in $U[x]$ with β monic. Then there is a unique ordered pair of polynomials γ and ρ in $U[x]$ such that $\alpha = \gamma\beta + \rho$ and either $\rho = 0$ or $\deg \rho < \deg \beta$. (γ is called the **quotient** and ρ the **remainder** in the division of α by β.)

Proof We first consider the case in which $\beta \mid \alpha$. Then there is a γ in $U[x]$ such that $\alpha = \gamma\beta = \gamma\beta + 0$, and we may choose this γ as the quotient and 0 as the remainder ρ.

Now let β not be a divisor of α and let S be the set of all polynomials of the form $\alpha - \delta\beta$, with δ in $U[x]$. We see that 0 is not in S, since otherwise $\beta \mid \alpha$. Hence every polynomial in S has a degree, and we let T be the nonempty set of all degrees of polynomials in S. It then follows from the well ordering principle that there is a least integer t in T. Now let γ be a polynomial in $U[x]$ such that $\alpha - \gamma\beta$ is a polynomial ρ in S of this minimal degree t. Let $\deg \beta = d$ and

$$\rho = r_t x^t + \cdots + r_0, \qquad r_t \neq 0.$$

We show that $t = \deg \rho < \deg \beta = d$ by assuming that $t \geq d$ and obtaining a contradiction. If $t \geq d$, then

$$\alpha - (\gamma + r_t x^{t-d})\beta$$

would be in S and would have degree less than t, contradicting the minimal nature of t. Therefore, $\deg \rho < \deg \beta$ and the existence of a quotient and remainder with the desired properties is established.

We complete the proof by showing that the ordered pair (γ, ρ) is unique. Let us assume that in $U[x]$

$$\alpha = \gamma\beta + \rho = \gamma'\beta + \rho'$$

with each of ρ and ρ' either 0 or of degree less than d. Then

$$(\gamma - \gamma')\beta = \rho' - \rho.$$

Since β is monic, either $\gamma - \gamma' = 0$ or $\deg[(\gamma - \gamma')\beta] \geq \deg \beta = d$. [See part (b) of Lemma 1 in Section 5.2.] On the other hand, $\rho' - \rho$ is either 0 or of degree less than d. Together these statements imply that $\gamma - \gamma'$ and $\rho - \rho'$ are both zero; that is, (γ, ρ) is unique. ∎

We next use Lemma 2 to obtain a more general form of the division algorithm.

THEOREM 2 The Division Algorithm in $U[x]$

Let α and β be in $U[x]$ and let β have an invertible v as its leading coefficient. Then there is a unique ordered pair γ and ρ in $U[x]$ such that $\alpha = \gamma\beta + \rho$ and either $\rho = 0$ or $\deg \rho < \deg \beta$.

Proof The polynomial $\beta' = v^{-1}\beta$ is monic, and hence it follows from Lemma 2 that there are polynomials γ' and ρ in $U[x]$ with $\alpha = \gamma'\beta' + \rho$ and either $\rho = 0$ or $\deg \rho < \deg \beta' = \deg \beta$. Now we let $\gamma = v^{-1}\gamma'$. Then $\gamma' = v\gamma$ and

$$\alpha = \gamma'\beta' + \rho = (v\gamma)(v^{-1}\beta) + \rho = \gamma\beta + \rho,$$

as desired. Uniqueness is left to the reader. ■

COROLLARY 1 In Theorem 2, $\beta \mid \alpha$ if and only if $\rho = 0$.

COROLLARY 2 In Theorem 2, if $\deg \beta = 1$, then ρ is in U.

In a ring $F[x]$ of polynomials over a field F, every nonzero β has an invertible as its leading coefficient and hence may be used as the divisor in the division algorithm of Theorem 2. Also, in $F[x]$ the quotient γ satisfies $\deg \gamma = \deg \alpha - \deg \beta < \deg \alpha$ when $\gamma \neq 0$ and $\deg \beta > 0$.

Our base 10 notation for the integers uses the fact that every positive integer is uniquely expressible as a polynomial

$$10^s c_s + 10^{s-1} c_{s-1} + \cdots + 10 c_1 + c_0, \quad c_s \neq 0,$$

with the c_j integers satisfying $0 \leq c_j < 10$. (There are similar expressions for other bases; see Problems *15 and *16 of Section 1.1.) An analogous result for polynomials over a field F is given in the following.

THEOREM 3 Polynomials in β

Let β be a polynomial in $U[x]$ with an invertible as its leading coefficient. Then in $U[x]$ every nonzero α is uniquely expressible in the form

$$\rho_s \beta^s + \cdots + \rho_1 \beta + \rho_0, \quad \rho_s \neq 0, \tag{1}$$

where each ρ_j either is 0 or has degree less than β.

Proof We assume the result false. This and the well ordering principle imply that, among the polynomials not of the form (1), there is an α with least degree d. Theorem 2 tells us that we can let this α equal $\gamma\beta + \rho_0$ where either $\rho_0 = 0$ or $\deg \rho_0 < \deg \beta$. We see that $\gamma \neq 0$, since otherwise $\alpha = \rho_0$ would be of form (1) with $s = 0$. Then $\gamma \neq 0$ implies that $\deg \gamma < \deg \alpha$. This and the minimal nature of $\deg \alpha$ mean that γ has an expression of the form (1), which we write as

$$\gamma = \rho_t \beta^{t-1} + \cdots + \rho_2 \beta + \rho_1.$$

Then

$$\alpha = \gamma\beta + \rho_0 = \rho_t \beta^t + \cdots + \rho_1 \beta + \rho_0$$

is of form (1). This contradiction completes the proof. ■

Problems

1. (i) In $\mathbf{Z}_{12}$, find all $\bar{a}$ such that $\bar{2} \mid \bar{a}$.
   (ii) In $\mathbf{Z}_{12}$, find all $\bar{b}$ such that $\bar{b} \mid \bar{2}$.
   (iii) In $\mathbf{Z}_{12}$, find all associates of $\bar{2}$.
   (iv) In $\mathbf{Z}_3[x]$, find all associates of $\bar{2}x^3 + x$.
2. Let p be a positive prime in $\mathbf{Z}$.
   (i) How many invertibles are there in $\mathbf{Z}_p$?
   (ii) How many associates does a nonzero α have in $\mathbf{Z}_p[x]$?
3. Let $\mathbf{Q}$ be the field of rational numbers. In $\mathbf{Q}[x]$, find the quotient and remainder in the division of each of the following by $x - 2$.
   (i) x^2; (ii) x^3; (iii) x^4; (iv) x^5.
4. In $\mathbf{Q}[x]$ find the quotient and remainder in the division of each of the following by $x + 3$.

   (i) $x^2 - 9$; (ii) $x^3 + 27$; (iii) $x^4 - 81$.

5. In $\mathbf{Z}_3[x]$, let $\alpha_0 = x^5 + x^4 + x^3 + x + \bar{2}$ and $\alpha_1 = x^4 + x^3 + \bar{2}x^2 + \bar{2}x$, and find polynomials $\gamma_1, \alpha_2, \gamma_2, \alpha_3$, and γ_3 such that the following are true.
   (i) $\alpha_0 = \gamma_1\alpha_1 + \alpha_2$ with $\deg \alpha_2 < \deg \alpha_1$;
   (ii) $\alpha_1 = \gamma_2\alpha_2 + \alpha_3$ with $\deg \alpha_3 < \deg \alpha_2$;
   (iii) $\alpha_2 = \gamma_3\alpha_3$.
6. (a) In $\mathbf{Q}[x]$, let $\alpha = x^3 - 5x + 8$ and $\beta = 2x^2 + 3x - 1$. Find γ and ρ such that $\alpha = \gamma\beta + \rho$ and $\deg \rho < 2$.
   (b) In $\mathbf{Z}_5[x]$, let $\alpha = x^3 + \bar{3}$ and $\beta = \bar{2}x^2 + \bar{3}x + \bar{4}$. Find γ and ρ such that $\alpha = \gamma\beta + \rho$ and $\deg \rho < 2$.
7. Let n be a positive integer and let a be in a commutative ring U with unity. In $U[x]$, find β such that $x^n - a^n = (x - a)\beta$.
8. Let $\alpha = a_d x^d + \cdots + a_1 x + a_0$ be in $U[x]$, let r be in U, and let $s = a_d r^d + \cdots + a_1 r + a_0$. Show that
$$(x - r) \mid (\alpha - s).$$
9. Let a and b be in $\mathbf{Q}$. Show that $(x + a + b) \mid (x^3 - 3abx + a^3 + b^3)$ in $\mathbf{Q}[x]$.
10. In the field $\mathbf{C}$ of complex numbers, let w be any number with $w^3 = 1$. Show that in $\mathbf{C}[x]$.
$$(x + aw + bw^2) \mid (x^3 - 3abx + a^3 + b^3).$$
11. Let R be a commutative ring. Show the following.

(a) If $d \mid a$ and $d \mid b$ in R, then $d \mid (ah + bk)$ for all h and k in R.
(b) Let $a = qb \pm c$ in R. Then $d \mid a$ and $d \mid b$ if and only if $d \mid b$ and $d \mid c$.

12. Let U be a commutative ring with unity. Show the following.
(a) In U, a is an associate of 1 if and only if a is an invertible.
(b) In $U[x]$, let β have an invertible as its leading coefficient $L(\beta)$, and let γ and ρ be the quotient and remainder in the division of α by β. If $\gamma \neq 0$, then

$$\deg \alpha = \deg \beta + \deg \gamma.$$

(c) Let v be an invertible in U. Then in U, $b \mid a$ if and only if $b \mid (av)$; also $b \mid a$ if and only if $(bv) \mid a$.

13. Let D be an integral domain. Show the following.
(a) α and β are associates in $D[x]$ if and only if $\alpha = v\beta$ with v an invertible in D. Also, associates in $D[x]$ have the same degree. (This proves the corollary to Lemma 1.)
(b) In $D[x]$, a nonzero α has a monic associate if and only if the leading coefficient $L(\alpha)$ is an invertible. If $L(\alpha)$ is an invertible, α has a unique monic associate.
(c) In $D[x]$, α is an associate of 1 if and only if α is an invertible in D.
(d) If $\beta \mid \alpha$ in $D[x]$, then $L(\beta) \mid L(\alpha)$ in D.
(e) In $D[x]$, if $\beta \mid \alpha$ and $\alpha \neq 0$, then $\deg \beta \leq \deg \alpha$.
(f) If $\beta \mid \alpha$ in $D[x]$ and α is monic, $L(\beta)$ is an invertible in D.
(g) If $\gamma \mid \beta$ in $D[x]$ and β is irreducible, then $\deg \gamma$ is either 0 or $\deg \beta$.
(h) In $D[x]$, every monic irreducible polynomial is a prime.
(i) In $D[x]$, the only monic divisors of a monic prime π are 1 and π.
(j) Let π be a prime with degree d in $D[x]$. Then every divisor of π has degree d or 0.

14. Let F be a field. Show the following.
(a) A nonzero β in $F[x]$ has exactly one monic associate.
(b) α is an associate of 1 in $F[x]$ if and only if α is a nonzero element of F.
(c) Let I be a principal ideal (β) in $F[x]$ with $\beta \neq 0$. Then there is a unique monic generator of I.

15. Give an example of a principal ideal in $\mathbf{Z}[x]$ that does not have a monic generator.

16. Let U be a commutative ring with unity, $c \in U$, $\alpha \in U[x]$, and $\deg \alpha = d$. Explain why there exist b_i in U such that

$$\alpha = b_0 + b_1(x - c) + b_2(x - c)^2 + \cdots + b_d(x - c)^d.$$

17. Let U be a commutative ring with unity. Let A be the relation in U for which rAs means that r and s are associates of each other. Is A an equivalence relation? Explain.

18. (i) Let α be in $D[x]$ with $\deg \alpha$ equal to 2 or 3. Explain why α is reducible if and only if it has a divisor of degree 1.
(ii) Show that $x^4 + x^2 + \bar{1}$ is reducible in $\mathbf{Z}_2[x]$ even though it has no divisor of degree 1. [*Hint:* See Problem 15(i) of Section 4.4.]

19. In $D[x]$, let $\alpha = \beta\gamma$ with α monic. Show that β and γ have monic associates β' and γ', respectively, and that $\alpha = \beta'\gamma'$.

20. Show that a monic α in $D[x]$ is reducible if and only if there exist monic polynomials β and γ in $D[x]$ such that

$$\alpha = \beta\gamma, \qquad \deg \beta < \deg \alpha, \qquad \deg \gamma < \deg \alpha.$$

21. Use Problem 20 to show the following.
(a) $x^3 - 3x + 1$ is irreducible (and prime) in $\mathbf{Q}[x]$.

(b) $x^3 + x + \bar{1}$ is irreducible (and prime) in $\mathbf{Z}_2[x]$.

22. Use Problem 20 to find three monic irreducible polynomials of degree 2 in $\mathbf{Z}_3[x]$. (The answer to Problem 3 of Section 5.2 may be helpful.)

23. Show that the only irreducible polynomials of degree 4 in $\mathbf{Z}_2[x]$ are $x^4 + x + \bar{1}$, $x^4 + x^3 + \bar{1}$, and $x^4 + x^3 + x^2 + x + \bar{1}$.

24. Find all the irreducible polynomials of degree 5 in $\mathbf{Z}_2[x]$.

25. Let $F = \{0, 1, c, c^2\}$ be a field with four elements. (In Problem 1 of Section 4.4, we saw that F has characteristic 2 and $c^2 = 1 + c$.) In $F[x]$, show the following.
(i) $x^4 - x = x(x - 1)(x - c)(x - c^2)$.
(ii) $x^{16} - x = (x^4 - x)\beta$, where

$$\beta = (x^2 + cx + 1)(x^2 + c^2x + 1)(x^2 + x + c^2)(x^2 + cx + c) \cdot (x^2 + c^2x + c^2)(x^2 + x + c).$$

(iii) The six second-degree factors of β in part (ii) are all the monic irreducible polynomials of degree 2 in $F[x]$.

26. In $\mathbf{Z}_2[x]$, show the following.
(i) $x^{16} - x = x(x - \bar{1})(x^2 + x + \bar{1})(x^4 + x + \bar{1})(x^4 + x^3 + \bar{1}) \cdot (x^4 + x^3 + x^2 + x + \bar{1})$.
(ii) The factors of $x^{16} - x$ on the right side of part (i) are all the irreducible polynomials of degree 1, 2, or 4.

27. Let F be a field with nine elements. Assume that in $F[x]$ the polynomial $x^{81} - x$ can be expressed as the product of all the monic irreducible polynomials of degree 1 or 2. How many of these factors are of degree 2?

28. Let R be a commutative ring. Let α and β be in $R[x]$ and let $\beta = b_d x^d + \cdots + b_1 x + b_0$ with $b_d \neq 0$. Show that there exist a nonnegative integer t and polynomials γ and ρ in $R[x]$ such that $(b_d)^t\alpha = \gamma\beta + \rho$ and either $\rho = 0$ or ρ has lower degree than β.

29. Let p be a positive prime in $\mathbf{Z}$ and let $a \mapsto \bar{a}$ give the natural map from $\mathbf{Z}$ to $\mathbf{Z}_p$. Let

$$\alpha = x^d + a_{d-1}x^{d-1} + \cdots + a_1x + a_0$$

be a monic polynomial in $\mathbf{Z}[x]$. Show the following.
(i) If α is reducible in $\mathbf{Z}[x]$, then

$$\bar{\alpha} = x^d + \bar{a}_{d-1}x^{d-1} + \cdots + \bar{a}_1x + \bar{a}_0$$

is reducible in $\mathbf{Z}_p[x]$.
(ii) $x^3 + (2a + 1)x + (2b + 1)$ is irreducible in $\mathbf{Z}[x]$ for all integers a and b.

30. Prove that $x^3 + (3a - 1)x + (3b + 1)$ is irreducible in $\mathbf{Z}[x]$ for all integers a and b. [*Hint:* Use Theorem 1 of Section 3.13 and Problem 29(i) above.]

31. For elements r, s, t of any ring R, let $r \equiv s \pmod{t}$ mean that $t \mid (r - s)$. Prove the following.
(a) $r \equiv r \pmod{t}$ for all r and t in R.
(b) If $r \equiv s \pmod{t}$, then $s \equiv r \pmod{t}$.
(c) If $a \equiv b \pmod{t}$ and $b \equiv c \pmod{t}$, then $a \equiv c \pmod{t}$.
(d) If $a \equiv b \pmod{t}$ and $c \equiv d \pmod{t}$, then $a \pm c \equiv b \pm d \pmod{t}$ and $ac \equiv bd \pmod{t}$.
(e) If $r \equiv s \pmod{t}$, then $r^n \equiv s^n \pmod{t}$ for all n in $\mathbf{Z}^+$.
(f) If $r \equiv s \pmod{mt}$ with m in R, then $r \equiv s \pmod{t}$.

32. Let F stand for any of the fields $\mathbf{Q}$, $\mathbf{R}$, and $\mathbf{C}$. Let $a_0, a_1, \ldots, a_{3e}$ be any elements of

F and $\alpha = x^3 - 1$. Show that in $F[x]$ one has

$$a_0 + a_1x + \cdots + a_{3e}x^{3e} \equiv (a_0 + a_3 + a_6 + \cdots + a_{3e}) + (a_1 + a_4 + \cdots + a_{3e-2})x + (a_2 + a_5 + \cdots + a_{3e-1})x^2 \pmod{\alpha}.$$

[See Problem 31 for the meaning of congruence modulo a polynomial.]

33. Let F and the a_i be as in Problem 32. Show that

$$(1 + x + x^2) \mid (a_0 + a_1x + a_2x^2 + \cdots + a_{3e}x^{3e})$$

in $F[x]$ if and only if

$$a_0 + a_3 + a_6 + \cdots + a_{3e} = a_1 + a_4 + a_7 + \cdots + a_{3e-2} = a_2 + a_5 + a_8 + \cdots + a_{3e-1}.$$

34. Generalize on Problem 33.

35. Let a, b, c, d, e be five consecutive terms from the sequence $1^5, 2^5, 3^5, \ldots$ of fifth powers. Show that $(1 + x + x^2 + x^3 + x^4) \mid (x^a + x^b + x^c + x^d + x^e)$ in $\mathbf{Q}[x]$, and hence in $\mathbf{R}[x]$ and in $\mathbf{C}[x]$.

5.4 Polynomial Functions

Associated with each polynomial

$$\alpha = a_d x^d + \cdots + a_1 x + a_0, \qquad a_j \text{ in } R,$$

over a commutative ring R is the mapping α^* from R to R for which the image of an r in R is

$$a_d r^d + \cdots + a_1 r + a_0. \tag{1}$$

Such a mapping α^* is called a ***polynomial function***. We let R^* represent the set of α^* for all α in $R[x]$.

Instead of designating the image of r under α^* given in display (1) as $\alpha^*(r)$, we shall use the simpler notation $\alpha(r)$. The sum σ^* and product π^* of polynomial functions α^* and β^* are defined, as in calculus, by

$$\sigma(r) = \alpha(r) + \beta(r), \qquad \pi(r) = \alpha(r)\beta(r). \tag{2}$$

Formulas (2) are equivalent to

$$(\alpha + \beta)^* = \alpha^* + \beta^*, \qquad (\alpha\beta)^* = \alpha^*\beta^*. \tag{3}$$

The operations defined in (2), or (3), make R^* into a commutative ring. (The details are tedious but straightforward and are left to the reader. See Problem 31 of Section 4.1.)

It also follows from (3) that the surjection f from $R[x]$ to R^* with $\alpha \mapsto \alpha^*$ is a ring homomorphism. This homomorphism need not be injective, as we see in the following example.

EXAMPLE 1 In $\mathbf{Z}_3[x]$, let $\alpha = x^3 + x$ and $\beta = \bar{2}x$. Here we show that $\alpha^* = \beta^*$ (although $\alpha \neq \beta$) and hence that the homomorphism f with $\alpha \mapsto \alpha^*$ is not an isomorphism from $\mathbf{Z}_3[x]$ to $\mathbf{Z}_3^*$. Since $\mathbf{Z}_3$ has only three elements, we solve our problem easily by noting that under either α^* or β^* we have $\bar{0} \mapsto \bar{0}$, $\bar{1} \mapsto \bar{2}$, and $\bar{2} \mapsto \bar{1}$.

Let R be a finite ring with m elements. Since there are m^m mappings from R to R (see Problem 10 of Section 3.1), there are at most m^m polynomial functions in R^*. If $m > 1$, there are an infinite number of polynomials in $R[x]$, and hence no mapping from $R[x]$ to R^* can be injective.

We show below that the homomorphism f from $D[x]$ to D^* with $\alpha \mapsto \alpha^*$ is an injection, and hence an isomorphism, for all infinite integral domains D. (See Corollary 2 to Theorem 3, below.)

DEFINITION 1 **Zero of a Function, Root of an Equation**

Let R be a subring of a commutative ring S and let α be in $R[x]$. An element s of S, such that $\alpha(s) = 0$, is a **zero** in S of the function α^* and of the polynomial α. Also, s is a **root** of the polynomial equation $\alpha(x) = 0$.

We next present some results linking zeros and first-degree divisors of a polynomial over a ring U with unity.

LEMMA 1 For each r in U and each positive integer n,

$$(x - r) \mid (x^n - r^n) \text{ in } U[x].$$

Proof This follows from the easily verified factorization

$$x^n - r^n = (x - r)(x^{n-1} + rx^{n-2} + \cdots + r^{n-2}x + r^{n-1}).$$ ■

LEMMA 2 Let α be in $U[x]$ and r be in U. Then, in $U[x]$,

$$(x - r) \mid [\alpha - \alpha(r)].$$

Proof The desired result is a consequence of Lemma 1 and the following:

$$\begin{aligned} \alpha - \alpha(r) &= (a_d x^d + \cdots + a_0) - (a_d r^d + \cdots + a_0) \\ &= a_d(x^d - r^d) + \cdots + a_1(x - r). \end{aligned}$$ ■

THEOREM 1 **The Remainder Theorem**

Let r be in U. Then in $U[x]$ the remainder in the division of α by $x - r$ is $\alpha(r)$.

Proof Lemma 2 tells us that $\alpha - \alpha(r) = (x - r)\gamma$ with γ in $U[x]$. Transposing, we have

$$\alpha = (x - r)\gamma + \alpha(r). \tag{4}$$

Since $\alpha(r)$ is either 0 or of degree 0, it is the unique remainder in the division of α by the first-degree $x - r$. ∎

THEOREM 2 The Factor Theorem

Let r be in U and α be in $U[x]$. Then r is a root of $\alpha(x) = 0$ (that is, a zero of α) if and only if $x - r$ is a divisor of α in $U[x]$.

Proof If r is a zero of α, that is, if $\alpha(r) = 0$, it follows immediately from (4) that $(x - r) \mid \alpha$. If $(x - r) \mid \alpha$, then (4) implies that

$$(x - r) \mid \alpha(r),$$

and this requires that $\alpha(r) = 0$, since a polynomial of degree 0 is not a multiple of a monic polynomial of degree 1. ∎

For polynomials over an integral domain D, we have the following result.

THEOREM 3 Linear Factors

Let α be in $D[x]$ and let $r_1, r_2, \ldots, r_n$ be n distinct zeros of α in D. Then there is a γ in $D[x]$ such that

$$\alpha = (x - r_1)(x - r_2) \cdots (x - r_n)\gamma.$$

Outline of Proof It follows from the Factor Theorem that $\alpha = (x - r_1)\beta$, with β in $D[x]$. Then

$$0 = \alpha(r_i) = (r_i - r_1)\beta(r_i).$$

For $2 \le i \le n$, $r_i - r_1 \ne 0$, since the zeros r_i are distinct. The fact that D has no 0-divisors then implies that, for $i > 1$, $\beta(r_i) = 0$; that is, r_i is a zero of β. An inductive argument may be used to complete a formal proof of this theorem. Informally, one gives β the treatment given to α above and continues the process until n first-degree divisors have been factored out of α. ∎

COROLLARY 1 If α has n distinct zeros and $\alpha \ne 0$, $\deg \alpha \ge n$. Hence a nonzero polynomial α has a finite number of zeros.

COROLLARY 2 If D is infinite and $\beta \ne \gamma$ in $D[x]$, then $\beta^* \ne \gamma^*$ in D^*; that is, the ring homomorphism $\alpha \mapsto \alpha^*$ is an injection, and hence it is an isomorphism.

The proofs of the corollaries are left to the reader as Problems 11 and 12 of this section.

Next we consider multiple roots of polynomial equations.

DEFINITION 2

Multiple Zero (or Root), Simple Zero

Let r be an element of U and let m be a positive integer. If $(x - r)^m$ is a divisor of α and $(x - r)^{m+1}$ is not a divisor of α in $U[x]$, one says that r is a zero of α [or a root of $\alpha(x) = 0$] with **multiplicity** m. A zero with multiplicity 1 is also called a **simple zero.** A zero with multiplicity greater than 1 is called a **multiple zero.**

THEOREM 4 **Test for Multiplicity**

Let m be a positive integer and let $\alpha = (x - r)^m\beta$ in $U[x]$. If r is not a zero of β, then r is a zero of α with multiplicity m.

Proof The hypothesis tells us that $(x - r)^m$ is a divisor of α; hence we need only show that $(x - r)^{m+1}$ is not a divisor of α. We do this by assuming that $\alpha = (x - r)^{m+1}\gamma$ in $U[x]$ and obtaining a contradiction.

This assumption leads to $(x - r)^m\beta = (x - r)^{m+1}\gamma$. The corollary to Lemma 1 of Section 5.2 tells us that the monic polynomial $x - r$ is not a 0-divisor in $U[x]$. Hence it may be canceled m times in the last equation, thus giving us $\beta = (x - r)\gamma$. Then the Factor Theorem implies that r is a zero of β. This contradicts the hypothesis and completes the proof. ∎

It follows from Problem *18 below that a zero s of a polynomial α in $U[x]$ has multiplicity greater than 1 if and only if s is a zero of the derivative α' of α.

EXAMPLE 2 Next we show that the polynomial $\alpha = x^3 - 3x + 2$ in $\mathbf{Q}[x]$ has 1 as a zero with multiplicity 2. We divide α by $(x - 1)^2$ and have

$$\alpha = (x - 1)^2(x + 2).$$

Since 1 is not a zero of $x + 2$, it follows from Theorem 4 that 1 is a zero with multiplicity 2 of α.

The two following results are needed in our impossibility proofs, in Section 6.3, concerning angle trisection and other classical problems on constructions with compass and straightedge. Also, Theorem 5 implies that the nonreal zeros of a polynomial with real coefficients occur in pairs of complex conjugates.

THEOREM 5 Pairs of Conjugate Roots

Let F be a subfield in a field E. Let t be an element of E such that t is not in F but t^2 is in F. If $r = a + bt$ is a zero in E of an α of $F[x]$, so is $\bar{r} = a - bt$.

Proof Let $t^2 = s$ and let $\beta = (x - r)(x - \bar{r}) = [x - (a + bt)][x - (a - bt)] = x^2 - 2ax + (a^2 - sb^2)$. Note that $\beta \in F[x]$. Let γ and ρ be the quotient and remainder, in $F[x]$, in the division of α by β.

In $E[x]$, we have $(x - r) \mid \beta$ by definition of β, and we have $(x - r) \mid \alpha$ from the hypothesis that r is a zero of α and the Factor Theorem. It follows from these facts and $\rho = \alpha - \gamma\beta$ that $(x - r) \mid \rho$ in $E[x]$. Since $\deg \beta = 2$, ρ does not have degree greater than 1. Hence, $(x - r) \mid \rho$ implies that $\rho = c(x - r)$, with c in E. Since ρ is in $F[x]$, its coefficients c and $-cr$ are in F.

We now consider two cases. If $c \neq 0$, then $r = -c^{-1}(-cr)$, and so r is in F. Since t is not in F, $r = a + bt$ can only be in F if $b = 0$; this would imply that $\bar{r} = a - bt = a = r$ is a zero of α.

If $c = 0$, then $\rho = 0$, $\beta \mid \alpha$, and hence $(x - \bar{r}) \mid \alpha$. It follows that, in this case also, $\bar{r}$ is a zero of α. ∎

COROLLARY Complex Zeros of a Real Polynomial

Let $a + b\mathbf{i}$ be a zero in $\mathbf{C}$ of an α in $\mathbf{R}[x]$. Then $a - b\mathbf{i}$ is also a zero of α.

The proof of the corollary is left to the reader as Problem 19 of this section.

THEOREM 6 Zeros of Cubics over Quadratic Extensions

Let K be an extension field over F. Let $t \in K$, $t^2 \in F$, and $E = \{a + bt : a,\ b \in F\}$. Then E is an extension field over F such that any polynomial α of degree 3 in $F[x]$ with a zero in E also has a zero in F.

Proof If t is in F, then $E = F$ and the desired result holds. So let t not be in F. One sees from

$$\frac{1}{a + bt} = \frac{a - bt}{(a + bt)(a - bt)} = \frac{a}{a^2 - b^2t^2} - \frac{b}{a^2 - b^2t^2}t$$

that an element $a + bt$ has as its inverse in E the element $c + dt$ with $c = a/(a^2 - b^2t^2)$ and $d = -b/(a^2 - b^2t^2)$ unless the denominator $a^2 - b^2t^2 = 0$. But $a^2 - b^2t^2 = 0$ implies that $b = 0$; and then $a = 0$ and $a + bt = 0$, since $b \neq 0$ and $a^2 - b^2t^2 = 0$ lead to the contradiction that $t = \pm a/b$ is in F. Hence every nonzero element of E has an inverse in E. It is now easy to see that E is a field with F as a subfield.

Let $r = a + bt$ be a zero in E of a polynomial α with degree 3 in $F[x]$. Now it follows from Theorem 5 above that $\bar{r} = a - bt$ is also a zero of α in E.

If $b = 0$, $r = a$ is the desired zero in F of α. If $b \neq 0$, then $\bar{r} \neq r$ and we can factor α in $E[x]$ as

$$\alpha = (x - r)(x - \bar{r})\beta.$$

Since $\deg \alpha = 3$, we have $\deg \beta = 1$; that is, $\beta = hx + k$ with h and k in F and $h \neq 0$. Let $u = -kh^{-1}$. Then u is a zero of β and hence is a zero of α. Also,

$$\alpha = h(x - r)(x - \bar{r})(x - u) = c_3 x^3 + c_2 x^2 + c_1 x + c_0$$

with the c_i in F and $c_3 \neq 0$. Expanding the left side and equating coefficients, we find that $c_3 = h$ and

$$c_2 = -h(r + \bar{r} + u) = -c_3(a + bt + a - bt + u) = -c_3(2 \cdot a + u).$$

Solving for u gives us $u = -c_2 c_3^{-1} - 2 \cdot a$. Since a, c_2, and c_3 are in F, this tells us that the zero u of α is in F, as desired. ■

Problems

Problem 20 below is cited in Section 5.8. Problems 37 and 38 are cited in Section 7.4

1. In $\mathbf{Q}[x]$, find the remainder in the division of $x^3 + 2x^2 - 5x - 7$ by $x - 4$ directly; then check using the Remainder Theorem.

2. In $\mathbf{Q}[x]$, find the remainder in the division of $4x^5 + 5x^4 - 1$ by $x + 1$ directly; then check using the Remainder Theorem.

3. Use the Remainder Theorem to show that $x - 1$ is a factor of $a_d x^d + \cdots + a_1 x + a_0$ in $U[x]$ if and only if $a_d + \cdots + a_1 + a_0 = 0$ in U.

4. Use the Remainder Theorem to find a necessary and sufficient condition on the coefficients a_i for $x + 1$ to be a factor of $a_d x^d + \cdots + a_1 x + a_0$.

5. In $\mathbf{Z}_5[x]$, let $\alpha = x^6 + \bar{2}x^5 - \bar{2}x$ and $\beta = x^2$. Tabulate $\alpha(r)$ and $\beta(r)$ for all r in $\mathbf{Z}_5$, and thus show that $\alpha^* = \beta^*$ in $\mathbf{Z}_5^*$ (even though $\alpha \neq \beta$ in $\mathbf{Z}_5[x]$).

6. Let $\alpha = x^7 - x$ in $\mathbf{Z}_7[x]$. Tabulate $\alpha(r)$ for all r in $\mathbf{Z}_7$, and thus show that α^* is the zero function of $\mathbf{Z}_7^*$.

7. Let F be a finite field with m elements. Explain why every element of F is a zero of $\alpha = x^m - x$, and hence α^* is the zero function of F^*.

8. Explain why $x^m - x + 1$ has no zero in a field F with m elements and hence has no first-degree divisor in $F[x]$.

9. (i) Show that $(x - \bar{2})(x - \bar{3}) = x(x - \bar{5})$ in $\mathbf{Z}_6[x]$.
(ii) How many zeros does $x^2 - \bar{5}x$ have in $\mathbf{Z}_6$?

10. How many zeros does $x^2 - \bar{7}x$ have in $\mathbf{Z}_{12}$?

11. Prove that an α with degree d in $D[x]$ has at most d zeros in the integral domain D (and thus prove Corollary 1 to Theorem 3 above).

12. Let D be an infinite integral domain. Prove that $\alpha \mapsto \alpha^*$ gives a ring isomorphism from $D[x]$ to D^*. (This proves Corollary 2 to Theorem 3 above.)

13. Let $\beta(h_i) = \gamma(h_i)$ for $i = 0, 1, \ldots, t$, where the h_i are $t + 1$ distinct elements of an integral domain D and $\beta \neq \gamma$ in $D[x]$. Explain why at least one of β and γ must have degree greater than t.

14. Let $s_1, s_2, \ldots, s_d$ be d distinct zeros in an integral domain D of an α with degree d in $D[x]$.
   (i) Use Theorem 3 above to show that
   $$\alpha = a(x - s_1)(x - s_2) \cdots (x - s_d),$$
   where a is a nonzero element of D.
   (ii) Explain why $a = 1$ in part (i) if and only if α is monic.

15. Let $F = \{b_1, b_2, \ldots, b_m\}$ be a field with m elements and let h be the ring homomorphism from $F[x]$ to F^* with $\alpha \mapsto \alpha^*$. Let K be the kernel of h; that is, let K consist of all the α of $F[x]$ such that α^* is the zero function of F^*. Let
$$\beta = (x - b_1)(x - b_2) \cdots (x - b_m).$$
Prove the following.
   (i) $\deg \alpha \geq m$ for every nonzero α in K.
   (ii) $\alpha^* = \rho^*$ in F^* if and only if $\alpha - \rho$ is in K.
   (iii) Let $\alpha = \gamma\beta + \rho$ in $F[x]$. Then $\alpha^* = \rho^*$ in F^*, and hence α is in K if and only if ρ is in K.
   (iv) $\beta \mid \alpha$ for every α in K; that is, $K = (\beta)$.
   (v) $\alpha^* = \rho^*$ in F^* if and only if $\beta \mid (\alpha - \rho)$ in $F[x]$.
   (vi) For every α in $F[x]$, there is a ρ in $F[x]$ such that $\alpha^* = \rho^*$ and either $\deg \rho < m$ or $\rho = 0$.

16. Let p be a positive prime in $\mathbf{Z}$ and let h be the ring homomorphism from $\mathbf{Z}_p[x]$ to $\mathbf{Z}_p^*$ with $\alpha \mapsto \alpha^*$. Let K be the kernel of h and let
$$\beta = x(x - \bar{1})(x - \bar{2}) \cdots (x - \overline{p-1}).$$
Explain why the following are true.
   (i) $K = (\beta)$.
   (ii) $\beta = x^p - x$.
   (iii) $(x - \bar{1})(x - \bar{2}) \cdots (x - \overline{p-1}) = x^{p-1} - 1$.
   (iv) $\bar{1} \cdot \bar{2} \cdot \bar{3} \cdots \overline{p-1} = -\bar{1}$ in $\mathbf{Z}_p$. [It may help to consider two cases: (a) $p = 2$; and (b) p an odd prime.]
   (v) $(p - 1)! + 1$ is a multiple of p in $\mathbf{Z}$.
   (vi) $(p - 1)! \equiv -1 \pmod{p}$. (This is known as Wilson's Theorem. See Problem 25 of Section 4.4 for an alternative proof.)

17. In each part below, α is in $F[x]$ and s is a zero of α. Find the multiplicity m of the zero s.
   (a) $\alpha = 3x^4 - 4x^3 + 1$, $s = 1$, $F = \mathbf{Q}$.
   (b) $\alpha = x^3 + x^2 - 2x - 2$, $s = -\sqrt{2}$, $F = \mathbf{R}$.
   (c) $\alpha = x^5 - x^4 + 2x^3 - 2x^2 + x - 1$, $s = \mathbf{i}$, $F = \mathbf{C}$.
   (d) $\alpha = x^9 + \bar{1}$, $s = \bar{2}$, $F = \mathbf{Z}_3$.
   (e) $\alpha = x^{10} - \bar{2}x^5 + \bar{1}$, $s = \bar{1}$, $F = \mathbf{Z}_5$.
   (f) $\alpha = nx^{n+1} - (n + 1)x^n + 1$, $s = 1$, $F = \mathbf{Q}$. (Here n is a positive integer.)

★ 18. In $F[x]$, let $\alpha = \pi_1\pi_2 \cdots \pi_d$, with $\deg \pi_i = 1$ for each i. Prove that α has no zero with multiplicity greater than 1 if and only if $\gcd(\alpha, \alpha') = 1$. (Here α' is the derivative of α specified in Definition 8 of Section 5.2.)

19. Let $a + b\mathbf{i}$ (with a and b real) be a zero of a polynomial α with real coefficients. Show that $a - b\mathbf{i}$ is a zero of α (and thus prove the corollary to Theorem 5 of this section).

20. (i) Let $\alpha \in F[x]$ and $\deg \alpha \in \{2, 3\}$. Explain why α is reducible (and composite) in $F[x]$ if and only if α has at least one zero in F.
(ii) Show that $\alpha = x^4 + x^2 + \bar{1}$ is reducible in $\mathbf{Z}_2[x]$ even though α has no zero in $\mathbf{Z}_2$. (See Problem 18 of Section 5.3.)

21. Use Problems 8 and 20(i) to prove that $x^3 - x + \bar{1}$ is irreducible (and prime) in $\mathbf{Z}_3[x]$.

22. Use Problem 20(i) to find all the monic irreducible polynomials of degree 3 in $\mathbf{Z}_3[x]$.

23. In a field F, let $a \neq 0$. In $F[x]$, let

$$a(x - r_1)(x - r_2)(x - r_3) = ax^3 + bx^2 + cx + d.$$

Show the following.
(a) $r_1 + r_2 + r_3 = -b/a$.
(b) $r_2 r_3 + r_1 r_3 + r_1 r_2 = c/a$.
(c) $r_1 r_2 r_3 = -d/a$.
(d) $r_1^2 + r_2^2 + r_3^2 = (b^2 - 2 \cdot ac)/a^2$.

24. Let U be a commutative ring with unity. In $U[x]$, let

$$(x - r_1)(x - r_2)(x - r_3)(x - r_4) = x^4 + a_3 x^3 + a_2 x^2 + a_1 x + a_0.$$

(a) Show that $r_1 + r_2 + r_3 + r_4 = -a_3$.
(b) Show that $r_1 r_2 r_3 r_4 = a_0$.
(c) Express each of a_1 and a_2 in terms of the r_i.

25. Let U be a commutative ring with unity. In $U[x]$, let

$$(x - r_1)(x - r_2) \cdots (x - r_d) = x^d + a_{d-1}x^{d-1} + \cdots + a_1 x + a_0.$$

(a) Express $r_1 + r_2 + \cdots + r_d$ in terms of the a_i.
(b) Express $r_1 r_2 \cdots r_d$ in terms of the a_i.

26. Let $a_d \neq 0$ in a field F. In $F[x]$, let

$$a_d(x - r_1)(x - r_2) \cdots (x - r_d) = a_d x^d + \cdots + a_1 x + a_0.$$

(a) Express $r_1 + r_2 + \cdots + r_d$ in terms of the a_i.
(b) Express $r_1 r_2 \cdots r_d$ in terms of the a_i.
(c) Show that $r_1^2 + r_2^2 + \cdots + r_d^2 = (a_{d-1}^2 - 2 \cdot a_d a_{d-2})/a_d^2$.

27. In $\mathbf{Q}[x]$, let

$$(x - 1)(x - 2)(x - 3) \cdots (x - d) = x^d - s_1 x^{d-1} + s_2 x^{d-2} - \cdots + (-1)^d s_d.$$

Show the following.
(a) $s_1 = d(d + 1)/2$.
(b) $s_d = 1 \cdot 2 \cdot 3 \cdots d = d!$
★ (c) $2s_2 = (1^3 + 2^3 + \cdots + d^3) - (1^2 + 2^2 + \cdots + d^2)$.

28. In $F[x]$, let

$$ax^2 + bx + c = a(x - r)(x - s).$$

For all nonnegative integers n, let $h_n = r^n + s^n$. Show the following.
(i) $ar^{n+2} + br^{n+1} + cr^n = 0$.
(ii) $as^{n+2} + bs^{n+1} + cs^n = 0$.
(iii) $ah_{n+2} + bh_{n+1} + ch_n = 0$.

29. Using the notation of Problem 28, express h_5 in terms of a, b, and c, assuming that $a \neq 0$.

30. In $F[x]$, let $(x - r)(x - s)(x - t) = x^3 + ux^2 + vx + w$. Express $r^5 + s^5 + t^5$ in terms of u, v, and w. Use the answer to Problem 29 as a partial check (by letting $t = 0 = w$, $a = 1$, $b = u$, and $c = v$).

31. Let **R** and $\mathbf{C} = \mathbf{R}[\mathbf{i}]$ be the real and complex fields, respectively. Let $\alpha = x^4 - 2x^3 + x^2$ and $\beta = x^2 + 1$.
   (i) Find the remainder ρ in the division of α by β in $\mathbf{R}[x]$.
   (ii) Considering α to be in $\mathbf{C}[x]$, evaluate $\alpha(\mathbf{i})$ in the form $a + b\mathbf{i}$.
   (iii) Explain the relation between the answers to parts (i) and (ii).

32. Let **R** be the field of real numbers. Describe an easy method for finding the remainder in the division of an α in $\mathbf{R}[x]$ by $x^2 + 1$. [*Hint:* See the preceding problem.]

33. Let **R** and **C** be the real and complex fields, respectively. Let I consist of all α in $\mathbf{R}[x]$ such that **i** is a zero in **C** of α. Prove that I is the principal ideal $(x^2 + 1)$.

★ 34. Let **Q** and **R** be the rational and real fields, respectively. Let K consist of all α in $\mathbf{Q}[x]$ such that $\sqrt{2} + \sqrt{3}$ is a zero in **R** of α. Prove that K is a principal ideal, and find a monic generator of K.

35. In $\mathbf{Q}[x]$, let $\alpha = 3x^4 - 7x^3 - 1$. Let c be in **Q** and

$$\beta = \alpha + c(x + 2)(x + 1)x(x - 1)(x - 2).$$

   (i) Explain why $\alpha(r) = \beta(r)$ for $r = -2, -1, 0, 1$, and 2.
   (ii) Given that $\beta(3) = 13$, find c.

36. In $\mathbf{Q}[x]$, let $\alpha = 2x^3 - 3x^2 + 5$. Find a polynomial β in $\mathbf{Q}[x]$ with $\deg \beta = 4$; $\beta(r) = \alpha(r)$ for $r = -2, -1, 0$, and 1; and $\beta(2) = -4$. (See Problem 35.)

37. Let V be the multiplicative group of invertibles in an integral domain D. Let ord V be an even integer $2t$ and W be the subset $\{v^2 : v \in V\}$ of squares in V. Show the following.
   (i) Each v of V is a root of $x^{2t} - 1 = 0$.
   (ii) Of the $2t$ elements of V, t elements are roots of $x^t - 1 = 0$, and the other t elements are roots of $x^t + 1 = 0$.
   (iii) There are t elements in W. [See Problem 23(v) of Section 4.4.]
   (iv) An element of V is in W if and only if it is a root of $x^t - 1 = 0$.
   (v) -1 is in W if and only if t is even.
   (vi) If w is in W, then $-w$ is in W if and only if t is even.

38. Let p be an odd positive prime in **Z** and let W_p be the subset of squares in the multiplicative group $\mathbf{V_p}$ of invertibles in $\mathbf{Z_p} = \mathbf{Z}/(p)$. Show the following.
   (i) -1 is in W_p if and only if $p \equiv 1 \pmod 4$.
   (ii) If w is in W_p, then $-w$ is in W_p if and only if $p \equiv 1 \pmod 4$.

39. In $\mathbf{C}[x]$, let $\alpha = a_d x^d + a_{d-1}x^{d-1} + \cdots + a_0$ and $\beta = x^2 + x + 1$. Use Problem 32 of Section 5.3 to find an easy method for obtaining the remainder in the division of α by β.

40. In $\mathbf{C}[x]$, let $\alpha = a_d x^d + a_{d-1}x^{d-1} + \cdots + a_0$. Find an easy method for obtaining the remainder ρ in the division of α by

$$\beta = x^4 + x^3 + x^2 + x + 1.$$

★ 41. In $\mathbf{C}[x]$, let $\alpha = a_d x^d + \cdots + a_0$ and $\beta = (x - r)^2(x - s)$, with $r \neq s$. Find the remainder in the division of α by β.

5.5 Integral and Rational Roots

Here we present methods of determining all the integers and all the rational numbers that are roots of a polynomial equation $\alpha(x) = 0$ in which the coefficients are integers or rational numbers.

If the polynomial equation under consideration has rational coefficients, it can be converted (without changing the roots) into one that has integer coefficients, by multiplying both sides of the equation by a common denominator for the original coefficients. Hence we let α be in $\mathbf{Z}[x]$; that is, we let

$$\alpha = a_d x^d + a_{d-1}x^{d-1} + \cdots + a_0, \qquad a_i \text{ in } \mathbf{Z}.$$

We first seek a means of discovering the integers, if any, that are zeros of α. We note that r is a zero of α if and only if

$$a_d r^d + a_{d-1}r^{d-1} + \cdots + a_1 r + a_0 = 0$$

or

$$r(a_d r^{d-1} + a_{d-1}r^{d-2} + \cdots + a_1) = -a_0. \tag{1}$$

If r is an integer, equation (1) implies that $r \mid a_0$. We now formally restate this result.

THEOREM 1 Integer Roots

The only integers that may be zeros of a polynomial

$$\alpha = a_d x^d + a_{d-1}x^{d-1} + \cdots + a_0,$$

with integer coefficients a_i, are the integral divisors of a_0.

If $a_0 \neq 0$, there are only a finite number of integral divisors d of a_0, and one can check to see which are zeros of α in a finite number of steps. If $a_0 = 0$, one modifies the procedure slightly, as indicated in the following example.

EXAMPLE 1 Integral Roots

Here we find all the integers that are zeros of

$$\alpha = x^7 + x^6 - 5x^5 + x^4 - 6x^3.$$

We note that $\alpha = x^3\beta$, where

$$\beta = x^4 + x^3 - 5x^2 + x - 6.$$

Clearly 0 is a zero of α, with multiplicity 3, and all the other zeros of α are zeros of β. Theorem 1 tells us that the set of integers that are zeros of β is a subset of $\{1, -1, 2, -2, 3, -3, 6, -6\}$, the set of integral divisors of -6. Evaluating $\beta(d)$ for each of these divisors d of a_0, we find that $\beta(2) = 0$, $\beta(-3) = 0$, and $\beta(d) \neq 0$ for the other choices of d. Hence the integral zeros of α are 0, 2, and -3.

Next we take up the question of rational zeros of a polynomial α in $\mathbf{Z}[x]$. Let

$$\alpha = a_d x^d + a_{d-1}x^{d-1} + \cdots + a_0, \qquad a_i \text{ in } \mathbf{Z}.$$

A rational number r is expressible as s/t, where s and t are relatively prime integers and $t \neq 0$. Then r is a zero of α if and only if

$$(a_d s^d/t^d) + (a_{d-1}s^{d-1}/t^{d-1}) + \cdots + (a_1 s/t) + a_0 = 0,$$

which is equivalent to each of the following:

$$a_d s^d + a_{d-1}s^{d-1}t + \cdots + a_1 st^{d-1} + a_0 t^d = 0,$$

$$a_d s^d = -t(a_{d-1}s^{d-1} + \cdots + a_1 st^{d-2} + a_0 t^{d-1}), \tag{2}$$

$$a_0 t^d = -s(a_d s^{d-1} + a_{d-1}s^{d-2}t + \cdots + a_1 t^{d-1}). \tag{3}$$

From (2) we see that $t \mid (a_d s^d)$. Since t and s are relatively prime, it follows that t and s^d are relatively prime and then that $t \mid a_d$. (See Problems 16 and 15 of Section 1.6.) Similarly, equation (3) implies that $s \mid a_0$. This leads to the following result.

THEOREM 2 Rational Roots

The only possibilities for rational zeros of a polynomial

$$a_d x^d + \cdots + a_1 x + a_0,$$

with integer coefficients a_i, are the numbers of the form s/t, where s and t are integers, $s \mid a_0$, and $t \mid a_d$. [One may restrict s and t so that they also satisfy $t > 0$ and $\gcd(s, t) = 1$.]

EXAMPLE 2 Applied to Nontrisection

Next we show that $\alpha = 8x^3 - 6x - 1$ is irreducible in $\mathbf{Q}[x]$. This fact will be used in Lemma 8 of Section 6.3 to prove that a 60° angle cannot be trisected with euclidean constructions. Since $\deg \alpha = 3$, the polynomial α is reducible if and only if it has a first-degree factor $ax + b$ and, therefore, a rational zero $-b/a$. Hence it will suffice to show that α has no rational zero.

Using Theorem 2 and the fact that α is also in $\mathbf{Z}[x]$, we see that the only possibilities for rational zeros are of the form s/t with s an integral divisor of 1 and t an integral divisor of 8; thus the only candidates are 1, -1, 1/2, $-1/2$, 1/4, $-1/4$, 1/8, and $-1/8$. Trial shows that none of these rational numbers is a zero of α. Therefore, α has no rational roots. Hence α has no factor of degree 1 in $\mathbf{Q}[x]$ and is irreducible.

EXAMPLE 3 Rational Roots

Here we find all the rational roots of $\alpha = 0$, where

$$\alpha = 6x^4 - 31x^3 + 25x^2 + 33x + 7,$$

and then we find the other roots of this equation. The possibilities for rational roots are the rational numbers s/t with $s \mid 7$ and $t \mid 6$; these numbers are ± 1, ± 7, $\pm 1/2$, $\pm 7/2$, $\pm 1/3$, $\pm 7/3$, $\pm 1/6$, and $\pm 7/6$. We substitute each of them into the given equation. The first root that we find is 7/2; this root and the Factor Theorem tell us that $x - (7/2)$ is a factor of α in $\mathbf{Q}[x]$. It is more convenient to use its associate $\beta = 2x - 7$ in seeking the complementary factor γ such that $\alpha = \beta\gamma$; then division by β leads to

$$\gamma = 3x^3 - 5x^2 - 5x - 1.$$

The roots of $\alpha = 0$ are $7/2$ and the roots of $\gamma = 0$. The possibilities for rational zeros of γ are ± 1 and $\pm 1/3$; trial shows that $-1/3$ is a zero. Then $3x + 1$ is a factor of γ, and division shows that

$$\gamma = (3x + 1)(x^2 - 2x - 1).$$

The zeros of $x^2 - 2x - 1$ are $1 \pm \sqrt{2}$. Hence the zeros of α are

$$\frac{7}{2}, -\frac{1}{3}, 1 + \sqrt{2}, 1 - \sqrt{2}.$$

Problems

1. Find an integer root of $x^3 + 3x^2 - 2 = 0$, and then use the Factor Theorem as an aid in finding the other roots in the field **C** of complex numbers.
2. Do as in Problem 1 for $x^3 - 6x + 4 = 0$.
3. Find all the roots of $x^4 + 2x^3 - x^2 + 4x - 6 = 0$ in **C**.
4. Find all the roots in **C** of
$$x^5 - 8x^4 + 15x^3 + 8x^2 - 64x + 120 = 0.$$
5. Find all the roots in **C** of each of the following equations.
   (a) $3x^3 + 4x^2 - 21x + 10 = 0.$
   (b) $12x^3 - 11x^2 - 13x + 10 = 0.$
   (c) $x^4 - x^3 - 3x^2 + 5x - 2 = 0.$
   (d) $81x^5 - 54x^4 + 3x^2 - 2x = 0.$
   (e) $x^4 + 3x^2 + 2 = 0.$
6. Find all the rational zeros of each of the following polynomials.
   (a) $2x^4 + 5x^3 - 5x^2 + 7x - 3.$
   (b) $x^5 - 6x^4 + 3x^3 - 3x^2 + 2x - 12.$
7. What are the possibilities for integral roots of
$$5x^4 + ax^3 + bx^2 + cx - 21 = 0,$$
given that a, b, and c are integers?
8. What are the possibilities for rational roots of the equation in the previous problem?
9. Given that a and b are integers, is it possible for $35/21$ to be a root of $9x^3 + ax^2 + bx - 10 = 0$? Explain.
10. Let a, b, and c be in **Z** and let $33/18$ be a root of
$$12x^3 + ax^2 + bx + c = 0.$$
Explain why $11 \mid c$ and why it is not necessarily true that $33 \mid c$.
11. Let r be a rational zero of a monic
$$\alpha = x^d + a_{d-1}x^{d-1} + \cdots + a_0, \qquad a_i \in \mathbf{Z}.$$
Explain why r must be an integer.

12. Let $\alpha = a_d x^d + \cdots + a_1 x$ be in $\mathbf{Z}[x]$. (Note that $a_0 = 0$.) State a necessary condition for an integer n to be a zero of α.

13. Let s and t be relatively prime integers. State a necessary condition for the rational number s/t to be a zero of the α of the preceding problem.

14. Let $a \in \mathbf{Z}$ and $\alpha = x^3 + x^2 + 2ax - 21$. Explain why $n^3 + n^2$ is even for every n in $\mathbf{Z}$ and use this to show that α has no integral zero. Can α have a rational zero? Explain.

15. Show the following.
(i) $x^5 - 7 = 0$ has no rational roots.
(ii) $\sqrt[5]{7}$ is not a rational number.

16. For each of the following polynomials α, show that α has no zero in $\mathbf{Q}$ and hence that α is irreducible in $\mathbf{Q}[x]$.
(a) $x^3 - 3x + 1$.
(b) $x^3 + x^2 - 2x - 1$.
(c) $8x^3 + 4x^2 - 4x - 1$.

17. Let

$$\alpha = (x - \sqrt{2} - \sqrt{3})(x - \sqrt{2} + \sqrt{3})(x + \sqrt{2} - \sqrt{3})(x + \sqrt{2} + \sqrt{3}).$$

Show the following.
(i) α is expressible as $x^4 + a_3 x^3 + a_2 x^2 + a_1 x + a_0$, with the a_i integers.
(ii) α has no rational zeros.
(iii) $\sqrt{2} + \sqrt{3}$ is irrational.

18. Prove that $\sqrt{5} - \sqrt{3}$ is irrational. [*Hint:* See Problem 17.]

19. Let α be in $\mathbf{Z}[x]$ and let r, s, and m be integers. Given that $r \equiv s \pmod{m}$, explain why $\alpha(r) \equiv \alpha(s) \pmod{m}$.

20. Let α be a polynomial in $\mathbf{Z}[x]$ such that $\alpha(0)$ and $\alpha(1)$ are both odd. Show that α has no zeros in $\mathbf{Z}$.

21. Let α be in $\mathbf{Z}[x]$ and let k be a positive integer such that $\alpha(j)$ is not an integral multiple of k for $j = 0, 1, 2, \ldots, k - 1$. Show that α has no zeros in $\mathbf{Z}$.

22. Let α be in $\mathbf{Z}[x]$ and let $\alpha(0)$, $\alpha(1)$, and the leading coefficient $L(\alpha)$ all be odd. Show that $\alpha = 0$ has no rational roots.

23. Let a, b, c, and d be distinct integers and let

$$\alpha = (x - a)(x - b)(x - c)(x - d) - 9.$$

If α has a zero s in $\mathbf{Z}$, show that $s = (a + b + c + d)/4$.

5.6 Polynomial Fitting, Finite Differences

Algebra is the intellectual instrument which has been created for rendering clear the quantitative aspect of the world.
— **A. N. Whitehead**

One cannot escape the feeling that these mathematical formulae have an independent existence and an intelligence of their own, that they are wiser than we are, wiser even than their discoverers, that we get more out of them than we originally put into them. — **Heinrich Hertz**

One of the most important applications of polynomials is to problems involving the fitting of an analytical formula to a finite table of values of a function. This arises in interpolation or extrapolation, approximating definite integrals, numerical solution of differential equations, and so on.

We now present one of Newton's techniques for fitting a table with a polynomial function. (See the biographical note on Newton on p. 330.)

Suppose that we are given a table

x	h_0	h_1	h_2	$\dots$	h_t
y	k_0	k_1	k_2	$\dots$	k_t

in which the h_i and k_i are elements of a field F. We seek a polynomial α, of lowest possible degree, in $F[x]$ such that

$$\alpha(h_i) = k_i \quad \text{for } i = 0, 1, \dots, t. \tag{1}$$

We assume that the h_i are $t + 1$ distinct elements of F, since otherwise there are conditions in (1) that are either redundant or contradictory.

Practically and theoretically, there are great advantages in looking for the polynomial α in the form

$$\alpha = c_0 + c_1(x - h_0) + c_2(x - h_0)(x - h_1) + \cdots + c_t[(x - h_0)(x - h_1) \cdots (x - h_{t-1})], \tag{2}$$

with the c_i elements of F.

THEOREM 1 Newton's Polynomial Fitting Formula

Let $h_0, h_1, \dots, h_t$ be $t + 1$ distinct elements of a field F and let $k_0, k_1, \dots, k_t$ be in F. Then in $F[x]$ there is a unique polynomial α of the form

$$\alpha = a_t x^t + a_{t-1} x^{t-1} + \cdots + a_0 \tag{3}$$

[or of the form (2)] such that

$$\alpha(h_i) = k_i \quad \text{for } i = 0, 1, \dots, t. \tag{4}$$

Proof We start by showing that the conditions in (4) determine one and only one sequence $c_0, c_1, \dots, c_t$ of coefficients in (2). In the evaluation of $\alpha(h_i)$, for a given i, all the products in (2) that have $x - h_i$ as a factor contribute zero. More specifically, we note that (2) and (4) imply that

$$\alpha(h_0) = c_0, \qquad \alpha(h_1) = c_0 + c_1(h_1 - h_0),$$

$$\alpha(h_2) = c_0 + c_1(h_2 - h_0) + c_2(h_2 - h_0)(h_2 - h_1),$$

and so forth. Then $\alpha(h_i) = k_i$ for $0 \le i \le t$ if and only if

$$c_0 = k_0, \qquad c_1 = (k_1 - c_0)/(h_1 - h_0),$$

and so forth. More precisely, we know that c_0 must equal k_0; and if we assume that $c_0, c_1, \ldots, c_{j-1}$ have been determined, then $c_j = u_j/v_j$, where

$$u_j = k_j - c_0 - c_1(h_j - h_0) - c_2(h_j - h_0)(h_j - h_1) - \cdots - c_{j-1}[(h_j - h_0)(h_j - h_1) \cdots (h_j - h_{j-2})]$$

and

$$v_j = (h_j - h_0)(h_j - h_1)(h_j - h_2) \cdots (h_j - h_{j-1}).$$

It is easily seen that either $\alpha = 0$ or $\deg \alpha \le t$, with $\deg \alpha$ the largest value of j for which $c_j \ne 0$.

The work above shows that the form (2) is unique for an α satisfying (4). To show uniqueness of the form (3), we let α satisfy (3) and (4) and let $\beta = b_t x^t + \cdots + b_0$ have $\beta(h_i) = k_i$ for $0 \le i \le t$. We also let $\gamma = \alpha - \beta$. Then

$$\gamma(h_i) = \alpha(h_i) - \beta(h_i) = k_i - k_i = 0,$$

and so the h_i are $t + 1$ distinct zeros of γ. It follows from Corollary 1 to Theorem 3 of Section 5.4 that either $\gamma = 0$ or $\deg \gamma \ge t + 1$. But the difference of two polynomials of the form (3) is either 0 or of degree at most t. Hence $\gamma = 0$ and $\alpha = \beta$. This proves uniqueness. ■

COROLLARY If $F = \{a_1, \ldots, a_m\}$ is a finite field, every mapping from F to F is a polynomial function in F^*, and so F^* has m^m elements.

Proof This follows from the Polynomial Fitting Theorem and the fact that a mapping θ from F to F is completely determined by the values of $\theta(a_1), \ldots, \theta(a_m)$. ■

EXAMPLE 1 Let us find the polynomial α of lowest possible degree in $\mathbf{Q}[x]$ such that the equation $y = \alpha(x)$ fits the data in the following table.

x	0	1	2	3	4	5
y	0	0	1	7	22	50

Theorem 1 tells us that the degree of α is 5 or less and that we may seek α in the form

$$\alpha = a + bx + cx(x - 1) + \cdots + fx(x - 1)(x - 2)(x - 3)(x - 4).$$

Using the fact that $\alpha(0) = 0$, we see that $a = 0$. Then $\alpha(1) = 0$ leads to $b = 0$. Next $\alpha(2) = 1$ implies that $1 = 2 \cdot 1 \cdot c$, or $c = 1/2$. Using these values of a, b, and c and $\alpha(3) = 7$ gives us

$$7 = \tfrac{1}{2} \cdot 3 \cdot 2 + 3 \cdot 2 \cdot 1 \cdot d \quad \text{or} \quad d = \tfrac{2}{3}.$$

Similarly, one finds that $e = 0$ and $f = 0$. Hence,

$$\begin{aligned} \alpha &= \tfrac{1}{2}x(x-1) + \tfrac{2}{3}x(x-1)(x-2) \\ &= \tfrac{1}{6}x(x-1)(4x-5). \end{aligned}$$

No polynomial of lower degree fits the data, since there is only one polynomial with degree 5 or less that meets the conditions.

EXAMPLE 2 **Interpolation**

Here we give a precise interpretation of the following question and then answer it: "What number y should be assigned to $x = 7/2$ so as to conform best with the table of Example 1?" We interpret this problem as asking for the value of $\alpha(7/2)$, where α is the unique polynomial of degree 5 or less that fits the data of Example 1; that is, α is the polynomial

$$x(x-1)(4x-5)/6$$

found in that example. With this interpretation, the answer is

$$\alpha(7/2) = (7/2)(5/2)(9)/6 = 105/8.$$

EXAMPLE 3 **Numerical Integration**

Let $y = f(x)$ have values as given in the table of Example 1. To approximate $I = \int_0^5 f(x)\,dx$, we let α be the polynomial $x(x-1)(4x-5)/6$ found in Example 1 and use $\int_0^5 \alpha(x)\,dx$ as the approximation.

In the remainder of this section, we let F be a field with characteristic 0; then the unity 1 of F generates an infinite cyclic subgroup in the additive group of F. We use a notation that allows us to think of this subgroup [1] as being $\mathbf{Z}$; specifically, we use the integer n as a symbol for $n \cdot 1$ in F.

NOTATION 1 **Factorials, Binomial Coefficients**

For n in $\{0, 1, 2, \ldots\}$, the symbol $n!$ is read as "n **factorial**" and is given by $0! = 1$, $1! = 1$, and $n! = 1 \cdot 2 \cdots n$ for $n \geq 2$. For n and k integers with $0 \leq k \leq n$, the **binomial coefficient** "n choose k" is given by

$$\binom{n}{0} = 1 \quad \text{and} \quad \binom{n}{k} = \frac{n(n-1)(n-2)\cdots(n-k+1)}{1 \cdot 2 \cdots k} = \frac{n!}{k!(n-k)!}.$$

For fixed k, we see that $\binom{n}{k}$ is a polynomial of degree k in n; this motivates the following introduction of a useful sequence of polynomials in x.

DEFINITION 1

Binomial Polynomials

For each positive integer k, the kth **binomial polynomial** is

$$B_k = \frac{1}{k!}[x(x-1)(x-2)\cdots(x-k+1)].$$

The zeroth binomial polynomial is $B_0 = 1$.

Let $k > 0$; then it is easily seen that for $n = 0, 1, 2, \ldots, k-1$ we have $B_k(n) = 0$, while for $n = k, k+1, k+2, \ldots$ the value of $B_k(n)$ is the binomial coefficient $\binom{n}{k} = \dfrac{n!}{k!(n-k)!}$.

Applying Theorem 1, or the technique of its proof, one easily obtains the following result.

THEOREM 2 **The Binomial Representation**

A polynomial α of degree d in $F[x]$ is uniquely expressible as

$$\alpha = b_0 + b_1 B_1 + b_2 B_2 + \cdots + b_d B_d, \qquad b_i \in F, \quad b_d \neq 0,$$

where the B_i are the binomial polynomials.

Explicit formulas for the coefficients b_i in this representation are given in Problem 23(ii) of this section. Also see Problems 25 and 26 for applications to polynomial fitting.

It is well known that the binomial coefficients $\binom{n}{k}$ satisfy

$$\binom{n}{k} + \binom{n}{k+1} = \binom{n+1}{k+1} \quad \text{or} \quad \binom{n}{k} = \binom{n+1}{k+1} - \binom{n}{k+1}.$$

In the notation of the binomial polynomials, the second of these becomes

$$B_k(n) = B_{k+1}(n+1) - B_{k+1}(n).$$

After we introduce some terminology, this formula will be generalized in Lemma 1 below.

NOTATION 2 **Composition of Polynomials**

Let R be a commutative ring. Let α and β be in $R[x]$, where

$$\alpha = a_d x^d + a_{d-1}x^{d-1} + \cdots + a_1 x + a_0.$$

Then $\alpha(\beta)$ denotes the polynomial

$$a_d \beta^d + a_{d-1}\beta^{d-1} + \cdots + a_1\beta + a_0.$$

Thus $\alpha(\beta)$ is the result of replacing x by β in the polynomial α. In particular, $\alpha(x)$ is the result of replacing x by x in α; that is, $\alpha(x)$ is another notation for α.

NOTATION 3 **The Difference Operator Δ**

The **difference** $\Delta\alpha$ of a polynomial α in $U[x]$ is given by

$$\Delta\alpha = \alpha(x+1) - \alpha. \tag{5}$$

We note that Δ is a mapping from $U[x]$ to itself and that the image of α under Δ is written as $\Delta\alpha$ rather than as $\Delta(\alpha)$. Since $\alpha(x) = \alpha$, formula (5) may be rewritten as

$$\Delta\alpha(x) = \alpha(x+1) - \alpha(x).$$

LEMMA 1 **Differencing a Binomial Polynomial**

$\Delta B_k = B_{k-1}$ for $k > 0$.

Proof We see that

$$\begin{aligned}
\Delta B_k &= B_k(x+1) - B_k(x)\\
&= [(x+1)x(x-1)(x-2)\cdots(x-k+2)/k!] -\\
&\quad [x(x-1)(x-2)\cdots(x-k+1)/k!]\\
&= [(x+1)-(x-k+1)][x(x-1)(x-2)\cdots(x-k+2)/k!]\\
&= kx(x-1)(x-2)\cdots(x-k+2)/k!\\
&= x(x-1)(x-2)\cdots(x-k+2)/(k-1)!\\
&= B_{k-1}.
\end{aligned}$$

■

THEOREM 3 **Differencing Using the Binomial Representation**

Let $\alpha = b_0 + b_1B_1 + \cdots + b_dB_d$, with b_i in F and with B_i the ith binomial polynomial. If $d > 0$, then

$$\Delta\alpha = b_1 + b_2B_1 + b_3B_2 + \cdots + b_dB_{d-1}.$$

Proof One easily sees that

$$\begin{aligned}
\Delta\alpha &= \Delta[b_0 + b_1B_1 + \cdots + b_dB_d]\\
&= \Delta b_0 + \Delta(b_1B_1) + \cdots + \Delta(b_dB_d)\\
&= 0 + b_1\Delta B_1 + b_2\Delta B_2 + \cdots + b_d\Delta B_d\\
&= b_1 + b_2B_1 + \cdots + b_dB_{d-1},
\end{aligned}$$

using Lemma 1.

■

COROLLARY If $\deg \alpha = d > 0$, then $\deg(\Delta\alpha) = d - 1$. If $\alpha = 0$ or $\deg \alpha = 0$ (that is, if α is in F), then $\Delta\alpha = 0$.

LEMMA 2 **Summing Values of a Binomial Polynomial**

For any $a \in F$ and any positive integers k and n,

$$B_{k-1}(a) + B_{k-1}(a+1) + \cdots + B_{k-1}(a+n) = B_k(a+n+1) - B_k(a). \quad (6)$$

Proof Lemma 1 tells us that $B_{k-1} = \Delta B_k$; hence the left side of (6) can be rewritten as

$$\Delta B_k(a) + \Delta B_k(a+1) + \cdots + \Delta B_k(a+n) = [B_k(a+1) - B_k(a)] + [B_k(a+2) - B_k(a+1)] + \cdots + [B_k(a+n+1) - B_k(a+n)].$$

When the terms that appear with both plus and minus signs are canceled, the only remaining terms are

$$-B_k(a) + B_k(a+n+1).$$

This equals the right side of (6), and so the lemma is proved. ∎

We now use Theorem 2 and Lemma 2 to obtain the following result.

THEOREM 4 **Summing Values of a Polynomial**

Let $a \in F$, $\alpha \in F[x]$, and $\deg \alpha = d$. Then there is a β of degree $d + 1$ in $F[x]$ such that

$$\alpha(a) + \alpha(a+1) + \cdots + \alpha(a+n) = \beta(n). \quad (7)$$

Proof Since $B_0 = 1$, Theorem 2 tells us that we can write

$$\alpha = b_0 B_0 + b_1 B_1 + \cdots + b_d B_d, \qquad b_d \neq 0. \quad (8)$$

Now let

$$\gamma = b_0 B_1 + b_1 B_2 + \cdots + b_d B_{d+1}.$$

We show below that the desired β is $\gamma(x + a + 1) - \gamma(a)$.

Let S be the sum on the left side of (7). Using (8) to express $\alpha(j)$ in terms of the $B_i(j)$, we find that

$$S = b_0[B_0(a) + B_0(a+1) + \cdots + B_0(a+n)] + \cdots + b_d[B_d(a) + B_d(a+1) + \cdots + B_d(a+n)].$$

Then Lemma 2 helps us to convert this to

$$\begin{aligned}
S &= b_0[B_1(a+n+1) - B_1(a)] + b_1[B_2(a+n+1) - B_2(a)] + \cdots \\
&\quad + b_d[B_{d+1}(a+n+1) - B_{d+1}(a)] \\
&= [b_0 B_1(a+n+1) + b_1 B_2(a+n+1) + \cdots + b_d B_{d+1}(a+n+1)] \\
&\quad - [b_0 B_1(a) + b_1 B_2(a) + \cdots + b_d B_{d+1}(a)] \\
&= \gamma(a+n+1) + \gamma(a).
\end{aligned}$$

Hence $S = \beta(n)$, where $\beta(x) = \gamma(x+a+1) - \gamma(a)$, as desired. ■

EXAMPLE 4 Next we find a closed form for the sum

$$C_n = 0^2 + 1^2 + 2^2 + \cdots + n^2.$$

We note that

$$C_n = \alpha(0) + \alpha(1) + \alpha(2) + \cdots + \alpha(n)$$

if α is chosen as x^2. Then it follows from Theorem 4 that $C_n = \beta(n)$ for some β of degree 3. Hence we may look for β in the form

$$\beta = a + bx + cx(x-1) + dx(x-1)(x-2). \tag{9}$$

One easily calculates that $\beta(0) = C_0 = 0$, $\beta(1) = C_1 = 1$, $\beta(2) = 1 + 4 = 5$, and $\beta(3) = 1 + 4 + 9 = 14$.

Successively letting $x = 0, 1, 2$, and 3 in (9), one then finds that $a = 0$, $b = 1$, $c = 3/2$, and $d = 1/3$. Hence,

$$\beta = x + \left(\frac{3}{2}\right)x(x-1) + \left(\frac{1}{3}\right)x(x-1)(x-2).$$

Expanding and collecting like terms, we have

$$\beta = \frac{2x^3 + 3x^2 + x}{6}.$$

Therefore,

$$1^2 + 2^2 + 3^2 + \cdots + n^2 = \frac{2n^3 + 3n^2 + n}{6}.$$

The ***Maclaurin Series Expansion*** of calculus for a polynomial $f(x)$ of degree d is the finite sum

$$f(x) = f(0) + f'(0)x + (1/2!)f''(0)x^2 + \cdots + (1/d!)f^{(d)}(0)x^d.$$

The analogue of this formula, with derivatives replaced by differences and each power x^n replaced by $x(x-1)\cdots(x-n+1)$ is in Problem 23 of this section.

Problems

1. Find the polynomial α of degree 4 in $\mathbf{Q}[x]$ such that 1, 2, 3, and 5 are zeros of α and $\alpha(4) = -18$.
2. Find the polynomial α of degree 3 in $\mathbf{Z}_7[x]$ such that $\bar{1}$, $\bar{3}$, and $\bar{4}$ are zeros of α and $\alpha(\bar{0}) = \bar{4}$.
3. Find the polynomial β of least degree in $\mathbf{Z}_5[x]$ such that $\beta(\bar{0}) = \bar{1}$, $\beta(\bar{1}) = \bar{0}$, $\beta(\bar{2}) = \bar{3}$, $\beta(\bar{3}) = \bar{2}$, and $\beta(\bar{4}) = \bar{4}$.
4. Find the polynomial β of least degree in $\mathbf{Z}[x]$ such that $\beta(-1) = -1$, $\beta(0) = -3$, $\beta(2) = 5$, and $\beta(5) = 47$.
5. Prove that there is no α of degree 3 in $\mathbf{Q}[x]$ to fit the following table.

x	1	2	3	4	5	6
$\alpha(x)$	1	2	6	20	76	312

6. Prove that $\deg \beta > 3$ for every β in $\mathbf{Q}[x]$ that fits the following table.

x	0	1	2	3	4
$\beta(x)$	6	5	6	13	32

7. Find the α of degree 3 in $\mathbf{Q}[x]$ that fits the following table.

x	0	1	4	6
$\alpha(x)$	1	0	117	415

8. Find the β of degree 3 in $\mathbf{Z}_7[x]$ that fits the following table.

x	$\bar{0}$	$\bar{1}$	$\bar{4}$	$\bar{6}$
$\beta(x)$	$\bar{1}$	$\bar{0}$	$\bar{5}$	$\bar{2}$

9. (a) Let $\alpha = x^3 - 3x^2 + 2x$. Express
$$\alpha(0) + \alpha(1) + \alpha(2) + \cdots + \alpha(n)$$
in the form $\beta(n)$ with β in $\mathbf{Q}[x]$.
(b) Let $\alpha = x^4 - 6x^3 + 11x^2 - 6x$. Express
$$\alpha(0) + \alpha(1) + \alpha(2) + \cdots + \alpha(n)$$
as a polynomial in n with coefficients in $\mathbf{Q}$.
10. (a) Express the following sum as $\beta(n)$ with β in $\mathbf{Q}[x]$:
$$1^3 + 2^3 + 3^3 + \cdots + n^3.$$
(b) Express the following sum as $\beta(n)$ with β in $\mathbf{Q}[x]$:
$$1^4 + 2^4 + 3^4 + \cdots + n^4.$$
11. Let A_n be the maximum number of regions into which a plane can be cut up by n straight lines.

(i) Find A_n for $n = 1, 2, 3$, and 4.
(ii) Prove that $A_n = A_{n-1} + n$ for $n > 1$.
(iii) Express A_n as a polynomial in n.

12. Let B_n be the maximum number of regions into which 3-space can be decomposed by n planes.
(i) Find B_n for $n = 1, 2, 3$, and 4.
★ (ii) Prove that $B_n = B_{n-1} + (n^2 - n + 2)/2$ for $n > 1$.
(iii) Find a closed form for B_n; that is, express B_n as a polynomial in n.

13. Let d be a nonnegative integer. For $k = 0, 1, \ldots, d$, let α_k be either 0 or a polynomial of degree k in $F[x]$ and let

$$\alpha = n^d\alpha_0 + n^{d-1}\alpha_1 + \alpha^{d-2}\alpha_2 + \cdots + n\alpha_{d-1} + \alpha_d.$$

Prove that there is a polynomial β in $F[x]$ with

$$\alpha(0) + \alpha(1) + \cdots + \alpha(n) = \beta(n)$$

and either $\beta = 0$ or $\deg \beta \leq d + 1$.

14. Let

$$A_n = 0 \cdot n + 1 \cdot (n-1) + 2(n-2) + 3(n-3) + \cdots + (n-1) \cdot 1 + n \cdot 0.$$

(i) Explain why there is a polynomial β of degree at most 3 in $\mathbf{Q}[x]$ such that $A_n = \beta(n)$.
(ii) Find a closed form for A_n. [*Hint:* Find β.]

15. For all nonnegative integers n, prove that

$$0^2 \cdot n + 1^2 \cdot (n-1) + 2^2 \cdot (n-2) + \cdots + (n-1)^2 \cdot 1 + n^2 \cdot 0 = n^2(n^2-1)/12.$$

16. Find a closed form for

$$0^2 \cdot n^2 + 1^2 \cdot (n-1)^2 + 2^2 \cdot (n-2)^2 + \cdots + (n-1)^2 \cdot 1^2 + n^2 \cdot 0^2.$$

17. In $\mathbf{Q}[x]$, let $\alpha = a_3x^3 + a_2x^2 + a_1x + a_0$ with $a_3 \neq 0$. Let $h = a_2/3a_3$. Prove that

$$\alpha(x-h) = a_3x^3 + b_1x + b_0$$

with b_1 and b_0 in $\mathbf{Q}$.

18. In $\mathbf{Q}[x]$, let $\alpha = a_dx^d + a_{d-1}x^{d-1} + \cdots + a_1x + a_0$ with $a_d \neq 0$. Find an h such that the coefficient of x^{d-1} in $\alpha(x-h)$ is 0.

19. Let the nth difference operator Δ^n be defined inductively for all positive integers n by

$$\begin{aligned} \Delta^1\alpha &= \Delta\alpha. \\ \Delta^2\alpha &= \Delta(\Delta\alpha) = \Delta[\alpha(x+1) - \alpha] = [\alpha(x+2) - \alpha(x+1)] - [\alpha(x+1) - \alpha] \\ &= \alpha(x+2) - 2\alpha(x+1) + \alpha, \\ &\ldots, \\ \Delta^{m+1}\alpha &= \Delta(\Delta^m\alpha). \end{aligned}$$

Let the binomial coefficient $\binom{n}{k} = n!/k!(n-k)!$. Prove that

$$\begin{aligned} \Delta^n\alpha &= \sum_{k=0}^{n}(-1)^k\binom{n}{n-k}\alpha(x+n-k) \\ &= \binom{n}{n}\alpha(x+n) - \binom{n}{n-1}\alpha(x+n-1) + \cdots + (-1)^n\binom{n}{0}\alpha(x). \end{aligned}$$

20. Let $\deg \alpha = d > 0$. Let Δ^n be as in the preceding problem. Prove the following.
(a) $\deg(\Delta^n\alpha) = d - n$ for $1 \leq n \leq d$.

(b) $\Delta^d\alpha = d!L(\alpha)$, where $L(\alpha)$ is the leading coefficient of α.
(c) $\Delta^n\alpha = 0$ for $n > d$.

21. Let K be a positive integer, let $\binom{n}{k} = n!/k!(n-k)!$, and let

$B_k = x(x-1)\cdots(x-k+1)/k!$ Show the following.
(a) $B_k(n) = 0$ for $n = 0, 1, 2, \ldots, k-1$.

(b) $B_k(n) = \binom{n}{k}$ for n an integer and $n \geq k$.

(c) $B_k(n) = (-1)^k\binom{k-n-1}{-n-1}$ for negative integers n.

22. Explain why $B_k(n)$ must be an integer for all integers n and all nonnegative integers k. (You may use the fact that $\binom{n}{k}$ is an integer for k and n integers with $0 \leq k \leq n$.)

23. Let $\alpha = b_0 + b_1B_1 + b_2B_2 + \cdots + b_dB_d$. Show the following.
(i) $\alpha(0) = b_0$ and $\Delta^j\alpha(0) = b_j$ for $j = 1, 2, \ldots, d$.

(ii) $b_j = \binom{j}{j}\alpha(j) - \binom{j}{j-1}\alpha(j-1) + \binom{j}{j-2}\alpha(j-2) - \cdots + (-1)^j\binom{j}{0}\alpha(0)$.

(iii) If each of $\alpha(0), \alpha(1), \ldots, \alpha(d)$ is an integer, then each of $b_0, b_1, \ldots, b_d$ is an integer.
(iv) If each of $\alpha(0), \alpha(1), \ldots, \alpha(d)$ is an integer or each of $b_0, b_1, \ldots, b_d$ is an integer, then $\alpha(n)$ is an integer for every n in $\mathbf{Z}$.

24. Express $\alpha = (x^4 - 2x^3 + 11x^2 + 14x)/24$ in the form

$$\alpha = b_0 + b_1B_1 + b_2B_2 + b_3B_3 + b_4B_4,$$

and also prove that $\alpha(n)$ is an integer for all integers n.

25. Given that $\deg \beta = 3$, complete the following table; then use the column for $x = 0$ and Problem 23(i) to express β in terms of the bionomial polynomials B_1, B_2, and B_3.

x	0	1	2	3
$\beta(x)$	1	1	7	25
$\Delta\beta(x)$	0	6	18	
$\Delta^2\beta(x)$	6			
$\Delta^3\beta(x)$				

26. Given that $\deg \alpha = 4$, use the table that follows to find $\Delta^j\alpha(0)$ for $j = 0, 1, 2, 3, 4$; then use Problem 23(ii) to express α in terms of the binomial polynomials B_k.

x	0	1	2	3	4
$\alpha(x)$	0	0	7	39	126

27. Let c be in F and α be in $F[x]$ with deg $\alpha = d$. Let α' be the derivative of α, $\alpha^{(2)}$ be the derivative of $\alpha', \ldots,$ and $\alpha^{(d)}$ be the derivative of $\alpha^{(d-1)}$. Show that

$$\alpha = \alpha(c) + (x - c)\alpha'(c) + (x - c)^2\alpha^{(2)}(c)/2! + \cdots + (x - c)^d\alpha^{(d)}(c)/d!.$$

(This is the ***Taylor Expansion*** of calculus for the special case of polynomials. The formula for the derivative of a polynomial is given in Definition 8 of Section 5.2.)

28. Use the notation of the preceding problem, and let k be an integer with $1 \le k \le d$. Prove that $(x - c)^k \mid \alpha$ if and only if

$$\alpha(c) = \alpha'(c) = \alpha^{(2)}(c) = \cdots = \alpha^{(k-1)}(c) = 0.$$

29. Let c be a zero with multiplicity m of an α in $F[x]$. Prove the following.
(a) If $m = 1$, c is not a zero of α' (the derivative of α).
(b) If $m > 1$, c is a zero of α' with multiplicity $m - 1$.

30. Let $\mathbf{C}$ be the field of complex numbers. In $\mathbf{C}[x]$, let

$$\alpha = x^3 + 3x^2 + 6x + 6.$$

Prove that each zero of α has multiplicity 1.

31. Let $\alpha \in \mathbf{Z}[x]$, $h \in \mathbf{Z}$, and $\alpha(h) = \alpha(h + 1) = \alpha(h + 2) = 0$. Show that $6 \mid \alpha(n)$ for all integers n.

32. Let $\alpha \in \mathbf{Z}[x]$, $h \in \mathbf{Z}$, $k \in \mathbf{Z}^+$, and

$$0 = \alpha(h) = \alpha(h + 1) = \alpha(h + 2) = \cdots = \alpha(h + k - 1).$$

Show that $k! \mid \alpha(n)$ for all integers n.

33. Let α and β be polynomials in $U[x]$. Show that

$$\alpha(x)\Delta\beta(x) + \beta(x + 1)\Delta\alpha(x) = \alpha(x + 1)\beta(x + 1) - \alpha(x)\beta(x).$$

34. Let α and β be in $\mathbf{Q}[x]$ and let $\gamma(x) = \alpha(x)\Delta\beta(x) + \beta(x + 1)\Delta\alpha(x)$. Show that

$$\gamma(0) + \gamma(1) + \cdots + \gamma(n) = \alpha(n + 1)\beta(n + 1) - \alpha(0)\beta(0).$$

35. In $\mathbf{C}[x]$, let $\alpha = a_d x^d + a_{d-1}x^{d-1} + \cdots + a_1 x + a_0$ and let $\alpha(n)$ be in $\mathbf{Z}$ for $d + 1$ consecutive integers n. Prove that $d!\alpha_i$ is in $\mathbf{Z}$ for $i = 0, 1, \ldots, d$ and that $\alpha(n)$ is in $\mathbf{Z}$ for all n in $\mathbf{Z}$.

5.7 Ideals in $F[x]$

In this section we note some properties of ideals in the integral domain $F[x]$ of polynomials over a field F; these properties are closely analogous to those for ideals in the integral domain $\mathbf{Z}$.

In $F[x]$ every nonzero β has an invertible as its leading coefficient. Therefore, the only restriction on the divisor β in the division algorithm is $\beta \neq 0$. Below, this is used to show that $F[x]$ shares with $\mathbf{Z}$ (but not with $\mathbf{Z}[x]$) the property that every ideal is a principal ideal. Then principal ideals help us to define the greatest common divisor and the least common multiple of polynomials in $F[x]$.

The division algorithm is also the basis for the proof of unique factorization into primes in $F[x]$ (see Section 5.8) and for the proof that $F[x]/(\beta)$ is a field whenever β is a prime (see Section 5.10).

We now go back to general commutative rings U with unity.

LEMMA 1 Linear Combinations in U

Let a and b be fixed elements of a commutative ring U with unity and let I consist of all the linear combinations

$$ax + by,$$

with x and y in U. Then I is an ideal in U, and both a and b are in I. (I is the smallest ideal containing both a and b in the ring U.)

Proof As in the Linear Combination Theorem for $\mathbf{Z}$ in Section 1.4, one easily sees that a and b are in I and that I is closed under subtraction. It is also clear that I is closed under left and right multiplication. Hence, I is an ideal ∎

Every ideal in the ring $\mathbf{Z}$ of the integers is a principal ideal. We next see that this is not true in $\mathbf{Z}[x]$.

EXAMPLE 1 Nonprincipal Ideal in $\mathbf{Z}[x]$

Let I consist of all the polynomials

$$\alpha = a_d x^d + \cdots + a_0, \qquad a_j \text{ in } \mathbf{Z},$$

such that $7 \mid a_0$ in $\mathbf{Z}$. Here we show that I is an ideal, but not a principal ideal, in $\mathbf{Z}[x]$. We see readily that α is in I if and only if $\alpha = 7\beta + x\gamma$, with β and γ in $\mathbf{Z}[x]$. It then follows from Lemma 1 that I is an ideal and that both 7 and x are in I.

Let us assume that I is a principal ideal (δ). This implies that $\delta \mid 7$ and $\delta \mid x$. Since $\mathbf{Z}$ is an integral domain, it follows from $\delta \mid 7$ that $\deg \delta \leq \deg 7 = 0$ and hence that $\deg \delta = 0$. However, the only polynomials of degree 0 in I are 7 and -7. But $I \neq (7) = (-7)$, since x is not a multiple of 7 (or -7) in $\mathbf{Z}[x]$. This contradiction shows that I is not a principal ideal.

Let α be a polynomial in a principal ideal (β) in $F[x]$ and let $\beta \neq 0$. Then $\beta \mid \alpha$ and $\deg \beta \leq \deg \alpha$. This shows that, if I is an ideal in $F[x]$ and $I \neq \{0\}$, a possible generator of I must be sought among the nonzero polynomials of least degree in I. The following result tells us that such a search will always be successful.

THEOREM 1 $F[x]$ Is a Principal Ideal Domain

Every ideal I in $F[x]$ is a principal ideal. If $I \neq \{0\}$, a nonzero β in I is a generator of I if and only if $\deg \beta \leq \deg \alpha$ for all nonzero α in I; also, I has a unique monic generator.

Proof The ideal $\{0\}$ is the principal ideal generated by 0; hence we may assume that there are nonzero polynomials in I. Then it follows from the well ordering principle that there is a β in I with $\deg \beta \leq \deg \alpha$ for all nonzero α in I.

Sir Isaac Newton
1642–1727

Nicolaus Copernicus, a Polish astronomer, published a book on the heliocentric theory in 1543. Tycho Brahe, a Dane, carefully recorded volumes of accurate observations of the positions of planets with the times of the observations. Galileo Galilei, an Italian, refined the telescope and saw Jupiter's orbiting moons. Johannes Kepler, a German, patiently tried pattern after pattern until he fitted Brahe's data into three laws of planetary motion which state that the path of a planet is an ellipse with the sun at one focus, that the line segment joining the planet and the sun sweeps out equal areas in equal times, and that the squares of the times of revolution of any two planets are proportional to the cubes of their distances from the sun.

Newton, an Englishman, went far beyond Kepler's tremendous feat. He created calculus and differential equations and used them to show that Kepler's laws were a consequence of Newton's universal law concerning the gravitational attraction between any two particles. One particular result of Newton's work was that the path of a rocket to the moon or to another body in space could be charted accurately. A general consequence is that, during the last three centuries, mathematics has been applied to an extremely wide range of problems.

From an inauspicious background Newton rose to become the towering figure of British science during his lifetime and one of the greatest men of science of all time. Contemporary portraits, busts, and medallions show him in almost godlike poses, one late portrait making him appear rather like a Roman senator. Recent studies, notably the biography by Frank Manuel, have focused on some flaws in Newton's character, such as a lack of generosity in failing to acknowledge work of his colleagues or his refusal to recognize ideas of his successors when they did not agree with his own. He controlled the Royal Society for many years and, partly because of his appointment as Director of the Royal Mint, he wielded considerable power. Nevertheless, his reputation as a great scientist has not been seriously tarnished by these revelations.

His greatest work was the *Philosophiae Naturalis Principia Mathematica* (1687). It contains some of his most startling discoveries concerning the applications of mathematics to physical and astronomical problems. Laplace said that the *Principia*, because of its profound and original ideas, was assured "a lasting pre-eminence over all other productions of the human mind." Some indication of the importance of this book might be the fact that a first edition would currently cost about $75,000 in the rare book market.

His *Arithmetica Universalis* (1707) contains many original results in the theory of algebraic equations, and his *Method of Fluxions* (1736) concerns not only an explanation of his discoveries in calculus but also his work on the approximation of roots of equations.

The question of whether Newton or Leibniz invented the calculus caused heated debate on both sides of the English Channel for many years. Newton's first ideas on the calculus apparently date from about 1665, but his first writings on the subject were the *De analysi per aequationes numero terminorum infinitas*, written in 1669 and first published in 1711, and his *Methodus fluxionum et serierum infinitorum*, written in 1671 and first published in English translation in 1736 and in the original Latin in 1742. Calculus methods were employed in the *Principia*, in 1687. So although Newton may have had the ideas of calculus prior to Leibniz, nevertheless Leibniz published first in 1684 with his famous paper "Nova methodus pro maximis et minimis" in the *Acta Eruditorum*. Furthermore, it is generally conceded today that both Newton and Leibniz discovered the calculus independently. This was not accepted generally during the height of the conflict between the two and their followers. Newton's friends accused Leibniz of plagiarism. In 1687 Newton had generously acknowledged in the first edition of the *Principia* that Leibniz had discovered a method similar to his. Following the development of the dispute over priority, Newton deleted this reference in the third edition of 1726. The conflict had the unfortunate effect of forcing mathematicians into two camps, the followers of Newton and the followers of Leibniz. British scientists used the notation of Newton, which is inferior to that of Leibniz; and following the deaths of Newton and Leibniz, mathematicians on the continent, using the notation of Leibniz, far surpassed the British in mathematical advances. The Leibniz notation is essentially that used in calculus courses today; that of Newton is reserved for certain applications in physics. Very recent scholarship on the work of Maclaurin has challenged the above conventional wisdom about the slow advance of British mathematics following the death of Newton. The controversy lingers on.

Whereas Leibniz was primarily a philosopher who created great mathematics as well, Newton was primarily a physicist who created the mathematics he needed. His *Opticks* (1704) is another great milestone in the history of science. Newton, who died a powerful and wealthy man, much honored in Britain and elsewhere, is buried in Westminster Abbey in a massive marble tomb prominently displayed in the nave. Leibniz died neglected; it is reported that his funeral was attended by a single mourner.

Using the division algorithm we note that for any α in I there exist γ and ρ in $F[x]$ such that $\alpha = \gamma\beta + \rho$ and either $\rho = 0$ or $\deg \rho < \deg \beta$. Then $\rho = \alpha - \gamma\beta$ is in I, since an ideal is closed under linear combinations. The inequality $\deg \rho < \deg \beta$ would contradict the minimal nature of the degree of β; hence, $\rho = 0$ and $\beta \mid \alpha$. This shows that $I = (\beta)$.

All generators of I are associates of each other, and just one of these associates is monic; thus the monic generator is unique. ∎

THEOREM 2 Common Divisors

Let α and β be in $F[x]$ with either $\alpha \neq 0$ or $\beta \neq 0$. Then there is a unique monic δ in $F[x]$ such that both of the following hold:

(i) $\delta \mid \alpha$ and $\delta \mid \beta$; that is, δ is a common divisor of α and β.
(ii) If $\gamma \mid \alpha$ and $\gamma \mid \beta$, then $\gamma \mid \delta$; that is, δ is a multiple of every common divisor γ of α and β.

This unique δ is expressible as $\delta = \zeta\alpha + \eta\beta$ with ζ and η in $F[x]$.

Proof Let I consist of all linear combinations

$$\xi\alpha + \eta\beta \tag{1}$$

with ξ and η in $F[x]$. Then I is an ideal and both α and β are in I by Lemma 1. Since either $\alpha \neq 0$ or $\beta \neq 0$, we have $I \neq \{0\}$. It follows from Theorem 1 that I is a principal ideal and has a unique monic generator, which we designate as δ.

Since α and β are in (δ), both $\delta \mid \alpha$ and $\delta \mid \beta$; thus we have proved part (i). Also, δ is of the form (1), since δ is in the set I. Then $\gamma \mid \alpha$ and $\gamma \mid \beta$ together imply $\gamma \mid (\xi\alpha + \eta\beta)$; that is, $\gamma \mid \delta$. This establishes part (ii).

To show uniqueness, we let each of δ and δ' be a monic polynomial satisfying (i) and (ii). It follows that both $\delta \mid \delta'$ and $\delta' \mid \delta$. But monic associates are equal; hence, $\delta = \delta'$ as desired. ■

DEFINITION 1

Greatest Common Divisor

If α and β are in $F[x]$ with either $\alpha \neq 0$ or $\beta \neq 0$, their **greatest common divisor** $\gcd(\alpha, \beta)$ is the unique monic δ in $F[x]$ with the properties that δ is a common divisor of α and β and that δ is a multiple of every common divisor γ of α and β. Also, we define $\gcd(0, 0)$ to be 0.

We have defined $\gcd(0, 0)$ to be 0, since the only linear combination $\xi \cdot 0 + \eta \cdot 0$ is 0 and $\{0\}$ is the principal ideal generated by 0.

For every α and β in $F[x]$, it follows from Theorem 2 that $\delta = \gcd(\alpha, \beta)$ exists and that $\delta = \xi\alpha + \eta\beta$ with ξ and η in $F[x]$. One can find δ (and also ξ and η, if one wishes) by the following analogue of Euclid's Algorithm for integers (which was described in Section 1.5). The work of the present algorithm can be tabulated as in Section 1.5.

ALGORITHM **Euclid's Algorithm for Polynomials**

Let α_0 and α_1 be in $F[x]$ with $0 \leq \deg \alpha_1 \leq \deg \alpha_0$. It follows from the division algorithm for polynomials (Theorem 2 of Section 5.3) that there exist γ_1 and α_2 in $F[x]$ with

$$\alpha_0 = \gamma_1\alpha_1 + \alpha_2 \quad \text{and either} \quad \alpha_2 = 0 \quad \text{or} \quad \deg \alpha_2 < \deg \alpha_1.$$

If $\alpha_2 \neq 0$, there are γ_2 and α_3 in $F[x]$ with

$$\alpha_1 = \gamma_2\alpha_2 + \alpha_3 \quad \text{and either} \quad \alpha_3 = 0 \quad \text{or} \quad \deg \alpha_3 < \deg \alpha_2.$$

The process is continued until one achieves a remainder α_{n+1} that is the zero polynomial. Then one has

$$\alpha_i = \gamma_{i+1}\alpha_{i+1} + \alpha_{i+2} \text{ with } \deg \alpha_{i+2} < \deg \alpha_{i+1} \qquad \text{for } i = 0, 1, \ldots, n-2;$$

$$\alpha_{n-1} = \gamma_n \alpha_n.$$

Let f be the leading coefficient of α_n. Then $\delta = f^{-1}\alpha_n$ is the unique monic associate of α_n and $\delta = \gcd(\alpha_0, \alpha_1)$.

If one wants polynomials ξ and η in $F[x]$ such that $\delta = \xi\alpha_0 + \eta\alpha_1$, one obtains polynomials $\xi_{n-1}, \xi_{n-2}, \ldots, \xi_1, \xi_0$ using

$$\xi_{n-1} = 1,\ \xi_{n-2} = -\gamma_{n-1}, \quad \text{and} \quad \xi_j = \xi_{j+2} - \xi_{j+1}\gamma_{j+1}$$
$$\text{for } j = n-3, n-4, \ldots, 0. \qquad (2)$$

As in Euclid's Algorithm for integers, it follows that

$$\alpha_n = \xi_1\alpha_0 + \xi_0\alpha_1.$$

Letting $\xi = f^{-1}\xi_1$ and $\eta = f^{-1}\xi_0$, we then have

$$\delta = f^{-1}\alpha_n = \xi\alpha_0 + \eta\alpha_1.$$

EXAMPLE 2 In $\mathbf{Z}_3[x]$, let $\alpha_0 = x^5 + x^4 + x^3 + x + \bar{2}$ and $\alpha_1 = x^4 + x^3 + \bar{2}x^2 + \bar{2}x$. We now try to find $\delta = \gcd(\alpha_0, \alpha_1)$ and ξ and η in $\mathbf{Z}_3[x]$ such that $\delta = \xi\alpha_0 + \eta\alpha_1$. Using long division (or the answers to Problem 5 of Section 5.3), we get

$$\alpha_0 = x\alpha_1 + \alpha_2 \quad \text{with} \quad \alpha_2 = \bar{2}x^3 + x^2 + x + \bar{2},$$
$$\alpha_1 = (\bar{2}x + \bar{1})\alpha_2 + \alpha_3 \quad \text{with} \quad \alpha_3 = \bar{2}x^2 + \bar{1},$$
$$\alpha_2 = (x + \bar{2})\alpha_3.$$

Since $\alpha_4 = 0$ and the leading coefficient of α_3 is $\bar{2}$,

$$\gcd(\alpha_0, \alpha_1) = (\bar{2})^{-1}\alpha_3 = \bar{2}(\bar{2}x^2 + \bar{1}) = x^2 + \bar{2}.$$

From (2) with $n = 3$, $\gamma_1 = x$, $\gamma_2 = \bar{2}x + \bar{1}$, and $\gamma_3 = x + \bar{2}$, we get

$$\xi_2 = 1,\ \xi_1 = -\gamma_2 = -(\bar{2}x + 1) = x + \bar{2},$$
$$\xi_0 = \xi_2 - \xi_1\gamma_1 = 1 - (x + \bar{2})x = -x^2 - \bar{2}x + 1 = \bar{2}x^2 + x + 1.$$

Finally, we let $\xi = (\bar{2})^{-1}\xi_1$ and $\eta = (\bar{2})^{-1}\xi_0$ and have

$$\gcd(\alpha_0, \alpha_1) = x^2 + \bar{2} = \xi\alpha_0 + \eta\alpha_1$$
$$= (\bar{2}x + \bar{1})\alpha_0 + (x^2 + \bar{2}x + \bar{2})\alpha_1.$$

DEFINITION 2

Relatively Prime Polynomials

If $\gcd(\alpha, \beta) = 1$, α and β are **relatively prime**.

The following result is easily obtained from Theorem 2 and the two definitions above.

THEOREM 3 Linear Combinations

Let α, β, and γ be in $F[x]$. Then:
(a) γ is expressible as $\xi\alpha + \eta\beta$ with ξ and η in $F[x]$ if and only if $\gcd(\alpha, \beta) \mid \gamma$.
(b) α and β are relatively prime if and only if $1 = \xi\alpha + \eta\beta$, with ξ and η in $F[x]$.
The proof is left to the reader.

Now let π be a prime in $F[x]$. Then the only monic divisors of π are 1 and the unique monic associate π_1 of π. Hence, $\gcd(\alpha, \pi) = \pi_1$ when $\pi \mid \alpha$, and $\gcd(\alpha, \pi) = 1$ when α is not a multiple of π.

LEMMA 2 Analogue of Euclid's Lemma

In $F[x]$, let π be a prime and let $\pi \mid (\alpha\beta)$. Then either $\pi \mid \alpha$ or $\pi \mid \beta$.

The proof is similar to that of Euclid's Lemma in Section 1.7 and is left to the reader as Problem 11 of this section.

Let α and β be in $F[x]$. Then the common multiples of α and β are the elements common to the principal ideals (α) and (β), that is, the elements of the intersection

$$I = (\alpha) \cap (\beta).$$

The intersection of ideals is also an ideal (see Problem 39 of Section 4.2). If $\alpha \neq 0$ and $\beta \neq 0$, I has nonzero elements; and it follows from Theorem 1 above that I is a principal ideal with a unique monic generator μ. It is natural to define this μ to be the ***least common multiple*** of α and β, which we denote by $\text{lcm}[\alpha, \beta]$. For completeness, we let $\text{lcm}[\alpha, 0] = 0 = \text{lcm}[0, \beta]$.

The following result establishes a mapping from any extension field E over F to the family of ideals I in $F[x]$ such that I has an irreducible generator.

THEOREM 4 Polynomials with a Fixed Zero

Let F be a subfield in E, let s be an element of E, and let I consist of all α in $F[x]$ such that s is a zero of α. Then I is a principal ideal. If $I \neq \{0\}$, the unique monic generator μ of I is irreducible.

Proof I is not empty, since 0 is in I. Let α and β be in I and let γ be in $F[x]$. Then it is clear that $\alpha - \beta$ and $\gamma\alpha$ are in I; that is, I is a nonempty set closed under subtraction and under left and right multiplication, and so I is an ideal. If

$I \neq \{0\}$, it follows from Theorem 1 above that I is a principal ideal and that its unique monic generator μ is also the unique monic polynomial of least degree in I. If μ were reducible in $F[x]$, we would have

$$\mu = \mu_1\mu_2, \quad \deg \mu_1 < \deg \mu, \quad \deg \mu_2 < \deg \mu$$

and $\mu_1(s)\mu_2(s) = \mu(s) = 0$. It would follow that s is a zero of either μ_1 or μ_2, contradicting the minimal nature of the degree of μ. This contradiction proves that μ is irreducible. ∎

COROLLARY Let α and a prime π be in $F[x]$. In an extension field E over F, let s be a zero of both α and π. Then $\pi \mid \alpha$ in $F[x]$.

The proof is left to the reader as Problem 12 below.

Problems

1. In $\mathbf{Z}_2[x]$, let $\alpha_0 = x^5 + x^4 + \bar{1}$ and $\alpha_1 = x^5 + x + \bar{1}$. Find the following.
   (i) γ_1 and α_2 such that $\alpha_0 = \gamma_1\alpha_1 + \alpha_2$ and $\deg \alpha_2 < \deg \alpha_1$.
   (ii) γ_2 and α_3 such that $\alpha_1 = \gamma_2\alpha_2 + \alpha_3$ and $\deg \alpha_3 < \deg \alpha_2$.
   (iii) γ_3 such that $\alpha_2 = \gamma_3\alpha_3$.
   (iv) $\delta = \gcd(\alpha_0, \alpha_1)$.
   (v) ξ and η such that $\delta = \xi\alpha_0 + \eta\alpha_1$ and $\deg \eta \le 2$.
2. In $\mathbf{Q}[x]$, let $\alpha_0 = x^5 + 2x^4 + x^3 + 2x^2 + x + 2$ and $\alpha_1 = x^4 - x$. Find the following.
   (i) γ_1 and α_2 such that $\alpha_0 = \gamma_1\alpha_1 + \alpha_2$ and $\deg \alpha_2 < \deg \alpha_1$.
   (ii) γ_2 and α_3 such that $\alpha_1 = \gamma_2\alpha_2 + \alpha_3$ and $\deg \alpha_3 < \deg \alpha_2$.
   (iii) γ_3 such that $\alpha_2 = \gamma_3\alpha_3$.
   (iv) $\delta = \gcd(\alpha_0, \alpha_1)$.
   (v) ξ and η such that $\delta = \xi\alpha_0 + \eta\alpha_1$ and $\deg \eta \le 2$.
3. (i) Find $\gcd(x - 1, x^2 + x + 1)$ in $\mathbf{Q}[x]$.
   (ii) Find the three smallest positive integers n such that $\gcd(n - 1, n^2 + n + 1) = 3$ in $\mathbf{Z}$.
   (iii) Find the three smallest positive integers n such that $\gcd(n - 1, n^2 + n + 1) = 1$ in $\mathbf{Z}$.
   (iv) In $\mathbf{Z}$, explain why $\gcd(n - 1, n^2 + n + 1)$ is in $\{1, 3\}$ for all n.
4. (i) Show that $\gcd(x + 1, x^2 - x + 1) = 1$ in $\mathbf{Q}[x]$.
   (ii) In $\mathbf{Z}$, show that there exist n such that $\gcd(n + 1, n^2 - n + 1) \neq 1$.
5. (i) Show that $\gcd(x + 1, 2x + 1) = 1$ in $\mathbf{Q}[x]$.
   (ii) In $\mathbf{Z}$, show that $\gcd(n + 1, 2n + 1) = 1$ for all n.
   (iii) Find $\gcd(x + 1, 3x + 1)$ in $\mathbf{Q}[x]$.
   (iv) In $\mathbf{Z}$, what are the possible values of $\gcd(n + 1, 3n + 1)$ with n in $\mathbf{Z}$?
6. (i) Find $\gcd(x - 1, 2x - 1)$ in $\mathbf{Q}[x]$.
   (ii) In $\mathbf{Z}$, what are the possible values of $\gcd(n - 1, 2n - 1)$ with n in $\mathbf{Z}$?
   (iii) Find $\gcd(x - 1, 3x - 1)$ in $\mathbf{Q}[x]$.
   (iv) In $\mathbf{Z}$, what are the possible values of $\gcd(n - 1, 3n - 1)$ with n in $\mathbf{Z}$?
7. Explain why distinct monic primes π and π' in $F[x]$ must be relatively prime.

8. Show with an example that distinct primes in $F[x]$ need not be relatively prime.
9. Explain why $x - a$ and $x - b$ are relatively prime in $F[x]$ if and only if $a \neq b$.
10. In $F[x]$, prove that $\gcd(\alpha, \beta) = 1$, $\alpha \mid \gamma$, and $\beta \mid \gamma$ together imply $(\alpha\beta) \mid \gamma$.
11. In $F[x]$, let π be a prime and let $\pi \mid (\alpha\beta)$. Prove that either $\pi \mid \alpha$ or $\pi \mid \beta$, and thus prove Lemma 2 above. (Also, see Euclid's Lemma in Section 1.7.)
12. Let α and a prime π be in $F[x]$. Let α and π have a common zero in an extension field E over F. Show that $\pi \mid \alpha$ (and thus prove the corollary to Theorem 4 above).
13. Let $\alpha = x + 1$ and I be the principal ideal (α^2) in $\mathbf{Q}[x]$. Prove that the quotient ring $\mathbf{Q}[x]/I$ is not a field.
14. Show that the quotient ring $\mathbf{Z}[x]/(x^2 + 1)$ is isomorphic to the ring of gaussian integers $m + n\mathbf{i}$ (with m and n in $\mathbf{Z}$).
15. Let s be a fixed element of a commutative ring R and h be the mapping from $R[x]$ to R with $\alpha \mapsto \alpha(s)$.
    (i) Explain why h is a ring homomorphism.
    (ii) What is the relation between s and the kernel K of h?
16. Let h be the mapping from $U[x]$ to U with
$$a_d x^d + \cdots + a_1 x + a_0 \mapsto a_d + \cdots + a_1 + a_0.$$
    (i) Explain why h is a ring homomorphism.
    (ii) Find the monic generator for the kernel of h.
17. Show that there exists an infinite field of characteristic p for every positive prime p in $\mathbf{Z}$.
18. Let x and y be independent indeterminates over $\mathbf{Q}$. Let I consist of all linear combinations $x\alpha + y\beta$, with α and β in $\mathbf{Q}[x, y]$. Show that I is an ideal in $\mathbf{Q}[x, y]$ but is not a principal ideal. (See Definitions 6 and 7 of Section 5.2.)

5.8 Factorizations of Polynomials

This section deals with analogues for the polynomial rings of the factorization theorems for the integers in Section 1.7.

THEOREM 1 Factorization into Irreducible Polynomials

Let R be a commutative ring. Then in $R[x]$ every α of positive degree is expressible in the form

$$\alpha = \beta_1\beta_2 \cdots \beta_n, \tag{I}$$

where the β_j are irreducible.

Proof We assume the result false and obtain a contradiction. Since the degree of a polynomial is a nonnegative integer, this assumption and the well ordering principle imply that, in the nonempty subset of $R[x]$ consisting of the polynomials that are neither irreducible nor the product of a finite number of irreducibles, there is an α with the minimum degree d for this subset.

Since this α is reducible, $\alpha = \alpha_1\alpha_2$ with $0 < \deg \alpha_1 < d$ and $0 < \deg \alpha_2 < d$. The minimal nature of d then implies that both α_1 and α_2 are expressible in the form (I); that is, we have

$$\alpha_1 = \beta_1 \cdots \beta_s, \qquad \alpha_2 = \beta_{s+1} \cdots \beta_t,$$

with the β_j irreducible polynomials in $R[x]$. Then

$$\alpha = \alpha_1\alpha_2 = \beta_1 \cdots \beta_s\beta_{s+1} \cdots \beta_t$$

is the desired contradiction, and the theorem is proved. ■

So far, the analogy between factorization into primes in **Z** and factorization into irreducible polynomials in $R[x]$ has been fairly close. When we consider uniqueness, some obstacles appear, as is seen in the following.

EXAMPLE 1 **Nonunique Factorization in $\mathbf{Z_m}$**
In $\mathbf{Z}_6[x]$, $x(x + \bar{5}) = (x + \bar{2})(x + \bar{3})$.

EXAMPLE 2 **Factorization in $\mathbf{Z_p}$**
In $\mathbf{Z}_5[x]$, $(\bar{2}x + \bar{1})(\bar{4}x + \bar{3}) = (x + \bar{3})(\bar{3}x + \bar{1})$.

Since all polynomials of degree 1 are irreducible, these examples show that a given polynomial may be factored into irreducible polynomials in more than one way. However, the factors in Example 1 are not primes in $\mathbf{Z}_6[x]$. (See Example 4 of Section 5.3, which shows that $x + \bar{5}$ is not a prime in $\mathbf{Z}_6[x]$.)

As a step toward resolving the lack of uniqueness in Example 2, we recall that, technically speaking, factorization of an integer into primes in **Z** is not unique; for example,

$$21 = 3 \cdot 7 = (-3)(-7).$$

But there is unique factorization into *positive* primes for an integer greater than 1.

In the ring $F[x]$ of polynomials over a field F, the irreducible polynomials are the same as the primes, and the analogue of a positive prime in **Z** is a monic prime polynomial. By restricting the ring R of coefficients to be a field F, we eliminate Example 1 from consideration, since $\mathbf{Z}_6$ is not a field.

We introduce the uniqueness into Example 2 by factoring out the leading coefficients from the first-degree factors and thus bringing monic primes into the picture. Then both sides of the equation in Example 2 become

$$\bar{3}(x + \bar{3})(x + \bar{2}).$$

We now present the general result.

THEOREM 2 **Unique Factorization in $F[x]$**

Every polynomial α of positive degree in $F[x]$ is expressible in the form

$$\alpha = \upsilon\pi_1^{n_1}\pi_2^{n_2} \cdots \pi_r^{n_r}, \qquad \text{(F)}$$

where v is a nonzero element of F, the π_j are distinct monic primes in $F[x]$, and the n_j are positive integers. This factorization is unique (except for the order of appearance of the powers of the π_j). If α is monic, $v = 1$ in display (F).

The proof is similar to that for unique factorization in the integers and is left to the reader.

Let **R** and **C** denote the fields of real and complex numbers. In $\mathbf{R}[x]$ and $\mathbf{C}[x]$ the above unique factorization results can be made more explicit by using the theorem of Gauss, stated in Section 4.9, that **C** is algebraically closed. We next restate Gauss's Theorem in the following form.

THEOREM 3 Fundamental Theorem of Algebra

If α is a polynomial of positive degree in $\mathbf{C}[x]$, $\alpha = 0$ has a root in **C**.

A proof requires advanced techniques of analysis or topology; we assume this result without proof.

The Fundamental Theorem of Algebra and the Factor Theorem (Theorem 2 of Section 5.4) together imply the following result.

THEOREM 4 Primes in C[x]

The primes (or irreducible polynomials) in $\mathbf{C}[x]$ are just the polynomials of degree 1.

Using the Unique Factorization Theorem for polynomials and Theorem 4, we obtain the result that follows.

THEOREM 5 Linear Factorization

Let α be a polynomial of positive degree d in $\mathbf{C}[x]$. Then α has d zeros $r_1, r_2, \ldots, r_d$ (not necessarily distinct) in **C**, and

$$\alpha = a(x - r_1)(x - r_2) \cdots (x - r_d), \tag{L}$$

with $a \neq 0$ in **C**.

Since the real numbers **R** form a subfield in **C**, a polynomial α in $\mathbf{R}[x]$ is also in $\mathbf{C}[x]$ and so is expressible in the form (L) if $\deg \alpha > 0$. Some or all of the zeros r_j in (L) may be real; the others come in pairs of complex conjugates. (See the corollary to Theorem 5 of Section 5.4.) We use this fact to characterize unique factorization in $\mathbf{R}[x]$.

THEOREM 6 Primes in $R[x]$

The primes (or irreducible polynomials) of $\mathbf{R}[x]$ are all the first-degree polynomials and those second-degree polynomials $ax^2 + bx + c$ (a, b, and c real) with $b^2 - 4ac < 0$.

Proof A polynomial α of positive degree in $\mathbf{R}[x]$ has a complex zero r, by the Fundamental Theorem. If r is real, $x - r$ divides α in $\mathbf{R}[x]$, while if $r = h + k\mathbf{i}$ is not real, $x^2 - (r + \bar{r})x + r\bar{r}$ divides α in $\mathbf{R}[x]$, where $\bar{r}$ is the complex conjugate $h - k\mathbf{i}$ of r. Hence an irreducible α must be of degree 1 or 2.

Let $\alpha = ax^2 + bx + c$. Then the (not necessarily distinct) zeros of α in $\mathbf{C}$ are

$$r_1 = \frac{-b + \sqrt{b^2 - 4ac}}{2a}, \qquad r_2 = \frac{-b - \sqrt{b^2 - 4ac}}{2a}.$$

A polynomial of degree 2 over the real numbers is composite if and only if it has a real root. [See Problem 20(i) of Section 5.4.] Hence α is prime in $\mathbf{R}[x]$ if and only if the roots r_1 and r_2 are not real; this is true if and only if $b^2 - 4ac < 0$. ■

THEOREM 7 Unique Factorization in R[x]

Every α of positive degree in $\mathbf{R}[x]$ is expressible as

$$\alpha = a(\pi_1)^{n_1}(\pi_2)^{n_2} \cdots (\pi_m)^{n_m}, \qquad a \text{ in } \mathbf{R},$$

where the π_j are distinct monic irreducible polynomials of degree 1 or 2 in $\mathbf{R}[x]$. The factorization is unique except for the order of appearance of the $(\pi_j)^{n_j}$.

Proof This is an immediate consequence of Theorems 6 and 2 above. ■

Problems

1. Explain why the following are true in $\mathbf{Z_{31}}[x]$.
   (i) There are 31 monic polynomials of degree 1.
   (ii) There are 961 monic polynomials of degree 2.
   (iii) Of the 961 monic second-degree polynomials, 496 are reducible (and composite) and 465 are irreducible (and prime).
2. Do the analogue of Problem 1 for $\mathbf{Z_p}[x]$, where p is a prime in $\mathbf{Z}$.
3. Let α_1, α_2, and α_3 denote distinct monic polynomials of degree 1 and let β denote a monic prime of degree 2 in $\mathbf{Z_{31}}[x]$. Explain why the following are true in $\mathbf{Z_{31}}[x]$.
   (i) There are 4495 polynomials of the form $\alpha_1\alpha_2\alpha_3$.
   (ii) There are 930 polynomials of the form $\alpha_1^2\alpha_2$.

(iii) There are 31 polynomials of the form α_1^3.
(iv) There are 14415 polynomials of the form $\alpha_1\beta$.
(v) There are 9920 monic primes of degree 3.

4. Do the analogue of Problem 3 for $\mathbf{Z_p}[x]$ where p is a prime in $\mathbf{Z}^+$.

5. Express each of the following irreducible polynomials as a product $\alpha\beta$ with α and β polynomials of degree 1 in $\mathbf{Z_6}[x]$.
(a) x; (b) $x + \bar{2}$; (c) $x + \bar{3}$.

6. Find first-degree polynomials α and β in $\mathbf{Z_{10}}[x]$ such that $\alpha\beta = x + \bar{3}$.

7. In $F[x]$, explain why $\gcd(\alpha, \beta) = 1$ if and only if there is no prime π with both $\pi \mid \alpha$ and $\pi \mid \beta$.

8. In $F[x]$, explain why $\gcd(\alpha_1\alpha_2 \cdots \alpha_m, \beta_1\beta_2 \cdots \beta_n) = 1$ if and only if $\gcd(\alpha_i, \beta_j) = 1$ for all i and j with $1 \le i \le m$ and $1 \le j \le n$.

9. Let $\pi_1, \pi_2, \ldots, \pi_n$ be n distinct monic primes in $F[x]$ and let $r_1, r_2, \ldots, r_n$ be (not necessarily distinct) positive integers. Explain why
$$\gcd(\pi_1^{r_1}, \pi_2^{r_2}\pi_3^{r_3} \cdots \pi_n^{r_n}) = 1.$$

10. Let $a_1, a_2, \ldots, a_n$ be n distinct elements of a field F and let $r_1, r_2, \ldots, r_n$ be (not necessarily distinct) positive integers. Explain why $(x - a_1)^{r_1}$ and $(x - a_2)^{r_2}(x - a_3)^{r_3} \cdots (x - a_n)^{r_n}$ are relatively prime.

5.9 Partial Fractions

If F is a field we can apply the construction outlined in Section 4.8 to the polynomial ring $F[x]$ and thus obtain the field of all quotients α/β, with α and β in $F[x]$ and $\beta \neq 0$.

NOTATION Field $F(x)$ of Quotients of Polynomials

If x is an indeterminate over a field F, the field
$$\{\alpha/\beta : \alpha, \beta \in F[x], \beta \neq 0\}$$
is denoted by $F(x)$.

It can be shown that the field $F(x)$ is essentially (in the sense of isomorphism) the smallest field in which x and all the elements of F are present. This justifies our calling $F(x)$ the ***field extension*** of F by x. In Section 5.10, field extensions of F by finite sets of elements (that may be algebraic or transcendental over F) will be discussed.

Unique factorization into primes in the polynomial ring $F[x]$ enables one to express any quotient of polynomials in $F(x)$ uniquely as a finite sum of special elements to be described below. This representation has important applications, one of which is in the calculus topic of integration of rational functions.

DEFINITION

Prime Power Denominator Term

A **ppd-term** is an element γ/π^k of $F(x)$ with k a positive integer, π a monic prime in $F[x]$, and $\deg \gamma < \deg \pi$.

Using the lemmas that follow, we shall show that every quotient of polynomials α/β of $F(x)$ is the sum of a polynomial in $F[x]$ and a finite number of ppd-terms.

LEMMA 1 Let α/β be in $F(x)$ and let $\beta = \beta_1\beta_2$ with $\gcd(\beta_1, \beta_2) = 1$. Then there exist α_1 and α_2 in $F[x]$ such that

$$\frac{\alpha}{\beta} = \frac{\alpha_1}{\beta_1} + \frac{\alpha_2}{\beta_2},$$

Proof Since $\beta_1$and β_2 are relatively prime, every α of $F[x]$ is a linear combination $\alpha_1\beta_2 + \alpha_2\beta_1$, with α_1 and α_2 in $F[x]$. Then

$$\frac{\alpha}{\beta} = \frac{\alpha_1\beta_2 + \alpha_2\beta_1}{\beta_1\beta_2} = \frac{\alpha_1}{\beta_1} + \frac{\alpha_2}{\beta_2}.$$ ■

LEMMA 2 Let β be a polynomial of positive degree in $F[x]$ and let s be a positive integer. Then an element α/β^s of $F(x)$ is expressible as

$$\frac{\alpha}{\beta^s} = \gamma_0 + \frac{\gamma_1}{\beta} + \frac{\gamma_2}{\beta^2} + \cdots + \frac{\gamma_s}{\beta^s} \tag{1}$$

with γ_i in $F[x]$ and either $\gamma_i = 0$ or $\deg \gamma_i < \deg \beta$ for $0 \le i \le s$.

Proof Theorem 3 of Section 5.3 tells us that there exist γ_i in $F[x]$ such that either $\gamma_i = 0$ or $\deg \gamma_i < \deg \beta$ and

$$\alpha = \gamma_0\beta^s + \gamma_1\beta^{s-1} + \cdots + \gamma_{s-1}\beta + \gamma_s. \tag{2}$$

Division of both sides of equation (2) by β^s leads to the desired equation (1). ■

We are now ready for the main theorem of this section.

THEOREM 1 **Partial Fraction Decomposition**

Every α/β of $F(x)$ is expressible as a sum of a polynomial in $F[x]$ and a finite number of ppd-terms γ/π^m (with π a monic prime and $\deg \gamma < \deg \pi$).

Proof If deg $\beta = 0$, α/β is a polynomial. Hence let deg $\beta > 0$. The leading coefficient of β is a nonzero element of the field F and can be canceled from both the numerator and the denominator of α/β; thus, we may assume that β is monic.

Then it follows from the Unique Factorization Theorem of Section 5.8 that

$$\beta = (\pi_1)^{n_1}(\pi_2)^{n_2} \cdots (\pi_r)^{n_r},$$

with the π_i distinct monic primes in $F[x]$.

If $r > 1$, we let $\beta_1 = (\pi_1)^{n_1}$ and $\beta' = (\pi_2)^{n_2} \cdots (\pi_r)^{n_r}$. Then β_1 and β' are relatively prime, and Lemma 1 tells us that $\alpha/\beta = (\alpha_1/\beta_1) + (\alpha'/\beta')$, with α_1 and α' in $F[x]$. If $r > 2$, α'/β' is broken up similarly. Proceeding in this way, we express α/β in the form

$$\frac{\alpha}{\beta} = \frac{\alpha_1}{\pi_1^{n_1}} + \frac{\alpha_2}{\pi_2^{n_2}} + \cdots + \frac{\alpha_r}{\pi_r^{n_r}}. \tag{3}$$

Then each term on the right side of (3) can be expressed as a sum of the form shown in (1), and this gives us the desired sum for α/β. ∎

In applications of this partial fraction decomposition (for example, to integration theory), the field F of coefficients is usually the field **R** of real numbers or the field **C** of complex numbers. Therefore, we make some further observations for these fields.

Theorem 4 of Section 5.8 tell us that deg $\pi = 1$ for every prime π in $\mathbf{C}[x]$ and hence in every ppd-term γ/π^m of $\mathbf{C}(x)$. Similarly, we see from Theorem 6 of Section 5.8 that deg π is 1 or 2 for every prime π in $\mathbf{R}[x]$ and in each ppd-term γ/π^m of $\mathbf{R}(x)$.

Hence every ppd-term of $\mathbf{R}(x)$ is of one of the forms

$$a(x+b)^{-m}, \qquad (ax+b)(x^2+cx+d)^{-m} \tag{4}$$

with a, b, c, and d real numbers and m a positive integer. In calculus texts it is shown that an integral of any function of one of these forms can be expressed in terms of polynomials, logarithms, and inverse trigonometric functions. It then follows from the partial fraction representation (Theorem 1) that an integral of each α/β of $\mathbf{R}(x)$ is expressible in this manner.

Problems

1. In $\mathbf{Q}(x)$, express each of the following as a finite sum of ppd-terms.
   (a) $(5x + 7)/(x - 2)(x + 3)$.
   (b) $(2x^3 - 5x^2 - 8)/(x + 1)^2(x^2 + 4)$.
2. Express $x^5/(x^2 + 1)(x^2 + 9)$ as the sum of a polynomial in $\mathbf{Q}[x]$ and a finite number of ppd-terms in $\mathbf{Q}(x)$.
3. Express $(x^3 + \bar{2}x)/(x - \bar{1})(x + \bar{1})$ as a sum of a polynomial in $\mathbf{Z}_5[x]$ and a finite number of ppd-terms in $\mathbf{Z}_5(x)$.

4. In $\mathbf{Z}_7(x)$, express $(\bar{2}x + \bar{1})/(x^2 + \bar{3})(x^2 + \bar{5})$ as a sum of ppd-terms. [*Warning:* Each of $x^2 + \bar{3}$ and $x^2 + \bar{5}$ is reducible in $\mathbf{Z}_7[x]$.]

5. Let r and s be distinct elements of a field F. In $F(x)$, show that
$$[(x - r)(x - s)]^{-1} = [(r - s)(x - r)]^{-1} + [(s - r)(x - s)]^{-1}.$$

6. Let r, s, and t be three distinct elements of a field F. In $F(x)$, show that
$$[(x - r)(x - s)(x - t)]^{-1} = a(x - r)^{-1} + b(x - s)^{-1} + c(x - t)^{-1},$$
where $a = [(r - s)(r - t)]^{-1}$, $b = [(s - r)(s - t)]^{-1}$, and $c = [(t - r)(t - s)]^{-1}$.

7. Generalize on Problems 5 and 6.

★ 8. Let $\alpha(x) = (x - r_1)(x - r_2) \cdots (x - r_n)$, with the r_i distinct elements of a field F, and let $\alpha'(x)$ be the derivative of $\alpha(x)$. (See Definition 8 of Section 5.2.) In the field $F(x)$, show that
$$\frac{1}{\alpha(x)} = \frac{1}{\alpha'(r_1)(x - r_1)} + \frac{1}{\alpha'(r_2)(x - r_2)} + \cdots + \frac{1}{\alpha'(r_n)(x - r_n)}.$$

★ 9. Let $\alpha(x)$ and $\alpha'(x)$ be as in Problem 8. Show that
$$\frac{\alpha'(x)}{\alpha(x)} = \frac{1}{x - r_1} + \frac{1}{x - r_2} + \cdots + \frac{1}{x - r_n}.$$

10. For each a in $\{1, 5, 7, 11, 13\}$, show that
$$\frac{a}{72} = \frac{b}{2} + \frac{c}{4} + \frac{d}{8} + \frac{e}{3} + \frac{f}{9}$$
with $|b|$, $|c|$, and $|d|$ in $\{0, 1\}$ and $|e|$ and $|f|$ in $\{0, 1, 2\}$.

11. Define an analogue for the rational numbers $\mathbf{Q}$ of a ppd-term in $F(x)$, and then state an analogue of Theorem 1 above in the form of a result on partial fraction decomposition of rational numbers.

5.10 Extension Fields

The mathematician, carried along on his flood of symbols, dealing apparently with purely formal truths, may still reach results of endless importance for our description of the physical universe. — **Karl Pearson**

In every field E, there is a smallest subfield P, the prime subfield. We know all the prime subfields, since P is isomorphic to the rational numbers $\mathbf{Q}$ when E has characteristic 0 and P is isomorphic to $\mathbf{Z_p}$ when E has prime characteristic p. This suggests that an approach to studying fields is to investigate methods of extending a given field F into a larger one.

There are two methods that use the integral domain $F[x]$ of polynomials over F. One is to apply the technique of Section 4.8 to $F[x]$ and thus obtain the field $F(x)$ of polynomial fractions α/β. Another method of extending F is to apply to $F[x]$ the technique used to construct the finite fields $\mathbf{Z_p}$ from the integers $\mathbf{Z}$. Specifically, we prove below that, when π is a prime in $F[x]$, the quotient ring $E = F[x]/(\pi)$ is a field having a subfield isomorphic to F.

Given a polynomial α of positive degree in $F[x]$, the latter technique can be used to extend F so that α factors into first-degree polynomials over the new field.

THEOREM 1 Quotient Rings $F[x]/(\beta)$

Let β be a polynomial of degree d in $F[x]$.

(a) The elements of $E = F[x]/(\beta)$ are the cosets $\bar{\rho} = \rho + (\beta)$ with either $\deg \rho < d$ or $\rho = 0$.

(b) If F has q elements, E has q^d elements.

(c) If β is a prime, E is a field that has a subfield $\bar{F}$ isomorphic to F.

Proof For each α in $F[x]$, let $\bar{\alpha}$ denote the ideal coset $\alpha + (\beta)$. The division algorithm gives us $\alpha = \gamma\beta + \rho$ with either $\deg \rho < \deg \beta = d$ or $\rho = 0$. Since the natural map $\alpha \mapsto \bar{\alpha}$ preserves additions and multiplications,

$$\bar{\alpha} = \overline{\gamma\beta + \rho} = \bar{\gamma}\bar{\beta} + \bar{\rho}.$$

Since $\bar{\beta} = \bar{0}$ in E, this means that $\bar{\alpha} = \bar{\rho}$ with

$$\rho = r_{d-1}x^{d-1} + \cdots + r_1 x + r_0, \qquad r_i \text{ in } F. \tag{1}$$

The representation (1) is unique, since $\bar{\rho} = \bar{\rho}_1$ if and only if $\beta | (\rho - \rho_1)$ and since the difference $\rho - \rho_1$ of two polynomials of the form (1) cannot be a multiple of β unless $\rho - \rho_1 = 0$.

If F is a finite field with q elements, there are q choices for each of the d coefficients $r_{d-1}, \ldots, r_0$ in (1) and hence q^d elements (cosets $\bar{\rho}$) in the ring E.

Now let β be a prime. If $\bar{\alpha} \neq 0$, α is not a multiple of the prime β, and it follows that $\gcd(\alpha, \beta) = 1$. Then

$$\xi\alpha + \eta\beta = 1, \tag{2}$$

with ξ and η in $F[x]$. Applying the natural map, we have

$$\bar{\xi}\bar{\alpha} + \bar{\eta}\bar{\beta} = \bar{1}.$$

But $\bar{\beta} = \bar{0}$ so $\bar{\xi}\bar{\alpha} = \bar{1}$. Therefore, every nonzero $\bar{\alpha}$ of E is an invertible and E is a field.

Now let a be in F. Then a is also in $F[x]$ and $\bar{a}$ is in E. Let $\bar{F}$ consist of the $\bar{a}$ for all a in F. It is easily seen that $\bar{F}$ is a subfield in E and that the mapping θ with $a \mapsto \bar{a}$ is an isomorphism from F to $\bar{F}$. (We note that θ is the modification of the natural map from $F[x]$ to E in which the domain is restricted to the subset F of $F[x]$.) ■

EXAMPLE 1 Let $\mathbf{R}$ be the field of real numbers, $\beta = x^2 + 1$, and $E = \mathbf{R}[x]/(\beta)$. Here we show that E is isomorphic to the complex numbers $\mathbf{C}$. Theorem 1 tells us that the elements of E are the cosets

$$\bar{\rho} = \overline{a + bx} = \bar{a} + \bar{b}\bar{x},$$

with a and b any real numbers. Since $\bar{\beta} = \bar{0}$,

$$\overline{x^2 + 1} = (\bar{x})^2 + \bar{1} = \bar{0}. \tag{3}$$

With the help of this equation, it is easy to see that the bijection θ from E to $\mathbf{C}$ with

$$\theta(\bar{a} + \bar{b}\bar{x}) = a + b\mathbf{i}$$

preserves additions and multiplications and hence is an isomorphism.

The polynomial equation $x^2 + 1 = 0$ does not have a root in the real numbers $\mathbf{R}$. Even if we did not know of the complex numbers, we could produce a field $E = \mathbf{R}[x]/(x^2 + 1)$ in which the coset $\bar{x} = x + (x^2 + 1)$ is a root. [See equation (3) in Example 1.] The polynomial $x^2 + 1$ is a prime in $\mathbf{R}[x]$, but in $E[x]$ it factors as a product $(x - \bar{x})(x + \bar{x})$ of first-degree polynomials.

We now generalize on this example. Let F be any field and let α be any polynomial of positive degree in $F[x]$. If α does not have a zero in F, we can choose a prime divisor π of α in $F[x]$ and construct the field $E = F[x]/(\pi)$, in which there is a subfield F_1 isomorphic to F. To avoid complicating the terminology, we act as if F_1 were F; that is, we assume that F is a subfield in E.

Then the coset $\bar{x} = x + (\pi)$ is a zero of π, and hence of α, in $E[x]$. The Factor Theorem then tells us that $\alpha = (x - \bar{x})\beta$, with β in $E[x]$. If β does not have a zero in E (or equivalently, a first-degree factor in $E[x]$) and $\deg \beta > 0$, one can give β the treatment accorded α. Continuing in this way, one shows that an extension of F exists over which α factors into first-degree polynomials.

We now introduce some terminology applying to extension fields.

DEFINITION

Field Extension

Let F be a subfield in E and let $s_1, s_2, \ldots, s_n$ be elements of E. Then $F(s_1, s_2, \ldots, s_n)$ denotes the smallest subfield S in E such that each s_i is in S and each element of F is in S. The field S is called the **field extension** of F by the s_i in E.

If x is an element of E that is transcendental over F (that is, if x is an indeterminate over F), then $F(x)$ consists of all polynomial fractions α/β, with α and $\beta \neq 0$ in $F[x]$; such an $F(x)$ is called a ***simple transcendental extension*** of F. This definition of $F(x)$ is consistent with Notation 1 in Section 5.9.

Let s be algebraic over F. We show below that the polynomial ring $F[s]$ is a field and hence is the same as the field extension $F(s)$; such an $F(s)$ is a ***simple algebraic extension*** of F.

If an α of $F[x]$ is the product

$$a(x - s_1)(x - s_2) \cdots (x - s_n),$$

with a in F and the s_i in E, the algebraic extension $F(s_1, s_2, \ldots, s_n)$ is called a ***splitting field*** (or a ***root field***) over F for α. It can be shown that any two splitting fields over F for α are isomorphic.

By definition, an element s of an extension E of F is algebraic over F if and only if s is a zero of a nonzero polynomial α in $F[x]$. If s is algebraic over F, it follows from Theorem 4 of Section 5.7 that there is a unique monic irreducible π in $F[x]$ with the property that s is a zero in E of an α in $F[x]$ if and only if $\pi | \alpha$; this π is called the ***minimal polynomial for s over F***. If $\deg \pi = d$, one says that s is ***algebraic of degree d over F***.

THEOREM 2 Simple Algebraic Extension

Let s be algebraic of degree d over F and let π be its minimal polynomial. Then:

(a) $F[s]$ is isomorphic to $F[x]/(\pi)$.
(b) The polynomial extension $F[s]$ is the same as the field extension $F(s)$.
(c) Every element of $F(s)$ is of the form

$$a_{d-1}s^{d-1} + \cdots + a_1 s + a_0, \qquad a_i \text{ in } F.$$

Proof The elements of $F[s]$ are the $\alpha(s)$ for all α in $F[x]$. Clearly, $\alpha_1(s) = \alpha_2(s)$ if and only if s is a zero of $\alpha_1 - \alpha_2$. By Theorem 4 of Section 5.7, s is a zero of $\alpha_1 - \alpha_2$ if and only if $\pi | (\alpha_1 - \alpha_2)$, that is, if and only if α_1 and α_2 are in the same coset of (π) in $F[x]$.

Then the mapping f from $F[s]$ to $F[x]/(\pi)$ with $\alpha(s) \mapsto \bar{\alpha} = \alpha + (\pi)$ is easily seen to be a bijection that preserves additions and multiplications; thus f is a ring isomorphism. Since π is a prime in $F[x]$, Theorem 1 tells us that $F[x]/(\pi)$ is a field. Hence the isomorphic $F[s]$ is also a field. This makes $F[s]$ the smallest extension field $F(s)$ of F in which s is present.

Part (c) follows from the fact that every $\alpha(s)$ of $F[s]$ may be rewritten as $\rho(s)$, where ρ is the remainder in the division of α by π. ■

COROLLARY 1 Simple Algebraic Extension of a Finite Field

Let F have q elements, let s be algebraic of degree n over F, and let $m = q^n$. Then $F[s]$ has m elements, and the minimal polynomial for s over F is a divisor of $x^m - x$.

COROLLARY 2 Number of Elements in a Finite Field

Let F have m elements and characteristic p. Then $m = p^n$ with n a positive integer.

COROLLARY 3 **Isomorphic Algebraic Extensions**

Let s and t (be algebraic and) have the same minimal polynomial over F. Then $F[s]$ and $F[t]$ are isomorphic.

The proofs of these corollaries are left to the reader as Problems 5 through 7 of this section.

Corollary 2 to Theorem 2 states that a finite field has p^n elements, where p and n are positive integers and p is a prime. Theorem 1 tells us how to construct such a field, if we can find an irreducible polynomial π of degree n in $\mathbf{Z_p}[x]$. Corollary 1 indicates that such a π should be sought among the factors of $x^m - x$, where $m = p^n$.

It can be proved that for every positive prime p in $\mathbf{Z}$ there exist prime polynomials of all positive degrees in $\mathbf{Z_p}[x]$ and hence fields with p^n elements for all positive integers n. [See G. J. Simmons, "The Number of Irreducible Polynomials of Degree n over $GF(p)$," *The American Mathematical Monthly* 77 (1970); 743–45.]

It can also be proved that any two finite fields with the same number of elements are isomorphic. A finite field is called a ***Galois field.*** If a Galois field has p^n elements, it is designated as $GF(p^n)$.

We now illustrate the construction of Galois fields.

EXAMPLE 2 **Field with 8 Elements**

Here we construct a field with 8 elements and discuss some of its properties. Since $8 = 2^3$, we start with a prime field $F = \{0, 1\}$ of characteristic 2 and look for an irreducible polynomial of degree 3 in $F[x]$. One sees that $\pi = x^3 + x + 1$ has no zeros in F, by noting that $\pi(0) = 1 = \pi(1)$. It follows that π has no first-degree divisors in $F[x]$. This and the fact that $\deg \pi = 3$ imply that π is irreducible.

Using Theorem 1, one now sees that $E = F[x]/(\pi)$ is a field with $2^3 = 8$ elements, each of which is a coset $\bar{\rho} = \rho + (\beta)$ with either $\deg \rho \le 2$ or $\rho = 0$.

We choose a notation for the elements of E that allows us to consider E to be an extension of F. Specifically, we let the elements $\bar{0}$, $\bar{1}$, and $\bar{x}$ of E be denoted by 0, 1, and c, respectively. Then the set for E is

$$\{0, 1, c, c + 1, c^2, c^2 + 1, c^2 + c, c^2 + c + 1\}.$$

Addition is performed easily in E using the fact that $y + y = 0$ for all y in E; that is, E has characteristic 2. For example,

$$(c^2 + c) + (c^2 + 1) = c + 1.$$

Since $\bar{\pi} = 0$ in E, we have $c^3 + c + 1 = 0$; that is, c is a zero in E of π. The multiplicative group V of E has 7 elements. Since 7 is a prime, V is cyclic with c one of its 6 generators. Using the fact that $c^3 = c + 1$, we express the elements of V as powers of c in the following table:

n	0	1	2	3	4	5	6
c^n	1	c	c^2	$c+1$	c^2+c	c^2+c+1	c^2+1

Lagrange's Theorem tells us that c^j is a zero of $x^7 - 1$ for $j = 0, 1, \ldots, 6$. Then it follows from Theorem 3 of Section 5.4 that in $E[x]$ we have

$$x^7 - 1 = (x-1)(x-c)(x-c^2)(x-c^3)(x-c^4)(x-c^5)(x-c^6). \tag{4}$$

Since π is the minimal polynomial for c over F and $x^7 - 1$ is a polynomial in $F[x]$ with c as a zero, it follows that π is a divisor of $x^7 - 1$. (See the corollary of Theorem 4 of Section 5.7.) In fact,

$$x^7 - 1 = (x-1)(x^3+x+1)(x^3+x^2+1).$$

This and the factorization (4) imply that c^j, for $1 \le j \le 6$, is a zero of either π or $\pi' = x^3 + x^2 + 1$. Substitution shows that c, c^2, and c^4 are zeros of π while c^3, c^5, and c^6 are zeros of π'.

Problems

1. In $\mathbf{Z}_3[x]$, let $\beta = x^3 + \bar{2}x + 1$ and $I = (\beta)$.
   (i) Show that β has no zero in $\mathbf{Z}_3 = \{\bar{0}, \bar{1}, \bar{2}\}$.
   (ii) Explain why $\mathbf{Z}_3[x]/I$ is a field F.
   (iii) How many elements are there in the F of (ii)?
2. Find a monic prime π in $\mathbf{Q}[x]$ such that $\mathbf{Q}[x]/(\pi)$ is isomorphic to the field extension $\mathbf{Q}(\sqrt{5})$.
3. Describe the construction of a field with 25 elements.
4. For which integers m in $\{1, 2, 3, \ldots, 20\}$ is there
   (i) a field with m elements?
   (ii) an integral domain with m elements?
5. Let s be algebraic of degree n over a field F with q elements. Prove that $F[s]$ has q^n elements and that the minimal polynomial for s over F is a divisor of $x^m - x$, where $m = q^n$. (This is the proof of Corollary 1 to Theorem 2.)
6. Prove that the number of elements in a finite field F is of the form p^n, where p is the characteristic of F and n is a positive integer (and thus prove Corollary 2 to Theorem 2).
7. Let s and t have the same minimal polynomial over F. Prove that $F[s]$ and $F[t]$ are isomorphic (and thus prove Corollary 3 to Theorem 2).
8. Let s be a zero in an extension field E over F of a monic prime π in $F[x]$. Show that π is the minimal polynomial for s over F.
9. Let F be a field with characteristic 5, and let θ be the mapping from F to itself with $\theta(x) = x^5$. Show the following.
   (a) θ is an endomorphism (that is, a homomorphism from F to itself).
   (b) If F is finite, θ is an automorphism and every element of F has a unique fifth root in F.

10. Do the analogue of Problem 9 in which F is a field of characteristic p and $\theta(x) = x^p$.

11. Let s and t be complex numbers. Use the fact that in $\mathbf{C}[x]$

$$(x + s + t) \mid (x^3 - 3stx + s^3 + t^3)$$

to find the minimal polynomial for $\sqrt[3]{2} + \sqrt[3]{4}$ over $\mathbf{Q}$.

12. Find the minimal polynomial for $\sqrt[3]{5} + \sqrt[3]{25}$ over $\mathbf{Q}$.

13. Show that the minimal polynomial for $\sqrt{7} + \mathbf{i}$ is
   (i) $x^2 - 2\sqrt{7}x + 8$ over the real numbers $\mathbf{R}$.
   (ii) $x^4 - 12x^2 + 64$ over the rational numbers $\mathbf{Q}$.

14. Find the minimal polynomial for $\sqrt{7} + \mathbf{i}$ over $\mathbf{Q}(\mathbf{i})$.

15. Let F be a field with 9 elements, let $P = \{0, 1, -1\}$ be its prime subfield, and let V be the multiplicative group of F. Explain why the following are true.
   (i) $x^8 = 1$ for each nonzero x in F.
   (ii) $x^9 = x$ and hence $x^9 - x = 0$ for all x in F.
   (iii) $x(x - 1)(x + 1)(x^2 + 1)(x^2 - x - 1)(x^2 + x - 1) = 0$ for all x in F.
   (iv) There is an element c in F such that $c^2 = -1$ and $F = \{0, 1, -1, c, -c, c - 1, -c + 1, -c - 1, c + 1\}$.
   (v) In V, the element $c + 1$ has order 8, and hence V is cyclic. [*Hint:* Find $(c + 1)^2$ and $(c + 1)^4$, using $c^2 = -1$.]

16. Let P be the prime subfield in a finite field F with m elements. Explain why F is the splitting field over P of $x^m - x$.

17. Let α be in $F[x]$ and let α' be the derivative of α. Prove that $\gcd(\alpha, \alpha') = 1$ if and only if there is no extension field E over F in which α has a zero with multiplicity greater than 1. (See Problem 29 of Section 5.6.)

18. Show that none of the following polynomials in $\mathbf{C}[x]$ has a zero of multiplicity greater than 1. [*Hint:* See the previous problem.]
   (i) $x^2 + 2x + 2$.
   (ii) $x^3 + 3x^2 + 6x + 6$.
   (iii) $x^4 + 4x^3 + 12x^2 + 24x + 24$.
   (iv) $x^5 + 5x^4 + 20x^3 + 60x^2 + 120x + 120$.

19. Show that $x^3 - 3x^2 + 6x - 6$ does not have a zero with multiplicity greater than 1.

20. Generalize on Problems 18 and 19.

★ 21. Let F be a finite field with 256 elements. How many monic irreducible polynomials of degree 8 are there with coefficients in F?

5.11 Equations of Degree 2, 3, or 4

For every positive integer n, a nonzero complex number

$$r(\cos a + \mathbf{i} \sin a), \qquad a \in \mathbf{R}, \quad r \in \mathbf{R}^+$$

has exactly n nth roots given by

$$\sqrt[n]{r}\{\cos[(a + 2k\pi)/n] + \mathbf{i} \sin[(a + 2k\pi)/n]\}; \quad k = 0, 1, \ldots, n - 1,$$

where $\sqrt[n]{r}$ denotes the unique positive nth root of r. The only nth root of 0 is 0 itself. (These facts are part of the well-known Theorem of DeMoivre; they also follow readily from Theorem 3 of Section 2.6.)

Hence the field $\mathbf{C}$ of complex numbers is closed under extraction of nth roots as well as under the four rational operations (addition, subtraction, multiplication, and division by nonzero elements.)

DEFINITION 1

Solution in Radicals

A polynomial equation $\alpha = 0$ is **solvable in radicals** if each root is expressible in terms of the coefficients, using only the rational operations and extraction of nth roots (with each n in $\mathbf{Z}^+$).

This section deals with methods for solution in radicals of polynomial equations with degree 2, 3, or 4. We have previously mentioned that it is not possible for such a method to apply to all equations with degree 5 or any higher degree. (See the biographical notes on Ruffini, Abel, and Galois and Section 5.12 below.)

In the rest of this section, F always designates a field that is closed under extraction of nth roots and has characteristic different from 2 and from 3. The methods given apply to all such fields F and hence to the complex numbers $\mathbf{C}$. Since the characteristic of F is not 2 or 3, we have $2 \cdot 1 \neq 0$ and $3 \cdot 1 \neq 0$ in F. Then $4 \cdot 1 = (1 + 1)(1 + 1) \neq 0$, since F has no 0-divisors. In F, we designate these nonzero elements $2 \cdot 1$, $3 \cdot 1$, and $4 \cdot 1$ by 2, 3, and 4, respectively.

If f is in F, the roots of

$$x^d + f = 0 \tag{1}$$

are the dth roots of $-f$; hence (1) is solvable in radicals. One approach to solving a general monic equation

$$x^d + a_{d-1}x^{d-1} + \cdots + a_1x + a_0 = 0, \qquad a_i \in F, \tag{2}$$

is to seek techniques for reducing (2) to (1). Such techniques are possible for $d \in \{2, 3, 4\}$, but we shall take only the first step in this direction.

For d in $\{2, 3, 4\}$, $h = a_{d-1}d^{-1}$ denotes an element of F. Then the substitution

$$x = y - h = y - a_{d-1}d^{-1}$$

in equation (2) leads to an equation

$$y^d + b_{d-2}y^{d-2} + \cdots + b_1y + b_0, \qquad b_i \in F, \tag{3}$$

in which the coefficient of y^{d-1} turns out to be 0. If the roots of (3) can be found, one subtracts h from each of them to obtain the roots of (2).

EXAMPLE 1 Let us find all roots of $x^2 + 6x + 13 = 0$ in $\mathbf{C}$. Here $d = 2$, $h = 6/2 = 3$, and the substitution $x = y - 3$ leads to

$$\begin{aligned} (y-3)^2 + 6(y-3) + 13 &= 0 \\ (y^2 - 6y + 9) + (6y - 18) + 13 &= 0 \\ y^2 + 4 &= 0 \\ y^2 &= -4. \end{aligned}$$

Hence, $y = \pm 2\mathbf{i}$ and $x = y - 3 = -3 \pm 2\mathbf{i}$.

EXAMPLE 2 Now we solve $ax^2 + bx + c = 0$, with $a \neq 0$, in F. Applying the technique of Example 1 to

$$x^2 + (b/a)x + (c/a) = 0,$$

we substitute $x = y - (b/2a)$ and find, after a few steps, that $y = \pm\sqrt{b^2 - 4ac}/2a$. Then

$$x = y - \left(\frac{b}{2a}\right) = \frac{-b \pm \sqrt{b^2 - 4ac}}{2a}.$$

Next we turn to monic cubic equations, that is, equations of the form

$$x^3 + a_2 x^2 + a_1 x + a_0 = 0.$$

The substitution $x = y - (a_2/3)$ leads to an equation of the form

$$y^3 + py + q = 0. \tag{4}$$

Since it took about 2000 years to go from solving quadratics to solving cubics, one should not be surprised that at this point we have to pull a rabbit out of a hat. Our approach is to factor the left side of (4) into first-degree polynomials with the help of the factorization

$$y^3 - 3rsy + (r^3 + s^3) = (y + r + s)(y + wr + w^2 s)(y + w^2 r + ws), \tag{5}$$

where $w = (-1 + \sqrt{-3})/2$. [In $\mathbf{C}$, one can rewrite w as $(-1 + \mathbf{i}\sqrt{3})/2$.] The factorization in (5) is easily verified by expanding the right side and using the fact that $w^2 + w + 1 = 0$.

Now we have to choose r and s in (5) so as to make the left side of (5) the same as the left side of (4). That is, we want r and s to satisfy

$$-3rs = p \quad \text{and} \quad r^3 + s^3 = q.$$

The first of these equations tells us that $s = -p/3r$; substituting this in the second equation leads to

$$\begin{aligned} r^3 + (-p/3r)^3 &= q \\ r^3 - (p^3/27r^3) &= q \\ r^6 - (p^3/27) &= qr^3 \\ (r^3)^2 - qr^3 - (p^3/27) &= 0. \end{aligned}$$

Now the quadratic formula (given in Example 2) leads to

$$r^3 = [q \pm \sqrt{q^2 + (4p^3/27)}]/2. \tag{6}$$

We do not need all the solutions for r and s; one pair will enable us to factor the left side of (4). Hence we let r be a cube root of the right side of (6), with one of the two choices in the $\pm$ sign. Then we let $s = -p/3r$ and equate to zero the first-degree factors on the right side of (5). In this way we find the roots of (4) in the form

$$-(r + s), \qquad -(wr + w^2s), \qquad -(w^2r + ws). \tag{7}$$

EXAMPLE 3 Next we find all the roots of $x^3 - 3x + 1 = 0$ in **C**. The technique used above on the general cubic tells us that we shall have the factorization

$$x^3 - 3x + 1 = (x + r + s)(x + wr + w^2s)(x + w^2r + ws)$$

if r and s satisfy

$$-3rs = -3, \; r^3 + s^3 = 1.$$

Substituting $s = -3/(-3r) = 1/r$ in $r^3 + s^3 = 1$ leads to

$$r^3 + (1/r^3) = 1$$
$$(r^3)^2 + 1 = r^3$$
$$(r^3)^2 - r^3 + 1 = 0.$$

This quadratic for r^3 is satisfied by $r^3 = (1 + \sqrt{1 - 4})/2$. Hence we may let

$$r = \sqrt[3]{(1 + \mathbf{i}\sqrt{3})/2}, \qquad s = 1/r,$$

and then the roots of $x^3 - 3x + 1 = 0$ are given by the expressions in (7).

Finally we consider monic fourth-degree polynomial equations, that is, equations of the form

$$x^4 + ax^3 + bx^2 + cx + d = 0. \tag{8}$$

Here we do not make the substitution $x = y - (a/4)$, which would knock out the y^3 term. Instead, our approach is to factor the left side of (8) into quadratics. The first step is to attempt to rewrite (8) in the form

$$(x^2 + hx + k)^2 - (ux + v)^2 = 0. \tag{9}$$

Then (9) can be put in the desired form

$$[(x^2 + hx + k) + (ux + v)][(x^2 + hx + k) - (ux + v)] = 0 \tag{10}$$

and solved by two applications of the quadratic formula.

What is necessary to convert (8) into (9)? Expanding the left side of (9) and collecting like terms, we see that (8) goes into (9) if we can choose h, k, u, and v so as to satisfy the simultaneous conditions

$$a = 2h, \quad b = h^2 + 2k - u^2, \quad c = 2hk - 2uv, \quad d = k^2 - v^2.$$

The first of these conditions shows that we should let $h = a/2$. We substitute $h = a/2$ in the other three conditions and rewrite them as

$$u^2 = (a^2/4) + 2k - b, \quad 2uv = ak - c, \quad v^2 = k^2 - d. \tag{11}$$

Since $4u^2v^2 - (2uv)^2 = 0$, the conditions in (11) can only be satisfied if

$$4[(a^2/4) + 2k - b][k^2 - d] - [ak - c]^2 = 0.$$

Simplifying this equation, we find that k has to satisfy

$$8k^3 - 4bk^2 + (2ac - 8d)k + (4bd - a^2d - c^2) = 0. \tag{12}$$

Equation (12) is called the ***resolvent cubic*** for the quartic (8).

Now one can show that the quartic (8) factors as in (10) if one chooses k to be a root of (12) and lets

$$h = a/2, \quad u = \sqrt{h^2 + 2k - b}, \quad v = \sqrt{k^2 - d}. \tag{13}$$

The details are left to the reader.

One may note that this solution for the quartic equation involves solutions of a cubic and two quadratics.

EXAMPLE 4 Let $\alpha = x^4 + 2x^3 + 3x^2 + 2x + 2$. Here we find all the roots of $\alpha = 0$ in the field $\mathbf{C}$ of complex numbers. First we try the easy (when it works) method of seeking rational roots by using Theorem 2 of Section 5.5. The result tells us that the only possibilities for rational roots are the integers -2, -1, 1, and 2. Substitution shows that none of these integers is a zero of α; hence we turn to the method for general quartics.

The resolvent cubic equation (12) for our given α is

$$2k^3 - 3k^2 - 2k + 3 = 0.$$

To solve $\alpha = 0$, we need only one root of this cubic. Hence we hope for a rational root. One of the possibilities for rational roots is 3/2, and we are fortunate that it actually checks out to be a root; so we let $k = 3/2$.

In equation (10) above, h is half of the coefficient of x^3 in α; thus $h = 2/2 = 1$. Using $h = 1$ and $k = 3/2$ in (13), we calculate $u = \sqrt{1 + 3 - 3} = 1$ and $v = \sqrt{(9/4) - 2} = 1/2$. Then we substitute these values for h, k, u, and v in (10), and $\alpha = 0$ is converted into

$$(x^2 + 1)(x^2 + 2x + 2) = 0.$$

Equating each of these quadratic factors to zero and solving the two quadratic equations that result, we find that the set of zeros of $\alpha = 0$ is $\{\mathbf{i}, -\mathbf{i}, -1 + \mathbf{i}, -1 - \mathbf{i}\}$.

In elementary algebra, the discriminant of the monic quadratic equation

$$x^2 + bx + c = 0 \tag{14}$$

is $D = b^2 - 4c$. It is left to the reader as Problem 5 below to show that D equals the square of the difference of the roots of equation (14) and that, when b and c are real, one obtains information concerning the roots by knowing whether $D = 0$, $D > 0$, or $D < 0$. We now generalize to monic polynomials of arbitrary degree that can be factored in the form

$$\alpha = (x - r_1)(x - r_2) \cdots (x - r_d). \tag{15}$$

DEFINITION 2

Discriminant

The **discriminant** of the α of (15) is

$$\left[\prod_{1 \le i < j \le d} (r_i - r_j)\right]^2,$$

that is, the square of the product of the differences of the roots of $\alpha = 0$.

For example, the discriminant of $(x - r_1)(x - r_2)(x - r_3)$ is

$$[(r_1 - r_2)(r_1 - r_3)(r_2 - r_3)]^2.$$

Discriminants of monic cubics are dealt with in Problems 7 through 10 below.

Problems

1. Let a and b be real numbers with $b \neq 0$. Let $r = \sqrt{a^2 + b^2}$. Show that the square roots of $a + b\mathbf{i}$ are

$$\pm[\sqrt{(r + a)/2} + \mathbf{i}b/2\sqrt{(r + a)/2}].$$

2. Let a and b be real numbers with $b \neq 0$. Let $r = \sqrt{a^2 + b^2}$. Show that the square roots of $a + b\mathbf{i}$ are

$$\pm[(b/2\sqrt{(r - a)/2}) + \mathbf{i}\sqrt{(r - a)/2}].$$

3. Solve $z^2 - 2\mathbf{i}\sqrt{7}\,z - 24\mathbf{i} = 0$ for z in the form $a + b\mathbf{i}$. [The result in Problem 1 (or Problem 2) may be helpful.]

4. Find the roots of $x^2 + 2\sqrt{5}\,x - 12\mathbf{i} = 0$ in the form $a + b\mathbf{i}$.

5. Let $x^2 + bx + c = (x - r)(x - s)$ in $\mathbf{C}[x]$. Let $D = (r - s)^2$. Show the following.
(i) $D = b^2 - 4c$.
(ii) $D = 0$ if and only if $r = s$.
In the parts that follow, let b and c be real.
(iii) $D > 0$ if and only if r and s are real and distinct.
(iv) $D < 0$ if and only if r and s are nonreal complex conjugates.

6. For each of the following monic quadratics α, use the discriminant tests of Problem 5 to see whether the roots of $\alpha = 0$ are equal, real and distinct, or nonreal complex conjugates.
(a) $x^2 + 3x + 1$.
(b) $x^2 + 3x + 3$.
(c) $x^2 + 14x + 49$.

7. Let $x^3 + px + q = (x - r)(x - s)(x - t)$ in $\mathbf{C}[x]$. Let $D = [(r - s)(r - t)(s - t)]^2$. Show the following.
(i) $D = -(4p^3 + 27q^2)$.

(ii) $D = 0$ if and only if two of r, s, and t are equal.
In the parts that follow, let p and q be real.
(iii) $D > 0$ if and only if r, s, and t are real and distinct.
(iv) $D < 0$ if and only if one of r, s, and t is real and the other two are nonreal complex conjugates.

8. Let $\alpha = x^3 + ax^2 + bx + c$ in $\mathbf{C}[x]$, let $h = a/3$, and $\alpha(x - h) = \beta$. Show the following.
(i) β is of the form $x^3 + px + q$.
(ii) The discriminants of α and β are equal.

9. Use the result in Problem 7(i) to find the discriminant of each of the following monic cubics, and then classify the roots as in parts (ii) through (iv) of Problem 7.
(a) $x^3 - x - 1 = 0$.
(b) $x^3 - 12x + 16 = 0$.
(c) $x^3 - 5x + 2 = 0$.

10. Use the result in Problem 8 to find the discriminant of

$$x^3 - 6x^2 + 7x + 9 = 0,$$

and then classify the roots as in parts (ii)–(iv) of Problem 7.

11. Find all the roots in $\mathbf{C}$ of $x^3 + 3x - 1 = 0$.

12. Find all the roots in $\mathbf{C}$ of $x^3 + 3x - 4 = 0$.

13. Use the triple-angle formula $\cos(3\theta) = 4\cos^3\theta - 3\cos\theta$ to show that $\cos 10°$, $\cos 110°$, and $\cos 130°$ are the three roots of

$$4x^3 - 3x - (\sqrt{3}/2) = 0.$$

14. Express the roots of $4x^3 - 3x = 1/2$ in terms of cosines. (See the preceding problem.)

15. Find all the zeros of α in $\mathbf{C}$, for each of the following α's.
(a) $x^4 - 4x^3 - 6x^2 - 12x + 9$.
(b) $x^4 - 10x^3 + 26x^2 - 5x - 2$.
(c) $x^4 + 4x^3 + 6x^2 + 4x + 1$.

16. Find all the zeros of α in $\mathbf{C}$, for each of the following α's.
(a) $x^4 - x^3 + x^2 - 3x + 2$.
(b) $x^4 + x^3 - 3x^2 - 5x - 2$.

17. Show that $c^2 = da^2$, given that, in $\mathbf{C}[x]$,

$$x^4 + ax^3 + bx^2 + cx + d = (x^2 + sx + p)(x^2 + tx + p).$$

18. Let $\alpha = x^4 + ax^3 + bx^2 + cx + d$ in $\mathbf{C}[x]$ with $c^2 = da^2 \neq 0$. Show that the roots of $\alpha = 0$, when suitably numbered, satisfy $r_1r_2 = r_3r_4$.

19. Let $\alpha = x^4 + ax^3 + bx^2 + cx + d$ in $\mathbf{C}[x]$ with $4ab = a^3 + 8c$. Show that the roots of $\alpha = 0$, when suitably numbered, satisfy $r_1 - r_2 = r_3 - r_4$.

20. Let $\alpha = x^4 + x^3 - x^2 - x + 1$. Factor α as in Problem 17, and thus find all the zeros of α in $\mathbf{C}$.

21. In $\mathbf{C}[x]$, let

$$x^3 - ax^2 + bx - c = (x - r)(x - s)(x - t)$$

and $N_k = r^k + s^k + t^k$. Show the following.
(i) $N_0 = 3$ if $c \neq 0$.
(ii) $N_1 = a$.
(iii) $N_2 = a^2 - 2b$.
(iv) $N_{k+3} - aN_{k+2} + bN_{k+1} - cN_k = 0$ for $k = 0, 1, 2, \ldots$.

22. In $\mathbf{C}[x]$, let

$$x^4 - ax^3 + bx^2 - cx + d = (x - r)(x - s)(x - t)(x - u)$$

and $N_k = r^k + s^k + t^k + u^k$. Show the following.

(i) $N_0 = 4$ if $d \neq 0$.

(ii) $N_1 = a$.

(iii) $N_2 = a^2 - 2b$.

(iv) $N_3 = a^3 - 3ab + 3c$.

(v) $N_{k+4} - aN_{k+3} + bN_{k+2} - cN_{k+1} + dN_k = 0$ for $k = 0, 1, 2, \ldots$.

(The results in Problems 21 and 22 are special cases of identities, known as ***Newton's Formulas***, on the sums of powers of the roots of a polynomial equation. The general case is given in Problem 30 of the Supplementary and Challenging Problems for Chapter 5.)

23. In $\mathbf{C}[x]$, let $\alpha(x) = ax^2 + bx + c$ with $a \neq 1$ and $\beta(x) = \alpha[\alpha(x)] - x$. If r and s are distinct complex roots of the equation $ax^2 + (b - 1)x + c = 0$, show that they are also roots of the fourth-degree equation $\beta(x) = 0$, and find a quadratic equation $\gamma(x) = 0$ satisfied by the other two roots of $\beta(x) = 0$.

24. Use the preceding problem to find all solutions in $\mathbf{C}$ of the equation

$$(x^2 - x - 2)^2 - (x^2 - x - 2) - 2 - x = 0.$$

Nicolò of Brescia (Tartaglia)
1499?–1557

Tartaglia (which means "the stammerer") was a teacher of mathematics in Brescia and Venice. In 1541 he discovered a general method for solving cubic equations. He made the mistake of delaying publication, however, and instead passed the method along to **Hieronimo Cardano** (1501–1576), a physician and professor of mathematics, who published the result (with due credit to Tartaglia) in his *Ars Magna* of 1545. The method was thereafter unfairly referred to by others as Cardano's method. Cardano's student, **Ludovico Ferrari** (1522–1565), discovered the general method for solving fourth-degree equations, a method also first described in Cardano's *Ars Magna.* It was the success of these men in solving by radicals equations of degrees 3 and 4 that encouraged many to seek solutions of equations of degree higher than 4. It was not until 1799 that Ruffini first indicated that this was a futile search.

★ 5.12 Automorphisms of *E* over *F*, Insolvability of a Quintic

Mathematicians are like Frenchmen: whatever you say to them they translate into their own language, and forthwith it is something entirely different. — **Johann Wolfgang von Goethe**

Galois reduced the problem of deciding whether a given polynomial equation can be solved in radicals to that of testing whether a group of permutations of the roots of the equations is a solvable group. Here we present the first small steps of this process. We also give an equation of the fifth degree that is not solvable in radicals, together with a number of theorems (some without proof) that imply this impossibility result.

We start by considering some special automorphisms of an extension field E over F.

DEFINITION

Automorphism of E over F

Let F be a subfield in E. An **automorphism** of E over F is an automorphism θ of E such that $\theta(a) = a$ for all a in F.

THEOREM 1 **Automorphisms over F Send Roots to Roots**

Let θ be an automorphism of E over F and let r be a zero in E of an α in $F[x]$. Then the image of r under θ is also a zero of α.

Proof Let the image of an element b of E under θ be denoted by $\bar{b}$. Let

$$\alpha = a_d x^d + \cdots + a_1 x + a_0, \qquad a_i \text{ in } F.$$

Since θ preserves additions and multiplications, applying θ to the given equation

$$a_d r^d + \cdots + a_1 r + a_0 = 0$$

leads to

$$\bar{a}_d \bar{r}^d + \cdots + \bar{a}_1 \bar{r} + \bar{a}_0 = \bar{0}.$$

Since θ leaves each element of F invariant, we then have

$$a_d \bar{r}^d + \cdots + a_1 \bar{r} + a_0 = 0;$$

that is, $\alpha(\bar{r}) = 0$ as desired. ∎

Let E be a splitting field for an α of positive degree in $F[x]$. In $E[x]$, let

$$\alpha = a(x - s_1)(x - s_2) \cdots (x - s_n),$$

where a is in F and the s_i are n distinct elements of E. Let θ be an automorphism of E over F. Then it follows from Theorem 1 that

$$s_i \mapsto \theta(s_i), \qquad i = 1, 2, \ldots, n$$

is a permutation on the set $\{s_1, s_2, \ldots, s_n\}$. This leads us to the following result.

THEOREM 2 Galois Group of a Polynomial

Let E be a splitting field for an α of degree n in $F[x]$. Let α have n distinct zeros in E. Then the automorphisms of E over F form a group G isomorphic to a subgroup G' in $\mathbf{S_n}$. (The operation in G is composition of mappings.)

The proof is left to the reader as Problem 1 of this section.

The group G' of Theorem 2 is called the ***Galois group*** for α over F.

EXAMPLE 1 Next we show that the Galois group G for $\pi = x^3 + x + 1$ over $F = \{0, 1\}$ is a cyclic group of order 3. In Example 2 of Section 5.10, we saw that π has a splitting field $E = \{0, 1, c, c^2, \ldots, c^6\}$ and that, in $E[x]$,

$$\pi = (x - c)(x - c^2)(x - c^4).$$

Let θ be an isomorphism of E over F; then

$$\theta(0) = 0, \ \theta(c^k) = [\theta(c)]^k \quad \text{for } 0 \le k \le 6.$$

This shows that θ is completely determined by the value of $\theta(c)$. Theorem 1 tells us that $\theta(c)$ must be in the set $\{c, c^2, c^4\}$ of zeros of π in E.

Let α be the mapping from E to E with

$$\alpha(0) = 0, \ \alpha(c^k) = c^{2k} \quad \text{for } 0 \le k \le 6.$$

It is now straightforward to verify that α, α^2, and the identity automorphism ε are the 3 automorphisms of E over F. (This is left to the reader.) Hence the Galois group of π is cyclic and of order 3, since it is isomorphic to the automorphism group $\{\varepsilon, \alpha, \alpha^2\}$.

We next state, but do not prove, an extremely difficult result that has greatly influenced abstract algebra.

THEOREM 3 A Galois Theorem

Let F be a field of characteristic 0 and let $\alpha \in F[x]$. Then $\alpha = 0$ is solvable in radicals if and only if the Galois group of α over F is a solvable group.

It can be shown that the Galois group of $x^5 - 4x^3 - 2$ over the rational numbers $\mathbf{Q}$ is the symmetric group $\mathbf{S_5}$. [This is not proved here. A proof is indicated in Exercise 8 of Chapter IX in Chih-Han Sah, *Abstract Algebra* (New York: Academic Press, 1967).]

Assuming this fact, it follows from Theorem 3 above and Theorem 3 of Section 2.13 that

$$x^5 - 4x^3 - 2 = 0$$

is not solvable in radicals; that is, its roots cannot be expressed in terms of the coefficients by using only the operations of addition, subtraction, multiplication, division, and extraction of nth roots (with n a positive integer).

Problems

1. Let E be a splitting field for an α of degree n in $F[x]$. Prove that the automorphisms of E over F form a group G isomorphic to a subgroup G' in $\mathbf{S_n}$ (and thus prove Theorem 2).
2. (a) Is every subgroup in $\mathbf{S_2}$ solvable?
   (b) Is every subgroup in $\mathbf{S_3}$ solvable?
3. How many automorphisms are there of
   (i) $\mathbf{Q}(\sqrt{2})$ over $\mathbf{Q}$? (ii) $\mathbf{Q}(\sqrt[3]{2})$ over $\mathbf{Q}$?
4. Explain why every automorphism θ of a field F is an automorphism of F over its prime subfield P.
5. Find the order of the Galois group of $x^3 - 5$ over
   (i) the rational numbers $\mathbf{Q}$; (ii) the real numbers $\mathbf{R}$.
6. Find the order of the Galois group of $x^8 - x$ over $\mathbf{Z_2}$.
7. Let s be a zero in $E = F(s)$ of a prime π in $F[x]$. Prove that there are at least as many automorphisms of E over F as there are zeros of π in E.

Review Problems

1. Show that $\sqrt{17} + 4\mathbf{i}$ is algebraic (over $\mathbf{Z}$).
2. Show that every complex number $a + b\mathbf{i}$ is algebraic over the field $\mathbf{R}$ of real numbers.
3. Explain why every element of $\mathbf{Q}[\sqrt[3]{5}]$ can be expressed in the form $a\sqrt[3]{25} + b\sqrt[3]{5} + c$, with a, b, and c in $\mathbf{Q}$.
4. List the 10 monic primes of degree 2 in $\mathbf{Z_5}[x]$.
5. Expand $(x + \bar{2})(x + \bar{1})x(x - \bar{1})(x - \bar{2})$ in $\mathbf{Z_5}[x]$.
6. In $\mathbf{Z_{11}}[x]$, express $x^{10} - \bar{1}$ as a product of monic first-degree polynomials.
7. Show that $(x^2 + x + 1) \mid (x^6 - 2x^3 + 1)$ in $\mathbf{Z}[x]$.
8. Let $\alpha = x^2 + 5$ and $\beta = 2x^2 + 10$.
   (i) Are α and β associates in $\mathbf{Q}[x]$? Explain.
   (ii) Are α and β associates in $\mathbf{Z}[x]$? Explain.
9. Find the quotient and remainder in the division of
   (i) $x^6 + 13x^5 + 5x^4 + 52x^3 + 3x^2 + 3x - 13$ by $x^2 + 4$ in $\mathbf{Z}[x]$.
   (ii) $x^6 + \bar{3}x^5 + \bar{2}x^3 + \bar{3}x^2 + \bar{3}x + \bar{2}$ by $x^2 - \bar{1}$ in $\mathbf{Z_5}[x]$.
10. Let c and k be complex numbers and let $\bar{c}$ and $\bar{k}$ be their complex conjugates. Show that the expansion of
$$(x^2 + cx + k)(x^2 + \bar{c}x + \bar{k})$$
    has all real coefficients.
11. How many roots are there of $x^2 - x = 0$
    (i) in $\mathbf{Z_{30}}$? (ii) in $\mathbf{Z_{29}}$?
12. How many zeros are there of $x^3 - x$
    (i) in $\mathbf{Z_{30}}$? (ii) in $\mathbf{Z_{31}}$?
13. Expand $(x^2 - \bar{2})^7$ in $\mathbf{Z_7}[x]$.
14. Expand $(x^5 + x^4 + \bar{1})^{81}$ in $\mathbf{Z_3}[x]$.
15. Let M consist of all α in $\mathbf{Z_{17}}[x]$ such that $\alpha^* = 0$ in $\mathbf{Z}_{17}^*$. Explain why M is an ideal in $\mathbf{Z_{17}}[x]$, and name the monic generator of M.

16. Show that $x^{11} - x^2 - x + \bar{2}$ has no first-degree factor in $\mathbf{Z}_{11}[x]$.

17. Find a rational root of $2x^3 - 5x^2 + 18x - 45 = 0$, and then find the other roots.

18. Give the complete set of possibilities for rational roots of

$$91x^3 + bx^2 + cx + 77 = 0$$

where b and c are integers. [*Hint:* gcd(91, 77) = 7.]

19. Find the α with deg $\alpha \leq 3$ in $\mathbf{Z}_7[x]$ that fits the following table.

x	$\bar{1}$	$\bar{4}$	$\bar{5}$	$\bar{6}$
$\alpha(x)$	$\bar{5}$	$\bar{6}$	$\bar{4}$	$\bar{1}$

20. Find the β in $\mathbf{Q}[x]$ such that

$$1 \cdot 2^2 + 2 \cdot 3^2 + 3 \cdot 4^2 + \cdots + n(n+1)^2 = \beta(n).$$

21. Let $\alpha = x^3 - x^2 - 8x + 12$. Express α in the form

$$a + b(x-2) + c(x-2)^2 + d(x-2)^3,$$

with a, b, c, and d in $\mathbf{Q}$, and thus find the multiplicity of 2 as a zero of α.

22. Show that 1 is a zero of $nx^{n+2} - (2n+1)x^{n+1} + (n+1)x^n + x - 1$ with multiplicity 3 for all positive integers n.

23. Show that $x^3 + x^2 - 2x - 1$ has no rational zeros.

24. Find all the roots in $\mathbf{C}$ of each of the following.
(a) $x^4 - 13x^2 + 36 = 0$.
(b) $x^4 + 4x^3 + 4x^2 - 4x - 5 = 0$.
(c) $x^4 - 4x^3 + 8x + 20 = 0$.

25. Let r be a fixed element of a commutative ring U with unity and let f be the mapping from $U[x]$ to U with $\alpha \mapsto \alpha(r)$.
(i) Explain why f is a ring homomorphism.
(ii) Is r a zero of every α in the kernel K of f?
(iii) Name the monic generator of the kernel K.

26. Which pair from the following three polynomials of $\mathbf{Q}[x]$ are relatively prime?
(i) $x^2 - 9$; (ii) $x - 3$; (iii) $x + 3$.

27. How many monic second-degree primes are there in $\mathbf{Z}_{29}[x]$?

28. Express $x^3/(x + \bar{1})(x + \bar{3})$ as the sum of a polynomial and a finite number of ppd-terms in $\mathbf{Z}_7(x)$; that is, find the partial fraction decomposition of this quotient of polynomials.

29. Is there a field with 99 elements? Explain.

30. Is there a field with 49 elements? Explain.

31. Let F be a finite field with n elements. Let π be a prime of degree d in $F[x]$ and $m = n^d$. Prove that $\pi \mid (x^m - x)$ in $F[x]$.

32. Let α and π be in $F[x]$, with π a prime. Prove that $\pi \mid \alpha$ in $F[x]$ if and only if there exists an extension field E over F in which α and π have a common zero.

33. Let F be a field with 81 elements. Is every automorphism θ of F an automorphism of F over its prime subfield $P = \{0, 1, 1 + 1\}$? Explain.

34. Is every automorphism θ of the complex numbers $\mathbf{C}$ an automorphism of $\mathbf{C}$ over $\mathbf{Q}$? Explain.

35. Find all the complex roots of $x^3 + x^2 - 2x - 1 = 0$. (We saw in Problem 23 that there are no rational roots).

CHAPTER 6

EUCLIDEAN CONSTRUCTIONS

Indeed, when in the course of a mathematical investigation we encounter a problem or conjecture a theorem, our minds will not rest until the problem is exhaustively solved and the theorem rigorously proved; or else, until we have found the reasons which made success impossible and, hence, failure unavoidable. Thus, the proofs of the impossibility of certain solutions play a predominant role in mathematics; the search for an answer to such questions has often led to the discovery of newer and more fruitful fields of endeavor. **—David Hilbert**

As early as the fifth century B.C., Greek mathematicians posed some geometric construction problems that have over the years inspired a great deal of interest. A euclidean construction is strictly defined; it must be done with only a straightedge (an unmarked ruler) and a compass (and in Greek times, this was a collapsing compass, but those have since been proved to be equivalent to a modern compass that does not collapse when lifted off the surface of the construction). Of course, in such constructions lines have no thickness and points have no length, breadth, or depth. What is sought is not just an excellent approximation to a geometrical figure but a theoretical method for obtaining the exact figure. Certainly there are constructions that yield figures so close to the desired one that the eye cannot perceive the difference; but these are not acceptable euclidean constructions. What is desired is a method of construction that would allow one to do the job exactly with the allowable tools if these were used perfectly. This problem is not one of applied science but of conceptual mathematics.

As we know, many euclidean constructions are possible. We can bisect a line segment; bisect an angle; construct an equilateral triangle, a square, or even a regular pentagon; and more. Unfortunately, given the constraints on the problem — the use of only straightedge and compass — many constructions are not possible. Included among these are some problems that date from antiquity and are easily understood.

One of these is the so-called Delian problem. The Athenians were told by the oracle of Delos that a plague wracking the country would end when they constructed a new altar to Apollo in the form of a cube with double the volume of the old altar, also in the form of a cube. The first solution, according to legend, involved doubling the edge of the cube, but that multiplied the volume by 8 instead of 2, of course. This and other false constructions failed to end the plague, and it continued to devastate the population. Rarely have there been stronger motivating factors in applied mathematics.

We recognize that the solution to the problem would have been to construct an edge of length equal to the cube root of 2, if the original edge had length 1. We shall prove that, starting with the rational numbers and the lengths corresponding to them, it is not possible to construct with straightedge and compass a length equal to the cube root of 2. In fact, the only lengths that we can construct are lengths expressible in terms of sums, differences, products, ratios, and square roots of given lengths. Our proof uses material, from Chapter 5, on field extensions.

This same technique is also used to show that we cannot trisect the angle — that is, construct an angle one-third of a general given angle. (Some specific angles, like 90°, can be trisected because we can construct a 30° angle.) The angle trisection, the "doubling of the cube," and the squaring of the circle (that is, finding a square with the area equal to the area of a given circle) are the three classical construction problems of antiquity. The impossibility proof for the last requires the more sophisticated methods mentioned in the biographical note on C. L. F. Lindemann in Chapter 5. These problems were so well-known in Greek times that they were even mentioned in a tragedy of Euripides, in one of the plays now lost.

The question of whether these three constructions are possible remained open until after the time of Galois. The techniques he developed to solve the problem of whether there is a method for solving the general fifth-degree polynomial equation were used by the French mathematician, Pierre Wantzel, to settle the question of angle trisection and cube doubling. He discovered that these constructions are simply not possible. This has not kept people from trying them, either through misunderstanding of the conditions set on the problem or through ignorance of the fact that the questions were settled once and for all by Wantzel. An amusing book on false attempts to trisect the angle is Underwood Dudley's *A Budget of Trisections*, published in 1989.

Of course, other famous constructibility problems have challenged mathematicians, most obviously the problem of the construction of regular n-sided polygons. Gauss proved that a regular polygon with n sides is constructible by straightedge and compass only when the number n is of the form $2^r p_1 p_2 \cdots p_k$, where the $p_1, p_2, \ldots, p_k$ are primes of the form $2^{2^t} + 1$, the so-called Fermat primes. Only five numbers of this form are known to be prime, namely, 3, 5, 17, 257, and 65,537. There may be other Fermat primes, but that is an open question at the present time. If more exist, they are very large.

6.1 Closure Under Euclidean Constructions

"When I use a word," Humpty Dumpty said in a rather scornful tone, "it means just what I choose it to mean— neither more nor less."

—**Charles Lutwidge Dodgson (Lewis Carroll)**

We deal with a fixed plane, which we call ***the plane***. We assume that we are given two fixed points π_0 and π_1 and that the distance between them is the unit of distance. We use ***line*** to mean "straight line."

DEFINITION

Collection Closed Under Euclidean Constructions

A collection Γ of points, lines, and circles of the plane is said to be **closed under euclidean constructions** if and only if the following three conditions are met:

(a) The line through any two distinct points of Γ is in Γ.

(b) Let α, β, and γ be points of Γ with $\alpha \neq \beta$. Let r be the distance between α and β. Then the circle with center at γ and radius r is in Γ.

(c) If α is a point of intersection of two distinct lines of Γ, or of two distinct circles of Γ, or of a line and a circle of Γ, then α is in Γ.

If Γ satisfies these three conditions, it is clear that Γ contains every point, line, and circle that might be obtained from those in Γ by constructions with straightedge and compass.

In Section 6.3, we shall describe a Γ that is closed under euclidean constructions (constructions with straightedge and compass) and in which there is a pair of lines forming a 60° angle. If it were possible to trisect every angle with euclidean constructions, there would have to be a pair of lines making a 20° angle in Γ, since Γ is closed under euclidean constructions. However, we shall prove that no lines of the specific Γ of Section 6.3 form a 20° angle; and thus we shall prove that a general method for trisecting all angles with straightedge and compass is impossible.

The same technique will be used to prove that some other classical constructions cannot be performed with straightedge and compass. Our advantage over the great mathematicians of ancient times is that now we can make use of analytic geometry and the theory of field extensions.

We start by listing the operations on lengths that have been known to be possible since Euclid's time. In elementary euclidean geometry one learns that, if given lengths of 1, a, and b, one can construct with straightedge and compass lengths of

$$a + b, \quad |a - b|, \quad ab, \quad a/b, \quad \sqrt{a}. \tag{1}$$

Figure 6.1 illustrates the construction of lengths $a + b$ and $a - b$ from given lengths a and b (with $a > b$). The techniques for constructing lengths ab and a/b using given lengths 1, a, and b are indicated in Figure 6.2. The mean proportional construction to obtain $\sqrt{a}$ from 1 and a is shown in Figure 6.3.

Since we intend to use both positive and negative numbers as coordinates (with the sign an indication of direction), we may replace $|a - b|$ with $a - b$ in (1) above. Thus we are motivated to consider subsets F, with at least two elements, in the real numbers $\mathbf{R}$ such that F is closed under addition, subtraction, multiplication, division by nonzero numbers, and extraction of

FIGURE 6.1

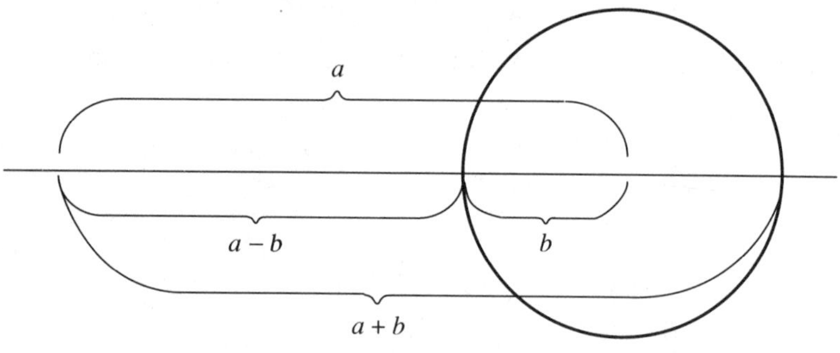

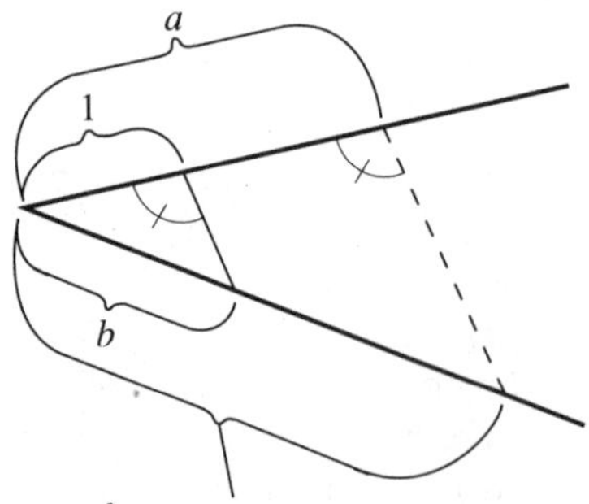

$\frac{x}{a} = \frac{b}{1}$ or $x = ab$

FIGURE 6.2a

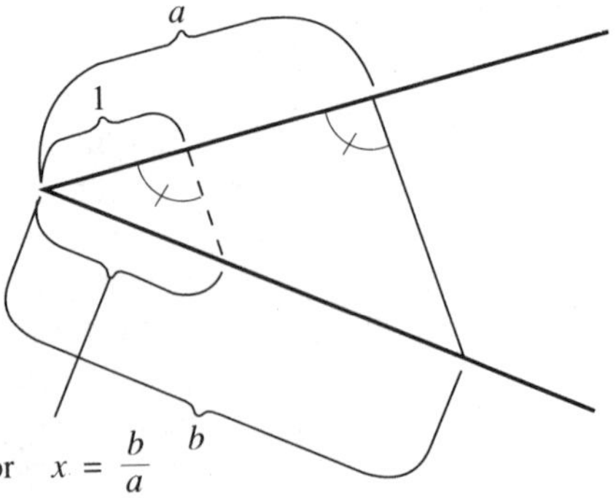

$\frac{x}{1} = \frac{b}{a}$ or $x = \frac{b}{a}$

FIGURE 6.2b

FIGURE 6.3

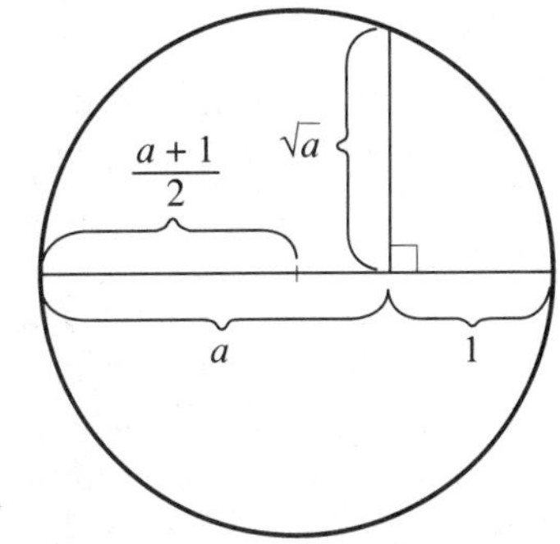

square roots of nonnegative numbers. A subset F with these properties is a subfield in $\mathbf{R}$ such that $\sqrt{a}$ is in F whenever a is in F and $a \geq 0$.

There are such subfields, since the field $\mathbf{R}$ of all the real numbers has this property. For our present purposes, the smallest F of this nature is most suitable; it will be characterized in the next section.

6.2 Closure Under Square Roots

The field $\mathbf{Q}$ of the rational numbers is the prime subfield in the real numbers $\mathbf{R}$; this means that $\mathbf{Q}$ is contained in every subfield in $\mathbf{R}$. We know that $\sqrt{a}$ need not be in $\mathbf{Q}$ when a is a positive rational number, since $\sqrt{2}$, $\sqrt{3}$, $\sqrt{5}$, and so on are not in $\mathbf{Q}$.

We seek the smallest extension F over $\mathbf{Q}$ such that F is closed under extraction of square roots of positive numbers in F. If a is a positive rational number, not only $\sqrt{a}$ but all the numbers in the field extension $\mathbf{Q}(\sqrt{a})$, must be in F. If b is a positive real number in $Q_1 = \mathbf{Q}(\sqrt{a})$, we similarly see that every number in $Q_1(\sqrt{b}) = \mathbf{Q}(\sqrt{a}, \sqrt{b})$ must be in F.

This motivates us to consider finite chains

$$Q_0, Q_1, Q_2, \ldots, Q_s, \tag{1}$$

such that $Q_0 = \mathbf{Q}$ and $Q_i = Q_{i-1}(r_i)$, with r_i a square root of a positive number in Q_{i-1}, for $i = 1, 2, \ldots, s$. We call the end result, $Q_s = \mathbf{Q}(r_1, r_2, \ldots, r_s)$, of such a chain a ***multiquadratic extension*** over $\mathbf{Q}$.

EXAMPLE 1 Let us find a multiquadratic extension $E = \mathbf{Q}(a, b, c)$ over $\mathbf{Q}$ such that all four roots of $x^4 - 8x^2 + 5 = 0$ are in E. Substituting u for x^2 converts the given equation into the quadratic $u^2 - 8u + 5 = 0$, which has $4 + \sqrt{11}$ and $4 - \sqrt{11}$ as its solutions. Then $x = \pm\sqrt{u}$ leads to the four solutions for x:

$$x_1 = \sqrt{4 + \sqrt{11}}, \quad x_2 = -x_1, \quad x_3 = \sqrt{4 - \sqrt{11}}, \quad x_4 = -x_3.$$

Now we let $a = \sqrt{11}$, $b = x_1$, and $c = x_3$. Then a^2, b^2, and c^2 are all positive real numbers with a^2 in $\mathbf{Q}$, b^2 in $\mathbf{Q}(a)$, and c^2 in $\mathbf{Q}(a, b)$. Thus $E = \mathbf{Q}(a, b, c)$ is a multiquadratic extension. It is also clear that all four roots x_i are in E.

NOTATION Special Field F

In the rest of this chapter, F denotes the set of all real numbers in all multiquadratic extensions E over $\mathbf{Q}$; that is, F is the union of all such fields E.

Next we show that this F is a field that is closed under extraction of square roots of positive numbers.

Let h and k be in F. Then h is in a multiquadratic extension $Q_s = \mathbf{Q}(a_1, a_2, \ldots, a_s)$ with each a_{i+1}^2 in $\mathbf{Q}(a_1, \ldots, a_i)$, and k is in a multiquadratic extension $Q_t = \mathbf{Q}(b_1, b_2, \ldots, b_t)$ with each b_{j+1}^2 in $\mathbf{Q}(b_1, \ldots, b_j)$. Then $h - k$ and hk^{-1} (if $k \neq 0$) are in F, since they are in the multiquadratic extension field

$$\mathbf{Q}(a_1, \ldots, a_s, b_1, \ldots, b_t).$$

Since F is nonempty, it now follows from Theorem 2 of Section 4.5 that F is a field.

Also, if g is a positive number in F, then g is in a multiquadratic extension $\mathbf{Q}(a_1, \ldots, a_s)$. Hence, $\sqrt{g}$ is in $\mathbf{Q}(a_1, \ldots, a_s, \sqrt{g})$ and therefore is in F. This means that F is closed under extraction of square roots of positive elements.

Problems

Here F always denotes the special field that is the union of all the multiquadratic extensions of the field $\mathbf{Q}$ of rational numbers.

1. Show that $\cos(\pi/3)$ is in $\mathbf{Q}$, that $\cos(\pi/4)$ is in $\mathbf{Q}(\sqrt{2})$, and hence that both $\cos(\pi/3)$ and $\cos(\pi/4)$ are in the field F.
2. Use the fact that the field F is closed under extraction of square roots of positive numbers to show that $\sin x$ is in F if and only if $\cos x$ is in F.
3. Given that $\cos x$ and $\cos y$ are in F, show that $\cos(x + y)$ and $\cos(x - y)$ are in F.
4. Given that $\cos(2x)$ is in F, show that $\cos x$ is in F.
5. Show that $\cos(\pi/12)$ is in F.
6. Show that $\sin(\pi/24)$ is in F.
7. Let $c = \cos(2\pi/5)$, $s = \sin(2\pi/5)$, and $z = c + \mathbf{i}s$. Show the following.
   (i) $z^5 - 1 = 0, z \neq 1$.
   (ii) $z^4 + z^3 + z^2 + z + 1 = 0$.
   (iii) $z^2 + z + 1 + z^{-1} + z^{-2} = 0$.
   (iv) $z^{-1} = c - \mathbf{i}s$, $z + z^{-1} = 2c$.
   (v) $z^2 + z^{-2} = 4c^2 - 2$.
   (vi) $4c^2 + 2c - 1 = 0$.
   (vii) $\cos(2\pi/5) = c = (-1 + \sqrt{5})/4$.
   (viii) $\cos(2\pi/5)$ is in the field F.
8. Explain why $\cos x$ is in F for each of the following values of x.
   (i) $\pi/5$; (ii) $(\pi/3) - (\pi/5)$; (iii) $\pi/15$; (iv) $\pi/30$; (v) $\pi/60$.
9. Show that all four roots of $x^4 - 6x^2 + 2 = 0$ are in a multiquadratic extension $E = \mathbf{Q}(a, b, c)$ over $\mathbf{Q}$.

6.3 Constructible Points, Lines, and Circles

Circles to square and cubes to double would give a man excessive trouble. —**Matthew Prior**

Let π_0 and π_1 be the two fixed points of the plane used in Section 6.1 to determine the unit of length. We set up a cartesian coordinate system in the plane with π_0 as the ***origin***, the line through π_0 and π_1 as the ***x-axis***, the perpendicular to the x-axis at the origin as the ***y-axis***, and the distance between π_0 and π_1 as the common unit of length on both axes. (See Figure 6.4.)

Then every point of the plane is uniquely represented by its coordinates (x, y), an ordered pair of real numbers. We shall sometimes use the ordered pair (x, y) as a name for the point.

We are now ready to define the collection Γ (of points, lines, and circles of the plane) that will enable us to prove certain classical constructions to be impossible.

DEFINITION

Special Collection Γ

The collection Γ consists of the following objects of the fixed plane:

(a) A point (u, v) is in Γ if and only if both u and v are in the special field F of the preceding section.

(b) A line is in Γ if and only if it has an equation $ax + by = c$ with each of a, b, and c in F.

(c) A circle is in Γ if and only if its radius d and the coordinates of its center (h, k) are in F.

We note that a circle in Γ has an equation of the form

$$x^2 + y^2 - rx - sy = t$$

with $r = 2h$, $s = 2k$, and $t = d^2 - h^2 - k^2$ (and hence with r, s, and t in F).

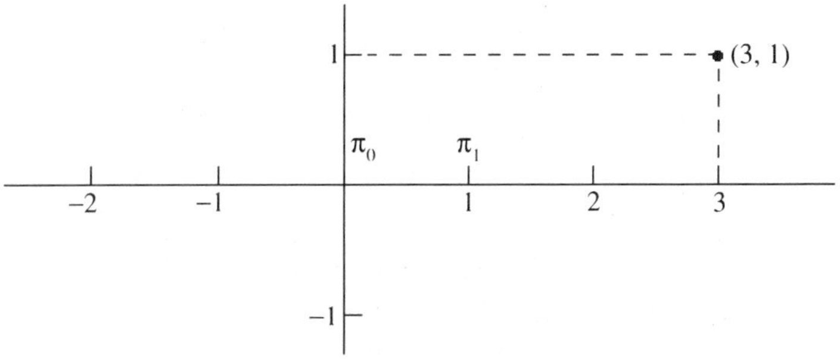

FIGURE 6.4

Next we give a number of lemmas that together show that this collection Γ is closed under euclidean constructions.

LEMMA 1 The line through two distinct points in Γ is also in Γ.

Proof Let the points have coordinates (u_1, v_1) and (u_2, v_2). Since the two points are distinct, either $u_1 - u_2 \neq 0$ or $v_1 - v_2 \neq 0$. Then the line through the two points has

$$(v_1 - v_2)(x - u_2) - (u_1 - u_2)(y - v_2) = 0$$

as an equation. This can be put in the form $ax + by = c$ with $a = v_1 - v_2$, $b = -(u_1 - u_2)$, and $c = v_1 u_2 - u_1 v_2$. Since F is a field, we see that a, b, and c are in F, and hence the line is in Γ. ∎

LEMMA 2 The point of intersection of two nonparallel lines in Γ is also in Γ.

Proof Let the lines have equations

$$a_1 x + b_1 y = c_1, \qquad a_2 x + b_2 y = c_2,$$

with a_1, b_1, c_1, a_2, b_2, and c_2 in the field F. Let $d = a_1 b_2 - a_2 b_1$. Since the lines are not parallel, $d \neq 0$. Solving the simultaneous linear equations, we find that the point of intersection is (u, v) with

$$u = (c_1 b_2 - c_2 b_1)/d, \qquad v = (a_1 c_2 - a_2 c_1)/d.$$

Now u and v are in the field F and the point (u, v) is in Γ. ∎

LEMMA 3 A point of intersection of a circle and line of Γ is also in Γ.

Proof Let the equation of the circle be

$$x^2 + y^2 - rx - sy = t \tag{1}$$

and that of the line be $ax + by = c$ with r, s, t, a, b, and c in F and either $a \neq 0$ or $b \neq 0$. For definiteness, let us assume that $a \neq 0$.

Then the y-coordinates of the points of intersection are found by solving $ax + by = c$ for x in terms of y, that is, as

$$\frac{c - by}{a}, \tag{2}$$

then substituting this for x in (1), and finally solving the resulting second-degree equation for y. The operations that have to be performed are additions, subtractions, multiplications, divisions by nonzero numbers, and extraction of a square root (of a nonnegative number when there is at least one actual point of intersection). Since F is closed under all these operations, the y for an intersection is in F. Substituting such a y in (2), we obtain the corresponding x coordinate and see that it is also in F. Hence the point of intersection is in Γ. ■

LEMMA 4 A point of intersection of two circles of Γ is also in Γ.

Proof Let the circles have equations

$$x^2 + y^2 - r_i x - s_i y = t_i, \qquad i = 1 \text{ and } 2.$$

Subtracting corresponding sides of these equations, we see that an intersection also satisfies

$$(r_1 - r_2)x + (s_1 - s_2)y = t_2 - t_1. \tag{3}$$

Since distinct intersecting circles cannot have the same center, either $r_1 - r_2 \neq 0$ or $s_1 - s_2 \neq 0$. Hence (3) is the equation of a line. Then a point of intersection of the two circles is a point of intersection of this line and either one of the circles and thus is in Γ by Lemma 3. ■

Summing up these four lemmas, we have the following.

THEOREM 1 **Closure Under Euclidean Constructions**

Γ is closed under euclidean constructions.

We need a few more lemmas.

LEMMA 5 There are lines in Γ that form a 60° angle.

Proof Clearly $\sqrt{3}$ is in F, and hence the lines with

$$y = 0, \qquad \sqrt{3}\,x - y = 0$$

as equations are in Γ. These lines form a 60° angle. ■

LEMMA 6 If a is an angle made by two lines of Γ, cos a is in F. (The proof of the converse of this lemma is left to the reader as Problem 1 of this section.)

Proof Let lines L and L' be in Γ and make an angle of a at their point of intersection α. (See Figure 6.5.) This implies that α is in Γ. Then it follows that the circle with α as center and 1 as radius is in Γ. Hence the intersections β and γ of this

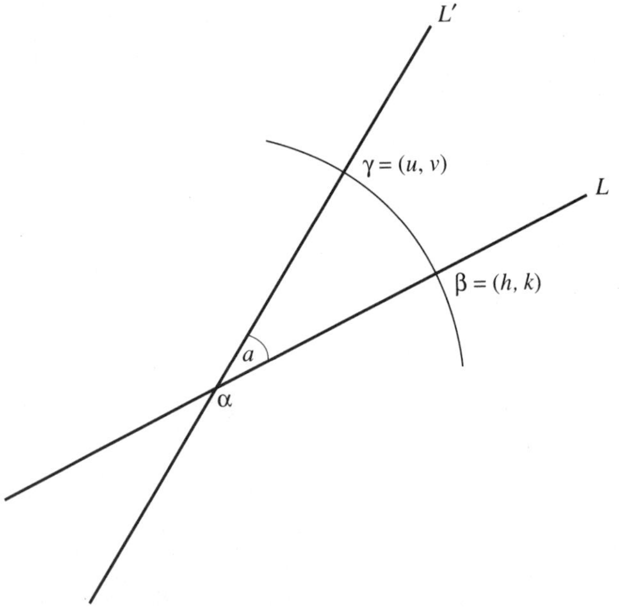

FIGURE 6.5

circle with the lines L and L' are in Γ. Let β and γ have (h, k) and (u, v) as coordinates. Then h, k, u, and v are in the field F.

Also, line segments $\alpha\beta$, $\alpha\gamma$, and $\beta\gamma$ have 1, 1, and $\sqrt{(h-u)^2+(k-v)^2}$, respectively, as their lengths. Using the Law of Cosines, one finds that

$$\cos a = \frac{1^2 + 1^2 - (h-u)^2 - (k-v)^2}{2 \cdot 1 \cdot 1}.$$

Since this expression for $\cos a$ involves operations under which the field F is closed, $\cos a$ is in F. ■

LEMMA 7 Let p, q, and r be rational numbers and let

$$\alpha = x^3 + px^2 + qx + r.$$

If $\alpha = 0$ has no root in $\mathbf{Q}$, it has no root in F.

Proof We give an indirect proof; that is, we assume that $\alpha = 0$ has a root in F but not in $\mathbf{Q}$, and then we show that this leads to a contradiction. From the assumption that there is a root of $\alpha = 0$ in F, it follows that there is a root in a multiquadratic extension $\mathbf{Q}(a_1, a_2, \ldots, a_n)$.

Let $Q_0 = \mathbf{Q}$ and $Q_i = \mathbf{Q}(a_1, \ldots, a_i)$ for $1 \le i \le n$. Also let h be the smallest integer in $\{0, 1, \ldots, n\}$ such that there is a root of $\alpha = 0$ in Q_h. By hypothesis, $h \neq 0$.

Since $Q_h = Q_{h-1}(a_h)$ with a_h^2 in Q_{h-1}, Theorem 6 of Section 5.4 tells us that if $\alpha = 0$ has a root in Q_h it has a root in Q_{h-1}. This contradicts the minimal nature of h and completes the proof. ■

LEMMA 8 No lines of Γ make a $20°$ angle.

Proof Let $s = \cos 20°$ and $\alpha = 8x^3 - 6x - 1$. Substituting $\theta = 20°$ in the triple-angle formula

$$4 \cos^3 \theta - 3 \cos \theta = \cos(3\theta),$$

we obtain $4s^3 - 3s = \cos 60° = 1/2$. Then $8s^3 - 6s - 1 = 0$; thus s is a root of $\alpha = 0$.

In Example 2 of Section 5.5, we showed that $\alpha = 0$ has no rational root. Then Lemma 7 above tells us that $\alpha = 0$ has no root in the special field F of this chapter. This means that the root $\cos 20°$ of $\alpha = 0$ is not in F. Finally, it follows from Lemma 6 above that no lines of Γ make a $20°$ angle. ∎

THEOREM 2 **No General Trisection Technique**

There is no general method for trisecting angles by euclidean constructions.

Proof The collection Γ of points, lines, and circles is closed under euclidean constructions and has lines making a $60°$ angle. (See Lemma 5.) If there were a method for trisecting all angles using euclidean constructions, a $20°$ angle would exist in Γ. But Lemma 8 states that Γ has no $20°$ angle. This proves the theorem. ∎

Another classical euclidean construction problem is that of "duplication of a cube." In this problem, one is given a length s and is asked to construct a length x such that a cube with edges of length x will have twice the volume of a cube with edges of length s, that is, with $x^3 = 2s^3$.

We assume that the given cube has sides of unit length. Then the length sought in the Delian problem is $\sqrt[3]{2}$. The following result is helpful.

LEMMA 9 The number $\sqrt[3]{2}$ is not in the special field F.

Proof It follows easily from unique factorization of positive integers as products of primes that $\sqrt[3]{2}$ is not a rational number. [See Problem 13(b) of Section 1.7.] Hence $\alpha = x^3 - 2$ has no zero in $\mathbf{Q}$. Then Lemma 7 tells us that α has no zero in F; hence $\sqrt[3]{2}$ is not in F. ∎

The point (0, 0) is in Γ. If a segment of length $\sqrt[3]{2}$ were constructible from a unit length with straightedge and compass, the point $(\sqrt[3]{2}, 0)$ would be in Γ, and hence $\sqrt[3]{2}$ would be in F. But since this would contradict Lemma 9, the Delian problem is not solvable by euclidean constructions.

Problems

1. Given that $\cos u$ is in the field F, show the following.
   (i) The point $(\cos u, \sin u)$ is in the collection Γ.
   (ii) In Γ, there is a line that makes an angle u with the x-axis.
2. Explain why $\cos u$ is in F if and only if $\cos(\pi - u)$ is in F.
3. Show that in Γ there are two lines forming an angle of 12° (that is, $\pi/15$).
4. Show that in Γ there are two lines forming an angle of 3°.
5. Show that Γ does not contain two lines forming an angle of 1°.
6. Show that Γ does not contain two lines forming an angle of 2°.
7. Show that cos 40° is not in F.
8. Show that cos 140° is not in F.
9. Let s be a positive real number in F. Show that in Γ there exist five points that are the vertices of a regular pentagon with sides of length s; that is, show that a regular pentagon with sides of given length s can be obtained by euclidean constructions. [See Problem 7(viii) of Section 6.2.]
10. Show that in Γ there do not exist nine points that are the vertices of a regular polygon (with nine sides); that is, a regular 9-gon is not constructible with straightedge and compass.
11. Show that $x^3 + x^2 - 2x - 1 = 0$ has no roots in F.
12. Show that $8y^3 + 4y^2 - 4y - 1 = 0$ has no roots in F.
13. Let $c = \cos(2\pi/7)$, $s = \sin(2\pi/7)$, and $z = c + \mathbf{i}s$. Show the following.
    (i) $z^7 - 1 = 0, \quad z \neq 1$.
    (ii) $z^6 + z^5 + z^4 + z^3 + z^2 + z + 1 = 0$.
    (iii) $z^3 + z^2 + z + 1 + z^{-1} + z^{-2} + z^{-3} = 0$.
    (iv) $z^{-1} = c - \mathbf{i}s, \quad z + z^{-1} = 2c$.
    (v) $z^2 + z^{-2} = 4c^2 - 2, \quad z^3 + z^{-3} = 8c^3 - 6c$.
    (vi) $8c^3 + 4c^2 - 4c - 1 = 0$.
    (vii) $\cos(2\pi/7)$ is not in the field F. (See Problem 12.)
14. Show that in Γ, there do not exist seven points that are the vertices of a regular polygon; that is, a regular 7-gon is not constructible with straightedge and compass. (See the preceding problem.)

Review Problems

1. Show that $\cos(\pi/20)$ is in the special field F.
2. Show that $\cos(19\pi/20)$ is in F.
3. Let $\cos a$ be in F and let n be a positive integer. Show that $\cos(na)$ is in F.
4. Let $\cos a$ be in F and let n be a positive integer. Show that $\cos(a/2^n)$ is in F.
5. Show that $\cos(\pi/7)$ is not in F.

Pierre Laurent Wantzel
1814–1848

Wantzel was the first to provide a rigorous proof of the impossibility of the trisection of general angles by straightedge and compass. He was répétiteur at the École Polytechnique in Paris and in 1845 provided a further proof of the insolvability of the general fifth-degree equation — a proof that in parts resembled those of Ruffini and Abel. He, like his contemporaries Galois and Abel, died young. His position at the École Polytechnique was taken, after his death, by the better-known mathematician, Charles Hermite.

6. Show that $\cos(\pi/180)$ is not in F.

7. Show that a regular 40-sided polygon, with sides of given length, is constructible with straightedge and compass.

8. Given a unit length, can one construct with straightedge and compass a regular 14-gon, with sides of unit length? Explain.

CHAPTER 7

MORE ON THE INTEGERS

Mathematics is the queen of the sciences and number theory the queen of mathematics. —**Carl Friedrich Gauss**

The study of the elementary theory of numbers usually takes place initially quite independently of abstract algebra. Here, however, we explore just a few topics from the theory of numbers and show some relationships between these topics and structures that we have studied in earlier chapters. For example, we see that the set of quadratic residues (squares in a residue system, mod a prime) and the quadratic nonresidues (nonsquares) can be looked at as the elements of a quotient group of order 2.

We begin with the Chinese Remainder Theorem, which gives a method for solving a system of simultaneous congruences. A problem of this type was included in the *Arithmetic* of Nicomachus around A.D. 100, and the solution was given at almost the same time in the form of a Chinese poem by Sun-Tsu. Problems of this sort appeared again in China and India in the seventh century A.D. and are the basis of many puzzles and problems in folklore.

We include some more general discussion of arithmetic functions, some of which were used earlier informally. And we close with a short discussion of Gauss's famous Law of Quadratic Reciprocity, a deep property of the primes.

7.1 Simultaneous Congruences — The Chinese Remainder Theorem

It seems safe to assume that the positive integers formed the first number system used by human beings. Therefore, it should not be surprising that certain types of mathematical puzzles, in which the answers are positive integers, go back to the beginnings of history. We present an example of such a problem (dressed in words of our times).

EXAMPLE 1 **Common Term in Several Arithmetic Progressions**

A child who can only count to 9 is given a batch of marbles. These are arranged in columns each having 9 marbles, but with 7 left over. So the child attempts to rearrange the batch in columns of 7 marbles each and finds that 6 are left over. Finally, an attempt to rearrange the batch in columns of 5 marbles leaves a remainder of 2. Let us find the two smallest positive integers that could be the total number of marbles in the batch. We seek the first two numbers common to all three arithmetic progressions:

$$7, \quad 16, \quad 25, \quad 34, \quad \ldots; \tag{1}$$

$$6, \quad 13, \quad 20, \quad 27, \quad \ldots; \tag{2}$$

$$2, \quad 7, \quad 12, \quad 17, \quad \ldots. \tag{3}$$

We see that 7 appears in both (1) and (3). It can also be seen that the other terms common to these two progressions are

$$7 + 45, \qquad 7 + 90, \qquad 7 + 135, \qquad \ldots.$$

Hence we seek the first two common terms of the two progressions

$$6, \quad 13, \quad 20, \quad 27, \quad \ldots; \tag{2}$$

$$7, \quad 52, \quad 97, \quad 142, \quad \ldots. \tag{4}$$

We test the first few numbers in (4) to see if any have a remainder of 6 when divided by 7, and we find that 97 is a solution to our problem. It is then fairly easy to see that the next solution is $97 + 7 \cdot 45 = 412$.

In the congruence notation, the problem of Example 1 could be stated as follows. Find the two smallest positive integers satisfying the three simultaneous congruences

$$x \equiv 7 \pmod 9, \qquad x \equiv 6 \pmod 7, \qquad x \equiv 2 \pmod 5.$$

DEFINITION 1 **Linear Congruence**

If a, b, and m are fixed integers, the congruence $ax \equiv b \pmod m$ is said to be **linear**; if $a = 1$, then the linear congruence is **monic**.

The results that follow enable one to tackle systems of any finite number of simultaneous congruences in which the one unknown x appears only to the first power.

LEMMA 1 **Making a Linear Congruence Monic**

(a) A congruence

$$ax \equiv c \pmod b \tag{5}$$

has solutions in **Z** if and only if $(a, b) \mid c$.

(b) If the set of solutions of (5) in $\mathbf{Z}$ is nonempty, it is also the set of solutions of a congruence

$$x \equiv k \pmod{s}. \tag{6}$$

Proof Part (a) is a restatement of Theorem 3 of Section 1.4.

If an integer x satisfies (5), there is an integer y such that

$$ax + by = c \tag{7}$$

and $d = \gcd(a, b)$ is an integral divisor of c. Then we can divide each of a, b, and c by d in (7) and obtain

$$rx + sy = t,$$

with r and s relatively prime. Hence (5) is equivalent to

$$rx \equiv t \pmod{s}, \qquad \gcd(r, s) = 1. \tag{8}$$

Translating (8) into the language appropriate to $\mathbf{Z_s} = \mathbf{Z}/(s)$ gives us

$$\bar{r} \cdot \bar{x} = \bar{t} \quad \text{in } \mathbf{Z_s}. \tag{9}$$

Since $\gcd(r, s) = 1$, $\bar{r}$ has a reciprocal in $\mathbf{Z_s}$ and (9) is equivalent to

$$\bar{x} = (\bar{r})^{-1}\bar{t} \quad \text{in } \mathbf{Z_s}. \tag{10}$$

Letting k be an integer such that $\bar{k} = (\bar{r})^{-1}\bar{t}$ in $\mathbf{Z_s}$, we can translate (10) back into the desired form (6). ■

Because of the result in Lemma 1(b), we restrict our consideration of linear congruences in what follows to those that are monic — that is, of the form $x \equiv k \pmod{s}$.

THEOREM 1 Simultaneous Congruences

In $\mathbf{Z}$, let $d = \gcd(s, t)$ and $m = \operatorname{lcm}[s, t]$.

(a) There is an integer x satisfying the two simultaneous congruences

$$\begin{aligned} x &\equiv a \pmod{s} \\ x &\equiv b \pmod{t} \end{aligned} \tag{11}$$

if and only if $a \equiv b \pmod{d}$.

(b) If $x = c$ satisfies the congruences (11), the set of all solutions of (11) in $\mathbf{Z}$ is the same as the set of all solutions of $x \equiv c \pmod{m}$ in $\mathbf{Z}$.

Proof Let c be a solution of (11). Then

$$a + hs = c = b + kt$$

for some integers h and k. Hence,

$$b - a = hs - kt.$$

Now it follows from Theorem 3 of Section 1.4 that $b - a$ is an integral multiple of $d = \gcd(s, t)$; that is, $a \equiv b \pmod{d}$.

Conversely, if $a \equiv b \pmod{d}$ it follows from Theorem 3 of Section 1.4 that there are integers u and v such that $b - a = us + vt$. Then $b - vt = a + us$, and $x = a + us$ is a solution of both congruences in (11). Hence part (a) is proved.

We start the proof of part (b) by letting c be a solution of the system (11). Then x is a solution of (11) if and only if $x \equiv c \pmod{s}$ and $x \equiv c \pmod{t}$ — that is, if and only if both $s \mid (x - c)$ and $t \mid (x - c)$. This means that x is a solution of (11) if and only if $x - c$ is a common multiple of s and t. Since the common multiples of s and t are the multiples of the least common multiple m (see Theorem 1 of Section 1.6), x is a solution of (11) if and only if $m \mid (x - c)$ — that is, if and only if $x \equiv c \pmod{m}$. ■

EXAMPLE 2 **No Solution**

We now consider the system of two simultaneous congruences

$$x \equiv 11 \pmod{15}, \qquad x \equiv 4 \pmod{21}.$$

We observe that $\gcd(15, 21) = 3$ and that 11 is not congruent to 4 modulo 3. Therefore, it follows from Theorem 1(a) that no integer x satisfies both congruences.

EXAMPLE 3 **One Congruence Replaces Two**

Let us find a single monic linear congruence $x \equiv c \pmod{m}$ that is equivalent to the system of two simultaneous congruences

$$x \equiv 14 \pmod{15}, \qquad x \equiv 5 \pmod{21}. \tag{A}$$

We use the technique of the proof of Theorem 1. An integer x satisfying both congruences of (A) exists if and only if there exist integers y and z such that

$$x = 14 + 15y = 5 + 21z. \tag{B}$$

The second equality in (B) shows that we have to solve

$$21z - 15y = 14 - 5 = 9$$

for integers y and z. Euclid's Algorithm of Section 1.5 provides a constructive method for doing this. However, the coefficients here are fairly small, so we can solve the equation by systematic trial.

First we divide by $3 = \gcd(21, 15)$ and convert the equation into

$$7z - 5y = 3 \quad \text{or} \quad z = (5y + 3)/7.$$

We try $0, 1, 2, \ldots$ for y until we find that $y = 5$ yields an integer, namely 4, for z. Substituting $y = 5$ or $z = 4$ in (B) results in $x = 89$ as a solution of the simultaneous congruences (A). Since $\operatorname{lcm}[15, 21] = 105$, Theorem 1(b) now tells us that the system (A) is equivalent to the single congruence $x \equiv 89 \pmod{105}$.

THEOREM 2 Relatively Prime Moduli

Let s and t be relatively prime integers. Then the system of simultaneous congruences

$$x \equiv a \pmod{s}$$
$$x \equiv b \pmod{t} \tag{12}$$

always has solutions. If c is one solution, the solutions of (12) are the solutions of

$$x \equiv c \pmod{st}. \tag{13}$$

Proof We are given that $\gcd(s, t) = 1$. Since any two integers a and b are congruent modulo 1, Theorem 1(a) tells us that there is a solution c of (12).

Using Theorem 2 of Section 1.6, we see that $\gcd(s, t) = 1$ implies $\operatorname{lcm}[s, t] = |st|$. Now it follows from Theorem 1(b) above that x satisfies (12) if and only if

$$x \equiv c \pmod{|st|}. \tag{14}$$

Since integers are congruent modulo $-st$ if and only if they are congruent modulo st, (14) and (13) are equivalent. ■

The solving of simultaneous congruences

$$x \equiv a \pmod{s}, \qquad x \equiv b \pmod{t}$$

involves the solution in integers y and z of

$$sy - tz = b - a.$$

We saw in Section 1.5 that such an equation can be solved with Euclid's Algorithm. We extend this technique to systems with any finite number of linear congruences after introducing some terminology.

DEFINITION 2

Set of Pairwise Relatively Prime Integers

The integers of a set S are **pairwise relatively prime** if $\gcd(m, n) = 1$ for all distinct m and n in S.

THEOREM 3 Chinese Remainder Theorem

Let $m_1, m_2, \ldots, m_r$ be pairwise relatively prime. Then the system of simultaneous congruences

$$x \equiv a_1 \pmod{m_1}, \ldots, x \equiv a_r \pmod{m_r} \tag{15}$$

always has solutions. If c is one solution, the set of solutions of (15) is the set of solutions of $x \equiv c \pmod{m_1 m_2 \cdots m_r}$.

Outline of Proof One uses Theorem 2 to replace two of the congruences of (15) with a single congruence. This process is continued until one has reduced the system to one congruence. (This is an inductive argument.)

It is left to the reader to show that the moduli are always relatively prime in the reduction process. ∎

EXAMPLE 4 Here we use Euclid's Theorem that there exist an infinite number of positive primes in **Z** to prove that there are a million consecutive positive integers

$$x, \quad x+1, \quad x+2, \quad \ldots, \quad x+999999,$$

each of which is an integral multiple of the cube of a prime.

Let the first million positive primes be

$$p_1, \quad p_2, \quad \ldots, \quad p_{1000000}.$$

Then we seek a positive integer x that satisfies the million simultaneous congruences:

$$x \equiv 0 \pmod{2^3}, \quad x \equiv -1 \pmod{3^3}, \quad \ldots, \quad x \equiv -999999 \pmod{p_{1000000}^3}.$$

Since $\gcd(p^3, q^3) = 1$ for distinct primes p and q, the existence of such an x is guaranteed by the Chinese Remainder Theorem. (The most modern electronic computer should have considerable difficulty in finding the smallest positive solution.)

Problems

1. Find a congruence $rx \equiv t \pmod{s}$, with $\gcd(r, s) = 1$, that has the same solutions as $56x \equiv 259 \pmod{987}$.
2. Find an integer c with $0 < c < 21$ such that $8x \equiv 15 \pmod{21}$ has the same solutions as $x \equiv c \pmod{21}$.
3. Find the two smallest positive integers x satisfying the simultaneous congruences
$$x \equiv 13 \pmod{14}, \qquad x \equiv 14 \pmod{15}.$$
4. Find the smallest positive integer and the largest negative integer satisfying the simultaneous congruences $3x \equiv 4 \pmod{5}$ and $8x \equiv 15 \pmod{21}$.
5. Find the two smallest positive integers having remainders 1, 2, 3, 4, and 5 when divided by 2, 3, 4, 5, and 6, respectively.
6. Find the two smallest positive integers having remainders 1, 2, 3, 4, and 5 when divided by 13, 14, 15, 17, and 19, respectively.
7. Find the smallest positive integer n satisfying the simultaneous conditions $4 \mid n$, $9 \mid (n+1)$, and $25 \mid (n+2)$.

8. An integer is called ***square-free*** if it is not an integral multiple of the square of a prime. Do there exist one million consecutive positive integers none of which is square-free? Justify your answer.
9. Let x and y be integers. Show that $19 \mid (3x + 5y)$ if and only if $19 \mid (13x + 9y)$.
10. Let x and y be integers. Find an integer c with $0 < c < 13$ such that $13 \mid (5x - 8y)$ if and only if $13 \mid (7x + cy)$.
11. Find the smallest positive integer s in $\{55, 56, 57, \ldots\}$ that satisfies both $x \equiv 0 \pmod 9$ and $x \equiv 5 \pmod{10}$.
12. Find the smallest positive integer t such that t is the sum of 9 consecutive positive integers and is also the sum of 10 consecutive positive integers.
13. (i) Find a and r in $\mathbf{Z}^+$ such that $y \equiv a \pmod r$ if and only if simultaneously $y \equiv 0 \pmod 7$ and $y + 1 \equiv 0 \pmod{11}$.
    (ii) Find b and s in $\mathbf{Z}^+$ such that $x \equiv b \pmod s$ if and only if simultaneously $x \equiv 0 \pmod{14}$ and $x + 2 \equiv 0 \pmod{22}$.
    (iii) Find c and t in $\mathbf{Z}^+$ such that $x \equiv c \pmod t$ if and only if simultaneously $x \equiv 0 \pmod{14}$, $x + 1 \equiv 0 \pmod{195}$, $x + 2 \equiv 0 \pmod{22}$.

★ 14. Explain why there must exist an integer y satisfying the simultaneous congruences $y \equiv 0 \pmod{10}$, $y + 1 \equiv 0 \pmod{231}$, and $y + 2 \equiv 0 \pmod{26}$.

★ 15. Explain why there exist integers x and y such that $\gcd(x + i, y + j) > 1$ for all i and j in $\{0, 1, 2\}$.

★ 16. Prove that for every positive integer n there exist integers x and y such that $\gcd(x + i, y + j) > 1$ for all i and j in $\{0, 1, 2, \ldots, n\}$.

7.2 More on Mathematical Induction

How often have I said to you that when you have eliminated the impossible, whatever remains, however improbable, must be the truth. —**Sir Arthur Conan Doyle**

Below we extend the problem set on mathematical induction of Section 1.1. First we present a result that is frequently used as a substitute for the well ordering principle, and then we state the related algorithm for a technique of proof that is more direct than well ordering proofs because it does not require proof by contradiction.

THEOREM 1 Condition for a Subset to Be $\mathbf{Z}^+$

Let S be a subset of $\mathbf{Z}^+ = \{1, 2, 3, \ldots\}$ with the following two properties:
(a) 1 is in S.
(b) Whenever $1, 2, \ldots, k - 1$ are in S, so is k.
Then $S = \mathbf{Z}^+$.

Proof Let T consist of the integers of $\mathbf{Z}^+$ that are not in S. We assume that T is nonempty and seek a contradiction. Under this assumption, it follows from the well ordering principle that T has a least integer t.

Since 1 is in S, $t \neq 1$. The fact that t is the least integer in T means that 1, 2, ..., $t - 1$ are in S. The hypothesis then implies that $(t - 1) + 1 = t$ is in S. Hence, t is in both S and T, which contradicts the definition of T. This contradiction shows that T is empty; that is, $S = \mathbf{Z}^+$. ■

ALGORITHM 1 **Proof by Strong Induction**

The following two steps constitute a proof by ***strong induction*** that an assertion $P(n)$ is true for all n in $X = \{a, a + 1, a + 2, \ldots\}$:

Basis. Show that $P(a)$ is true.

Inductive Part. Assume that $P(a)$, $P(a + 1)$, ..., $P(h - 1)$ are all true for some h in $\{a + 1, a + 2, \ldots\}$, and show that this ***inductive hypothesis*** implies that $P(h)$ is true.

It is left to the reader as Problem 5 below to prove that Algorithm 1 is a valid technique for showing that a proposition $P(n)$ is true for all n in

$$X = \{a, a + 1, a + 2, \ldots\}.$$

EXAMPLE 1 Let $P(n)$ be the assertion that n is expressible in the form $3x + 22y$, with x and y in $\mathbf{N} = \{0, 1, 2, \ldots\}$. We prove as follows that $P(n)$ is true for all n in $X = \{42, 43, \ldots\}$.

It is helpful to consider the first three cases. The equations

$$42 = 3 \cdot 14 + 22 \cdot 0, \quad 43 = 3 \cdot 7 + 22 \cdot 1, \quad 44 = 3 \cdot 0 + 22 \cdot 2 \tag{1}$$

show that each of 42, 43, and 44 is expressible in the desired form. Now we are ready for the parts of Algorithm 1.

Basis. Since $42 = 3 \cdot 14 + 22 \cdot 0$, $P(42)$ is true.

Inductive Part. Let h be in $\{43, 44, \ldots\}$ and assume that

$$P(42), \quad P(43), \ldots, \quad P(h - 1) \quad \text{are all true.} \tag{2}$$

We wish to show that this assumption (2) implies that $P(h)$ is true. Since display (1) shows that $P(42)$, $P(43)$, and $P(44)$ are true, we can restrict our attention to h's with $45 \leq h$. For such an h, we have $42 \leq h - 3 < h$, and it follows from the inductive hypothesis (2) that $P(h - 3)$ is true; that is,

$$h - 3 = 3x + 22y \quad \text{for some } x \text{ and } y \text{ in } \mathbf{N}.$$

Then $h = 3(x + 1) + 22y$ with $x + 1$ and y in $\mathbf{N}$, and so $P(h)$ is true. This completes the proof by strong induction that $P(n)$ is true for all n in X.

The reader may find any one of the inductive techniques given in Section 1.1 or here to be most convenient for a given problem in the set that follows.

Problems

Some of the following problems do not require an inductive proof.

1. Prove that there exists an infinite sequence of positive primes

$$p_1, p_2, p_3, \ldots$$

in $\mathbf{Z}$ such that $p_{n+1} > p_n$ for all positive integers n.

2. In a round robin tennis tournament, each player had exactly one match with every other contestant, and there were no ties. Prove that it is possible to list the players' names in a column so that each contestant beat the one listed immediately below him (except that the last one listed need not have won any match).

3. In the "Hit One Hundred" game, the two players alternate in announcing numbers. The first number must be in $\{1, 2, 3\}$. Then each player must announce a number that is 1, 2, or 3 more than the last number given by his opponent. A player wins if he calls out the number 100. Prove that the second player can always win if he uses perfect strategy.

4. Generalize on the game of Problem 3.

5. (i) Let S be a subset of $X = \{a, a+1, a+2, \ldots\}$ with a in S and such that h is in S whenever $a, a+1, \ldots, h-1$ are in S. Prove that $S = X$.
(ii) Prove that Algorithm 1 is a valid technique for establishing that a proposition $P(n)$ is true for all n in $X = \{a, a+1, a+2, \ldots\}$.

6. For all n in $X = \{336, 337, 338, \ldots\}$, prove that n cents worth of postage can be assembled using only 17-cent stamps and 22-cent stamps. [*Hint:* See Example 1.]

7. Let a sequence $u_0, u_1, u_2, \ldots$ be defined by $u_0 = 1$ and $u_{n+1} = 4u_n - 1$. Prove that $u_n = (2 \cdot 4^n + 1)/3$.

8. Let a sequence $v_0, v_1, \ldots$ be defined by $v_0 = 0$, $v_1 = 1$, and $v_n = v_{n-1} - v_{n-2}$ for $n \geq 2$. Prove that $v_{n+3} = -v_n$ and $v_{n+6} = v_n$ for all $n \geq 0$.

9. Let $0! = 1$; and for every n in $\mathbf{Z}^+$, let $n! = 1 \cdot 2 \cdot 3 \cdots n$. Let $\binom{n}{k} = n!/[k!(n-k)!]$ for integers k and n with $0 \leq k \leq n$. Prove the following.

(a) $\binom{n}{k-1} + \binom{n}{k} = \binom{n+1}{k}$ for $1 \leq k \leq n$.

(b) $(1+x)^n = \binom{n}{0} + \binom{n}{1}x + \binom{n}{2}x^2 + \cdots + \binom{n}{n}x^n$.

(c) $\binom{n}{k}$ is the number of subsets with k elements in a set with n elements.

10. For $k = 1, 2, \ldots, n$, let

$$u^{(k)} = \frac{d^k u}{dx^k} \quad \text{and} \quad v^{(k)} = \frac{d^k v}{dx^k}$$

exist. Prove ***Leibniz's Rule*** — that is,

$$\frac{d^n(uv)}{dx^n} = \binom{n}{0}uv^{(n)} + \binom{n}{1}u^{(1)}v^{(n-1)} + \binom{n}{2}u^{(2)}v^{(n-2)} + \cdots + \binom{n}{n}u^{(n)}v.$$

11. Let $F_1, F_2, \ldots$ be the Fibonacci sequence $1, 1, 2, 3, \ldots$. Prove that

$$\binom{n}{0}F_m + \binom{n}{1}F_{m+1} + \binom{n}{2}F_{m+2} + \cdots + \binom{n}{n}F_{m+n} = F_{m+2n}.$$

12. Prove that the number of terms in $(x_1 + x_2 + \cdots + x_k)^n$ is

$$\binom{n+k-1}{k-1}.$$

13. Let S be a set $\{p_1, p_2, \ldots, p_k\}$ of k distinct positive primes in $\mathbf{Z}$. Let $A(k, n)$ be the number of positive integers s such that s is a product $q_1 q_2 \cdots q_n$ of n (not necessarily distinct) primes with each q_i in S. Find a simple closed form for $A(k, n)$, and prove it correct for all positive integers k and n.

14. Prove the following.
(a) $1 + 2 + 3 + \cdots + n = n(n + 1)/2$.
(b) $1 \cdot 2 + 2 \cdot 3 + 3 \cdot 4 + \cdots + n(n + 1) = n(n + 1)(n + 2)/3$.
(c) $1 \cdot n + 2(n - 1) + 3(n - 2) + \cdots + (n - 1) \cdot 2 + n \cdot 1 = n(n + 1)(n + 2)/6$.
(d) $1^3 + 2^3 + 3^3 + \cdots + n^3 = (1 + 2 + 3 + \cdots + n)^2$.

15. A permutation θ in $\mathbf{S}_n$ is a ***derangement*** if $\theta(x) \neq x$ for all x in $\{1, 2, \ldots, n\}$. For $n \geq 1$, let d_n be the number of derangements in $\mathbf{S}_n$. Also let $d_0 = 1$. Prove the following.

(a) $d_n = (n - 1)(d_{n-1} + d_{n-2})$ for $n \geq 2$.
(b) $d_n = nd_{n-1} + (-1)^n$ for $n \geq 1$.
(c) $d_n = n!\left[\frac{1}{0!} - \frac{1}{1!} + \frac{1}{2!} - \cdots + (-1)^n \frac{1}{n!}\right]$.
(d) $\binom{n}{0}d_0 + \binom{n}{1}d_1 + \binom{n}{2}d_2 + \cdots + \binom{n}{n}d_n = n!$.
(e) $n!\binom{n}{n} - (n-1)!\binom{n}{n-1} + (n-2)!\binom{n}{n-2} - \cdots + (-1)^n 0!\binom{n}{0} = d_n$.
(f) $\binom{n}{1}d_{n-1} + 2\binom{n}{2}d_{n-2} + 3\binom{n}{3}d_{n-3} + \cdots + n\binom{n}{n}d_0 = n!$.

16. Prove that all the subsets of a set X with n elements can be numbered as $\beta_1, \beta_2, \ldots, \beta_m$ (with $m = 2^n$) such that β_1 is the empty set, β_m has a single element, and for $1 \leq i < m$ the set β_{i+1} is obtained either by adjoining a single element of X to β_i or by deleting a single element from β_i.

17. Let A_n be the maximum number of regions into which a plane can be cut by $n + 5$ straight lines, of which 5 are parallel to each other. Find and prove a simple closed form for A_n.

18. Let $B(m, n)$ be the maximum number of 3-dimensional regions into which 3-space can be cut using $m + n$ planes of which m are parallel to each other. Find and prove a closed form for $B(m, n)$.

19. Let U be a noncommutative ring with unity. Let r and v be in U with v an invertible. Prove that there exists an element s in U such that $vr^n = s^n v$ for all positive integers n.

20. Prove the following for all integers n; it may sometimes be helpful to give a proof first for all nonnegative n.
(a) $n^5 + 4n \equiv 0 \pmod 5$.
(b) $n(n^2 + 1)(n^2 + 4) \equiv 0 \pmod{10}$.
(c) $n^5 - 5n^3 + 4n \equiv 0 \pmod{120}$.
(d) $n^5 + 11n^3 + 4n \equiv 0 \pmod 8$.

21. Prove that $n^5 - 5n^3 + 4n \equiv 0 \pmod{360}$ for all integers n such that $n \not\equiv 0 \pmod 3$.

22. Let a be in $\mathbf{Z}$, let n be in $\mathbf{Z}^+$, and let $m = 2^n$.
(i) Prove that $(2a - 1)^m \equiv 1 \pmod{4m}$.
(ii) Explain why the multiplicative group $\mathbf{V_{4m}}$ of invertibles in $\mathbf{Z_{4m}}$ is not cyclic.

23. Let a be in $\mathbf{Z}$, let n be in $\mathbf{Z}^+$, and let $m = 3^n$. Prove the following.
(i) $(3a \pm 1)^m \equiv \pm 1 \pmod{3m}$. (The $\pm$ signs must be chosen to be the same.)
(ii) Let $q = (3m - 1)/2$. Then there do not exist $x_1, x_2, \ldots, x_{q-1}$ in $\mathbf{Z}$ such that

$$x_1^m + x_2^m + \cdots + x_{q-1}^m = 3ma \pm q.$$

24. Let D be an ordered integral domain. Prove the following.
(i) If $0 < a < 1$ in D, then $0 < a^{n+1} < a^n < 1$ for all n in $\mathbf{Z}^+$.
(ii) If $1 < b$ in D, then $1 < b^n < b^{n+1}$ for all n in $\mathbf{Z}^+$.
(iii) If v is an invertible in D and the multiplicative order of v is finite, v must be in $\{-1, 1\}$.

25. Let D be an ordered integral domain. Given that $1 + a \geq 0$ in D, prove that $(1 + a)^n \geq 1 + n \cdot a$ for all n in $\mathbf{Z}^+$.

26. Prove that $3^n \geq n^3$ for all n in $\mathbf{Z}^+$.

27. Prove that $2n\sqrt{n}/3 < \sqrt{1} + \sqrt{2} + \sqrt{3} + \cdots + \sqrt{n}$.

28. Prove that $\sqrt{1} + \sqrt{2} + \cdots + \sqrt{n} < (4n + 3)\sqrt{n}/6$.

29. Let F_n be the nth Fibonacci number and let $R_n = F_{n+1}/F_n$. Prove that $1 \leq R_{2n-1} < R_{2n+1} < R_{2n+2} < R_{2n}$.

30. Let F_n and L_n be the nth Fibonacci and nth Lucas numbers, respectively. Let $a = (1 + \sqrt{5})/2$ and $b = (1 - \sqrt{5})/2$. Prove the following.
(i) $F_n = (a^n - b^n)/(a - b)$.
(ii) $L_n = a^n + b^n$.
(iii) $a^n = (L_n + F_n\sqrt{5})/2$.
(iv) $b^n = (L_n - F_n\sqrt{5})/2$.
(v) $a^n = aF_n + F_{n-1}$ for $n \geq 2$.

31. Let F be a field with m elements. Let d be a fixed positive integer. How many distinct polynomials α are there such that

$$\alpha = (x - r_1)(x - r_2)(x - r_3) \cdots (x - r_d)$$

with the r_i (not necessarily distinct) elements of F?

32. Complete the proof of the Chinese Remainder Theorem (Theorem 3 of Section 7.1).

33. Let $x = t - t^{-1}$ and $y_n = t^n - t^{-n}$ for all positive integers n. Prove that y_n is a polynomial in x with rational coefficients if and only if n is odd.

34. Let $u_1, u_2, u_3, \ldots$ be defined by $u_1 = 2$ and $u_{n+1} = u_n(u_n - 1) + 1$ for $n \geq 1$. In $\mathbf{Q}[x]$ let

$$(x - u_1)(x - u_2) \cdots (x_1 - u_n) = x^n - s_{n-1}x^{n-1} + \cdots + (-1)^{n-1}s_1 x + (-1)^n s_0.$$

Prove that $s_1 = s_0 - 1$.

35. Let the u_i be as in Problem 34. Let $v_i = u_i$ for $1 \leq i < n$ and $v_n = u_n - 2$. In $\mathbf{Q}[x]$, let

$$(x - v_1)(x - v_2) \cdots (x - v_n) = x^n - t_{n-1}x^{n-1} + \cdots + (-1)^{n-1}t_1 x + (-1)^n t_0.$$

Prove that $t_1 = t_0 + 1$.

36. Let n be a positive integer; $m = 2^n$; and $a_1, a_2, \ldots, a_m$ be positive elements of an ordered integral domain D. Prove that

$$m^m \cdot (a_1 a_2 \cdots a_m) \leq (a_1 + a_2 + \cdots + a_m)^m,$$

with equality holding if and only if $a_1 = a_2 = \cdots = a_m$.

37. Let $b_1, b_2, \ldots, b_n$ be positive elements of an ordered integral domain D. Prove that

$$n^n \cdot (b_1 b_2 \cdots b_n) \leq (b_1 + b_2 + \cdots + b_n)^n,$$

with equality holding if and only if $b_1 = b_2 = \cdots = b_n$.

38. Let $a_1, a_2, \ldots, a_n$ be nonnegative real numbers;

$$A = (a_1 + a_2 + \cdots + a_n)/n; \quad \text{and} \quad G = \sqrt[n]{a_1 a_2 \cdots a_n}.$$

Prove that $A \geq G$ and that $A = G$ if and only if

$$a_1 = a_2 = \cdots = a_n.$$

(This is known as the ***Inequality on the Means***.)

39. Let $a_1, a_2, \ldots, a_n$ be positive real numbers;

$$G = \sqrt[n]{a_1 a_2 \cdots a_n}; \quad \text{and} \quad H = \frac{n}{\dfrac{1}{a_1} + \dfrac{1}{a_2} + \cdots + \dfrac{1}{a_n}}.$$

Prove that $G \geq H$ and that $G = H$ if and only if

$$a_1 = a_2 = \cdots = a_n.$$

40. Let $n > 1$; let $r_1, r_2, \ldots, r_n$ be nonnegative real numbers; and let

$$(x - r_1)(x - r_2) \cdots (x - r_n)$$
$$= x^n - s_{n-1}x^{n-1} + s_{n-2}x^{n-2} - \cdots + (-1)^{n-1}s_1 x + (-1)^n s_0.$$

Prove that $s_{n-j}s_j \geq \binom{n}{j}^2 s_0$ for $0 < j < n$ $\left[\text{where } \binom{n}{j} \text{ is the binomial coefficient } n(n-1)\cdots(n-j+1)/1 \cdot 2 \cdots j\right]$.

7.3 Some Number Theoretic Functions

Arithmetic is where the answer is right and everything is nice and you can look out of the window and see the blue sky — or the answer is wrong and you have to start all over and try again. —**Carl Sandburg**

Here we derive explicit formulas for some of the more frequently encountered functions from the Theory of Numbers.

Let ϕ be the Euler ϕ-function; then $\phi(n)$ is the number of integers in $\{1, 2, \ldots, n\}$ that are relatively prime to n. (See Problem 38 of Section 4.3.) We now define two other number-theoretic functions (and later in this section introduce one more, the partition function).

NOTATION **The σ and τ Functions**

Given a positive integer n, $\sigma(n)$ denotes the sum of the positive integral divisors of n, and $\tau(n)$ denotes the number of such divisors.

First let us consider the problem of calculating $\phi(n)$.

EXAMPLE 1 **Inclusion–Exclusion Calculation of $\phi(28)$**

The set of positive integers less than or equal to 28 and relatively prime to 28 is

$$A = \{1, 3, 5, 9, 11, 13, 15, 17, 19, 23, 25, 27\}.$$

Let us count these numbers in a way that will suggest a method to handle the general case. Since $28 = 2^2 \cdot 7$, a number is in A if and only if it is in $X = \{1, 2, \ldots, 28\}$ and is not an integral multiple of 2 or of 7. In X there are 28/2 multiples of 2 and 28/7 multiples of 7.

When we delete the multiples of 2 and the multiples of 7 from X, are we left with $28 - 14 - 4$ integers in A? No, since our tally of the integers deleted has counted the integers 14 and 28 twice, once as a multiple of 2 and once as a multiple of 7. We add on a correction term and have

$$\begin{aligned}\phi(28) &= 28 - \frac{28}{2} - \frac{28}{7} + \frac{28}{2 \cdot 7}\\ &= 28\left(1 - \frac{1}{2} - \frac{1}{7} + \frac{1}{2 \cdot 7}\right)\\ &= 28\left(1 - \frac{1}{2}\right)\left(1 - \frac{1}{7}\right) = 28 \cdot \frac{1}{2} \cdot \frac{6}{7} = 12.\end{aligned}$$

Next we take up the general case. In the following, we let n be in $\{2, 3, \ldots\}$ and let its standard factorization be

$$n = (p_1)^{r_1}(p_2)^{r_2} \cdots (p_m)^{r_m}. \tag{S}$$

Note that here $p_1, p_2, \ldots, p_m$ are distinct positive primes.

THEOREM 1 **Euler ϕ-Function**

Let the standard factorization of n be as in (S). Then

$$\phi(n) = n\left(1 - \frac{1}{p_1}\right)\left(1 - \frac{1}{p_2}\right) \cdots \left(1 - \frac{1}{p_m}\right).$$

Proof As in Example 1, we first count all the positive integers less than or equal to n and then subtract away the number of multiples of p_1, of p_2, of p_3, and so on to p_m, since these are clearly the integers not relatively prime to n. But we have deleted multiples of p_1p_2 twice from our list, for example, once as multiples of p_1 and once as multiples of p_2. We must, therefore, add back to our sum the number of multiples of p_1p_2, p_1p_3, p_2p_3, and so forth. Now we consider multiples of $p_1p_2p_3$ and other such triple products. They have been deleted three times (once for each prime factor) and put back in three times (once for each combination of two prime factors). But they must be deleted, so we must subtract away the number of multiples of three prime factors. We continue this process and obtain the following value for $\phi(n)$:

$$\phi(n) = n - \sum_{1 \le i \le m} \frac{n}{p_i} + \sum_{1 \le i < j \le m} \frac{n}{p_i p_j} - \sum_{1 \le i < j < k \le m} \frac{n}{p_i p_j p_k} + \cdots$$

$$+ (-1)^m \frac{n}{p_1 p_2 \cdots p_m}$$

$$= n\left(1 - \frac{1}{p_1}\right)\left(1 - \frac{1}{p_2}\right) \cdots \left(1 - \frac{1}{p_m}\right). \qquad \blacksquare$$

The technique of proof above, of counting the elements of a set by alternately deleting from and adding to a set, is known as ***Sylvester's Principle*** or the ***inclusion–exclusion principle***. (See the biographical note on Sylvester at the close of this section.)

THEOREM 2 **σ-Function**

If the standard factorization of n is

$$n = (p_1)^{r_1}(p_2)^{r_2} \cdots (p_m)^{r_m},$$

then

$$\sigma(n) = \frac{p_1^{r_1+1} - 1}{p_1 - 1} \cdot \frac{p_2^{r_2+1} - 1}{p_2 - 1} \cdots \frac{p_m^{r_m+1} - 1}{p_m - 1}.$$

Proof Let us consider the product

$$(1 + p_1 + \cdots + p_1^{r_1})(1 + p_2 + \cdots + p_2^{r_2}) \cdots (1 + p_m + \cdots + p_m^{r_m}). \qquad \text{(P)}$$

When multiplied out, this product will consist of all possible terms of the form

$$p_1^{s_1} p_2^{s_2} \cdots p_m^{s_m},$$

where $0 \le s_i \le r_i$ for $i = 1, 2, \ldots, m$ and each term will appear only once. But such products are, in fact, the divisors of n; and hence the product (P) is the sum of all the divisors of n. Using the formula for the sum of a geometric progression to sum each factor, we have the formula for $\sigma(n)$ given above. ■

THEOREM 3 **τ-Function**

If the standard factorization of n is

$$n = (p_1)^{r_1}(p_2)^{r_2} \cdots (p_m)^{r_m},$$

then

$$\tau(n) = (r_1 + 1)(r_2 + 1) \cdots (r_m + 1).$$

Proof In the multiplied-out expansion of the product (P) given in Theorem 2, if each term (that is, positive integral divisor of n) is replaced by 1, then the sum of the divisors is changed into the number of divisors. Thus, $\tau(n)$ may be calculated by letting $p_1 = p_2 = \cdots = p_m = 1$ in (P); then $\tau(n) = (r_1 + 1)(r_2 + 1) \cdots (r_m + 1)$. ∎

Each of the functions ϕ, σ, and τ has the following property.

DEFINITION 1

Multiplicative Function

A function f from $\mathbf{Z}^+$ to $\mathbf{Z}^+$ is said to be **multiplicative** if $f(rs) = f(r)f(s)$ whenever $\gcd(r, s) = 1$.

The reader is asked in Problems 8, 24, and 32 of this section to show that each of ϕ, σ, and τ is multiplicative.

We next use the σ-function to introduce a concept that is involved in a famous unsolved problem.

DEFINITION 2

Perfect Numbers

A positive integer n is **perfect** if $\sigma(n) = 2n$.

One easily sees that 6 is perfect since $\sigma(6) = 1 + 2 + 3 + 6 = 12 = 2 \cdot 6$. The next perfect integer is 28. Each of these perfect numbers is of the form $2^s p$, where s and p are positive integers with p prime and

$$p = 1 + 2 + 2^2 + \cdots + 2^s = 2^{s+1} - 1.$$

It is left to the reader, as Problem 19 of this section, to prove that all numbers of this form are perfect. The much harder converse — that all even perfect numbers are of this form — was proved by Euler. It has been conjectured that no odd perfect numbers exist, but no proof or counterexample has been obtained yet. It is known that an odd perfect number would have to be greater than 10^{36}, so one should not expect to find a counterexample easily.

A somewhat related concept is given in the following.

DEFINITION 3

Amicable Pair

Positive integers a and b form an **amicable pair** if

$$\sigma(a) = a + b = \sigma(b).$$

In Problem 21 of this section, the reader is asked to show that 220 and 284 form an amicable pair.

In Section 2.13 we used the calculation of the number of cycles and products of disjoint cycles of various types to obtain important results concerning the symmetric group $\mathbf{S_5}$. One may note that every permutation in $\mathbf{S_5}$ is representable in one of the following ways:

$$(abcde),\ (abcd)(e),\ (abc)(de),\ (abc)(d)(e),\ (ab)(cd)(e),\ (ab)(c)(d)(e),\ (a)(b)(c)(d)(e),$$

where a, b, c, d, and e are distinct integers in $\{1, 2, 3, 4, 5\}$.

These types of representations correspond to the ways in which 5 can be written as a positive integer or a sum of positive integers:

$$5,\ 4 + 1,\ 3 + 2,\ 3 + 1 + 1,\ 2 + 2 + 1,\ 2 + 1 + 1 + 1,\ 1 + 1 + 1 + 1 + 1. \qquad (1)$$

We see that there are seven such partitions of the positive integer 5.

In classifying the permutations in $\mathbf{S_n}$, and in other applications, it is helpful to be able to generalize from the integer 5 to a general positive integer n, as we now do.

DEFINITION 4

The Partition Function p

An expression for a positive integer n as (a positive integer or) a sum of positive integers will be called a **partition of n**. The number of such partitions, where the order of the terms is not taken into account, is denoted by $p(n)$, and p is called the **(unordered) partition function**.

We note that $3 + 1 + 1$, $1 + 3 + 1$, and $1 + 1 + 3$ are all considered to be the same partition of 5 and that 5 itself is a partition of 5. Display (1) above shows that $p(5) = 7$.

Euler examined this partition function in his *Introductio in Analysin Infinitorum* (1748) and obtained the following remarkable result.

THEOREM 4 **Generating Function for p**

$$\frac{1}{(1 - x)(1 - x^2)(1 - x^3)(1 - x^4)\cdots} = p(0) + p(1)x + p(2)x^2 + \cdots.$$

This means that, if the function on the left is expanded as a power series in x, we can read off the values of $p(n)$ as the coefficients of the powers of x in the series; such a function is called a ***generating function***. [For uniformity, we let $p(0)$ be 1.]

We do not prove Theorem 4 or even define precisely what is meant by the infinite product in the denominator of the generating function; instead we give an intuitive argument to show that the result is plausible. If we were to expand

$$\frac{1}{1-x}\cdot\frac{1}{1-x^2}\cdot\frac{1}{1-x^3}\cdot\frac{1}{1-x^4}\cdots$$

factor by factor into series of powers of x, we would have

$$(1+x+x^2+\cdots)(1+x^2+x^4+\cdots)(1+x^3+x^6+\cdots)(1+x^4+x^8+\cdots)\cdots$$
$$=(1+x^1+x^{1+1}+\cdots)(1+x^2+x^{2+2}+\cdots)(1+x^3+x^{3+3}+\cdots)\cdots. \quad (2)$$

For example, what is the coefficient of x^5 when this is multiplied out and like terms collected? First we note that we can ignore series $(1+x^m+x^{2m}+\cdots)$ in which $m>5$. Then Table 7.1 shows all the ways of selecting terms x^j from the series that are the first five factors in (2) so as to obtain x^5 in the expansion. Thus we see that x^5 appears precisely 7 times in the product; the coefficient of x^5 is therefore 7, which is $p(5)$.

The values of $p(n)$ can also be calculated from a recursion formula, which we now give without proof:

$$\begin{aligned}p(n) &= p(n-1)+p(n-2)-p(n-5)-p(n-7)\\ &\quad + p(n-12)+p(n-15)-p(n-22)-p(n-26)\\ &\quad +\cdots\\ &\quad +(-1)^k p\left(n-\frac{k(3k-1)}{2}\right)+(-1)^k p\left(n-\frac{k(3k+1)}{2}\right)+\cdots\end{aligned}$$

In 1917 S. Ramanujan discovered an explicit formula for $p(n)$. Unfortunately it involved transcendental numbers and, on the face of it, gave those who saw it little confidence that it would work. Since it is clear that the values of $p(n)$ must be integers, and since a function involving transcendental

TABLE 7.1

x^5 term in expansion	Term chosen from: 1st factor	2nd factor	3rd factor	4th factor	5th factor
x^5	1	1	1	1	x^5
x^{1+4}	x^1	1	1	x^4	1
x^{2+3}	1	x^2	x^3	1	1
x^{1+1+3}	x^{1+1}	1	x^3	1	1
x^{1+2+2}	x^1	x^{2+2}	1	1	1
$x^{1+1+1+2}$	x^{1+1+1}	x^2	1	1	1
$x^{1+1+1+1+1}$	$x^{1+1+1+1+1}$	1	1	1	1

numbers will — without using the greatest integer function or some such device — usually not give an integer value, the formula did not look promising. So, as a test, the value of $p(200)$ was found using the formula, and simultaneously P. A. MacMahon was put to work calculating $p(200)$ using the recursion formula. The two results agreed: $p(200) = 3{,}972{,}999{,}029{,}388$. Ramanujan was vindicated and Hardy set about finding a proof of the formula. It is evident that $p(n)$ grows rapidly. The explicit formula is quite complicated and is useful mainly in estimating the size of $p(n)$ for large values of n. Readers interested in pursuing this matter further should read lecture VIII in G. H. Hardy's book *Ramanujan: Twelve Lectures on Subjects Suggested by His Life and Work* (Cambridge University Press, 1940). In this lecture Hardy also gave the value of $p(14031)$, a formidable number indeed and an impressive calculation in the days prior to electronic computers.

Problems

1. Calculate the following.
(a) $\phi(60)$; (b) $\sigma(728)$; (c) $\tau(728)$.

2. Calculate the following.
(a) $\phi(360)$; (b) $\sigma(7280)$; (c) $\tau(7280)$.

3. Find all the positive integers n such that
(a) $\phi(n) = 4$; (b) $\phi(n) = 6$.

4. Find all the positive integers n such that
(a) $\phi(n) = 8$; (b) $\phi(n) = 10$.

5. Explain why $\phi(n) = \phi(2n)$ for all odd positive integers n.

6. Show the following.
(i) If $\gcd(a, n) = 1$, then $\gcd(n - a, n) = 1$.
(ii) For $n \geq 2$, the sum of all the integers a such that both $1 \leq a \leq n$ and $\gcd(a, n) = 1$ is equal to $n\phi(n)/2$.

7. Let $p, q, \ldots, r$ be distinct positive primes and let $a, b, \ldots, c$ be any positive integers. Explain why

$$\phi(p^a q^b \cdots r^c) = \phi(p^a)\phi(q^b) \cdots \phi(r^c).$$

8. Prove that ϕ is a multiplicative function; that is, prove that $\phi(rs) = \phi(r)\phi(s)$ whenever r and s are relatively prime positive integers.

9. Let a and b be positive integers and let $d = \gcd(a, b)$. Prove that

$$\phi(ab) = d\phi(a)\phi(b)/\phi(d).$$

10. Let p and n be positive integers with p prime. Prove the following.
(a) $p \mid n$ implies that $(p - 1) \mid \phi(n)$.
(b) $(p - 1) \mid \phi(n)$ does not imply that $p \mid n$.

11. Let d and n be positive integers with $d \mid n$. Show that $\phi(d) \mid \phi(n)$.

12. (i) Find the smallest positive prime p with $7 \mid \phi(p)$.
(ii) Prove that there is no positive integer n with $\phi(n) = 14$.

13. (i) Let $\phi(n) = 48$, let p be a positive prime, and let $p \mid n$. Prove that p is 2, 3, 5, 7, or 13.
(ii) Given that $\phi(n) = 48$, prove that n must be an integral multiple of 7 or 9 or 13.
(iii) Find the eleven positive integers n satisfying $\phi(n) = 48$.

★ **14.** How many rational numbers are expressible as a/b with a and b integers satisfying both $0 \le a < b \le n$ and $\gcd(a, b) = 1$?

15. Show that, for every positive integer n,

$$\frac{\sigma(n)}{n} = \sum_{d|n} \frac{1}{d},$$

where $\sum_{d|n}$ means the sum over all positive integral divisors d of n.

$$\left[\text{For example, } \sum_{d|6} \frac{1}{d} = \frac{1}{1} + \frac{1}{2} + \frac{1}{3} + \frac{1}{6} = \frac{12}{6} = \frac{\sigma(6)}{6}.\right]$$

16. Let $\sigma_2(n)$ denote the sum of the squares of the positive integral divisors of n; that is, let

$$\sigma_2(n) = \sum_{d|n} d^2.$$

Also, let the standard factorization of n be

$$n = (p_1)^{r_1}(p_2)^{r_2} \cdots (p_m)^{r_m}.$$

Show that

$$\sigma_2(n) = \sigma(n) \prod_{i=1}^{m} \frac{(p_i)^{r_i+1} + 1}{p_i + 1}.$$

17. Prove that 6 and 28 are the two smallest perfect numbers.

18. Show that 496 is a perfect number.

19. Let s be a positive integer such that $2^{s+1} - 1$ is a prime p. Prove that $2^s p$ is perfect.

20. Given that n is perfect, show that $\sum_{d|n} \frac{1}{d} = 2$.

21. Show that 220 and 284 are an amicable pair.

22. (i) Find $\sigma(1184) - 1184$.
(ii) Find $\sigma(1210) - 1210$.
(iii) Find an amicable pair of positive integers whose difference is 26.

23. Let $p, q, \ldots, r$ be distinct positive primes and let $a, b, \ldots, c$ be positive integers. Explain why

$$\sigma(p^a q^b \cdots r^c) = \sigma(p^a)\sigma(q^b) \cdots \sigma(r^c).$$

24. Prove that σ is multiplicative; that is, prove that $\sigma(mn) = \sigma(m)\sigma(n)$ whenever m and n are relatively prime positive integers.

25. (i) Tabulate $\tau(n)$ for $n = 1, 2, 3, \ldots, 20$.
(ii) For which numbers n in $\{1, 2, 3, \ldots, 20\}$ is $\tau(n)$ odd?

26. Make a conjecture concerning the positive integers n such that $\tau(n)$ is odd, and then prove that the conjecture is correct.

27. Characterize the positive integers n such that $\tau(n)$ is a prime p.

28. Find the smallest positive integer n such that $\tau(n) = 5$.

29. Find the smallest positive integer n such that $\tau(n) = 10$.

30. (a) Find the smallest positive integer that has at least 10 positive integral divisors.
(b) Find the smallest positive integer with exactly 28 positive integral divisors.

31. Do the analogue of Problems 7 and 23 for the τ-function.

Godfrey Harold Hardy
1877–1947

G. H. Hardy was at one time Savilian Professor of Geometry at Oxford and later Sadleirian Professor of Pure Mathematics at Cambridge. He was certainly one of the most eminent mathematicians of the twentieth century, and he made, in collaboration with his colleagues, J. E. Littlewood (1885–1977) and S. Ramanujan (see separate biographical note), profound contributions to several areas of number theory.

Hardy was unusually colorful, and much has been written about his eccentricities. There are many wonderful stories about his life, far more than we could hope to recount here. We refer the reader to Hardy's own account of his life in mathematics, *A Mathematician's Apology* (Cambridge University Press, 1969), and an account by another of Hardy's collaborators, George Pólya, "Some Mathematicians I Have Known," *American Mathematical Monthly* 76 (7) (1969): 746–53.

Srinivasa Ramanujan
1887–1920

Ramanujan was born in the Tanjore district of southern India and received only a minimal formal education in mathematics. He nevertheless had a phenomenal talent for discovering profound and complex formulas, many of which he was able to prove to the satisfaction of other mathematicians of the day. In many cases, however, he was unable to explain to contemporaries how he came up with the formulas. Much effort has been expended in the intervening years trying to prove some of these. With the discovery of some of his lost notebooks within the last decade, the effort continues. With the achievement of each proof, Ramanujan's stature grows.

Before coming to England, he corresponded with G. H. Hardy, who recognized his extraordinary genius and arranged to have him come to Cambridge. He arrived in England in 1914 and proceeded to fill in gaps in his understanding of function theory and modern concepts of what constituted proof.

He became ill in 1917 and died three years later. During those years, he continued to make dramatic new discoveries; and it was during this period, part of which he spent in the hospital, that he gave us one of the best-known anecdotes passed along in mathematics classrooms. On one visit to Ramanujan in the hospital, Hardy mentioned that he had arrived in taxi number 1729. Ramanujan immediately commented that 1729 is indeed an interesting number since it is the smallest positive integer that can be represented in two different ways as the sum of two cubes: $1729 = 9^3 + 10^3 = 1^3 + 12^3$.

In 1991 a very interesting and popular biography of Ramanujan appeared: *The Man Who Knew Infinity*, by Robert Kanigel. A few years earlier, a television program of his life appeared on the PBS series *Nova*.

James Joseph Sylvester
1814–1897

Sylvester was born in London and was educated in England at St. John's College, Cambridge. He was made a member of the Royal Society at the age of 25, but he was denied a degree at Cambridge because of his religion. He spent a considerable amount of his adult life in the United States. In 1841 he was appointed professor of mathematics at the University of Virginia, but there too he met with hostility because of his nationality, his religion, and his opposition to slavery. After an encounter in which two students attacked him with a club (and he defended himself with a sword-cane), Sylvester returned to England. He came back to the United States in 1876, when he joined the faculty of Johns Hopkins University in Baltimore. There he founded the first graduate program in mathematics and the first mathematical research journal in the United States, the *American Journal of Mathematics.*

Between his two stays in the United States, he worked with Cayley in England and with him developed much of their important work on the theory of invariants. He was the first to use a number of words currently common in mathematics, including *invariant*, *Jacobian*, *discriminant*, and *Hessian.*

Sylvester, in addition to doing very good mathematics, wrote poetry. Unfortunately, the quality of his poetry did not measure up to that of his mathematics. He tended to write verse in which every line ended with the same sound. Two such poems were "Rosalind" and "Spring's Debut." Florian Cajori, the mathematical historian, wrote that "at the reading, at the Peabody Institute in Baltimore, of his Rosalind poem, consisting of about 400 lines all rhyming with 'Rosalind,' he first read all his explanatory footnotes, so as not to interrupt the poem; these took one hour and a-half. Then he read the poem itself to the remnant of his audience."

32. Prove that τ is multiplicative.

33. Tabulate the partition function $p(n)$ for $n = 1, 2, \ldots, 8$.

34. Classify the permutations in $\mathbf{S}_6$ according to the type of the standard form, and count the number of each type. Check using ord $\mathbf{S}_6 = 6! = 720$.

35. Find the positive integer n, given that the product of all the positive integral divisors of n is 2,560,000.

36. Prove that, for every n in $\mathbf{Z}^+$, the product P of all the positive integral divisors of n is $n^{\tau(n)/2}$.

★ **37.** For n in $\mathbf{Z}^+$, let $\gamma(n)$ be the smallest k in $\mathbf{Z}^+$ such that $\tau(k) = n$. For m in $\mathbf{Z}^+$, prove that $\gamma(2^m)$ is the product of the m smallest integers of the form p^e, with p a positive prime and e in the set $\{1, 2, 4, 8, \ldots\}$ of powers of 2.

★ **38.** State and prove a rule for obtaining $\gamma(3^m)$, where γ is as defined in Problem 37.

39. (a) Does $\phi(1)\tau(6) + \phi(2)\tau(3) + \phi(3)\tau(2) + \phi(6)\tau(1) = \sigma(6)$?
(b) Does $\phi(1)\tau(12) + \phi(2)\tau(6) + \phi(3)\tau(4) + \phi(4)\tau(3) + \phi(6)\tau(2) + \phi(12)\tau(1) = \sigma(12)$?

7.4 Quadratic Residues

The moving power of mathematical invention is not reasoning but imagination. —**Augustus De Morgan**

In the Theory of Numbers, an integer a is called a ***quadratic residue*** modulo a positive integer m if $\gcd(a, m) = 1$ and there exist integers x that satisfy

$$x^2 \equiv a \pmod{m}. \tag{1}$$

If $\gcd(a, m) = 1$ and there are no integer solutions of the congruence (1), a is a ***quadratic nonresidue*** modulo m.

If $a \equiv b \pmod{m}$, then b is a quadratic residue modulo m if and only if a is a quadratic residue; similarly for nonresidues. Hence we may restrict ourselves to integers a satisfying $0 < a < m$. We also note that a is a quadratic residue modulo m if and only if there is an integer x such that $\bar{x}^2 = \bar{a}$ in the multiplicative group $\mathbf{V_m}$ of invertibles of $\mathbf{Z_m} = \mathbf{Z}/(m)$.

For small positive integers m, one can easily find the quadratic residues modulo m by squaring m consecutive integers, as one can see in Examples 3 and 4 of Section 4.2 and Example 3 of Section 4.3. (In those examples, we used quadratic nonresidues to prove nonexistence of integer solutions for certain quadratic equations.)

Let us examine the pattern formed by the quadratic residues among 1, 2, 3, ..., $p - 1$ for some small odd primes p. In Table 7.2, an entry "R" means that n is a quadratic residue modulo p while an entry "—" means that it is not.

We note that precisely half of the considered integers are quadratic residues in each case. This is proved for all odd primes p in Theorem 1 below.

One might also see that the entries read the same from left to right as from right to left when $p \equiv 1 \pmod 4$. When $p \equiv 3 \pmod 4$, there is antisymmetry; that is, if one reverses each entry symbol as one reads from left to right, one obtains the entries from right to left. Each of these statements is true for general odd positive primes p; a proof is outlined in Problems 37 and 38 of Section 5.4.

Carl Friedrich Gauss derived a number of results on quadratic residues in his 1801 *Disquisitiones Arithmeticae*. We state the most famous as Theorem 2 below but first note that among these results is the proof that, for a fixed odd

TABLE 7.2

p \ n	1	2	3	4	5	6	7	8	9	10	11	12
3	R	—										
5	R	—	—	R								
7	R	R	—	R	—	—						
11	R	—	R	R	R	—	—	—	R	—		
13	R	—	R	R	—	—	—	—	R	R	—	R

positive prime p, the product of two quadratic residues or of two nonresidues is a quadratic residue, while the product of a quadratic residue and a nonresidue is a nonresidue. We shall obtain this as a special case using the following results.

THEOREM 1 Squares of Invertibles in D

Let V be the multiplicative group of invertibles of an integral domain D whose characteristic is not 2. Let W be the set of squares of elements of V. Then W is a subgroup in V and has index 2 when V is finite.

Proof Let θ be the mapping from V to itself with $\theta(v) = v^2$. Since D is commutative,

$$\theta(ab) = abab = aabb = a^2b^2 = \theta(a)\theta(b);$$

that is, θ is a group homomorphism. It then follows from Theorem 2(c) of Section 3.3 that the image set W is a subgroup in the codomain V.

Next we find the kernel K of θ. We note that k is in K if and only if $k^2 = 1$, which is equivalent to

$$(k-1)(k+1) = 0.$$

Since D has no 0-divisors, this implies that k is 1 or -1. Also, 1 and -1 are distinct, since the characteristic of D is not 2. Hence, $K = \{1, -1\}$.

If a^2 is in W, the complete inverse image $\theta^{-1}(a^2)$ is the coset $aK = \{a, -a\}$, and the collection of all these cosets partitions V into r pairs, where $r = \text{ord } W$. Therefore, ord $V = 2r$, and W has index 2 in V. ∎

If p is an odd prime, $\mathbf{Z_p}$ is a finite integral domain with characteristic different from 2. Then Theorem 1 tells us that the set R of squares in $\mathbf{V_p}$ is a subgroup with index 2 in $\mathbf{V_p}$. If N denotes the other coset of R in $\mathbf{V_p}$ (that is, the coset of nonsquares), then $\mathbf{V_p}/R = \{R, N\}$ has the multiplication table

	R	N
R	R	N
N	N	R

(2)

Translating these facts into congruence language, we see that half of the integers in $\{1, 2, \ldots, p-1\}$ are quadratic residues and that table (2) sums up concisely in group theoretic terms the content of Gauss's theorem on products of quadratic residues and nonresidues.

The following definition will help us state a famous and deep property of quadratic residues that was first discovered independently by Euler and Legendre in 1785 but not proved by them. Gauss discovered it in 1795, proved it in several ways, and published a proof in his *Disquisitiones Arithmeticae* of 1801.

NOTATION Legendre Symbol

If p is an odd prime and $\gcd(n, p) = 1$, the Legendre symbol $\left(\frac{n}{p}\right)$ is defined as follows:

$$\left(\frac{n}{p}\right) = 1 \quad \text{if } n \text{ is a quadratic residue modulo } p,$$

$$\left(\frac{n}{p}\right) = -1 \quad \text{if } n \text{ is a quadratic nonresidue modulo } p.$$

THEOREM 2 Law of Quadratic Reciprocity

If p and q are distinct odd positive primes in $\mathbf{Z}$,

$$\left(\frac{p}{q}\right)\left(\frac{q}{p}\right) = (-1)^{(p-1)(q-1)/4}.$$

Since all the known proofs are somewhat involved, we do not present a proof; instead we refer the reader to one of the standard works on number theory listed in the bibliography. This law is useful in determining the quadratic residues and nonresidues for large primes.

Problems

1. Find the quadratic residues in $\{1, 2, 3, \ldots, p-1\}$ modulo
   (a) $p = 17$; (b) $p = 19$.
2. Find the quadratic residues modulo 23 in $\{1, 2, \ldots, 22\}$.
3. Show that the Law of Quadratic Reciprocity can be restated as follows. If p and q are distinct odd positive primes, then

$$\left(\frac{p}{q}\right) = \left(\frac{q}{p}\right)(-1)^{(p-1)(q-1)/4}.$$

4. Let p and q be distinct odd positive primes. Show that the Law of Quadratic Reciprocity is equivalent to the two following statements.
   (a) $\left(\frac{p}{q}\right) = \left(\frac{q}{p}\right)$ if either $p \equiv 1 \pmod 4$ or $q \equiv 1 \pmod 4$.
   (b) $\left(\frac{p}{q}\right) = -\left(\frac{q}{p}\right)$ if $p \equiv 3 \equiv q \pmod 4$.
5. Let p be an odd positive prime with $p \neq 5$. Use the preceding problem to show that 5 is a quadratic residue modulo p if and only if p is a quadratic residue modulo 5 — that is, if and only if $p \equiv \pm 1 \pmod 5$.

6. Let p be an odd prime with $p > 3$. Find $\left(\frac{p}{3}\right)$ and $\left(\frac{3}{p}\right)$ under each of the following assumptions.

   (a) $p \equiv 1 \pmod{12}$; (b) $p \equiv 5 \pmod{12}$; (c) $p \equiv 7 \pmod{12}$;
   (d) $p \equiv 11 \pmod{12}$.

7. Let p be an odd positive prime. Use Problem 38 of Section 5.4 to show the following.

   (a) $\left(\frac{-1}{p}\right) = 1$ and $\left(\frac{-n}{p}\right) = \left(\frac{n}{p}\right)$ if $p \equiv 1 \pmod 4$.

   (b) $\left(\frac{-1}{p}\right) = -1$ and $\left(\frac{-n}{p}\right) = -\left(\frac{n}{p}\right)$ if $p \equiv 3 \pmod 4$.

8. Use Problems 6 and 7 to find $\left(\frac{-3}{p}\right)$ under each of the following assumptions. (Here, p is a positive prime.)
   (a) $p \equiv 1 \pmod 6$; (b) $p \equiv 5 \pmod 6$.

9. Let p be an odd positive prime. Then it is a theorem of Euler that $\gcd(p, n) = 1$ implies that

$$\left(\frac{n}{p}\right) \equiv n^{(p-1)/2} \pmod p.$$

   Use this result to show the following.

   (i) $\left(\frac{p-1}{p}\right) = \left(\frac{-1}{p}\right) = (-1)^{(p-1)/2}$.

   (ii) $p - 1$ is a quadratic residue modulo p if and only if $p \equiv 1 \pmod 4$.

   (iii) If m is an integer greater than 1 and $p \mid (m^2 + 1)$, then $p \equiv 1 \pmod 4$.

Review Problems

1. Find the two smallest positive integers x satisfying both

$$x \equiv 0 \pmod 5 \quad \text{and} \quad x \equiv 4 \pmod 8.$$

2. Find the two smallest positive integers x satisfying the simultaneous congruences

$$x \equiv 4 \pmod 5, \quad x \equiv 3 \pmod 6, \quad \text{and} \quad x \equiv 2 \pmod 7.$$

3. Find all n in $\{0, 1, 2, \ldots, 105\}$ such that n is congruent to either 1 or -1 modulo p for every p in $\{2, 3, 5, 7\}$.

4. Find all n in $\{0, 1, 2, \ldots, 71\}$ such that n is congruent to either 1 or -1 modulo p for both $p = 11$ and $p = 13$.

5. Let R be a commutative ring. Prove that every α with $\deg \alpha > 0$ in $R[x]$ has an irreducible divisor in $R[x]$.

6. Let $p, a_1, a_2, \ldots, a_n$ be integers with p a prime. Also, let $p \mid (a_1 a_2 \cdots a_n)$. Prove that $p \mid a_i$ for at least one i in $\{1, 2, \ldots, n\}$ (and thus prove Lemma 2 of Section 1.7).

7. Let D be an integral domain with prime characteristic p. Prove that

$$(x + y)^{p^n} = x^{p^n} + y^{p^n}$$

   for all x and y in D and all nonnegative integers n.

8. Let D be an integral domain with prime characteristic p. Prove that

$$(x_1 + x_2 + \cdots + x_m)^{p^n} = (x_1)^{p^n} + (x_2)^{p^n} + \cdots + (x_m)^{p^n}$$

for all integers $m \geq 1$ and $n \geq 0$ and all x_i in D.

9. Let $u_0, u_1, u_2, \ldots$ be the sequence with $u_0 = u_1 = u_2 = 0$, $u_3 = 1$, and $u_n = 3u_{n-1} + 6u_{n-2} - 3u_{n-3} - u_{n-4}$ for $n > 3$. For $n > 1$, prove the following.
(a) $u_2 + u_4 + \cdots + u_{2n} = (5u_{2n} - 3u_{2n-1} - u_{2n-2})/4$.
(b) $u_3 + u_5 + \cdots + u_{2n+1} = (5u_{2n+1} - 3u_{2n} - u_{2n-1} - 1)/4$.

10. Let $F_0, F_1, F_2, \ldots$ be the Fibonacci sequence with $F_0 = 0$, $F_1 = 1$, and $F_n = F_{n-1} + F_{n-2}$ for $n > 1$. Let $v_n = F_n F_{n-1} F_{n-2}/2$. Show that v_n equals the u_n of the preceding problem for $n > 1$.

11. Let ϕ be the Euler ϕ-function. Find all a in $\{1, 2, \ldots, 25\}$ such that $\phi(p^n) = a$ for some positive prime p and positive integer n.

12. (i) Find all positive primes p such that $3 \mid (p - 1)$ and $\phi(p) \mid 96$.
(ii) Find all the positive integers n such that $\phi(n) = 96$.

13. Explain why $\phi(n)$ is an even integer for $n > 2$.

14. Let a be a positive integer. Explain why $\{n : \phi(n) = a\}$ is a finite set of positive integers.

15. For n in $\mathbf{Z}^+ = \{1, 2, 3, \ldots\}$, let $\sigma(n)$ be the sum of the positive integral divisors of n. Find all n in $\mathbf{Z}^+$ with $\sigma(n) = 18$.

16. Is there a maximum value of $\sigma(n)$? Explain.

17. Show that $\sigma(n) = 2n - 1$ for an infinite number of n's in $\mathbf{Z}^+$.

18. Let p be a prime in $\mathbf{Z}^+$ and $u_n = p^n$. Show that

$$\sigma(u_n) = u_n + \sigma(u_{n-1}) \quad \text{for } n \in \mathbf{Z}^+.$$

19. For n in $\mathbf{Z}^+$, let $\tau(n)$ be the number of d's in $\mathbf{Z}^+$ with $d \mid n$. Find the smallest n in $\mathbf{Z}^+$ with $\tau(n) = 14$.

20. Characterize the positive integers n such that $\tau(n) = 2$.

21. Characterize the positive integers n such that $\tau(n) = 3$.

22. For n in $\mathbf{Z}^+$, let $A(n)$ be the number of ordered pairs (x, y) of positive integers such that

$$\frac{1}{x} + \frac{1}{y} = \frac{1}{n}.$$

Tabulate $A(n)$ and $\tau(n^2)$ for $n = 1, 2, 3, 4, 5, 6$, and 7.

23. Find $p(9)$ (that is, the number of unordered partitions of 9).

24. Find $p(10)$.

25. Let p and q be odd positive primes in $\mathbf{Z}$, with $q \equiv 1 \pmod 4$ and $q \not\equiv 1 \pmod 8$. Show that p is a quadratic residue modulo q if and only if q is a quadratic residue modulo p.

26. For each of the following, find the three smallest odd positive primes p in $\mathbf{Z}$ satisfying the condition.

(i) $\left(\dfrac{p}{19}\right)\left(\dfrac{19}{p}\right) = 1.$

(ii) $\left(\dfrac{p}{19}\right)\left(\dfrac{19}{p}\right) = -1.$

27. Let $I = (r)$ and $J = (s)$ be principal ideals in **Z**. Show that there exist x in I and y in J with $x + y = 1$ if and only if r and s are relatively prime.

28. Let $I = (r)$ and $J = (s)$ be principal ideals in **Z** and let $\gcd(r, s) = 1$. Prove that the intersection

$$(a + I) \cap (b + J)$$

of a coset of I and a coset of J is nonempty and is a coset of $I \cap J$.

29. Let p be an odd positive prime in **Z**. Prove without group theory the following results that are established, using group theory, after Theorem 1 of Section 7.4.

(i) Modulo p, the product of two quadratic residues or of two quadratic nonresidues is a quadratic residue.

(ii) Modulo p, the product of a quadratic residue and a quadratic nonresidue is a quadratic nonresidue.

CHAPTER 8

SOME APPLICATIONS TO CODING

The word *code* is often associated with clandestine activities, the world of spies and intrigue. In that context, a code provides the form in which messages are transmitted in order to make them more difficult (or impossible) for anyone other than the intended recipient to read. The science of developing or breaking such codes is called *cryptography*. We describe one code of this type in Section 8.3. It was created for purposes such as secure electronic transmission of funds from one financial institution to another.

But first we discuss codes of another kind. These detect or both detect and correct certain transmission errors in sending messages. The errors can be caused by weather conditions, great distances, and other factors that result in static, fading signals, and so on. But the ability of these codes to detect and correct errors has allowed us to receive messages from our probes to other planets as well as to develop more prosaic uses in telephone communication, in the networking of computers, and the like.

Figure 8.1 is a simplified diagram depicting a message going from the source to its destination via an encoder, a transmission channel that is subject to noise, and a decoder. If the code is good enough, errors introduced by noise in the channel have a high probability of being detected and corrected.

The channel can be anything from the space between Saturn and Earth to a magnetic tape used for computer storage. Messages may be in various

FIGURE 8.1

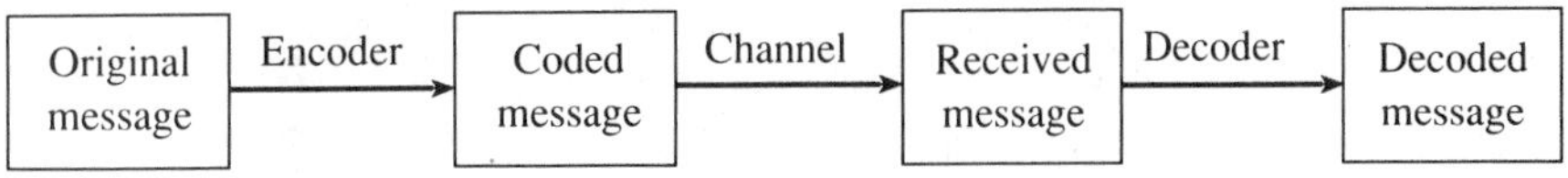

forms, but a very common form (especially where computers are involved) is as a bit string — that is, a finite sequence of elements of $\mathbf{Z_2} = \{0, 1\}$. We assume that, when a string goes through the channel, a string of equal length is received, although some of the bits of the string may be turned into their opposites by the noise.

8.1 Binary Codes

We recall from Definition 2 of Section 3.9 that a ***bit*** is a digit in the set $\{0, 1\}$ and that a string $a_1 a_2 \ldots a_n$, with each a_i a bit, is an ***n-bit string***. We use $\mathbf{W_n}$ to denote the set of all n-bit strings. For example,

$$\mathbf{W_3} = \{000, 001, 010, 011, 100, 101, 110, 111\}.$$

DEFINITION 1

Weight of a Bit String

The **weight** of a string $\alpha = a_1 a_2 \ldots a_n$ in $\mathbf{W_n}$ is the number of 1's among the bits a_i of α; we let **wgt $\boldsymbol{\alpha}$** denote the weight of α.

Thus, wgt 01010 = 2 and wgt 1110 = 3. If $\alpha = a_1 a_2 \ldots a_n$ is a bit string, it is clear that wgt $\alpha = a_1 + a_2 + \cdots + a_n$.

EXAMPLE 1 **Detecting an Error in One Bit**

Let the set of possible messages be

$$\{\text{no, maybe, await later message, yes}\}.$$

We use the bit strings of $\mathbf{W_2} = \{00, 01, 10, 11\}$ to represent these messages. To aid in detecting transmission errors, we then replace a string $\alpha = ab$ in $\mathbf{W_2}$ by the string abc in $\mathbf{W_3}$, in which $c = 0$ when wgt α is even and $c = 1$ when wgt α is odd. That is, we send the string $E(\alpha)$ of the following table.

α	00	01	10	11
$E(\alpha)$	000	011	101	110

as a codeword for α. Since the four codewords are the four strings of even weight in $\mathbf{W_3}$, this code has the nice property that an error in only one bit changes a codeword β into a string γ that has odd weight and hence is not a codeword. Thus the code enables the receiver to detect single-bit errors. In particular, if the string $\gamma = 010$ of odd weight is received, we would know that a transmission error had occurred. As γ could be the result of a single-bit error in any one of the codewords 000, 011, or 110, this code does not provide an infallible way of correcting one-bit errors.

DEFINITION 2

Encoding Function, Messageword, Codeword

In the codes of this section, an **encoding function** is an injective mapping $E\colon \mathbf{W_r} \to \mathbf{W_s}$, where r and s are fixed positive integers with $r < s$. A **messageword** is a string α in the domain $\mathbf{W_r}$. The image $E(\alpha)$ is the **codeword** for α.

The encoding function must be injective, since distinct messagewords should have distinct codewords to allow the receiver to decode them unambiguously.

DEFINITION 3

Decoding Function

Let $E\colon \mathbf{W_r} \to \mathbf{W_s}$ be an encoding function. A **decoding function** for E is a mapping $D\colon \mathbf{W_s} \to \mathbf{W_r} \cup \{\text{error}\}$ with the following properties:

(a) If γ is a codeword (that is, if $E(\alpha) = \gamma$ for some α in $\mathbf{W_r}$), then $D(\gamma) = \alpha$.

(b) If γ is not a codeword but there is good reason to believe that γ is a mistransmission of a codeword β, then $D(\gamma) = D(\beta)$.

(c) If neither (a) nor (b) is applicable, then $D(\gamma) =$ "error."

A decoding function for the encoding function E of Example 1 is the $D\colon \mathbf{W_3} \to \mathbf{W_2} \cup \{\text{error}\}$ such that $D(abc) = ab$ when abc has even weight and $D(abc) =$ "error" when abc has odd weight.

NOTATION 1

Juxtaposition of Strings, $\alpha 0$, $\alpha 1$

Let $\alpha = a_1 \ldots a_t$ and $\beta = b_1 \ldots b_u$ be bit strings. Then $\alpha\beta$ stands for the $(t + u)$-bit string $a_1 \ldots a_t b_1 \ldots b_u$. In particular, $\alpha 0$ denotes $a_1 \ldots a_t 0$, and $\alpha 1$ denotes $a_1 \ldots a_t 1$. Also, $\alpha\alpha\alpha$ stands for $(\alpha\alpha)\alpha = a_1 \ldots a_t a_1 \ldots a_t a_1 \ldots a_t$.

For example, if $\alpha = 10$ and $\beta = 001$, then $\alpha 0 = 100$, $\alpha 1 = 101$, $\alpha\beta = 10001$, $\beta\alpha = 00110$, and $\alpha\alpha\alpha = 101010$.

Using this notation, the encoding function $E\colon \mathbf{W_2} \to \mathbf{W_3}$ of Example 1 can be given by

$$E\colon \alpha \mapsto \alpha 0 \text{ when wgt } \alpha \text{ is even and } \alpha \mapsto \alpha 1 \text{ when wgt } \alpha \text{ is odd.}$$

DEFINITION 4

(r, s)-Code

If r and s are integers with $0 < r < s$, an **(r, s)-code** is an ordered pair $[E, D]$ in which $E: \mathbf{W_r} \to \mathbf{W_s}$ is an encoding function such that the codeword for each α in $\mathbf{W_r}$ is of the form $\alpha\delta$, with δ an $(s - r)$-bit string that may vary with α, and D is a decoding function for E.

Note that one of the requirements in Definition 4 is that a messageword $\alpha = a_1 \ldots a_r$ has a codeword $E(\alpha) = b_1 \ldots b_r b_{r+1} \ldots b_s$ in which $b_i = a_i$ for $1 \le i \le r$. (This follows from the fact that $E(\alpha)$ is of the form $\alpha\delta$, with δ an $(s - r)$-bit string.) If a received string γ in an (r, s)-code is known to be a codeword, it is decoded by dropping its rightmost $s - r$ bits; if γ is not a codeword, decoding is usually more difficult.

NOTATION 2

Modulo 2 Addition

We let $\oplus$ stand for the addition operation of the field $\mathbf{Z_2} = \{0, 1\}$; that is, $\oplus$ is specified by

$$0 \oplus 0 = 0 = 1 \oplus 1, \qquad 0 \oplus 1 = 1 = 1 \oplus 0.$$

EXAMPLE 2

The $(r, r + 1)$ Parity Codes

We use modulo 2 addition of bits to describe a family of codes that includes the code of Example 1 as a special case. For a fixed positive integer r, the **$(r, r + 1)$ parity code** is the $[E, D]$ with

$$E: \mathbf{W_r} \to \mathbf{W_{r+1}};\ a_1 \ldots a_r \mapsto a_1 \ldots a_r d \quad \text{with } d = a_1 \oplus a_2 \oplus \cdots \oplus a_r$$

and with a string $\gamma = c_1 \ldots c_r c_{r+1}$ in $\mathbf{W_{r+1}}$ decoded as $c_1 \ldots c_r$ when γ has even weight and decoded as "error" when γ has odd weight. In these codes, $E(\alpha) = \alpha 0$ when wgt α is even, $E(\alpha) = \alpha 1$ when wgt α is odd, and the codewords $E(\alpha)$ are the strings of even weight in $\mathbf{W_{r+1}}$. A codeword is decoded by dropping its rightmost bit, and the noncodeword is decoded as "error."

The parity codes detect single-bit errors because codewords have even weight and an error in just one bit changes a codeword into a string of odd weight. These codes do not correct errors, since we have no way to tell which bits are incorrect, nor do they detect a mistransmission with an even number of wrong bits.

EXAMPLE 3

Application of the (5, 6) Parity Code

In the (5, 6) parity code, we can let the 32 messagewords of $\mathbf{W_5}$ represent the letters $A, B, \ldots, Z$ and 6 punctuation marks. In particular, we let 10010 stand for S and 01110 stand for 0. Then

$$E(10010) = 100100, \text{ since } 10010 \text{ has even weight,}$$

$$E(01110) = 011101, \text{ since } 01110 \text{ has odd weight.}$$

Hence this code sends *SOS* as

$$100100/011101/100100.$$

Now let us look at cases of decoding in the (5, 6) parity code. If $\beta = 101011$, we note that wgt β is even and so obtain $D(\beta) = 10101$ by dropping the last bit of β. If $\gamma = 010101$, we note that wgt γ is odd and that γ is consequently not a codeword; therefore, γ is the result of a transmission error, and we have $D(\gamma) =$ "error."

Next we describe a family of codes that both detect and correct single-bit errors — in contrast to the parity codes, which only detect such errors.

EXAMPLE 4 **The $(r, 3r)$ Triple Repetition Codes**

Here we define, for each positive integer r, a code $[E, D]$ called the **$(r, 3r)$ triple repetition code**. Encoding is given by

$$E\colon \mathbf{W}_r \to \mathbf{W}_{3r}; \quad \alpha \mapsto \alpha\alpha\alpha.$$

In particular, for $r = 2$ the encoding function has the table

α	00	01	10	11
$E(\alpha)$	000000	010101	101010	111111

As an aid in decoding a $3r$-bit string $\gamma = c_1 \ldots c_{3r}$, we break γ into three r-bit strings and write them in the following form:

$$\begin{array}{ccccccc} c_1 & c_2 & \cdots & c_j & \cdots & c_r \\ c_{r+1} & c_{r+2} & \cdots & c_{r+j} & \cdots & c_{2r} \\ c_{2r+1} & c_{2r+2} & \cdots & c_{2r+j} & \cdots & c_{3r} \end{array}$$

Now we let $D(\gamma) = a_1 \ldots a_j \ldots a_r$, where a_j is the bit that appears at least twice in the ordered triple (c_j, c_{r+j}, c_{2r+j}). This decoding process corrects transmission errors in which there is at most one error in the bits of each such triple.

With $r = 4$, the (4, 12) triple repetition code decodes a received string $\gamma = 111000100100$ by breaking it into thirds as 1110/0010/0100 and placing these parts as follows:

$$\begin{array}{cccc} 1 & 1 & 1 & 0 \\ 0 & 0 & 1 & 0 \\ 0 & 1 & 0 & 0 \end{array}$$

Then $D(\gamma) = 0110$, since 0 appears a majority of the times in the first column, 1 appears a majority of the times in the second column and also in the third column, and 0 all three times in the last column.

Since the thirds are not identical, we know that transmission errors have occurred. If there is at most one error in the three bits of any column, the decoding corrects the errors.

Problems

1. In the (4, 5) parity code, encode (that is, find the codewords for) the following.
(a) 1110; (b) 1010.

2. Find the parity codewords for the following.
(a) 0000; (b) 0001.

3. In the (4, 5) parity code, decode the following.
(a) 10101; (b) 01010.

4. In the (4, 5) parity code, find the following.
(a) $D(11110)$; (b) $D(00001)$.

5. In the (6, 7) parity code, let the messageword be $\alpha = 110110$. If its codeword 1101100 is mistransmitted as 1000100, could the receiver detect the error? Explain.

6. If the codeword 1101100 in the (6, 7) parity code comes through as 1101000, could the decoder detect the error? Explain.

7. In the (3, 9) triple repetition code $[E, D]$, find the codewords $E(\alpha)$ for the following.
(a) $\alpha = 000$; (b) $\alpha = 110$; (c) $\alpha = 100$.

8. In the (3, 9) triple repetition code, find the following.
(a) $E(111)$; (b) $E(001)$; (c) $E(011)$.

9. In the (4, 12) triple repetition code, decode the following.
(a) 101000111111; (b) 000000000000; (c) 101011111010.

10. In the (4, 12) triple repetition code, find the following.
(a) $D(010011000000)$; (b) $D(010100001010)$; (c) $D(111111111111)$.

11. In the (2, 6) triple repetition code, let the messageword be $\alpha = 00$.
(a) If its codeword 000000 is received incorrectly as 100100, will it be decoded correctly?
(b) If 000000 is mistransmitted as 000101, will it be decoded correctly?

12. In the (2, 6) triple repetition code, let the messageword be $\alpha = 11$. Will the decoding be correct if α's codeword 111111 is received incorrectly as:
(a) 110101? (b) 101101?

13. Let $E\colon \mathbf{W}_4 \to \mathbf{W}_7$; $abcd \mapsto abcdefg$ with

$$e = a \oplus b \oplus c, \quad f = a \oplus b \oplus d, \quad g = a \oplus c \oplus d.$$

Find the following.
(a) $E(0000)$; (b) $E(1000)$; (c) $E(0100)$; (d) $E(0001)$.

14. Using the encoding function E of Problem 13, find the following.
(a) $E(1100)$; (b) $E(1001)$; (c) $E(0101)$.

15. For the E of Problem 13, find the following.
(a) $E(1110)$; (b) $E(1101)$; (c) $E(1011)$.

16. For the E of Problem 13, find the following.
(a) $E(0111)$; (b) $E(1111)$.

17. Which of the following strings is a codeword for the E of Problem 13?
(a) 1110100; (b) 1110011.

18. Which of the following strings is a codeword for the E of Problem 13?
(a) 1100101; (b) 1100001.

8.2 Matrix Codes, A Hamming Code

Encoding and decoding for the codes of the previous section and for many more complicated codes can be accomplished by using matrix multiplication over the field $\mathbf{Z}_2 = \{0, 1\}$, and hence these operations are easily performed by electronic computers. To aid in describing matrix codes, we now summarize some material from Section 4.7.

We recall that the ***product*** AB of matrices $A = (a_{ij})_{r\times s}$ and $B = (b_{ij})_{s\times t}$ is the matrix $C = (c_{ij})_{r\times t}$ with

$$c_{ij} = a_{i1}b_{1j} + a_{i2}b_{2j} + \cdots + a_{is}b_{sj}; \quad 1 \le i \le r,\ 1 \le j \le t.$$

The $m \times n$ ***zero matrix*** $0_{m\times n}$ has all entries equal to zero. The $m \times m$ ***unity matrix*** is the square matrix $I_m = (\delta_{ij})_{m\times m}$, where

$$\delta_{ij} = \begin{cases} 1 & \text{if } i = j, \quad \text{and} \\ 0 & \text{if } i \ne j. \end{cases}$$

The ***transpose*** of $A = (a_{ij})_{m\times n}$ is $A^T = (a'_{ij})_{n\times m}$, where $a'_{ij} = a_{ji}$. Thus, taking the transpose of a matrix changes rows into columns and columns into rows. Transposes play an important role in matrix theory, but here we will use the notation for transposes mainly for convenience in printing.

Next we slightly change our notation for messagewords and codewords.

NOTATION 1 Words as Single-Row Matrices

Here we use one-rowed matrices $(a_1 \ \ a_2 \ \ \ldots \ \ a_r)$ and $(b_1 \ \ b_2 \ \ \ldots \ \ b_s)$ as messagewords and codewords, instead of using the bit strings $a_1 \ldots a_r$ and $b_1 \ldots b_s$.

EXAMPLE 1 The (2, 3) Parity Code as a Matrix Code

In our new notation, the encoding function for the (2, 3) parity code is

$$E\colon (a \quad b) \mapsto (a \quad b \quad a \oplus b),$$

where $\oplus$ is the addition operation of the field $\mathbf{Z}_2 = \{0, 1\}$ given by

$$0 \oplus 0 = 0 = 1 \oplus 1, \qquad 0 \oplus 1 = 1 = 1 \oplus 0.$$

If we let

$$\alpha = (a \quad b) \quad \text{and} \quad G = \begin{pmatrix} 1 & 0 & 1 \\ 0 & 1 & 1 \end{pmatrix},$$

then the codeword $E(\alpha)$ for α is the matrix product

$$\alpha G = (a \quad b)\begin{pmatrix} 1 & 0 & 1 \\ 0 & 1 & 1 \end{pmatrix} = (a \quad b \quad a \oplus b)$$

over $\mathbf{Z}_2$.

EXAMPLE 2 **Matrix for the (2, 6) Triple Repetition Code**

In matrix notation, the encoding function for the (2, 6) triple repetition code is $E: (a \quad b) \mapsto (a \quad b \quad a \quad b \quad a \quad b)$. We let

$$\alpha = (a \quad b) \quad \text{and} \quad G = \begin{pmatrix} 1 & 0 & 1 & 0 & 1 & 0 \\ 0 & 1 & 0 & 1 & 0 & 1 \end{pmatrix}.$$

Then the codeword $E(\alpha)$ for α is the matrix product αG over $\mathbf{Z_2}$.

DEFINITION 1

Matrix Code, Generator Matrix

If the encoding function for an (r, s)-code can be given by

$$E: (a_1 \ \dots \ a_r) \mapsto (a_1 \ \dots \ a_r)G = (b_1 \ \dots \ b_s),$$

with G a fixed $r \times s$ matrix over $\mathbf{Z_2}$, then the code is an $\boldsymbol{r \times s}$ **matrix code** with G as its **generator**.

NOTATION 2 **Horizontal Matrix Juxtaposition**

For matrices $A = (a_{ij})_{m \times r}$, $B = (b_{ij})_{m \times s}$, and $C = (c_{ij})_{m \times t}$, we let $(A : B : C)$ denote the $m \times (r + s + t)$ matrix $Q = (q_{ij})$ with

$$q_{ij} = a_{ij} \quad \text{for } 1 \le j \le r,$$
$$q_{i,\,r+j} = b_{ij} \quad \text{for } 1 \le j \le s,$$
$$q_{i,\,r+s+j} = c_{ij} \quad \text{for } 1 \le j \le t.$$

Similarly, one can define $(A : B)$, $(A : B : C : D)$, and so on.

With this notation, the generator matrix for the $(r, r + 1)$ parity code is $(I_r : F)$, where $F = (1 \quad 1 \quad \dots \quad 1)^T$ is the $r \times 1$ matrix with each entry equal to 1. In particular, the generator matrix for the (3, 4) parity code is

$$\begin{pmatrix} 1 & 0 & 0 & 1 \\ 0 & 1 & 0 & 1 \\ 0 & 0 & 1 & 1 \end{pmatrix} = (I_3 : F) \quad \text{with } F = \begin{pmatrix} 1 \\ 1 \\ 1 \end{pmatrix} = (1 \quad 1 \quad 1)^T.$$

Also, one easily sees that the generator matrix for the $(r, 3r)$ triple repetition code is $(I_r : I_r : I_r)$.

In an (r, s)-code, the first r bits of the codeword $E(\alpha)$ are the bits of α. (This is part of Definition 4 of Section 8.1.) In matrix notation, this forces the generator for an $r \times s$ matrix code to be of the form $(I_r : F)$, with F an $r \times (s - r)$ matrix.

The triple repetition codes detect and correct single-bit errors. In 1950, Richard W. Hamming, then at Bell Laboratories, developed a sequence of single-bit-error correcting codes that use shorter codewords than the triple repetition codes. These save transmission time and also cut down on errors by decreasing the number of bits sent.

The Hamming codes use multiplication of matrices over the mod 2 ring $\mathbf{Z}_2$ both for encoding and for decoding, as we shall see below.

The first of Hamming's sequence of codes is identical to the (1, 3) triple repetition code. The second of Hamming's codes is a 4×7 matrix code [as compared with the (4, 12) triple repetition code]. We devote the rest of this section to this 4×7 Hamming Code.

EXAMPLE 3 **Generator Matrix for the 4 × 7 Hamming Code**

Let $\alpha = (a \quad b \quad c \quad d)$ be a 1×4 matrix over $\mathbf{Z}_2$,

$$F = \begin{pmatrix} 1 & 1 & 1 \\ 1 & 1 & 0 \\ 1 & 0 & 1 \\ 0 & 1 & 1 \end{pmatrix}, \quad \text{and} \quad G = (I_4 : F) = \begin{pmatrix} 1 & 0 & 0 & 0 & 1 & 1 & 1 \\ 0 & 1 & 0 & 0 & 1 & 1 & 0 \\ 0 & 0 & 1 & 0 & 1 & 0 & 1 \\ 0 & 0 & 0 & 1 & 0 & 1 & 1 \end{pmatrix} \tag{1}$$

Then G is the generator for the **4 × 7 Hamming Code**; that is, the codeword for α in this code is the matrix product

$$\alpha G = (a \quad b \quad c \quad d \quad a \oplus b \oplus c \quad a \oplus b \oplus d \quad a \oplus c \oplus d).$$

For example, if the messageword is $\alpha = (0 \quad 0 \quad 1 \quad 0)$, then its codeword is

$$\begin{aligned} \alpha G &= (0 \quad 0 \quad 1 \quad 0 \quad 0 \oplus 0 \oplus 1 \quad 0 \oplus 0 \oplus 0 \quad 0 \oplus 1 \oplus 0) \\ &= (0 \quad 0 \quad 1 \quad 0 \quad 1 \quad 0 \quad 1). \end{aligned}$$

We note that encoding here is the same as in Problem 13 of Section 8.1.

Example 3 tells us how to encode in the 4×7 Hamming Code. How do we decode? Let $\gamma = (a \quad b \quad c \quad d \quad e \quad f \quad g)$ be a received 1×7 matrix. If

$$e = a \oplus b \oplus c, \quad f = a \oplus b \oplus d, \quad \text{and} \quad g = a \oplus c \oplus d, \tag{2}$$

then γ is a codeword, and we decode it by dropping its last three bits. If at least one of the equations in (2) is not satisfied, γ is not a codeword. When γ results from a transmission error changing only one bit, we can still decode correctly, as described below.

A 1×7 matrix $\gamma = (a \quad b \quad c \quad d \quad e \quad f \quad g)$ is a codeword if and only if the system of three equations

$$\begin{aligned} a \oplus b \oplus c \qquad &= e \\ a \oplus b \oplus \qquad d &= f \\ a \qquad \oplus c \oplus d &= g \end{aligned}$$

is satisfied. Since $x \oplus x = 0$ for each of the bits x in $\{0, 1\}$, we can add e (modulo 2) to both sides of the first equation, f to both sides of the second equation, and g to both sides of the third equation. The system of three equations then becomes

$$\begin{array}{lllllll} a \oplus b \oplus c & & \oplus e & & & = 0 \\ a \oplus b & \oplus d & & \oplus f & & = 0 \\ a & \oplus c \oplus d & & & \oplus g & = 0 \end{array}$$

This system of three simultaneous equations is equivalent to the single matrix equation

$$\gamma K = 0_{1 \times 3},$$

where

$$K = \begin{pmatrix} 1 & 1 & 1 \\ 1 & 1 & 0 \\ 1 & 0 & 1 \\ 0 & 1 & 1 \\ 1 & 0 & 0 \\ 0 & 1 & 0 \\ 0 & 0 & 1 \end{pmatrix} \qquad (3)$$

and $0_{1 \times 3}$ is the 1×3 zero matrix $(0 \ \ 0 \ \ 0)$.

DEFINITION 2

Check Matrix, Syndrome

The 7×3 matrix K given in (3) is called the **check matrix** for the 4×7 Hamming Code. If γ is a 1×7 matrix, the matrix product γK (over $\mathbf{Z}_2$) is called the **syndrome** for γ.

We note that a syndrome γK is a 1×3 matrix over $\mathbf{Z}_2 = \{0, 1\}$, since γ is a 1×7 matrix and K is a 7×3 matrix. There are eight 1×3 matrices over $\mathbf{Z}_2$, because each of the three entries is in $\{0, 1\}$. If we look at the check matrix K, we see that the eight possibilities for a syndrome are the zero matrix $(0 \ \ 0 \ \ 0)$ and the seven rows of K.

NOTATION 3 δ_i

Here we use δ_i for the 1×7 matrix $(d_1 \ d_2 \ \dots \ d_7)$ with $d_i = 1$ and $d_j = 0$ for $j \neq i$.

For example, $\delta_1 = (1 \ \ 0 \ \ 0 \ \ 0 \ \ 0 \ \ 0 \ \ 0)$ and $\delta_5 = (0 \ \ 0 \ \ 0 \ \ 0 \ \ 1 \ \ 0 \ \ 0)$.

Let $\beta = (b_1 \ \ b_2 \ \ \dots \ \ b_7)$ and $\beta + \delta_i = (c_1 \ \ c_2 \ \ \dots \ \ c_7)$, where δ_i is as in Notation 3. Then $c_i = b_i \oplus 1$, and so c_i is the bit in $\{0, 1\}$ different from b_i, while for $j \neq i$ we have $c_i = b_i \oplus 0 = b_i$. Hence, $\beta + \delta_i$ is the received matrix resulting from a transmission error that changes only the ith bit.

LEMMA 1 Syndrome for a Codeword or Its Slight Modification

Let $\beta = (b_1 \ \ b_2 \ \ \ldots \ \ b_7)$ be a codeword, and let δ_i be as in Notation 3. Then:
(a) The syndrome βK is the zero matrix $(0 \ \ 0 \ \ 0)$.
(b) The syndrome $(\beta + \delta_i)K$ equals the ith row of the check matrix K.

Proof We have already established part (a) since we showed above that βK is the zero matrix if and only if β is a codeword. Part (b) follows from the definition of matrix multiplication and the fact that the only nonzero entry of δ_i is the 1 in the ith column. ■

Lemma 1 provides the justification for the following.

DECODING RULES FOR THE 4 × 7 HAMMING CODE

Let γK be the syndrome for $\gamma = (c_1 \ \ \ldots \ \ c_7)$. Then:
(a) If $\gamma K = (0 \ \ 0 \ \ 0)$, γ is a codeword; and we decode it by dropping its last three bits.
(b) If γK equals one of the bottom three rows of the matrix K, we also decode γ by dropping its last three bits. [In this case, γ is not a codeword, but it is probably the result of a transmission error in one of the three bits that we are discarding.]
(c) If γK equals the ith row of K with i in $\{1, 2, 3, 4\}$, we decode γ by changing its ith bit and then dropping its last three bits. [Such a received transmission is most likely the result of an error in the ith bit, since changing the ith bit will make γ into a codeword.]

EXAMPLE 4 Illustrations of Hamming Decoding

The accompanying table gives three 1×7 received matrices γ, each with its syndrome γK and its decoded message matrix $D(\gamma)$.

Received γ	γK	$D(\gamma)$
$\gamma_1 = (0 \ \ 0 \ \ 1 \ \ 0 \ \ 1 \ \ 0 \ \ 1)$	$(0 \ \ 0 \ \ 0)$	$(0 \ \ 0 \ \ 1 \ \ 0)$
$\gamma_2 = (1 \ \ 1 \ \ 1 \ \ 1 \ \ 1 \ \ 1 \ \ 0)$	$(0 \ \ 0 \ \ 1)$	$(1 \ \ 1 \ \ 1 \ \ 1)$
$\gamma_3 = (0 \ \ 0 \ \ 1 \ \ 0 \ \ 0 \ \ 1 \ \ 1)$	$(1 \ \ 1 \ \ 0)$	$(0 \ \ 1 \ \ 1 \ \ 0)$

We see that the syndrome $\gamma_1 K = (0 \ \ 0 \ \ 0)$, and this tells us that γ_1 is a codeword; therefore, we use rule (a) and decode γ_1 by dropping its last three bits. Since the syndrome $\gamma_2 K$ is the seventh row of K, we use rule (b) and decode γ_2 by dropping its last three bits. The syndrome $\gamma_3 K$ is the second row of K;

hence we use rule (c) and decode γ_3 by changing its second bit (from 0 to 1) and then dropping the last three bits.

The 4×7 Hamming Code has the largest r for all the $(r, r+3)$ codes that both detect and correct single-bit errors. It thus maximizes the number of message words for all such codes. The 4×7 code described above is the case $t = 3$ of a sequence of Hamming codes; these are $r \times (r+t)$ codes with $r = 2^t - 1 - t$. For fixed t, they have the largest r of the $(r, r+t)$ codes that both detect and correct single-bit errors. More sophisticated multiple-error-correcting codes of similar efficiency were developed independently by A. Hoquenghem in 1959 and by R. C. Bose & D. V. Ray-Chanduri in 1960; they are called ***BCH codes***.

Problems

1. Let G be as in display (1). Find the codewords αG for each of the following message-words α.
   (a) $(1\ \ 0\ \ 0\ \ 0)$; (b) $(1\ \ 0\ \ 1\ \ 0)$; (c) $(1\ \ 1\ \ 0\ \ 1)$; (d) $(0\ \ 0\ \ 0\ \ 0)$.
2. Do as in Problem 1 for the following.
   (a) $(0\ \ 1\ \ 0\ \ 0)$; (b) $(0\ \ 1\ \ 0\ \ 1)$; (c) $(0\ \ 1\ \ 1\ \ 1)$; (d) $(1\ \ 1\ \ 1\ \ 1)$.
3. How many messagewords are there in the 4×7 Hamming Code?
4. How many codewords are there in the 4×7 Hamming Code?
5. Let K be as in display (3). For each of the following 1×7 matrices γ, find the syndrome γK and then decode γ.
   (a) $(0\ \ 0\ \ 1\ \ 0\ \ 1\ \ 1\ \ 1)$; (b) $(1\ \ 1\ \ 1\ \ 1\ \ 1\ \ 1\ \ 1)$;
   (c) $(0\ \ 1\ \ 1\ \ 0\ \ 1\ \ 0\ \ 1)$.
6. Do as in Problem 5 for the following.
   (a) $(0\ \ 0\ \ 0\ \ 1\ \ 0\ \ 1\ \ 1)$; (b) $(1\ \ 0\ \ 1\ \ 1\ \ 1\ \ 0\ \ 0)$;
   (c) $(0\ \ 0\ \ 1\ \ 1\ \ 0\ \ 1\ \ 0)$.
7. Let $\alpha = (0\ \ 0\ \ 0\ \ 1)$ and G be as in display (1).
   (a) Find the codeword $\beta = \alpha G$.
   (b) Let β be received (incorrectly) as $\gamma = (0\ \ 0\ \ 1\ \ 1\ \ 0\ \ 1\ \ 0)$. Find the syndrome γK, and then use the decoding rules to get $D(\gamma)$. Does $D(\gamma) = \alpha$?
8. Do as in Problem 7, but here let $\alpha = (1\ \ 0\ \ 0\ \ 1)$ and $\gamma = (1\ \ 1\ \ 0\ \ 1\ \ 0\ \ 0\ \ 0)$.
9. If two of the bits of a received matrix are incorrect, will the 4×7 Hamming Code always correct the error? Explain.
10. Let F be the transpose of the matrix

$$F^T = \begin{pmatrix} 1 & 1 & 1 & 1 & 0 & 1 & 1 & 1 & 0 & 0 & 0 \\ 1 & 1 & 1 & 0 & 1 & 1 & 0 & 0 & 1 & 1 & 0 \\ 1 & 1 & 0 & 1 & 1 & 0 & 1 & 0 & 1 & 0 & 1 \\ 1 & 0 & 1 & 1 & 1 & 0 & 0 & 1 & 0 & 1 & 1 \end{pmatrix}$$

and let $G = (I_{11} : F)$ be the generator matrix for the *11 × 15* ***Hamming Code*** $[E, D]$ with

$$E: (a_1\ \ \dots\ \ a_{11}) \mapsto (a_1\ \ \dots\ \ a_{11})G.$$

Find a check matrix K such that $\gamma = (c_1 \ \dots \ c_{15})$ is a codeword if and only if γK is the 1×4 zero matrix, and give rules for the decoding function D.

★ **11.** Do Problem 9 with "correct" replaced by "detect."

8.3 Modular Codes, Trapdoor Functions

Now we consider codes designed to maintain privacy of the messages rather than to combat transmission errors. Here elements of the multiplicative group V_m of invertibles in $\mathbf{Z_m} = \{\bar{0}, \bar{1}, \dots, \overline{m-1}\}$, for a suitable m, are used for messagewords and codewords instead of the binary strings of previous sections.

The modulus m is chosen to be a product pq of distinct positive primes (which should be very large in real-life encryption). For $m = pq$, let $f = \phi(m) = (p-1)(q-1)$, where ϕ is the Euler ϕ-function. Let e be in $\mathbf{Z}^+$ and be relatively prime to f. Then there exist d and g in $\mathbf{Z}$ with $ed - fg = 1$. [See Corollary (b) to Theorem 2 of Section 1.4.] Thus $ed = fg + 1$, and for $\bar{x}$ in V_m one has

$$\bar{x}^{ed} = \bar{x}^{fg+1} = (\bar{x}^f)^g \bar{x} = \bar{1}^g \bar{x} = \bar{x},$$

since $\bar{x}^f = \bar{x}^{\phi(m)} = \bar{1}$ in V_m by Euler's Theorem (Theorem 1 of Section 4.3). The equation $\bar{x}^{ed} = \bar{x}$ for all $\bar{x}$ in V_m means that the mappings

$$E: \bar{x} \mapsto \bar{x}^e, \qquad D: \bar{x} \mapsto \bar{x}^d, \tag{1}$$

from V_m to itself, are inverses of each other. Hence these mappings are bijections. In particular, E is injective and so is available as an encoding function.

DEFINITION

Modular Code

The ordered pair $[E, D]$, with E and D as in (1), is the **modular code** with **modulus** m, **encoding exponent** e, **decoding exponent** d, and **modular factors** p and q.

Such codes are important in encryption for the following reasons. If the modular prime factors p and q are large enough (say, each with about 100 digits), we can publish (at least for a limited time) the values of the modulus m and the encoding exponent e, and thus make it possible for anyone to send us a coded message. However, we keep p, q, and d secret so that only authorized persons can decode. This is based on the practical impossibility of factoring a very large number m, in a limited time, even when one knows that it is the product of two primes. To create one's own modular code, one needs two large primes, but modern electronic computers can obtain sufficient numbers of such primes.

We now illustrate the principles of modular codes by using a case with fairly small modular factors; this code could be "cracked" rather easily.

EXAMPLE 1 We announce that anyone wishing to send us a message should use the elements $\bar{2}, \bar{3}, \ldots, \overline{27}$ of the group V_{899} for the letters $a, b, \ldots, z$ and that the message should be encoded by replacing each element $\bar{x}$ by $\bar{x}^{11}$. Thus the letter o would be represented by the word $\overline{16}$ and encoded as $\overline{16}^{11}$. We use the fact that $\overline{900} = \bar{1}$ in V_{899} to calculate $\overline{16}^{11}$ as follows:

$$\overline{16}^2 = \overline{256},\ \overline{16}^4 = \overline{256}^2 = \overline{65536} = \overline{72 \cdot 900 + 736} = \overline{72 + 736} = \overline{808},$$

$$\overline{16}^8 = \overline{808}^2 = \overline{652864} = \overline{725 \cdot 900 + 364} = \overline{725 + 364} = \overline{1089} = \overline{190},$$

$$\overline{16}^{10} = \overline{16}^8 \cdot \overline{16}^2 = \overline{190 \cdot 256} = \overline{48640} = \overline{54 \cdot 900 + 40} = \overline{54 + 40} = \overline{94},$$

$$\overline{16}^{11} = \overline{16}^{10} \cdot \overline{16} = \overline{94 \cdot 16} = \overline{1504} = \overline{900 + 604} = \overline{605}.$$

This tells us that the codeword to be sent is $\overline{605}$.

We as receivers of the codeword would know that the modular factors are 29 and 31. Hence, 2, 3, ..., 27 are all relatively prime to $m = 29 \cdot 31$ and the "words" $\bar{2}, \bar{3}, \ldots, \overline{27}$ are all in V_m. We would also know that $\phi(m) = 28 \cdot 30 = 840 = f$, and we could use Euclid's Algorithm to obtain the equation $11 \cdot 611 - 840 \cdot 8 = 1$. This would tell us that $d = 611$ and that the decoding function is $D: \bar{x} \mapsto \bar{x}^{611}$.

Even for the relatively small p, q, and e of Example 1, encoding and decoding are tedious. For very large p and q, these calculations are still practical with computers; but factoring m and thus being able to find the decoding function are not possible in any reasonable amount of time with present-day techniques and computers. As computing equipment gets faster, one might be able to maintain the qualities of the code by increasing the size of p and q or by decreasing the time interval for which the code is in use.

An encoding function E that can be described exactly while keeping its inverse D secret is called a ***trapdoor function***. The first practical construction of trapdoor functions was by R. L. Rivest, L. Shamir, and L. M. Adelman in their paper "A method for obtaining digital signatures and public-key cryptosystems," *Communications of the Association for Computing Machinery* 21 (1978): 120–26.

The application of this section depends on the lack of a practical computer algorithm for factoring a number m that is known to be the product of two large primes. If such an algorithm were developed, the modular codes would cease to have their desirable "trapdoor" property and would have to be replaced by codes using other trapdoor functions. Information on such codes is given by M. E. Hellman in the paper "An overview of public key cryptography," *IEEE Communications Society Magazine* 16 (1978): 24–32. It is very probable that algebra and number theory will continue to be important in such applications.

Problems

1. Use the encoding function $E: \bar{x} \mapsto \bar{x}^3$ to encode the following elements of V_{15}.
   (a) $\bar{2}$; (b) $\bar{4}$; (c) $\bar{7}$.
2. Use the encoding function $E: \bar{x} \mapsto \bar{x}^5$ to encode the following elements of V_{21}.
   (a) $\bar{2}$; (b) $\bar{4}$; (c) $\bar{5}$.
3. Let $E: \bar{x} \mapsto \bar{x}^{13}$ be the encoding function on V_{77}. Find the decoding exponent d for the decoding function $D: \bar{x} \mapsto \bar{x}^d$.
4. Let $E: \bar{x} \mapsto \bar{x}^7$ be the encoding function on V_{143}. Find the decoding exponent d.
5. Which numbers in $\{2, 3, \ldots, 76\}$ are possible as encoding exponents e or decoding exponents d for a modular code using the elements of V_{77}?
6. How many numbers in $\{2, 3, \ldots, 142\}$ are possible as encoding exponents e for a modular code on V_{143}?
7. In Example 1, five multiplications (most of them followed by reductions modulo m) were required to calculate $\bar{x}^{11}$ in V_{899}. How many multiplications are needed to find $\bar{x}^{101}$ in V_{1133}?
8. How many multiplications are needed to find $\bar{x}^{10247}$ in V_{643063}?
9. ★ Crack the code on V_{1133} with $E(\bar{x}) = \bar{x}^{101}$; that is, find D.
10. ★ Crack the code on V_{643063} with $E(\bar{x}) = \bar{x}^{10247}$.
11. If one knew that the prime factors p and q of 643063 are greater than 700, could one use an element of V_{643063} to represent an ordered pair of letters in a modular code? Explain.
12. How large would the modular factors p and q have to be to allow elements $\bar{2}, \bar{3}, \ldots, \bar{r}$ of V_m to represent ordered triples of letters in a modular code?

Review Problems

1. In the (5, 6) parity code, suppose that the received string is 011001. Assuming a 1-digit error, what are all the possibilities for the message word in W_5?
2. Do as in Problem 1 for the (4, 5) parity code and the received string 01000.
3. For the (2, 6) triple repetition code, find the following.
   (a) $E(01)$; (b) $D(110110)$; (c) $D(100010)$.
4. For the (3, 9) triple repetition code, find the following.
   (a) $E(101)$; (b) $D(011010110)$; (c) $D(111001011)$.
5. In the 4×7 Hamming Code, find the codewords for each of the following message-words.
   (a) (0 1 0 0); (b) (1 0 0 1); (c) (1 0 1 1).
6. Do as in Problem 5 for the following.
   (a) (0 0 1 0); (b) (0 1 1 0); (c) (1 1 1 0).
7. For each of the following 1×7 matrices γ, find the syndrome γK, and then decode γ in the 4×7 Hamming Code.
   (a) (0 0 1 0 0 1 1).
   (b) (0 1 1 0 0 1 1).
   (c) (0 1 1 0 0 1 0).

8. Do as in Problem 7 for each of the following.
(a) (1 1 0 0 1 0 0).
(b) (1 1 1 0 1 1 0).
(c) (1 1 1 0 1 0 0).

9. (i) Identify the possible values for the encoding exponent e in a modular code encoding function

$$E: V_{35} \to V_{35}; \qquad \bar{x} \mapsto \bar{x}^e.$$

(ii) For each e in the answer to part (i), give the decoding exponent d for the decoding function

$$D: V_{35} \to V_{35}; \qquad \bar{x} \mapsto \bar{x}^d.$$

10. Find the decoding function for the modular code encoding function

$$E: V_{187} \to V_{187}; \qquad \bar{x} \mapsto \bar{x}^7.$$

Computer Programming Projects

Projects Related to Chapter 1

1. Write a program for calculating gcd(a, b), by Euclid's Algorithm, for a and b integers with $a \geq b > 0$.
2. Use the formula in Problem 18(iv) of Section 1.6 to write a program for calculating lcm[a, b] for integers a and b with $a \geq b > 0$.

Projects Related to Chapters 2 and 3

1. Find all the subgroups in $\mathbf{S_5}$.
2. Characterize all groups of order 12, in a manner similar to the characterization of all groups of order 8 in Problems 17 through 21 of Section 2.12.

Projects Related to Chapter 4

1. Find all the primes in
   (a) $(1 + 6\mathbf{Z}^+) \cap \{1, 2, 3, \ldots, 10^6\}$;
   (b) $(-1 + 6\mathbf{Z}^+) \cap \{1, 2, 3, \ldots, 10^6\}$.
2. Program multiplication of 5×5 matrices over $\mathbf{Z_2} = \{0, 1\}$.

Projects Related to Chapter 5

1. Let $\alpha = a_d x^d + a_{d-1}x^{d-1} + \cdots + a_0$ and $\beta = x^e + b_{e-1}x^{e-1} + \cdots + b_0$ be in $\mathbf{Z}[x]$ (with β monic). Program the calculations of the coefficients of polynomials γ and ρ such that $\alpha = \gamma\beta + \rho$ and either $\rho = 0$ or deg $\rho <$ deg β.
2. The cyclotomic polynomials are the sequence $C_1, C_2, C_3, \ldots$ of monic polynomials in $\mathbf{Z}[x]$ such that $x^n - 1$ equals the product of the C_d for all the positive integral divisors d of n. Show that every coefficient of C_n is in $\{1, 0, -1\}$ for $n = 1, 2, \ldots, 104$ but not for $n = 105$.
3. Let p be a prime in $\mathbf{Z}^+$. For a finite number of values of d in $\mathbf{Z}^+$, find the number of monic prime polynomials of degree d in $\mathbf{Z_p}[x]$. (See Problems 1 through 4 of Section 5.8.)
4. Let $F = \{0, 1\}$ be a field with two elements and let s be a zero, in an extension field E over F, of the polynomial $\beta = x^p + x + 1$. For $p = 5$, 7, and/or 11, tabulate

integer coefficients c_{ij} such that

$$s^n = c_{n,\,p-1}s^{p-1} + c_{n,\,p-2}\,s^{p-2} + \cdots + c_{n1}s + c_{n0}\,.$$

for $0 \le n \le 2^p - 2$. (See Example 2 in Section 5.10.)

Projects Related to Chapter 7

1. Find the smallest integer n with $n > 1$ and $n \equiv \pm 1 \pmod{p}$ for all p in $\{2, 3, 5, 7, 11, 13, 17, 19, 23, 29\}$.
2. For n in $\mathbf{Z}^+$, let $p(n)$ be the number of partitions of n (into positive integer parts). Tabulate $p(n)$ for $n = 1, 2, 3, \ldots, 100$. [See Theorem 4 of Section 7.3.]

Projects Related to Chapter 8

1. Program coding and decoding for the 4×7 Hamming Code.
2. Find some primes greater than 10^{50}.

Supplementary and Challenging Problems

Supplementary and Challenging Problems for Chapter 1

1. Let a and b be integers with b odd and $\gcd(a, b) = 1$. Prove that either $\gcd(a, 2b) = 1$ or $\gcd(a + b, 2b) = 1$.
2. Show that for every integer n there exist unique integers q and r such that $n = 15q + r$ and $|r| \le 7$.
3. Let a and b be integers with $b \neq 0$. Show that there exist unique integers q and r with $a = qb + r$ and $2|r| \le |b|$, or give an example of lack of uniqueness.
4. Let a, b, c, and q be integers with $c \neq 0$ and $a = bq + c$. Let $m = \text{lcm}[b, c] = hb = kc$. Prove that $ka = \text{lcm}[a, b]$.
5. With the help of Problem 4, explain how the division algorithm could be used to calculate the least common multiple of two integers.
6. Let a, b, and c be integers. Let $d_1 = \gcd(a, b)$ and $d_2 = \gcd(b, c)$. Prove that $\gcd(d_1, c) = \gcd(a, d_2)$.
7. Generalize the result in Problem 6.
8. Do the problem analogous to Problem 6 in which gcd is replaced by lcm.
9. Generalize the result in Problem 8.
10. Let $p_1, p_2, \ldots, p_r$ be r distinct positive primes.
    (i) Explain why the standard factorization for
    $$1 + p_1 p_2 \cdots p_r$$
    involves only primes that differ from all these p_j.
    (ii) Explain why there must be an infinite number of primes.
11. Let $(p_1)^{e_1}(p_2)^{e_2} \cdots (p_r)^{e_r}$ be the standard factorization of the positive integer n. Express each of the following in terms of the e_i.
    (a) The number of ordered pairs of positive integers (x, y) such that $\text{lcm}[x, y] = n$.
    (b) The number of ordered pairs of positive integers (x, y) such that
    $$\frac{1}{x} + \frac{1}{y} = \frac{1}{n}.$$
12. Prove a relationship between the answers for the preceding problem and that for Problem 8(ii) of Section 1.7.

Supplementary and Challenging Problems for Chapter 2

1. Let γ be an s-cycle $(a_1 a_2 \ldots a_s)$ and θ be any permutation in $\mathbf{S_n}$. Let $\theta(a_i) = b_i$ for $i = 1, 2, \ldots, s$. Show that
$$\theta^{-1}\gamma\theta = (b_1 b_2 \ldots b_s).$$

2. Let γ and γ' be s-cycles in $\mathbf{S_n}$. Show that there exists at least one permutation θ in $\mathbf{S_n}$ such that

$$\gamma' = \theta^{-1}\gamma\theta.$$

3. Let the permutation α of $\mathbf{S_n}$ be a product $\gamma_1\gamma_2 \cdots \gamma_t$ of disjoint cycles $\gamma_1, \gamma_2, \ldots, \gamma_t$ of lengths $k_1, k_2, \ldots, k_t$, respectively. Show that, for every θ of $\mathbf{S_n}$, $\theta^{-1}\alpha\theta$ is also a product of t disjoint cycles whose lengths are $k_1, k_2, \ldots, k_t$.

4. (i) Let a and b be two distinct elements of order 2 in a finite group G of order m, and let $ab = ba$. Show that $4 \mid m$.
   (ii) Explain why a finite group of even order must have an element of order 2.
   (iii) Let n be in $\mathbf{Z}^+$ and let G be an abelian group of order $4n + 2$. Prove that G has exactly one element of order 2.

5. Show that the multiplicative group $\mathbf{R}^{\pm}$ of the nonzero real numbers has only one subgroup N with index 2. (You may use the fact that every positive real number has real square roots.)

6. Let H be a subgroup in G. Let K consist of all the elements of a of G such that $aH = Ha$. Prove that K is a subgroup in G and that H is a normal subgroup in K.

7. Let a be the only element of order 2 in a group G. Prove that a is in the center of G.

8. Let G be a group of finite order u. Show the following.
   (i) The number of elements of G with order greater than 2 is even.
   (ii) If u is odd, every element of G is a square of an element of G.
   (iii) If u is even, there are at least two elements g of G with $g^2 = \mathbf{e}$, and hence at least one element of G is not a square.

9. Let $a^2b^2 = b^2a^2$ for all a and b in a group G. Let H consist of the elements of odd order in G. Prove that H is a commutative subgroup in G.

10. Let r be a positive integer and let $a^rb^r = b^ra^r$ for all a and b in a group G. Let H be the subset of G consisting of the elements with order relatively prime to r. Prove that H is a commutative subgroup in G.

11. Give the set of positive integers that are orders of subgroups in $\mathbf{S_5}$.

12. Prove that no 3-cycle (abc) is the cube of a permutation in $\mathbf{S_n}$.

13. Does there exist a nonabelian group G whose center C has index 9 in G?

14. Prove that a group G that has a finite number of subgroups must be finite.

15. In a finite group G, let T be a subset with more than half of the elements of G. Prove that every element of G is a product tt', with t and t' in T.

16. For every positive integer n, let m_n be the smallest positive integer m such that $\theta^m = (1)$ for all θ in $\mathbf{S_n}$. Prove the following.
   (i) $m_n = m_{n-1}$ if n is an integral multiple of two distinct primes.
   (ii) $m_n = pm_{n-1}$ if n is a power of a prime p.

17. Let ord $G = n$ and let $g_1, g_2, \ldots, g_n$ be n not necessarily distinct elements of G. Show that there exist integers i and j such that $1 \le i \le j \le n$ and $g_ig_{i+1} \cdots g_j = \mathbf{e}$, the identity of G.

18. Let $T = \{t_1, t_2, \ldots, t_m\}$ be a subset with m elements in an abelian group G. Let t^{-1} not be in T whenever t is in T. Show that the product t_it_j is in T for at most $m(m-1)/2$ of the m^2 choices of i and j with $1 \le i \le m$ and $1 \le j \le m$.

19. Give an example of a noncommutative group G in which $(ab)^3 = a^3b^3$ for all a and b in G.

20. Prove that every subgroup in $\mathbf{S_4}$ is solvable.

Supplementary and Challenging Problems for Chapter 3

1. Let β be a fixed odd permutation in $\mathbf{S_n}$.
(i) Show that $\beta^{-1}\alpha\beta$ is in $\mathbf{A_n}$ for every α in $\mathbf{A_n}$.
(ii) Show that the mapping f with $f(\alpha) = \beta^{-1}\alpha\beta$ is an automorphism of $\mathbf{A_n}$.
(iii) Explain why f is not an inner automorphism.

2. Show the following.
(i) A homomorphism f from $\mathbf{A_4}$ to some group G is completely determined once one knows the images of (12)(34) and (123) under f.
(ii) $\mathbf{A_4}$ has exactly 24 automorphisms, of which 12 are inner automorphisms.

3. (i) How many automorphisms are there of $\mathbf{S_4}$?
(ii) How many inner automorphisms are there of $\mathbf{S_4}$?

4. A mapping θ from a group G to a group G' with

$$\theta(xy) = \theta(y)\theta(x) \quad \text{for all } x \text{ and } y \text{ in } G$$

is a group ***antihomomorphism***. Tabulate all the antihomomorphisms from the symmetric group $\mathbf{S_3}$ to a group $\{\mathbf{e}, b\}$ of order 2.

5. An ***antiautomorphism*** of a group G is a bijective antihomomorphism from G to itself. (See Problem 4.) Tabulate all the antiautomorphisms of the symmetric group $\mathbf{S_3}$.

6. Let $\mathbf{A}(G)$ and $A'(G)$ consist of all the automorphisms and all the antiautomorphisms, respectively, of a group G. (See Problem 5.) Let $A''(G) = \mathbf{A}(G) \cup A'(G)$. Show the following.
(i) $A'(G)$ is not the empty set.
(ii) If either $\mathbf{A}(G)$ or $A'(G)$ is finite, there are the same number of mappings in each of these sets.
(iii) $A''(G)$ is a group, with $\mathbf{A}(G)$ as a normal subgroup, under composition of mappings.

7. Let M and N be normal subgroups in G such that $M \cap N = \{\mathbf{e}\}$. Prove that $ab = ba$ for all $a \in M$ and $b \in N$.

8. Let H and K be subgroups in G and let $D = H \cap K$. Let $h_1k_1 = h_2k_2$ with h_1 and h_2 in H and k_1 and k_2 in K. Show that h_1 and h_2 are in the same left coset of D in H.

9. Let θ be a surjective group homomorphism from G to G'. Let H' be a subgroup in G' and let H be the subset of G consisting of all the elements h of G such that $\theta(h)$ is in H'. (H is the union of the complete inverse images of the elements of H'.) Show the following.
(a) H is a subgroup in G.
(b) H is normal in G if H' is normal in G'.

10. Let G be a group of order p^m, where p and m are positive integers with p prime and $m \geq 2$. Explain why the order of the center C in G is neither p^{m-1} nor 1. What is the order when $m = 2$?

11. Let G be a nonabelian group of order pq, where p and q are distinct primes. Prove that the center C in G is the trivial subgroup $\{\mathbf{e}\}$.

12. (i) Prove that every group of order 15 is cyclic.
(ii) Explain why any two groups of order 15 are isomorphic.

13. Let G be a group of order 10. Prove the following.

(i) Either G is cyclic or G is isomorphic to the dihedral group $\mathbf{D_5}$ of symmetries of a regular pentagon (that is, to the group generated by $\{(12345), (25)(34)\}$).
(ii) G has exactly one subgroup of order 5.
(iii) G has 4 elements of order 5.
(iv) The number of elements of order 2 in G is either 1 or 5.

14. Generalize on Problem 13 to groups of order $2p$, where p is an odd prime.

15. Let G be a group of order 8 and let f_j designate the number of elements of order j in G (for $j = 1, 2, 4, 8$). Show the following.
(i) If G is abelian, (f_1, f_2, f_4, f_8) is (1, 1, 2, 4), (1, 3, 4, 0), or (1, 7, 0, 0).
(ii) If G is not abelian, (f_1, f_2, f_4) is (1, 1, 6) or (1, 5, 2).

16. What is the maximum number of groups of order 12 that one can have without two of them being isomorphic?

17. Let T be a set $\{\gamma_1, \gamma_2, \ldots, \gamma_s\}$ of s disjoint transpositions in $\mathbf{S_n}$. Let H be the subgroup in $\mathbf{S_n}$ generated by T.
(i) Show that H consists of the permutations of the form

$$\gamma_1^{e_1}\gamma_2^{e_2} \cdots \gamma_s^{e_s},$$

where e_j is 0 or 1 for $j = 1, 2, \ldots, s$.
(ii) What is the order of H?

18. In Problem 17, let γ_j be the transposition $(a_j b_j)$ with $a_j = 2j - 1$ and $b_j = 2j$ for $j = 1, 2, \ldots, s$. Explain why each left coset of the subgroup H (of Problem 17) in $\mathbf{S_{2s}}$ has exactly one permutation

$$\alpha = \begin{pmatrix} 1 & 2 & \cdots & 2s \\ c_1 & c_2 & \cdots & c_{2s} \end{pmatrix}$$

such that in the listing $c_1, c_2, \ldots, c_{2s}$ one has $2j - 1$ to the left of $2j$ for $j = 1, 2, \ldots, s$.

19. Let Γ be the collection of all the right cosets Hg of a subgroup H in G. For every a in G, let $\theta(a)$ be the mapping from Γ to Γ with $Hg \mapsto H(ga)$. Let M consist of the $\theta(a)$ for all a in G. Prove the following.
(i) M is a subgroup in the group $\mathbf{B}(\Gamma)$ of bijections from Γ to Γ.
(ii) The mapping f with $a \mapsto \theta(a)$ is a homomorphism from G to M.
(iii) Let K be the kernel of f. Then K is a subgroup in H, K is a normal subgroup in G, and K is the union of all the subgroups in H that are normal in G.
(iv) Let H and K have indices t and u, respectively, in G. Then $t \mid u$ and $u \mid (t!)$.

20. Let H be a subgroup of order 7 in a group G of order 21. Show that H is normal in G.

21. Prove that a group G of order pq is solvable if p and q are distinct primes in $\mathbf{Z}$.

22. Let θ be a mapping from an additive group G to the real numbers $\mathbf{R}$. Show that there exist mappings α and β from G to $\mathbf{R}$ such that all of the following hold for every g in G.
(a) $\theta(g) = \alpha(g) + \beta(g)$.
(b) $\alpha(-g) = \alpha(g)$.
(c) $\beta(-g) = -\beta(g)$.

23. Let T be a set of $mn + 1$ positive integers, where m and n are positive integers. Let D be the relation on T for which aDb if and only if $a \mid b$. Also, let E be the relation on T such that aEb if and only if both $a\bar{D}b$ and $b\bar{D}a$. Prove that in T there

SUPPLEMENTARY AND CHALLENGING PROBLEMS

exists either a subset $\{a_1, a_2, \ldots, a_{m+1}\}$ with $a_i D a_{i+1}$ for $1 \le i \le m$ or a subset $\{b_1, b_2, \ldots, b_{n+1}\}$ with $b_j E b_k$ for $j \ne k$.

24. Do the analogue of Problem 23 for a collection Γ of $mn + 1$ subsets of a set S and the relation Δ in Γ such that $\alpha\Delta\beta$ if and only if $\alpha \subset \beta$.

25. Let a boolean algebra have a finite carrier X. Prove that the number of elements in X is a positive integral power of 2.

26. Let α be an injection from X to Y and γ be an injection from Y to X. Prove that there exists a bijection β from X onto Y. (This is known as the ***Cantor-Bernstein Theorem.***)

27. Prove that there exists a bijection β from $\mathbf{Z}^+ \times \mathbf{Z}^+$ to $\mathbf{Z}^+$.

28. Prove that there exists a bijection from $\mathbf{Q}^+$ to $\mathbf{Z}^+$. ($\mathbf{Q}^+$ is the set of positive rational numbers.)

Supplementary and Challenging Problems for Chapter 4

1. Let R be a ring whose additive group is cyclic. Prove that R is commutative (under multiplication).

2. Let R be a ring with p elements, where p is prime. Prove that either R and $\mathbf{Z_p}$ are isomorphic rings or $rs = 0$ for all r and s in R.

3. In a ring U with unity $1 \ne 0$, let $ab = 1$ and let a not be a 0-divisor. Prove that $ba = 1$.

4. Let a be an element that is neither 0 nor a 0-divisor in a finite ring R. Show that R has a unity and that a is an invertible.

5. Let $\mathbf{V_m}$ be the multiplicative group of invertibles in $\mathbf{Z_m}$. Let m be odd and $\mathbf{V_m}$ be a cyclic group $[\bar{a}]$. Prove that $\mathbf{V_{2m}}$ is cyclic and that either $\bar{a}$ or $\overline{a+m}$ is a generator of $\mathbf{V_{2m}}$. Given that $\mathbf{V_{11}} = [\bar{2}]$, find a generator for $\mathbf{V_{22}}$.

6. Given that $a \equiv 7 \pmod 8$, prove that there do not exist integers x, y, and z such that $x^2 + y^2 + z^2 = a$.

7. Use the fact that no integer x satisfies $x^2 \equiv 3 \pmod 4$ to show that no integer y satisfies $y^2 \equiv 12 \pmod{16}$ and no integer z satisfies $z^2 \equiv 4444 \pmod{10^4}$.

8. Let a, b, and m be in $\mathbf{Z}$ with $m > 0$. Prove that $a \equiv b \pmod{7^m}$ implies that $a^7 \equiv b^7 \pmod{7^{m+1}}$.

9. Let a, b, and c be integers with

$$bc \equiv 1 \pmod a, \qquad ac \equiv 1 \pmod b, \qquad ab \equiv 1 \pmod c.$$

Show that $bc + ac + ab \equiv 1 \pmod{abc}$.

10. Find positive integers x and y, with x as small as possible, such that $x^2 + (x+1)^2 + (x+2)^2 + \cdots + (x+10)^2 = y^2$.

11. Let p be an odd positive prime in $\mathbf{Z}$ and let $r = (p-1)/2$. Show the following.
(a) In $\mathbf{Z_p}$, $\overline{2r+1} = 0$, $\overline{-2r} = \bar{1}$, and $(\overline{p-2})^{-1} = (\overline{-2})^{-1} = \bar{r}$.
(b) In $\mathbf{V_p}$, $\bar{1} \cdot \bar{2} \cdot \bar{3} \cdots \overline{p-3} = \bar{r}$.
(c) $(p-3)! - r$ is a multiple of p in $\mathbf{Z}$.
(d) $(p-3)! \equiv r \pmod p$.

12. Let p be a positive prime in $\mathbf{Z}$ of the form $p = 6s + 1$, with s an integer. Show the following.

SUPPLEMENTARY AND CHALLENGING PROBLEMS

(a) In $\mathbf{Z_p}$, $\overline{6s+1} = 0$, $\overline{-6s} = \bar{1}$, $(\overline{p-2})^{-1} = (\overline{-2})^{-1} = \overline{3s}$, and $(\overline{p-3})^{-1} = (\overline{-3})^{-1} = \overline{2s}$.
(b) In $\mathbf{V_p}$, $\bar{1} \cdot \bar{2} \cdot \bar{3} \cdots \overline{p-4} = \overline{6s^2}$.
(c) $(p-4)! - 6s^2$ is a multiple of p in $\mathbf{Z}$.
(d) $(p-4)! \equiv 6s^2 \pmod{p}$.

13. Let p be a positive prime in $\mathbf{Z}$ of the form $p = 6t - 1$, with t in $\mathbf{Z}$. Show the following.
(a) In $\mathbf{Z_p}$, $(\overline{p-2})^{-1} = \overline{-3t}$ and $(\overline{p-3})^{-1} = \overline{-2t}$.
(b) In $\mathbf{V_p}$, $\bar{1} \cdot \bar{2} \cdot \bar{3} \cdots \overline{p-4} = \overline{6t^2}$.
(c) $(p-4)! - 6t^2$ is a multiple of p in $\mathbf{Z}$.
(d) $(p-4)! \equiv 6t^2 \pmod{p}$.

14. Let p be a prime in the integers with $p \geq 7$. Prove the following.
(a) $(p^2 - p) \mid [(p-2)! + p - 1]$.
(b) $(p^2 + p) \mid [(p-2)! - p - 1]$.

15. Prove that $[n!/(n^2 + 3n + 2)]$ is an even integer for all positive integers n. Here $[x]$ denotes the greatest integer in x (that is, the integer such that $[x] \leq x < [x] + 1$).

16. Let S be the subring in the real numbers $\mathbf{R}$ consisting of all $a + b\sqrt{3}$, with a and b integers. Prove that S is not an ideal in $\mathbf{R}$.

17. An ideal I in a commutative ring R is said to be a ***prime ideal*** if $a \in R$, $b \in R$, and $ab \in I$ together imply that either $a \in I$ or $b \in I$. Prove that an ideal I in R is prime if and only if the quotient ring R/I has no 0-divisors.

18. Let I be an ideal in R. If $a - b \in I$, we say that a is ***congruent*** to b ***modulo*** I and write $a \equiv b \bmod I$.
(a) Prove that the relation in R of congruence modulo I is an equivalence relation.
(b) Prove that congruence modulo I is preserved under addition, subtraction, and multiplication; that is, $a \equiv b \bmod I$ and $c \equiv d \bmod I$ imply that $a \pm c \equiv b \pm d \bmod I$ and $ac \equiv bd \bmod I$.
(c) Prove that $a \equiv b \bmod I$ implies $a^n \equiv b^n \bmod I$ for all positive integers n.

19. Describe an infinite field with only a finite number of subfields.

20. Describe an infinite collection of subfields in the field $\mathbf{R}$ of real numbers.

21. Can an infinite integral domain D that is not a field have only a finite number of subdomains? Explain.

22. Let $\mathbf{A}(R)$ be the set of automorphisms of a ring R. Prove that $\mathbf{A}(R)$ is a subgroup in the group $\mathbf{B}(R)$ of bijections from R to R.

23. Define "characteristic of a group" in such a way that the previously defined characteristic of an integral domain D is always equal to the characteristic of the additive group of D. What is the characteristic of a Klein 4-group under this definition?

24. Let R be a commutative ring and let I consist of all the elements r of R such that $r^n = 0$ for some integer n. Prove the following.
(i) I is an ideal in R.
(ii) In the quotient ring R/I, if $\bar{r}^n = \bar{0}$ with n an integer, then $\bar{r} = \bar{0}$.

25. Let V and $\mathbf{R}^+$ be the multiplicative groups of nonzero quaternions and positive real numbers, respectively. Let S be the subgroup of all α in V with $\|\alpha\| = 1$. Show that V is isomorphic to the direct product $S \times \mathbf{R}^+$.

26. Let W be the ring of hamiltonian integers $a + b\mathbf{i} + c\mathbf{j} + d\mathbf{k}$ (with $a, b, c, d \in \mathbf{Z}$). Let $\lambda = (1 + \mathbf{i} + \mathbf{j} + \mathbf{k})/2$ and let W' consist of all α such that either $\alpha \in W$ or $\alpha = \beta + \lambda$ with $\beta \in W$. Show the following.

(a) W' is a subring in the skew-field H of quaternions.
(b) $\|\alpha\|$ is an integer for all α in W'.

27. Show that every quaternion γ is a sum $\gamma_1 + \gamma_2$ with γ_1 in the ring W' of Problem 26 and $\|\gamma_2\| < 1$.

28. Show that every complex number z is a sum $z_1 + z_2$, where z_1 is a gaussian integer $m + n\mathbf{i}$, with m and n integers, and $|z_2| \leq \sqrt{2}/2$.

29. Let α and β be in the ring W' of Problem 26 and let $\beta \neq 0$. Show the following.
(i) $\alpha = \gamma\beta$, with γ a quaternion.
(ii) $\alpha = \gamma_1\beta + \gamma_2\beta$, with γ_1 in W' and $\|\gamma_2\| < 1$.
(iii) $\alpha = \gamma_1\beta + \rho$, with γ_1 and ρ in W' and $\|\rho\| < \|\beta\|$.

30. Let z_1 and z_2 be gaussian integers with $z_2 \neq 0$. Show that there exist gaussian integers z_3 and z_4 with $z_1 = z_2 z_3 + z_4$ and $|z_4| < |z_2|$.

31. Let L be a subring in the W' of Problem 26 with the property that $\alpha\beta$ is in L for every α in W' and β in L. (This makes L a ***left ideal*** in W'.)
(a) Explain why L is not necessarily an ideal in W'.
★ (b) Show that L consists of all products $\alpha\gamma$, where α may be any element of W' and γ is some fixed element of L (that is, every left ideal in W' is a ***principal left ideal***).

32. Prove that, in the ring of gaussian integers, every ideal I is a principal ideal. (See Problem 31.)

33. Let S consist of all the real numbers of the form $(m + n\sqrt{3})/2$ with m and n in $\mathbf{Z}$ and $m \equiv n \pmod 2$.
(i) Show that S is a subring in the ring $\mathbf{R}$ of real numbers.
(ii) Let α and β be in S with $\beta \neq 0$. Show that there exist γ and ρ in S such that $\alpha = \gamma\beta + \rho$ and $|\rho| < |\beta|$.

34. Let a, d, and n be positive integers. Prove that the infinite arithmetic progression

$$a, \quad a + d, \quad a + 2d, \quad a + 3d, \ldots$$

either has no exact nth powers r^n of integers r or has infinitely many nth powers.

Supplementary and Challenging Problems for Chapter 5

1. Let z and w be complex numbers with z transcendental (over $\mathbf{Z}$). Prove the following.
(i) There is exactly one homomorphism θ from $\mathbf{Z}[z]$ to $\mathbf{Z}[w]$ with $\theta(1) = 1$ and $\theta(z) = w$.
(ii) The θ of (i) is an isomorphism if and only if w is transcendental (over $\mathbf{Z}$).

2. Let α and β be in $\mathbf{Q}[x]$. In $\mathbf{C}[x]$, let

$$\alpha = (x - r)(x - s), \qquad \beta = (x - a)(x - b)(x - c),$$

Prove the following.
(i) $r - a$ is algebraic over $\mathbf{Q}$ (and hence over $\mathbf{Z}$).
(ii) ra is algebraic over $\mathbf{Q}$.
(iii) $1/a$ is algebraic over $\mathbf{Q}$ if $a \neq 0$.

3. Let A be the set of all complex numbers a that are algebraic (over $\mathbf{Z}$). Prove that A is a subfield in the complex numbers $\mathbf{C}$.

4. (i) Let n be an integer and let $\beta \mid \alpha$ in $\mathbf{Z}[x]$. Show that $\beta(n) \mid \alpha(n)$ in $\mathbf{Z}$.

(ii) Find all the integers m such that $x^2 - x + m$ is a divisor of $x^{13} + x + 90$ in $\mathbf{Z}[x]$.

5. Let L_n be the nth Lucas number defined by

$$L_1 = 1, \qquad L_2 = 3, \qquad L_n = L_{n-1} + L_{n-2} \quad \text{for } n = 3, 4, 5, \ldots.$$

Prove that, for all positive integers n, $x^2 - x - 1$ is a divisor of $x^{2n} - L_n x^n + (-1)^n$ in $\mathbf{Z}[x]$.

6. Let $\alpha = u_0 x^d + u_1 x^{d-1} + \cdots + u_d$ and $\beta = v_0 x^{d-1} + v_1 x^{d-2} + \cdots + v_{d-1}$ in $F[x]$ with the coefficients satisfying

$$v_j = u_0 + u_1 + \cdots + u_j \quad \text{for } 0 \le j < d.$$

Also, let α and β have a common zero in an extension field E over F. Prove that $\alpha = \beta\gamma$ with γ in $F[x]$, and find γ.

7. (i) Let $p\alpha = \beta\gamma$ with p a prime in $\mathbf{Z}$ and α, β, and γ in $\mathbf{Z}[x]$. Also, let γ have a coefficient that is not a multiple of p. Prove that every coefficient of β is a multiple of p.

(ii) Prove that α is reducible in $\mathbf{Z}[x]$ if and only if it is reducible (and composite) in $\mathbf{Q}[x]$.

8. Let $\alpha = a_d x^d + \cdots + a_1 x + a_0$ be in $\mathbf{Z}[x]$. Let p be a prime in $\mathbf{Z}$ and let

$$a_d \not\equiv 0 \pmod{p}, \qquad a_j \equiv 0 \pmod{p} \quad \text{for } 0 \le j < d,$$
$$a_0 \not\equiv 0 \pmod{p^2}.$$

Prove that α is irreducible (and a prime) in $\mathbf{Q}[x]$. (This is known as ***Eisenstein's Irreducibility Criterion***.)

9. Prove that $x^5 - 4x^3 - 2$ is a prime in $\mathbf{Q}[x]$.

10. (i) Prove that $x^4 + 5x^3 + 10x^2 + 10x + 5$ is a prime in $\mathbf{Q}[x]$.

(ii) Use the substitution $x = y + 1$ to prove that

$$x^4 + x^3 + x^2 + x + 1$$

is a prime in $\mathbf{Q}[x]$.

11. Let m, n, and p be positive integers with p prime. Let α_n be the complex number $\cos(2\pi/n) + \mathbf{i}\sin(2\pi/n)$. The minimal polynomial $C_n(x)$ for α_n over $\mathbf{Q}$ is called the nth ***cyclotomic polynomial***. Prove the following.

(a) $C_n(x)$ is the minimal polynomial for each primitive nth root of unity — that is, for each

$$\cos(2k\pi/n) + \mathbf{i}\sin(2k\pi/n) \quad \text{with } \gcd(k, n) = 1.$$

The degree of $C_n(x)$ is $\phi(n)$, where ϕ is the Euler ϕ-function.

(b) $x^n - 1$ is the product of the $C_d(x)$ for all positive integral divisors d of n.

(c) $C_p(x) = x^{p-1} + x^{p-2} + \cdots + x + 1$. (Recall that p is a prime.)

(d) If $p \mid n$, then $C_{pn}(x) = C_n(x^p)$.

(e) If p is not an integral divisor of n, then

$$C_{pn}(x)C_n(x) = C_n(x^p).$$

★ (f) If p is not an integral divisor of n, then

$$C_{pn}(x) = C_n(\alpha_p x)C_n(\alpha_p^2 x)C_n(\alpha_p^3 x) \cdots C_n(\alpha_p^{p-1} x)$$

and $C_{pn}(x)C_n(x) = C_n(x^p) = D(p, n)$, where $D(p, n)$ is the determinant of the p by p matrix (a_{ij}) in which a_{ij} is the sum of all the terms $c_k x^k$ of $C_n(x)$ for which $k \equiv i - j + 1 \pmod{p}$.

(g) $C_1(1) = 0$, $C_n(1) = p$ if n is a power p^m, and $C_n(1) = 1$ if n is neither 1 nor a power of a prime.

★ (h) Each coefficient of $C_n(x)$ is in $\{-1, 0, 1\}$ if n is not a multiple of three distinct odd primes.

12. Prove that $x^{n-1} + x^{n-2} + \cdots + x + 1$ is a prime in $\mathbf{Q}[x]$ if and only if n is a prime in $\mathbf{Z}^+$.

13. In $\mathbf{Z_2}[x]$, let $\beta_n = x^{n-1} + x^{n-2} + \cdots + x + \bar{1}$.
(i) Prove that β_n is reducible whenever n is a composite integer.
(ii) Show that β_p may or may not be reducible when p is a prime in $\mathbf{Z}$.

14. Let $\beta = (x^3 - 1)(x^2 - 1)(x - 1)$ and $\alpha_n = (x^{n+2} - 1)(x^{n+1} - 1)(x^n - 1)$. Prove that $\beta \mid \alpha_n$ in $\mathbf{Z}[x]$ for all positive integers n. Generalize.

15. Let $p = 2q + 1$, where q is an odd positive integer and p is a prime. Prove that $q! \equiv \pm 1 \pmod{p}$.

16. Let p and d be positive integers with p a prime and $d \mid (p - 1)$. Show that there are at most $\phi(d)$ elements with order d in the multiplicative group $\mathbf{V_p}$ of invertibles in $\mathbf{Z_p}$. [Here ϕ is the Euler ϕ-function. It can be shown that there are exactly $\phi(d)$ elements with order d in $\mathbf{V_p}$. See Problem 5 of the Supplementary and Challenging Problems for Chapter 7.]

17. Let $r = \cos(2\pi/n) + \mathbf{i}\sin(2\pi/n)$, where n is an integer greater than 1. Show the following.
(i) $x^n - 1 = (x - 1)(x - r) \cdots (x - r^{n-1})$ in $\mathbf{C}[x]$.
(ii) $1 + r + r^2 + \cdots + r^{n-1} = 0$.
(iii) $1 \cdot r \cdot r^2 \cdots r^{n-1} = (-1)^{n-1}$.
(iv) Let $p_n = (1 + r)(1 + r^2) \cdots (1 + r^{n-1})$. Then $p_n = 0$ when n is even, and $p_n = 1$ when n is odd.
(v) $\{1, r, r^2, \ldots, r^{n-1}\}$ is the only subgroup with order n in the multiplicative group of the nonzero complex numbers.

18. Let $r_1, r_2, \ldots, r_n$ be the n nth roots of 1 in $\mathbf{C}$. Evaluate.

$$\prod_{i<j}(r_i - r_j)^2;$$

that is, the product of the squares of the differences of the nth roots of unity.

19. Let $\alpha = a + b(x + 3) + c(x + 3)(x - 4)$, where a, b, and c are in $\mathbf{Q}$. Find a, b, and c given that, in $\mathbf{Q}[x]$,

$$(x + 3) \mid (\alpha - 6),\ (x - 4) \mid (\alpha + 7),\ \text{and}\ (x - 6) \mid (\alpha - 10).$$

Note that here the parentheses are symbols of aggregation and do not denote principal ideals.

20. Find the smallest positive integer a such that there is a polynomial $ax^2 + bx + c$ in $\mathbf{Z}[x]$ with real zeros r and s satisfying $0 < r < s < 1$.

21. For every positive integer n, let

$$(x + 1)^n(x^2 - 1.998x + 1) = x^{n+2} + a_{n+1}x^{n+1} + a_n x^n + \cdots + a_0.$$

Find the smallest n such that $a_j > 0$ for $0 < j < n + 1$.

22. Let F be an extension field over $\mathbf{Q}$ and let n denote $n \cdot 1$ in F. In $F[x]$, let

$$x_a^{(d)} = (x + a)(x + a - 1)(x + a - 2) \cdots (x + a - d + 1).$$

Prove that for every α of degree d in $F[x]$, there are c_i in F such that

$$\alpha = c_0 x_0^{(d)} + c_1 x_1^{(d)} + \cdots + c_d x_d^{(d)}.$$

23. In $\mathbf{Q}[x]$, let $\alpha = x^4 + x^3 - x^2 - x + 1$.
(i) Show that if r is a zero in $\mathbf{C}$ of α, so is $-1/r$.

(ii) Find the monic second-degree β in $\mathbf{Q}[x]$ such that $s = r - r^{-1}$ is a zero of β when r is a zero of α.

(iii) Find all the roots of $\alpha = 0$ in $\mathbf{C}$.

24. Generalize on parts (i) and (ii) of Problem 23.

25. Let $\alpha = x^4 + ax^3 + bx^2 + cx + d$ be in $\mathbf{C}[x]$. Use the substitution $x = \sqrt{y}$ to show that the square of a root of $\alpha = 0$ satisfies

$$y^4 + (2b - a^2)y^3 + (b^2 + 2d - 2ac)y^2 + (2bd - c^2)y + d^2 = 0.$$

26. (i) In $\mathbf{C}[x]$, show that $(x + a + b) \mid (x^3 - 3abx + a^3 + b^3)$ for all a and b in $\mathbf{C}$.

(ii) Let a, b, c, and z be complex numbers such that $z + az^{2/3} + bz^{1/3} + c = 0$. Explain why $(z + c)^3 + a^3z^2 + b^3z - 3abz(z + c) = 0$.

(iii) In $\mathbf{C}[x]$, find the monic third-degree polynomial whose zeros are the cubes of the zeros of $x^3 + ax^2 + bx + c$.

27. Assume the fact that, if $r_1, r_2, \ldots, r_n$ are nonnegative real numbers and

$$(x - r_1)(x - r_2) \cdots (x - r_n) = x^n - s_1x^{n-1} + s_2x^{n-2} - \cdots + (-1)^n s_n,$$

then $s_1/n \geq \sqrt[n]{s_n}$. Use this result to prove that the roots of

$$x^n + a_1x^{n-1} + a_2x^{n-2} + \cdots + a_{n-1}x + a_n = 0$$

cannot all be real when each a_i is in $\{-1, 1\}$ and $n > 3$.

28. Prove that there is no choice of k as a real number such that

$$x^4 + x^3 - x^2 + kx + 1 = 0$$

has 4 real roots.

29. Let $\mathbf{C}$ be the field of complex numbers. Let α and β be in $\mathbf{C}[x]$ and have the same set of zeros in $\mathbf{C}$ (but possibly with different multiplicities). Let the same be true for the polynomials $\alpha + 1$ and $\beta + 1$. Prove that $\alpha = \beta$ or α has degree zero.

30. Let F be an extension field over $\mathbf{Q}$. In $F[x]$, let

$$(x - r_1)(x - r_2) \cdots (x - r_n) = x^n - s_1x^{n-1} + s_2x^{n-2} - \cdots + (-1)^n s_n.$$

Also, let $N_k = (r_1)^k + (r_2)^k + \cdots + (r_n)^k$. Show the following.

(i) $N_0 = n$ if $s_n \neq 0$.

(ii) $N_k - s_1N_{k-1} + s_2N_{k-2} - \cdots + (-1)^{k-1}s_{k-1}N_1 + (-1)^k s_k k = 0$ for $1 \leq k < n$.

(iii) $N_k - s_1N_{k-1} + s_2N_{k-2} - \cdots + (-1)^n s_n N_{k-n} = 0$ for $k = n, n + 1, \ldots$.

(These results are known as the ***Newton formulas*** on the sums of the powers of the roots of a polynomial equation.)

31. Let $w = (-1 + \mathbf{i}\sqrt{3})/2$. Find all the invertibles in $\mathbf{Z}[w]$.

32. Find all the invertibles in $\mathbf{Z}[\mathbf{i}\sqrt{5}]$.

33. In the integral domain $\mathbf{Z}[\mathbf{i}]$ of the gaussian integers, let

$$\alpha_1\alpha_2 \cdots \alpha_m = \beta_1\beta_2 \cdots \beta_n,$$

with the α's and β's primes in $\mathbf{Z}[\mathbf{i}]$. Prove the following.

(i) Each α_i has as many associates among the α's as it has among the β's.

(ii) $m = n$.

(This is an analogue for $\mathbf{Z}[\mathbf{i}]$ of the unique factorization theorem for $\mathbf{Z}$ given in Theorem 2 of Section 1.7.)

34. Let $D = \mathbf{Z}[\mathbf{i}\sqrt{5}]$. For all $\alpha = a + b\mathbf{i}\sqrt{5}$ (with a and b in $\mathbf{Z}$) in D, let $\bar{\alpha} = a - b\mathbf{i}\sqrt{5}$ and $\|\alpha\| = \alpha\bar{\alpha}$. Prove the following.

(i) $\|\alpha\beta\| = \|\alpha\| \cdot \|\beta\|$ in D.

(ii) D is an integral domain.

(iii) $6 = 2 \cdot 3 = (1 + \mathbf{i}\sqrt{5})(1 - \mathbf{i}\sqrt{5})$ in D.

(iv) 2, 3, $1 + \mathbf{i}\sqrt{5}$, and $1 - \mathbf{i}\sqrt{5}$ are primes in D.

(v) No two of the primes in part (iv) are associates. (This shows that $\mathbf{Z}[\mathbf{i}\sqrt{5}]$ does not have the type of unique factorization into primes described for $\mathbf{Z}[\mathbf{i}]$ in the preceding problem.)

35. Let F be a finite field with m elements and let q be a positive prime in $\mathbf{Z}$. Show that there are exactly $(m^q - m)/q$ monic prime polynomials of degree q in $F[x]$.

36. Let θ be a mapping from $\mathbf{C}$ to $\mathbf{C}$ with $\Delta^2\theta(x) = 2\theta(x)$; that is,

$$\theta(x + 2) - 2\theta(x + 1) + \theta(x) = 2\theta(x).$$

Explain why there cannot be any polynomial α in $\mathbf{C}[x]$ such that the polynomial function $\alpha^* = \theta$.

37. For each integer $s > 1$, let d_s be the lowest degree of a monic polynomial α in $\mathbf{Z}[x]$ with the property that $s \mid \alpha(n)$ for all integers n. Prove that d_s is the smallest positive integer k such that $s \mid k!$.

38. Find all the primes $\pi = m + n\mathbf{i}$ in the gaussian integers $\mathbf{Z}[\mathbf{i}]$ such that π is an associate of its conjugate $\bar{\pi} = m - n\mathbf{i}$.

39. Find a positive prime p in $\mathbf{Z}$ and a prime $\pi = m + n\mathbf{i}$ in $\mathbf{Z}[\mathbf{i}]$ such that $\pi^2 \mid p$ in $\mathbf{Z}[\mathbf{i}]$.

40. In $\mathbf{C}[x]$, let α' be the derivative of α and let $(\alpha')^r \mid \alpha^s$ for some positive integers r and s. Prove that $\alpha' \mid \alpha$ and that α has only one zero (and that its multiplicity is $\deg \alpha$).

41. The sequence of ***Chebyshev Polynomials*** is defined by

$$\gamma_0 = 1, \qquad \gamma_1 = x, \qquad \text{and} \qquad \gamma_n = 2x\gamma_{n-1} - \gamma_{n-2} \quad \text{for} \quad n \geq 2.$$

Show that both -1 and $\cos[\pi/(2n + 1)]$ are real zeros of $\gamma_{n+1} + \gamma_n$ for $n \geq 0$.

42. For n in $\mathbf{Z}^+$, show that $\cos[\pi/(2n + 1)]$ is a zero of the polynomial β_n defined by $\beta_1 = 2x - 1$, $\beta_2 = 4x^2 - 2x - 1$, and the recursion relation $\beta_n = 2x\beta_{n-1} - \beta_{n-2}$.

Supplementary and Challenging Problems for Chapter 6

1. Prove that a regular polygon with 17 sides, each of given length, is constructible with straightedge and compass.

2. Prove that one cannot construct with straightedge and compass a regular 11-gon having sides of a given length.

Supplementary and Challenging Problems for Chapter 7

1. Let U be a commutative ring with unity. We say that ideals I and J in U are ***relatively prime*** if there exist an h in I and a k in J such that $h + k = 1$. Given that I and J are relatively prime ideals in U, prove that the intersection

$$(a + I) \cap (b + J),$$

of any cosets of I and J, respectively, is nonempty and is a coset of $I \cap J$. Also, explain why this result is a generalization of Theorem 2 of Section 7.1.

2. State and prove a generalization of the Chinese Remainder Theorem that deals with a set $\{I_1, I_2, \ldots, I_r\}$ of r pairwise relatively prime ideals in U. (See the preceding problem.)

3. Let $\alpha : \mathbf{Z}^+ \to N = \{0, 1, 2, \ldots\}$ be a mapping with the following three properties:
(a) $\alpha(1) = 0$;
(b) $\alpha(p) = 1$ if p is prime;
(c) $\alpha(mn) = n\alpha(m) + m\alpha(n)$ for all m and n in $\mathbf{Z}^+$.
Prove that one and only one such α exists, and express $\alpha(n)/n$ in terms of the standard factorization $n = (p_1)^{e_1}(p_2)^{e_2} \cdots (p_r)^{e_r}$.

4. Prove that a prime $p > 3$ is expressible as $p = m^2 + 3n^2$, with m and n in $\mathbf{Z}$, if and only if $p \equiv 1 \pmod{6}$.

5. Let F be a finite field with m elements. Prove the following.
(i) If $d > 0$ and $d \mid (m - 1)$, there are $\phi(d)$ elements with order d in the multiplicative group V of nonzero elements of F.
(ii) V is cyclic.

6. Let p be an odd positive prime in $\mathbf{Z}$ and let

$$1 + \frac{1}{2} + \frac{1}{3} + \cdots + \frac{1}{p-1} = \frac{r}{s},$$

with r and s in $\mathbf{Z}$. Prove the following.
(i) $p \mid r$.
(ii) If $p > 3$, then $p^2 \mid r$.

7. For all positive integers n, let $A(n)$ be the number of ordered pairs (x, y) of positive integers such that

$$\frac{1}{x} + \frac{1}{y} = \frac{1}{n}.$$

Prove that $A(n) = \tau(n^2)$.

8. For all positive integers n, let $B(n)$ be the number of ordered pairs (x, y) of positive integers such that $\operatorname{lcm}[x, y] = n$.
(i) Prove that $B(n)$ is a multiplicative function; that is, $B(rs) = B(r)B(s)$ whenever $(r, s) = 1$.
(ii) Give a formula for $B(p^m)$, where p and m are positive integers with p a prime.

9. State and solve the analogue of Problem 8 dealing with ordered triples (x, y, z) of positive integers having n as their least common multiple.

10. Use Sylvester's Inclusion-Exclusion Principle to find the one millionth term in the sequence

$$2, 3, 5, 6, 7, 10, \ldots, 26, 28, 29, 30, 31, 33, \ldots$$

that results when the perfect squares, cubes, and fifth powers are removed from the positive integers $1, 2, 3, \ldots$.

11. Let θ be a permutation on $\{1, 2, \ldots, n\}$. Show that

$$n\prod_{i=1}^{n}[i - \theta(i)] = n[1 - \theta(1)][2 - \theta(2)] \cdots [n - \theta(n)]$$

is always an even integer.

12. Let α and β be permutations in $\mathbf{S}_{400}$. For $1 \le i \le 400$, let c_i be the product $\alpha(i)\beta(i)$ in $\mathbf{Z}$. Prove that

$$c_1c_2 \cdots c_{400} \equiv 1 \pmod{401}.$$

13. Let $T = \{1, 51, 2551, 1252551, 6251252551, \ldots\}$ consist of all the numbers formed by placing the digits of $5^n, 5^{n-1}, \ldots, 5^2, 5, 1$ side by side, where n is a nonnegative integer. Find all the integers t in T such that $t = x^2 + y^2 + z^2$ with x, y, and z in $\mathbf{Z}$.

14. Characterize the positive integers n that are not expressible as the sum, $a + (a + 1) + \cdots + (a + b)$, of two or more consecutive positive integers.

15. Let $a \equiv b \pmod{p}$, with p a prime. For all n in $\mathbf{Z}^+$, prove that

$$a^{p^n} \equiv b^{p^n} \pmod{p^{n+1}}.$$

16. Let $a_1 = 2$ and $a_{n+1} = a_n^2 - a_n + 1$ for $n \in \mathbf{Z}^+$. Prove that a_s and a_t are relatively prime whenever $s \neq t$.

17. Let a, b, c, and d be distinct integers and let

$$\alpha = x^4 - (a + b + c + d)x^3 + (ab + ac + ad + bc + bd + cd)x^2$$
$$- (bcd + acd + abd + abc)x + (abcd - 49).$$

Show that the only possibility for an integral zero of α is $(a + b + c + d)/4$.

18. Prove that among any 10 consecutive integers there is at least one that is relatively prime to each of the others.

19. Show that for every positive integer n there is a positive integer m such that $(\sqrt{2} - 1)^n = \sqrt{m + 1} - \sqrt{m}$.

20. Show that, for each positive integer n, there is an integer a_n with both

$$0 < a_n - (\sqrt{3} + 1)^{2n} < 1 \quad \text{and} \quad 2^{n+1} \mid a_n.$$

21. Let n be an integer such that

$$n^2 = 10^d c_d + 10^{d-1} c_{d-1} + \cdots + 10c_1 + c_0,$$

with each c_j in $\{1, 4, 9\}$. What are the possible values of the tens digit c_1 of n^2?

22. For each of the following, find all solutions in $\mathbf{Z}$.
(i) $x^2 + 3 = u^2$.
(ii) $4y^2 + 4y + 4 = u^2$.
(iii) $y^2 + y + 1 = v^2$.
(iv) $z^2 - z + 1 = v^2$.

23. Find all triples of integers x, y, and z such that

$$x = 3y^2 + 1 \quad \text{and} \quad x^2 + x + 1 = 3z^2.$$

24. Find all solutions in integers x and y of $x^3 - 1 = y^2$.

25. Find all solutions in integers x and y of $x^3 + 1 = y^2$.

26. Prove that the product of four consecutive positive integers is never a perfect square or cube.

27. Prove that $2^n - 1$ is not an integral divisor of $3^n - 1$ when n is an integer greater than 1.

Supplementary and Challenging Problems for Chapter 8

1. Give the generator matrix G and the check matrix K for a (26, 31) code that corrects one-digit errors. State the decoding rules for this code.

2. Let r and s be in $\mathbf{Z}^+$ with $2^{r-1} \leq s < 2^r$. Show that x^s can be calculated with no more than $2(r - 1)$ multiplications.

3. Describe a practical method for computer search for a prime with 50 digits.

4. In $\mathbf{Z}_2[x]$ let π be a prime polynomial of degree d and γ be relatively prime to π. Let $E : a_1 a_2 \ldots a_d \mapsto b_1 b_2 \ldots b_d$ be the mapping from W_d to itself with

$$\gamma(a_1 + a_2 x + \cdots + a_d x^{d-1}) - (b_1 + b_2 x + \cdots + b_d x^{d-1})$$

in the principal ideal (π). Is E bijective? If so, describe its inverse function.

5. Is Problem 4 the basis for a code? If so, is it error-detecting, or does it serve for encryption?

ANNOTATED BIBLIOGRAPHY

General

Albert, A. Adrian, ed. *Studies in Modern Algebra* (MAA Studies in Mathematics, Vol. 2). Washington, D.C.: Mathematical Association of America, 1963. A collection of essays on advances in algebra by Mac Lane, Bruck, C. W. Curtis, Kleinfeld, and Paige.

Aleksandrov, A. D.; A. N. Kolmogorov; and M. A. Lavrent'ev, eds. *Mathematics: Its Content, Methods, and Meaning*, 2nd ed. (3 vols.). Cambridge, Mass.: MIT Press, 1969. (Paperback available.) An excellent general view of mathematics by a group of Russian mathematicians. The section on groups is good for its discussion of symmetry and the applications of group theory to other parts of mathematics — for example, topology.

Bourbaki, N. *Éléments de mathématiques.* Paris: Hermann, 1942–1958. Although the *Éléments* covers a wide range of mathematical ideas, the material on algebra has been particularly important in influencing terminology and notation. For experienced readers.

Courant, Richard, and Herbert Robbins. *What Is Mathematics?* Oxford: Oxford University Press, 1978. (Paperback.) A beautifully written, lucid account of some of the most interesting and important problems of mathematics, designed for the reader without much mathematical background. Excellent sections on number theory and on constructibility.

Fraleigh, J. B. *A First Course in Abstract Algebra*, 4th ed. Reading, Mass.: Addison-Wesley, 1989. A standard text in abstract algebra with many sections indicating the ways in which abstract algebra is used in the study of topology.

Herstein, I. N. *Topics in Algebra*, 2nd ed. New York: Wiley, 1975. A popular intermediate-level text in abstract algebra.

Jacobson, Nathan. *Basic Algebra I*, 2nd ed. New York: W. H. Freeman, 1985. An important text that will strike some readers as something beyond the basic level.

———. *Lectures in Abstract Algebra* (3 vols.). New York: Springer, 1976, 1975, 1964. A classic treatment of abstract algebra written at a fairly sophisticated level.

Lang, Serge. *Algebra*, 2nd ed. Menlo Park, Calif.: Benjamin-Cummings, 1984. A widely used graduate-level text. Very terse.

Mac Lane, Saunders, and Garrett Birkhoff. *Algebra*, 2nd ed. New York: Macmillan, 1979. A modern successor to the authors' early and pace-setting classic, *A Survey of Modern Algebra.* Fairly sophisticated.

van der Waerden, B. L. *Modern Algebra* (2 vols.). New York: Ungar, 1949, 1950. Originally published in German, this great classic was one of the first texts in abstract algebra and was used to introduce the subject into graduate curricula in the United States in the 1930s.

Group Theory

Alperin, J. "Groups and Symmetry." In *Mathematics Today* (L. A. Steen, ed.). New York: Springer-Verlag, 1978. A popular account of the search for the finite simple groups, with amazing tables showing the orders of some of the large sporadic groups.

Aschbacher, Michael. "The Classification of the Finite Simple Groups," *Mathematical Intelligencer* 3(2) (1981): 59–64. An expository article on the history of the classification theorem for finite simple groups.

Gallian, J. A. "The Search for Finite Simple Groups," *Mathematics Magazine* 49 (1976): 163–179. A good expository account of the monumental effort to finish the classification of the finite simple groups.

Gorenstein, Daniel. *Finite Groups*, 2nd ed. New York: Chelsea, 1980. An advanced treatise on group theory.

Hall, Marshall, Jr. *The Theory of Groups*, 2nd ed. New York: Chelsea. An advanced, modern treatment of group theory. A good reference.

Kleiner, Israel. "The Evolution of Group Theory: A Brief Summary," *Mathematics Magazine* 59 (1986): 195–213. A very readable survey of the history of group theory. This article won the Carl B. Allendoerfer Award of the Mathematical Association of America for expository writing. Highly recommended.

Rotman, Joseph. *An Introduction to the Theory of Groups*, 3rd ed. Boston: Allyn & Bacon, 1984. A standard graduate text in group theory.

Schattschneider, Doris. "The Plane Symmetry Groups: Their Recognition and Notation." *American Mathematical Monthly* 85 (1978): 439–450. A charming article on transformations that leave invariant designs or patterns in the plane. It includes many patterns and a discussion of connections with the art of M. C. Escher.

Thompson, J. G., and W. Feit. "Solvability of Groups of Odd Order." *Pacific Journal of Mathematics* 13 (1963): 775–1029. This is the now-famous paper containing the results mentioned in the biographical note on Burnside at the close of Section 2.13.

Symmetry

Coxeter, H. S. M. *Introduction to Geometry*, 2nd ed. New York: Wiley, 1969. This standard work on geometry by probably the most renowned geometer of the mid-twentieth century contains chapters on symmetry, isometries, and the various groups involved, always from the point of view of a geometer.

MacGillavry, Caroline H. *Fantasy and Symmetry: The Periodic Drawings of M. C. Escher*. New York: Harry Abrams, 1976. (An earlier version, published in the Netherlands in 1965, was titled *Symmetry Aspects of M. C. Escher's Periodic Drawings*.) A beautifully produced book showing the extraordinary drawings of Escher that illustrate plane symmetry.

Pólya, George. "Über die Analogie der Kristallsymmetrie in der Ebene," *Zeitschrift für Kristallographie* 60 (1924); 278–282, and Paul Niggli, "Die Flächensymmetrien homogener Diskontinuen," *Zeitschrift für Kristallographie* 60 (1924): 283–298. The original 1924 papers enumerating the 17 symmetries in the plane.

Schattschneider, Doris. *Visions of Symmetry: Notebooks, Periodic Drawings, and Related Work of M. C. Escher*. New York: W. H. Freeman, 1990. A scholarly book dressed up as a lavishly illustrated and printed art book. The treatment of plane symmetries is detailed and erudite. The book contains over 350 illustrations, 180 never published before.

Schattschneider, Doris, and W. Walker. *M. C. Escher Kaleidocycles.* New York: Ballantine Books, 1977. A book with cutouts so that you can construct your own polyhedra with faces showing Escher drawings that illustrate plane symmetries.

Weyl, Hermann. *Symmetry.* Princeton, N.J.: Princeton University Press, 1952. The classic book on symmetry for a general audience, written by one of the finest mathematicians of the twentieth century.

Rings

Burton, David M. *A First Course in Rings and Ideals.* Reading, Mass.: Addison-Wesley, 1970. An advanced undergraduate or beginning graduate text in rings and ideals, containing a great deal of material on the subject beyond the present text.

Divinsky, N. J. *Rings and Radicals.* Toronto: University of Toronto Press, 1965. A graduate-level treatment of rings, including fairly recent developments with an emphasis on the theory of radicals (Jacobson, Brown-McCoy, Levitski, and others).

McCoy, Neal H. *Rings and Ideals* (Carus Monograph No. 8). Washington, D.C.: Mathematical Association of America, 1948. A classic work on rings in the well-known series of monographs written for the college-level audience.

———. *The Theory of Rings.* New York: Chelsea, 1964. A text for students with basic knowledge of abstract algebra. It contains some results that are comparatively recent in the development of the subject.

Fields

Artin, Emil. *Galois Theory* (Notre Dame Mathematical Lecture No. 2), 2nd ed. Notre Dame, Ind.: University of Notre Dame Press, 1944. A classic view of Galois theory by a great teacher and mathematician.

Lieber, Lillian R. *Galois and the Theory of Groups: A Bright Star in Mathesis.* Brooklyn, N.Y.: Galois Institute of Mathematics and Art, 1956. An unconventional free-verse treatment of Galois theory that conveys the principal ideas of the subject, if not the details. Amusing illustrations are provided by the author's husband.

Pollard, Harry, and H. G. Diamond. *The Theory of Algebraic Numbers* (Carus Monograph No. 9). Washington, D.C.: Mathematical Association of America, 1975. An elementary treatment of algebraic number theory in the well-known Carus series of books aimed at undergraduates.

Theory of Numbers

Grosswald, Emil. *Topics from the Theory of Numbers*, 2nd ed. Boston: Birkhäuser, 1981. Many of the usual topics of elementary number theory as well as an introduction to algebraic number theory and analytic number theory. Excellent treatment of the theory of partitions. Some work in abstract algebra included. Unusual for its treatment of many topics not included in other elementary texts.

Hardy, G. H., and E. M. Wright. *An Introduction to the Theory of Numbers*, 5th ed. Oxford: Clarendon Press, 1979. The greatest classic of them all — at least in the theory of numbers and in the English language. The footnotes are almost as exciting as the text, and the text is superb.

Khinchin, Aleksander Y. *Three Pearls of Number Theory*. Rochester, N.Y.: Graylock Press, 1957. An elegant treatment of three proofs, by youthful mathematicians, of conjectures in number theory that resisted the efforts of mature mathematicians for some time.

LeVeque, W. J. *Topics in Number Theory* (2 vols.). Reading, Mass.: Addison-Wesley, 1956. A collection of topics in number theory, some not commonly found in elementary texts.

———, ed. *Studies in Number Theory* (MAA Studies in Mathematics, Vol. 6). Washington, D.C.: Mathematical Association of America, 1969. A collection of essays on developments in number theory by well-known mathematicians: LeVeque, Lewis, J. Robinson, D. H. Lehmer, P. T. Bateman, and H. G. Diamond.

Niven, Ivan, and Herbert S. Zuckerman. *An Introduction to the Theory of Numbers*, 4th ed. New York: Wiley, 1980. An important text with some unusual topics and many challenging, interesting problems.

Rademacher, Hans. *Lectures on Elementary Number Theory*. Melbourne, Fla.: Krieger, 1977. A beautifully written collection of classic theorems in number theory with comparatively straightforward proofs. A good reference for solutions of some of the famous and difficult problems of number theory.

Sierpiński, Wacław. *250 Problems in Elementary Number Theory*. New York: American Elzevier, 1970. A beautiful set of problems by a great mathematician. Some of the problems have been solved only in recent years.

Uspensky, J. V., and M. A. Heaslet. *Elementary Number Theory*. New York: McGraw-Hill, 1939. This remains, in spite of its age, a standard work in number theory. It contains many derivations that show the real reasons why various formulas work, rather than clever tricks that work but do not aid one's intuition. Many nontrivial problems.

Weil, André. *Number Theory/An Approach through History from Hammurabi to Legendre*. Boston: Birkhäuser, 1984. Written for a general audience, this survey of number theory spanning 36 centuries is scholarly and pleasant to read. The author is one of the giants of twentieth-century mathematics.

Theory of Equations

Uspensky, J. V. *Theory of Equations*. New York: McGraw-Hill, 1948. A very good text on the classical topics of the theory of equations, including methods for solving the cubic and quartic equations and methods of calculating or approximating roots of higher-degree equations. A discussion of symmetric functions.

Geometric Constructions

Dudley, Underwood. *A Budget of Trisections*. New York: Springer-Verlag, 1987. An amusing book on the folly of attempting to trisect the angle. The title is a clever play on that of DeMorgan's *Budget of Paradoxes*, which included among other things the work of circle-squarers. This book contains a collection of failed attempts at trisecting the angle, a collection gathered by the author over many years.

History of Mathematics

Albers, Donald J.; G. L. Alexanderson; and Constance Reid. *International Mathematical Congresses/An Illustrated History 1893–1986*. New York: Springer-Verlag, 1986. A detailed account of the establishment of the Fields Medals, along with pictures of all the medalists up through the year 1986. (Pictures of the medalists up through 1990 appear in the Japanese edition).

———. *More Mathematical People*. Boston: Harcourt, Brace, Jovanovich, 1990. Interviews or biographical essays of eighteen prominent mathematicians, including two eminent American algebraists, Saunders Mac Lane and Irving Kaplansky. Mac Lane's interview includes the words of a song about group theory, written by Mac Lane.

Alexanderson, Gerald L; Leonard F. Klosinski; and Loren C. Larson, eds. *The William Lowell Putnam Mathematical Competition Problems and Solutions: 1965–84*. Washington, D.C.: Mathematical Association of America, 1985. The problems and solutions of 20 Putnam competitions; a follow-up of the Gleason-Greenwood-Kelly volume listed below, without the extensive discussion of subsequent work the problems stimulated.

Bell, Eric Temple. *The Development of Mathematics*, 2nd ed. New York: McGraw-Hill, 1945. More or less chronological accounts of the growth and development of important areas of mathematics by one of the most entertaining and popular of mathematical writers.

———. *Men of Mathematics*. New York: Simon & Schuster, 1961. (Available in paperback.) An entertaining, anecdotal account of the lives of prominent mathematicians. For a popular treatment of the lives of Galois, Abel, and others, this is highly recommended. For reservations about the historical accuracy of Bell's account of the life of Galois, see the paper by Tony Rothman below.

Boyer, Carl B., and Uta C. Merzbach. *A History of Mathematics*, 2nd ed. New York: Wiley, 1989. One of the few popular sources to cover (in a limited manner) results of the twentieth century.

Cajori, Florian. *A History of Mathematics*, 4th ed. New York: Chelsea, 1985. An excellent, detailed history of mathematics by a recognized authority. Unfortunately, for twentieth-century mathematics and twentieth-century discoveries concerning earlier results, one must go elsewhere. Nevertheless, it remains a very valuable source on the subject.

Campbell, Paul, and L. Grinstein. *Women of Mathematics*. New York: Greenwood Press, 1987. A counter to Bell's classic *Men of Mathematics*, this book gives biographies and brief surveys of the contributions of many women in mathematics, several of whom have been prominent algebraists (Emmy Noether, Käte Fenchel, and Olga Taussky-Todd).

Dickson, Leonard Eugene. *History of the Theory of Numbers* (3 vols.). New York: Chelsea, 1952. A comprehensive history of the literature on the theory of numbers up to approximately 1920.

Eves, Howard. *An Introduction to the History of Mathematics*, 5th ed. New York: Saunders, 1983. A popular, well-written history, widely used as a textbook.

———. *In Mathematical Circles* (Vols. I and II); *Mathematical Circles Revisited: Mathematical Circles Squared.* Boston: Prindle, Weber & Schmidt, 1969, 1971, 1972. A very entertaining collection of anecdotes and historical accounts of mathematics and mathematicians. Delightful reading for the layman as well as the professional.

Gleason, A. M.; R. E. Greenwood; and L. M. Kelly. *The William Lowell Putnam Mathematical Competition Problems and Solutions: 1938–1964.* Washington, D.C.: Mathematical Association of America, 1980. The problems of the first 25 competitions with solutions and references to related research in mathematics.

Hardy, G. H. *Ramanujan: Twelve Lectures on Subjects Suggested by His Life and Work,* 3rd ed. New York: Chelsea, 1978. A great twentieth-century mathematician develops some topics associated with his remarkable collaborator.

Infeld, Leopold. *Whom the Gods Love: The Story of Évariste Galois.* Reston, Va.: National Council of Teachers of Mathematics, 1978. A biography of Galois by one of his modern admirers.

Kanigel, Robert. *The Man Who Knew Infinity/A Life of the Genius Ramanujan.* New York: Charles Scribner's Sons, 1991. A superbly written and entertaining account of the life of the Indian mathematician, Srinivasa Ramanujan. No mathematical background is required.

Kline, Morris. *Mathematical Thought from Ancient to Modern Times.* New York: Oxford University Press, 1972. A detailed, up-to-date history of mathematics. If one is to own one history of mathematics, this is it.

Kramer, Edna. *The Nature and Growth of Modern Mathematics.* Princeton, N.J.: Princeton University Press, 1982. A general book on the development of mathematical ideas. Excellent for biographical data on some twentieth-century mathematicians. The accounts of the growth of algebra are easy to read and interesting. Also of interest is the chapter on Klein and his *Erlanger Programm.*

Newman, James R. *World of Mathematics* (4 vols.). New York: Simon & Schuster, 1956. A popular collection of excerpts on mathematics from many sources. A good set for reference as well as reading for pleasure.

Nový, Luboš. *Origins of Modern Algebra.* Leyden: Noordhof International Publishing, 1973. This is a detailed history of abstract algebra between 1770, the year of Lagrange's important work, and 1870, the end of a very active period in the development of algebra in England.

Ore, Oystein. *Niels Henrik Abel, Mathematician Extraordinary.* New York: Chelsea, 1974. A biography of Abel by a leading twentieth-century mathematician.

———. *Number Theory and Its History.* New York: McGraw-Hill, 1948. An excellent, easy-to-read book on number theory with historical accounts and portraits of mathematicians.

Phillips, Esther R., ed. *Studies in the History of Mathematics.* Washington, D.C.: Mathematical Association of America, 1987. This collection contains an excellent article by Harold M. Edwards on the invention of ideals by Richard Dedekind.

Rothman, Tony. "Genius and Biographers: The Fictionalization of Évariste Galois," *American Mathematical Monthly* 89 (1982): 84–106. A thorough reevaluation of the biographical accounts of Galois, with much newly examined evidence to indicate that earlier accounts were overly romantic and misleading.

———. "The Short Life of Évariste Galois," *Scientific American* (April 1982): 136–149. A popular account of recent research into the life of Galois.

Smith, David E. *A Source Book in Mathematics* (2 vols.). New York: Dover, 1984. Here one can find the original papers on the cubic, quartic, and quintic equations by Cardano, Ferrari (Cardano), Abel, and Galois, translated into English.

Struik, Dirk J. *A Concise History of Mathematics.* New York: Dover, 1967. (Paperback.) A good introduction to the history of mathematics, but very concise. There is nothing on twentieth-century developments.

———. ed. *A Source Book in Mathematics, 1200–1800.* Cambridge, Mass.: Harvard University Press, 1969. A collection of original papers by great mathematicians up to 1800, along with comments and background by the editor. Much of abstract algebra is missing here because of the 1800 terminal data. However, it is a good source if one wishes to see excerpts from the original publications on the solution of equations by Cardano and Lagrange, for example.

Wussing, Hans. *The Genesis of the Abstract Group Concept.* Cambridge, Mass.: MIT Press, 1984. An excellent description of the work of Lagrange, Ruffini, Cauchy, Galois, Kronecker, Jordan, Klein, Netto, Hölder, and others in group theory and the subjects in which group theory grew.

Applications

Beckenbach, Edwin F., ed. *Applied Combinatorial Mathematics.* Melbourne, Fla.: Krieger, 1981. A collection of papers of which one is especially interesting in the present context: "Pólya's Theory of Counting," by N. G. de Bruijn. Group theory is used to develop a fundamental theorem in combinatorial mathematics.

Berlekamp, Elwyn R. *Algebraic Coding Theory*, rev. ed. Laguna Hills, Calif.: Aegean Park Press, 1984. One of the standard works on the subject, often cited in the literature.

Birkhoff, Garrett, and Thomas C. Bartee. *Modern Applied Algebra.* New York: McGraw-Hill, 1970. Applications of abstract algebra to problems of computer design, programming languages, coding and decoding, mathematical linguistics, and radar.

Dornhoff, Larry L., and Franz E. Hohn. *Applied Modern Algebra.* New York: Macmillan, 1978. An accessible treatment of a wide range of algebraic topics with applications, including boolean algebra and coding theory. Some topics are not strictly algebraic. For example, there are discussions of some topics from graph theory (including the four-color problem and the traveling salesperson problem) and from combinatorics (the Pólya enumeration theorem).

Dyson, Freeman J. "Mathematics in the Physical Sciences." In *The Mathematical Sciences: A Collection of Essays.* Cambridge, Mass.: MIT Press, 1969. An easy-to-read essay on the uses of group theory in the discovery of elementary particles in physics.

Gell-Mann, Murray, and Yuval Ne'eman. *The Eightfold Way.* New York: W. A. Benjamin, 1965. Applications of group theory to the present-day study of elementary particles.

Hohn, Franz E. *Applied Boolean Algebra*, 2nd ed. New York: Macmillan, 1966. This is a simple, easy-to-read introduction to the applications of boolean algebra.

Kohavi, Zvi. *Switching and Finite Automata Theory*, 2nd ed. New York: McGraw-Hill, 1978. A popular text by an engineer. It starts with error-correcting codes, moves on to lattices and boolean algebra with applications to switching, and then to a number of more advanced topics.

Levinson, Norman. "Coding Theory: A Counterexample to G. H. Hardy's Conception of Applied Mathematics." *American Mathematical Monthly* 77 (1970): 249–258. Applications of finite fields and number theory to coding.

Loebl, Ernest M. ed. *Group Theory and Its Applications* (3 vols.). New York: Academic Press, 1968–1975. A collection of essays on applications of group theory to quantum mechanics, solid-state physics, atomic spectroscopy, nuclear theory, elementary particle theory, and relativity.

MacWilliams, F. J., and N. J. A. Sloane. *The Theory of Error-Correcting Codes.* New York: Elzevier, 1978. An encyclopedic treatment of error-correcting codes by authors who have made major contributions to the field. Although the book covers very sophisticated topics, the early chapters are easy to read.

Thompson, Thomas M. *From Error-Correcting Codes through Sphere Packings to Simple Groups.* (Carus Monograph No. 21). Washington, D.C.: Mathematical Association of America, 1984. A prize-winning account of the connections between the pure mathematics of group theory and sphere packing and applications of error-correcting codes. A good historical account; easy to read.

Answers, Hints, or Solutions for Most Odd-Numbered Problems

Section 1.1, page 8

1. *Hint:* See Example 2 of this section.

3. For all m in $\mathbf{Z}^+$, L_{3m} is even and L_{3m-2} and L_{3m-1} are odd. The proof is similar to that of Example 1.

5. (ii) $T_n = (n + 1)/2n$. Yes. Yes.
   (iii) **Basis.** The formula holds for $n = 2$, since $T_2 = 3/4$.
   Inductive Part. Assume that $T_k = (k + 1)/2k$. Then
$$T_{k+1} = T_k\left[1 - \frac{1}{(k+1)^2}\right] = \frac{k+1}{2k} \cdot \frac{(k+1)^2 - 1}{(k+1)^2} = \frac{k^2 + 2k + 1 - 1}{2k(k+1)}$$
$$= \frac{k(k+2)}{2k(k+1)} = \frac{k+2}{2(k+1)}.$$
   Since $T_{k+1} = (k + 2)/2(k + 1)$ is the claimed formula when $n = k + 1$, this completes the proof.

7. (ii) $A_n = n^2$.

9. (i) $v_n = (-1/2)^{n-1}$.
   (ii) *Hint:* Show that $v_{k+1} = -v_k/2$, and then use mathematical induction.
   (iv) $u_n = (2/3)[1 - (-1/2)^n]$.

11. $B_n = (n + 1)! - 1$.

13. (i) $P(n)$ is true for all n in $\mathbf{Z}^+$.
   (ii) $P(4) \Rightarrow P(5)$ is false.
   (iii) $P(1)$ is false.

Section 1.2, page 13

1. 12, -12, 4002, and -4002.

3. (a) -1, 1. (b) All integers. (c) 4. (d) 6.
   (e) ± 1, ± 2, ± 3, ± 6.

5. Let a, b, d, h, and k be integers with d an integral divisor of a and b. Then d is an integral divisor of $ha + kb$.

7. (a) Yes. Yes. (b) Yes, since $0 = d \cdot 0$ for all d in $\mathbf{Z}$.
   (c) $0\mathbf{Z} = \{0\}$. (d) 0. (e) Yes, since $d = d \cdot 1$ with 1 in $\mathbf{Z}$.
   (f) Yes, since $n = n \cdot 1$ with n and 1 in $\mathbf{Z}$.
   (g) Yes; $d \mid m$ implies that $m = dn$ with n in $\mathbf{Z}$, and $m = (-d)(-n)$ with $-n$ in $\mathbf{Z}$ implies that $(-d) \mid m$. (h) Yes, if $m = 9n$ is in $9\mathbf{Z}$, then $m = 3(3n)$ is in $3\mathbf{Z}$.

9. (i) $A = \{1, 2, 3, 5, 6, 10, 15, 30\}$.
   (ii) $B = \{1, 3, 7, 21\}$. (iii) $A \cap B = \{1, 3\}$.

11. Since $d \mid b$ and $d \mid c$, we have $b = ds$ and $c = dt$ with s and t in **Z**. Then $a = qb + c = qds + dt = d(qs + t)$ with $qs + t$ in **Z**; hence, $d \mid a$.

13. (a) Let a and b be in $8\mathbf{Z}$. Then $a = 8m$ and $b = 8n$ with m and n in **Z**; hence, $a - b = 8m - 8n = 8(m - n)$ with $m - n$ in **Z** and $a - b$ is in $8\mathbf{Z}$.
(b) Yes.

15. (a) $0, \pm 3, \pm 6, \pm 9, \pm 12, \pm 15, \pm 18$.
(b) Yes; the set **Z**, of all the integers, is one of the possibilities for the set T.

17. 0, 2, 4, 6. These are the only possibilities, since $10x + 14y$ must be even. They are of this form because $0 = 10 \cdot 0 + 14 \cdot 0$, $2 = 10 \cdot 3 + 14(-2)$, $4 = 10 \cdot 6 + 14(-4)$, and $6 = 10 \cdot 9 + 14(-6)$.

19. Let $a \mid b$ and $b \mid c$. Then $c = bm$ and $b = an$ with m and n in **Z**. Hence, $c = bm = (an)m = a(nm)$ with nm in **Z**; so $a \mid c$.

21. (ii) 4, 6, 8, 9, 10, 12, 14, 15, 16, 18, 20, 21, 22, 24, 25, 26, 27, 28, 30, 32.

23. $x = k$, $y = h - qk$.

25. (a) 6. (b) 2. (c) 6. (d) 3.

27.

x	−31	−5	5	7	9	11	21	47
y	−6	−8	−18	−44	34	8	−2	−4

29. Yes. If $c = ab$ with a and b in $1 + \mathbf{Z}^+$, then $c = xy + xz + yz + 1$ with $x = a - 1$, $y = b - 1$, and $z = 1$.

Section 1.3, page 19

1. (a) $q = 11, r = 16$;
(b) $q = -12, r = 1$;
(c) $q = 0 = r$.

3. (i) $-60, 0, \pm 6, \pm 12, \pm 18, \pm 24, \pm 30, \pm 36, \pm 42, \pm 48, \pm 54$.
(ii) $-59, -53, -47, \ldots, -11, -5, 1, 7, 13, \ldots, 49, 55$.
(iii) $-55, -49, -43, \ldots, -13, -7, -1, 5, 11, \ldots, 53, 59$.
(iv) $-57, -51, \ldots, -3, 3, 9, \ldots, 51, 57$.

5. Yes. **7.** Yes.

9. (a) No; 3 and 6 are in $3\mathbf{Z}$, and they have opposite parity.
(b) Yes; they are both odd.

11. (i) Yes. If $2a$ and $2b$ are in $2\mathbf{Z}$, so is $2a - 2b = 2(a - b)$.
(ii) No; 1 and 3 are in $1 + 2\mathbf{Z}$, but $3 - 1 = 2$ is not.

13. (i) Yes. (ii) Yes.

15. (i) If a is even, $a + 2d$ is the sum of two even integers and so is also even. If a is odd, $a + 2d$ is the sum of an odd integer and an even integer and so is odd, like a.
(ii) This follows from (i), since $m + n = (m - n) + 2n$.

17. (i) $2 \cdot 44$, $4 \cdot 22$.
(ii) Letting $x - y = 2$ and $x + y = 44$, we get $(x, y) = (23, 21)$. Also, with $x - y = 4$ and $x + y = 22$, we get $(x, y) = (13, 9)$.

19. Since $2 \mid a$, we have $2 \mid (x - y)(x + y)$; and thus at least one of $x - y$ and $x + y$ is even. So is the other factor, by Problem 15(ii).

21. Letting $x - y = 1$ and $x + y = a$, we get $x = (a + 1)/2$ and $y = (a - 1)/2$. [x and y are integers, since a is odd; they are positive, since $a > 1$.]

Section 1.4, page 23

1. $d = 11, e = 7, f = 1, g = 1$.

3. (i) 14. (ii) 7. (iii) 7. (iv) 7.
(v) Yes, $s = 7 = u = \gcd(14, 42, 35)$.

5. (a) 1, 2, 3, and 4. (b) 1 and 5.

7. (a) 1 or -1. (b) 2 or 4.

9. (i) 11. This follows from $(22, 55) = 11$ and Theorem 2(a).
(ii) $x = 3, y = -1$.

11. 1, 2, 3, 4, 6, 8, 9, 12, 18, 24, 36, 72.

13. (i) 1, 19.

23. (a) By Theorem 3(e) of Section 1.3, S consists of all the (integral) multiples of the smallest integer t of S. Hence, t must be one of the positive common divisors 1, 2, 17, and 34 of 170 and -102. In each case, the multiples of t include the multiples of 34.
(b) Yes. S might be the set of all integers.

25. By Lemma 1 of Section 1.2, a common (integral) divisor of n and $n + 1$ must be a divisor of their difference 1.

27. The positive integral divisors 1, 2, 5, and 10 of the difference 10 of n and $n + 10$.

29. *Hint:* $n^2 + n + 1 = (n + 2)(n - 1) + 3$.

31. It is true when u or v is even; also, if u and v are odd, then $u + v$ is even.

Sections 1.5 and 1.6, page 33

1.

	(c, d)	$[c, d]$	$(c, d)[c, d]$	cd
(a)	4	24	96	96
(b)	7	35	245	245
(c)	1	42	42	42

3. $d = 907, x = -257$, and $y = 302$.

5. $u = 9x = -2313$, and $v = 9y = 2718$.

7. (i) 14; (ii) 14; (iii) 14; (iv) 14.

11. (i) $(a, b) = 3$ and $(a, c) = 3$. (ii) 3. (iii) 630.
(iv) 3. (v) Yes.

13. (i) 30. (ii) 210. (iii) 105. (iv) 210. (v) Yes.

15. The hypothesis implies that there are integers m, h, and k with $bc = am$ and $1 = ah + bk$. Then $c = ach + bck = ach + amk = a(ch + mk)$. Hence, $a \mid c$.

Section 1.7, page 37

1. (i) $2^4 \cdot 3^2 \cdot 5 \cdot 7$. (ii) $2^8 \cdot 3^4 \cdot 5^2 \cdot 7^2$. (iii) $2^{12} \cdot 3^6 \cdot 5^3 \cdot 7^3$.

3. (i) 1, 2, 4, 8. (ii) 1, 2, 4, 8, 16. (iii) 1, 2, 4, 8, 16, 32. (iv) 1, 2, 4, 8, 16, 32, 64.

7. $(a + b)(b + 1)(c + 1)$.

11. All the exponents e_i must be even.

21. $(a, b) = 2^6 \cdot 19 \cdot 23^3$, $[a, b] = 2^{30} \cdot 3 \cdot 5^{21} \cdot 7^4 \cdot 11^2 \cdot 19^5 \cdot 23^7$.

27. 2^4 (or 16).

Review Problems (Chapter 1), page 39

1. (i) 97. (iii) $x = -15$, $y = 14$.

3. Yes.

5. The gcd is a positive integral divisor of every linear combination of n and $n + 4$, and hence their difference 4.

7. (ii) 12. (iii) Yes. See Theorem 3(e) of Section 1.3.

Section 2.1, page 43

1. (a) Not a permutation on $\mathbf{X}_5$, since 1 appears twice (and 2 does not appear) in the listing 1, 3, 4, 5, 1.
(b) A permutation on $\mathbf{X}_5$. (c) A permutation on $\mathbf{X}_5$.

3. $\begin{pmatrix} 1 & 2 \\ 1 & 2 \end{pmatrix}$, $\begin{pmatrix} 1 & 2 \\ 2 & 1 \end{pmatrix}$.

5. (a) $4! = 4 \cdot 3 \cdot 2 \cdot 1 = 24$. (b) $5! = 120$. **7.** $3! = 6$.

9. (i) $\gamma : 1 \mapsto 3, 2 \mapsto 2, 3 \mapsto 1$. (ii) $\delta : 1 \mapsto 1, 2 \mapsto 3, 3 \mapsto 2$.

11. $\begin{pmatrix} 1 & 6 & 3 & 2 & 4 & 5 \\ 6 & 3 & 1 & 4 & 2 & 5 \end{pmatrix}$.

13. 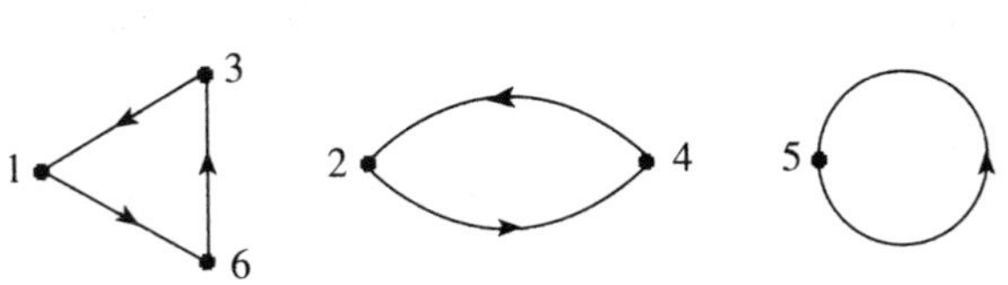

15. (i) 1. (ii) 1, 3, 6. (iii) Yes. (iv) 2, 4. (v) Yes.

17. $1 \mapsto 1, 2 \mapsto 2, 3 \mapsto 3, 4 \mapsto 4$; $1 \mapsto 2, 2 \mapsto 1, 3 \mapsto 3, 4 \mapsto 4$;
$1 \mapsto 3, 2 \mapsto 2, 3 \mapsto 1, 4 \mapsto 4$; $1 \mapsto 4, 2 \mapsto 2, 3 \mapsto 3, 4 \mapsto 1$;
$1 \mapsto 1, 2 \mapsto 3, 3 \mapsto 2, 4 \mapsto 4$; $1 \mapsto 1, 2 \mapsto 4, 3 \mapsto 3, 4 \mapsto 2$;
$1 \mapsto 1, 2 \mapsto 2, 3 \mapsto 4, 4 \mapsto 3$; $1 \mapsto 2, 2 \mapsto 1, 3 \mapsto 4, 4 \mapsto 3$;
$1 \mapsto 3, 2 \mapsto 4, 3 \mapsto 1, 4 \mapsto 2$; $1 \mapsto 4, 2 \mapsto 3, 3 \mapsto 2, 4 \mapsto 1$.

Section 2.2, page 49

1. $\varepsilon\varepsilon = \varepsilon = \theta\theta$; $\varepsilon\theta = \theta = \theta\varepsilon$.

3. (i) $\rho^2 = \begin{pmatrix} 1 & 2 & 3 \\ 3 & 1 & 2 \end{pmatrix}$. (ii) $\phi\rho = \begin{pmatrix} 1 & 2 & 3 \\ 2 & 1 & 3 \end{pmatrix} = \rho^2\phi$.

(iv)

	ε	ρ	ρ^2	ϕ	$\rho\phi$	$\rho^2\phi$
ρ^2	ρ^2	ε	ρ	$\rho^2\phi$	ϕ	$\rho\phi$
ϕ	ϕ	$\rho^2\phi$	$\rho\phi$	ε	ρ^2	ρ
$\rho\phi$	$\rho\phi$	ϕ	$\rho^2\phi$	ρ	ε	ρ^2
$\rho^2\phi$	$\rho^2\phi$	$\rho\phi$	ϕ	ρ^2	ρ	ε

5. (i) $\alpha^2 : 1 \mapsto 3, 2 \mapsto 2, 3 \mapsto 6, 4 \mapsto 4, 5 \mapsto 5, 6 \mapsto 1$.
(ii) $\alpha^2\alpha : 1 \mapsto 1, 2 \mapsto 4, 3 \mapsto 3, 4 \mapsto 2, 5 \mapsto 5, 6 \mapsto 6$.
(iii) $\alpha\alpha^2 = \alpha^2\alpha$. [See part (ii).]
(iv) $\alpha^{-1} : 1 \mapsto 3, 2 \mapsto 4, 3 \mapsto 6, 4 \mapsto 2, 5 \mapsto 5, 6 \mapsto 1$.

7. (a) 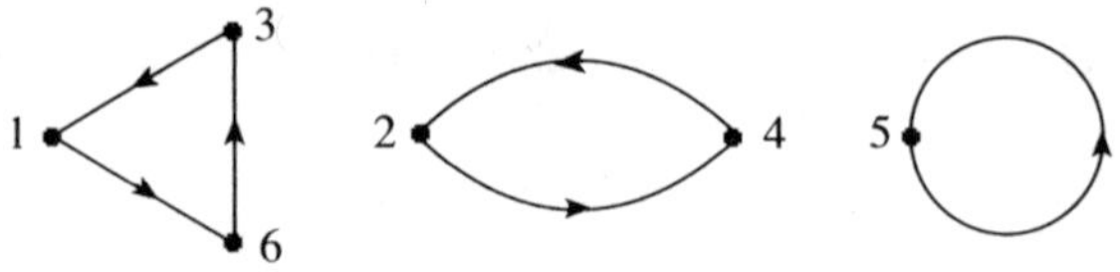

(b) 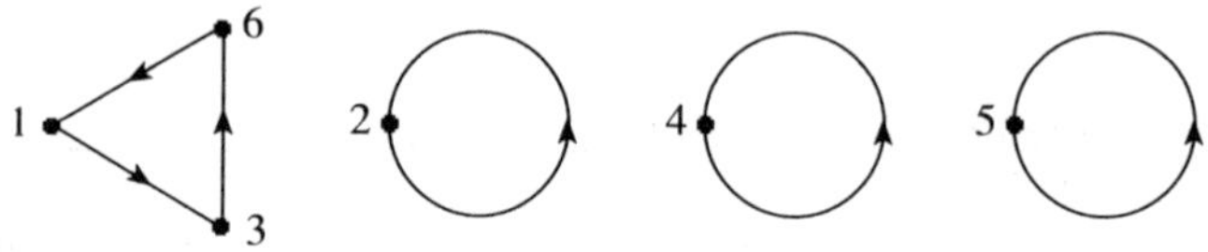

(c) 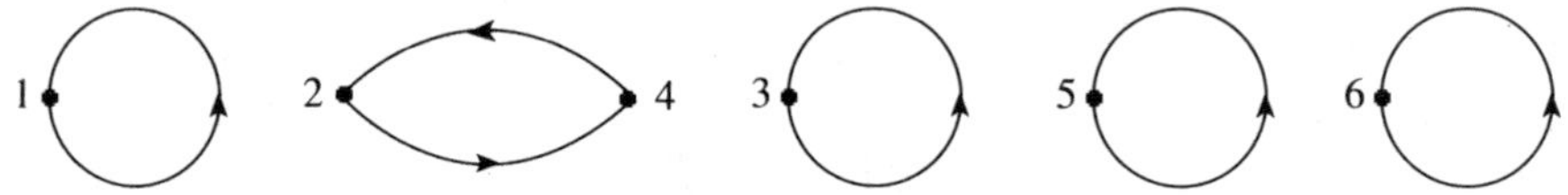

(d) 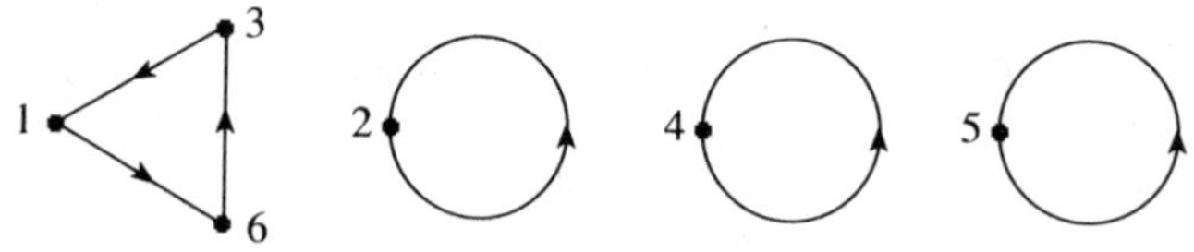

(e) 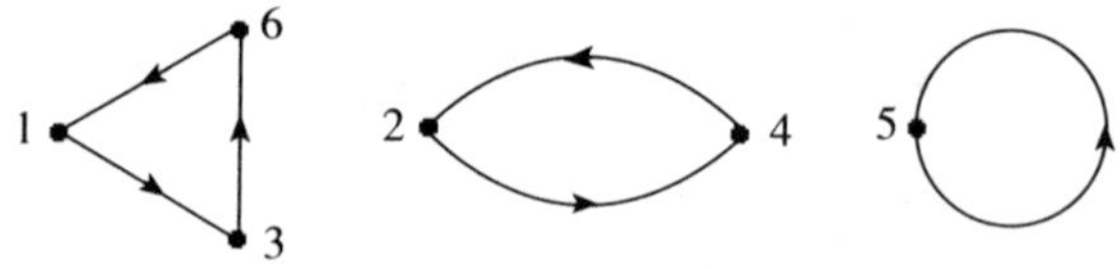

(f) 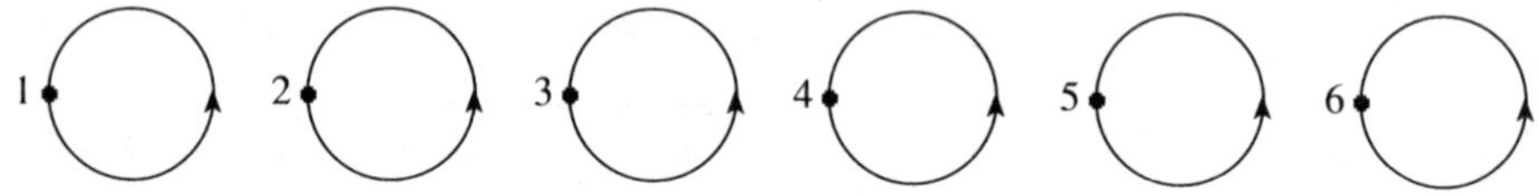

(g) Same as (e).

9. Yes.

Section 2.3, page 57

1. (a) $\theta^{-1} = \theta^3 = \theta^{101} = \theta$; $\theta^2 = \theta^4 = \theta^{100} = \varepsilon$, where $\varepsilon : 1 \mapsto 1, 2 \mapsto 2$.
(b) $\theta^{-1} : 1 \mapsto 3,\ 2 \mapsto 1,\ 3 \mapsto 2$; $\theta^2 = \theta^{-1}$; $\theta^3 : 1 \mapsto 1,\ 2 \mapsto 2,\ 3 \mapsto 3$; $\theta^4 = \theta^{100} = \theta$; $\theta^{101} = \theta^{-1}$.
(c) $\theta^{-1} = \theta^3 = \theta^{101} = \theta$; $\theta^2 = \theta^4 = \theta^{100} = \varepsilon : 1 \mapsto 1, 2 \mapsto 2, 3 \mapsto 3, 4 \mapsto 4$.

3. (i) $(\alpha\beta)^2 : 1 \mapsto 1, 2 \mapsto 2, 3 \mapsto 3$; $\alpha^2\beta^2 : 1 \mapsto 3, 2 \mapsto 1, 3 \mapsto 2$.
(ii) $(\alpha\gamma)^2 = \alpha^2\gamma^2 : 1 \mapsto 1, 2 \mapsto 2, 3 \mapsto 3$.

5. (i) Given that $ab = ba$, we have $(ab)^{-1} = (ba)^{-1} = a^{-1}b^{-1}$ by Theorem 4.
(ii) Given that $(ab)^{-1} = a^{-1}b^{-1}$, we have $[(ab)^{-1}]^{-1} = [a^{-1}b^{-1}]^{-1}$, and so $ab = (b^{-1})^{-1}(a^{-1})^{-1} = ba$.

7. (i) Under multiplication, H is closed and associative, with 1 as the identity and with each of 1 and -1 as its own inverse.
(ii) $\mathbf{ee} = \mathbf{e}$ and $\mathbf{e}a = a = a\mathbf{e}$, since $\mathbf{e}$ is the identity. Then $aa = \mathbf{e}$ because no element is repeated in a row (or column) of the table.
(iii) G is commutative if and only if the table is symmetric about the main diagonal (the diagonal from the top left corner to the bottom right corner).
(iv) Yes. Part (ii) gives us the entries of the table for G, and with part (iii) we see that G is commutative.

	e	a
e	e	a
a	a	e

9. (i)

·	e	b	c	d
e	e	b	c	d
b	b	e	d	c
c	c	d	e	b
d	d	c	b	e

(ii) Yes.

11. For all g in G, one has $g^{-1} = g$ since $gg = \mathbf{e}$. For a and b in G, $a^{-1} = a$, $b^{-1} = b$, $(ab)^{-1} = ab$, and hence $ab = (ab)^{-1} = b^{-1}a^{-1} = ba$.

13. One choice is $\alpha : 1 \mapsto 2,\ 2 \mapsto 1,\ 3 \mapsto 3$; $\beta : 1 \mapsto 2,\ 2 \mapsto 3,\ 3 \mapsto 1$; and $\gamma : 1 \mapsto 3,\ 2 \mapsto 1,\ 3 \mapsto 2$. Then $\alpha\beta$ and $\gamma\alpha$ have $1 \mapsto 3, 2 \mapsto 2, 3 \mapsto 1$, but $\beta \neq \gamma$.

15. (b) $(ab)^{-1} = b^{-1}a^{-1}$. (c) $(abc)^{-1} = c^{-1}b^{-1}a^{-1}$.

17. First we note that $a \neq a^2$, since $a = a^2$ would imply that $a\mathbf{e} = aa$ and then $\mathbf{e} = a$, using left cancellation. Hence $\mathbf{e}$, a, and a^2 are three distinct elements.
(i) For n in $\mathbf{Z}$, we have $n = 3q + r$ with q in $\mathbf{Z}$, and r in $\{0, 1, 2\}$. Then $a^n = a^{3q+r} = (a^3)^q a^r = \mathbf{e}^q a^r = \mathbf{e}a^r = a^r$. Part (i) implies that

$$\cdots = a^{-6} = a^{-3} = \mathbf{e} = a^3 = a^6 = \cdots,$$
$$\cdots = a^{-5} = a^{-2} = a = a^4 = a^7 = \cdots,$$
$$\cdots = a^{-4} = a^{-1} = a^2 = a^5 = a^8 = \cdots.$$

Parts (ii) and (iii) now follow.

21. If $xx \neq \mathbf{e}$, then x is not its own inverse. Such elements come in pairs $\{x, x^{-1}\}$, and hence there are an even number of them.

23. $bca = a^{-1}(abc)a = a^{-1}\mathbf{e}a = \mathbf{e}$, $cab = b^{-1}(bca)b = b^{-1}\mathbf{e}b = \mathbf{e}$.

25. The entry c appears more than once in the row for c.

Section 2.4, page 66

1. (a) (15432); (b) (156)(234); (c) (12345);
(d) (123), (132), (124), (142), (134), (143), (234), (243);
(e) (12)(34), (13)(24), (14)(23).

3.

	γ^{-1}	γ^2	γ^3	γ^4	γ^5	γ^6
(a)	(132)	(132)	(1)	(123)	(132)	(1)
(b)	(1432)	(13)(24)	(1432)	(1)	(1234)	(13)(24)
(c)	(15432)	(13524)	(14253)	(15432)	(1)	(12345)
(d)	(165432)	(135)(246)	(14)(25)(36)	(153)(264)	(165432)	(1)

5. (i) (123); (ii) (1234); (iii) (12345); (iv) (123456).

7. $(12)(13)(14)\cdots(1s)$ or $(1s)(2s)(3s)\cdots(\overline{s-1}s)$.

9. (i) (acb); (ii) $(xyz) = (xzy)^2$; (iii) $(abc)^5 = (acb)$; (iv) $(xyz) = (xzy)^5$.

11. $(a_1 a_3 a_5 \ldots a_{2r-1})(a_2 a_4 a_6 \ldots a_{2r})$.

13. Only for $n = 1$ or $n = 2$.

15. (i) $(1) = (12)(12)$, $(123) = (12)(13)$, $(132) = (13)(12)$.
(ii) The only interesting cases are $(123)(132) = (132)(123) = (1)$.
(iii) (1) is its own inverse, while (123) and (132) are inverses of each other.

17. Yes, as $\gamma_1\gamma_2 \cdots \gamma_{2h}\delta_1\delta_2 \cdots \delta_{2k}$. The integer $2h + 2k$ is even.

19. (i) Yes. (ii) Yes. (iii) 4.

21.

	(1)	θ	θ^2	θ^3	θ^4
(1)	(1)	θ	θ^2	θ^3	θ^4
θ	θ	θ^2	θ^3	θ^4	(1)
θ^2	θ^2	θ^3	θ^4	(1)	θ
θ^3	θ^3	θ^4	(1)	θ	θ^2
θ^4	θ^4	(1)	θ	θ^2	θ^3

23. $r = 6 = m$. **27.** 12.

29.

θ	(1)	(132)	(23)	(12)
$(12)\theta(12)$	(1)	(123)	(13)	(12)

31. (i) θ cannot have a cycle $(abc \ldots)$ in its standard form, since that would imply that θ^2 sends a to c and hence $\theta^2 \neq (1)$.

Section 2.5, page 72

1.

	(1)	(123)	(132)
(1)	(1)	(123)	(132)
(123)	(123)	(132)	(1)
(132)	(132)	(1)	(123)

The table shows that A is closed under multiplication. Also, the inverse of each element of A is in A, since (1) is its own inverse and (123) and (132) are inverses of each other. Now Lemma 2 tells us that A is a subgroup in $\mathbf{S}_3$.

3. (a) $(123)(23) = (13)$, which is not in D.
(b) $(123)^{-1} = (132)$, which is not in F.

7. (iii) No; if β is in a subgroup, so are β^2, β^3, and $\beta^4 = (1)$ by closure under multiplication.
(iv) $\{(1), \beta^2\}$ is a subgroup in H, and in $\mathbf{S}_4$, since it is closed under multiplication and each of its elements is its own inverse.

9. (i) Every group of order 2 consists of the identity $\mathbf{e}$ and another element b such that $b^2 = \mathbf{e}$. [See Problem 7(ii) of Section 2.3.]
(ii) $\{(1), (23)\}$, $\{(1), (13)\}$, $\{(1), (12)\}$.

11. (i) This part is similar to Problem 7(iv).
(ii) $\{1, \mathbf{i}\}$ is not closed under multiplication, and hence is not a subgroup, since $\mathbf{ii} = -1$ is not in this set. Alternatively, $\mathbf{i}^{-1} = -\mathbf{i}$ is not in $\{1, \mathbf{i}\}$.

13. Yes; $a^i a^j = a^{i+j} = a^{j+i} = a^j a^i$ for $0 \le i < s$ and $0 \le j < s$. Commutativity in G follows from commutativity of addition of integers.

15. (i) Yes; if h and k are in H, then $hk = kh$, since multiplication in H is the same as in G and G is abelian.
(ii) No; $\mathbf{S}_3$ is not abelian even though its subgroup $\{(1)\}$ is.

17. The desired results follow from Lemma 2 or Theorem 1.

19. *Hint:* Use Lemma 2.

25. (ii) When $G = \mathbf{S}_3$ and $s = 2$, $H = \{(1), (123), (132)\}$ is a subgroup in $\mathbf{S}_3$. (See Problem 1.)
(iii) When $G = \mathbf{S}_3$ and $s = 3$, $H = \{(1), (23), (13), (12)\}$ is not a subgroup $\mathbf{S}_3$, since it is not closed under multiplication; for example, $(12)(13) = (123)$, which is not in H.

27. (i) G, $\{1, -1\}$, $\{1\}$. (ii) Yes.

Section 2.6, page 79

1. (a) It is a group.
(b) It is not a group, since it does not satisfy Axiom 2 (Identity).
(c) It is a group.

3. (a) It is not a group, since it does not satisfy Axiom 3 (Inverses).
(b) It is not a group, since it does not satisfy Axiom 1 (Closure).
(c) It is a group.

5. The positive integers.

7. Assume that there exists an additive group $G = \{z, b, c\}$ of order 3 with z as the identity. Show that $b + c = z = c + b$, $2 \cdot b = c$, $2 \cdot c = b$, and $3 \cdot b = z = 3 \cdot c$ in such a group.

9. $-(-a) = a$ for every a of an additive group.

11. (i)

$\bar{a}$	$\bar{3}$	$\bar{4}$	$\bar{5}$	$\bar{6}$	$\bar{7}$	$\bar{8}$	$\bar{9}$
$-\bar{a}$	$\bar{7}$	$\bar{6}$	$\bar{5}$	$\bar{4}$	$\bar{3}$	$\bar{2}$	$\bar{1}$
$\bar{a}+\bar{3}$	$\bar{6}$	$\bar{7}$	$\bar{8}$	$\bar{9}$	$\bar{0}$	$\bar{1}$	$\bar{2}$
$\bar{2}\cdot\bar{a}$	$\bar{6}$	$\bar{8}$	$\bar{0}$	$\bar{2}$	$\bar{4}$	$\bar{6}$	$\bar{8}$
$\bar{3}\cdot\bar{a}$	$\bar{9}$	$\bar{2}$	$\bar{5}$	$\bar{8}$	$\bar{1}$	$\bar{4}$	$\bar{7}$

(ii) $\bar{7}$.
(iii) No.
(iv) Yes, $\bar{5}$.
(v) No.

13. $\begin{pmatrix} 2 & 4 & 4 \\ 8 & -12 & 6 \end{pmatrix}$.

15. (a) $6\sqrt{2}[\cos(3\pi/4) + \mathbf{i}\sin(3\pi/4)]$;
(b) $9[\cos(\pi/2) + \mathbf{i}\sin(\pi/2)]$;
(c) $5(\cos\pi + \mathbf{i}\sin\pi)$.

17. (a) $4\sqrt{3} + 4\mathbf{i}$; (b) $-1 - \mathbf{i}$.

19. $-2^{27} + 2^{27}\sqrt{3}\mathbf{i}$.

Section 2.7, page 86

1. (i) 1; (ii) 2; (iii) 3; (iv) 4; (v) 5.

3. (1243); it has order 4.

5. (i) 2; (ii) 6; (iii) 4; (iv) 10.

7. (i) 6; (ii) 4; (iii) 20; (iv) 12; (v) 60; (vi) 60.

9. (i) $r = 4$. (ii) $s = 4$. (iii) 4. (iv) 3.

11. (a) 3; (b) 2; (c) 3; (d) 6; (e) 1.

13. (a) 7; (b) 7; (c) 7; (d) 7; (e) 7; (f) 1.

15. (i) $[\mathbf{i}] = \{1, \mathbf{i}, -1, -\mathbf{i}\}$. (ii) 4. (iii) Yes. (iv) $\mathbf{i}$ and $-\mathbf{i}$.

17. (i) $[(1)] = \{(1)\}$, $[(123)] = [(132)]$
$= \{(1), (123)\,(132)\}$, $[(23)] = \{(1), (23)\}$,
$[(13)] = \{(1), (13)\}$, $[(12)] = \{(1), (12)\}$.

θ	(1)	(123)	(132)	(23)	(13)	(12)
order of $[\theta]$	1	3	3	2	2	2

(ii) Yes. (iii) Yes; 6 is not the order of any θ in $\mathbf{S}_3$.
(iv) Yes.

19. (a) $\{\mathbf{e}, a^{15}\}$; (b) $\{\mathbf{e}, a^{10}, a^{20}\}$; (c) $\{\mathbf{e}, a^6, a^{12}, a^{18}, a^{24}\}$;
(d) $\{\mathbf{e}, a^5, a^{10}, a^{15}, a^{20}, a^{25}\}$.

21. (a) a^{15}; (b) a^{10}, a^{20}; (c) a^3, a^9, a^{21}, a^{27};
(d) $a, a^7, a^{11}, a^{13}, a^{17}, a^{19}, a^{23}, a^{29}$.

25. The order of a is 1, 3, 5, or 15.

27. (i) 1; (ii) $m - 1$.

29. Yes. In Section 2.3, see Problem 7(iv) and Example 2.

31.

	0	1	2	3	4
0	0	1	2	3	4
1	1	2	3	4	0
2	2	3	4	0	1
3	3	4	0	1	2
4	4	0	1	2	3

41. (a) 1, 2. (b) 1, 3. (c) 1, 2, 3, 4. (d) 1, 5. (e) 1, 2, 3, 4, 5, 6.
(f) 1, 3, 5, 7. (g) 1, 2, 4, 5, 7, 8.

45. If $(ab)^q = \mathbf{e}$, then $(ba)^q = b(ab)^q b^{-1} = beb^{-1} = \mathbf{e}$. Similarly, if $(ba)^q = \mathbf{e}$ then $(ab)^q = \mathbf{e}$.

47. $[1] = \{n \cdot 1 : n \in \mathbf{Z}\} = \mathbf{Z} = \{n \cdot (-1) : n \in \mathbf{Z}\} = [-1]$. No other generators.

49. (iv) $[w] = \{1, w, w^2\}$. (v) 3. (vi) Yes. (vii) w and w^2.

51. (a) 22. (b) 11. (c) 44.

Section 2.8, page 93

1. (a) $x_2^2 + x_3^2 - x_1^2$.
(b) $x_2^2 + x_1^2 - x_3^2$.
(c) $x_2^2 + x_3^2 + x_4^2$.

3. (1), (14).

5. (a) even; (b) odd; (c) even; (d) even.

7. γ_i is a product of $s_i - 1$ transpositions; hence, θ is a product of

$$(s_1 - 1) + (s_2 - 1) + \cdots + (s_r - 1) = s_1 + s_2 + \cdots + s_r - r$$

transpositions.

9. (i) We have $\alpha = \gamma_1\gamma_2 \cdots \gamma_{2s}$ and $\alpha' = \delta_1\delta_2 \cdots \delta_{2t}$, with the γ_i and δ_i transpositions. Then each of $\alpha\alpha' = \gamma_1 \cdots \gamma_{2s}\delta_1 \cdots \delta_{2t}$ and $\alpha^{-1} = \gamma_{2s} \cdots \gamma_1$ is a product of an even number of transpositions and hence is an even permutation. Similarly, $\beta\beta'$ is even.
(ii) This is similar to part (i).

13. {(1), (123), (132)}.

15. We can let $\alpha = \gamma_1\gamma_2 \cdots \gamma_h$ and $\beta = \delta_1\delta_2 \cdots \delta_k$ with the γ_i and δ_j transpositions. Then

$$\alpha^{-1}\beta^{-1}\alpha\beta = \gamma_h \cdots \gamma_1\delta_k \cdots \delta_1\gamma_1 \cdots \gamma_h\delta_1 \cdots \delta_k$$

is a product of $2(h + k)$ transpositions and so is even.

Section 2.9, page 98

1. (i) $\rho^2 = (13)(24)$, $\rho^3 = (1432)$, $\rho\phi = (14)(23)$, $\rho^2\phi = (13)$, $\rho^3\phi = (12)(34)$, and these are the remaining elements of the octic group.
(ii) $\phi\rho = (12)(34) = \rho^3\phi$.

3. See the table on the inside back cover.

5. {(1), (13)(24)} or $\{\varepsilon, \rho^2\}$.

7. (i) [(1234)] = {(1), (1234), (13)(24), (1432)}.

(ii) $\{(1), (13), (24), (13)(24)\}$ and $\{(1), (12)(34), (13)(24), (14)(23)\}$.
(iii) $\{\varepsilon, \rho^2\}, \{\varepsilon, \phi\}, \{\varepsilon, \rho\phi\}, \{\varepsilon, \rho^2\phi\}, \{\varepsilon, \rho^3\phi\}$.

11. (i) Yes. It is a finite subset closed under multiplication.
(ii) $(1), \rho^2, \rho^3\phi, \rho\phi$.
(iii) Yes. $K = H$.

13. (1) and (23).

15. (a) 1 or 3. (b) $a_3 = 5, a_4 = 1, a_5 = 2$. Yes, $\alpha = (12345)^2$.
(c) $a_3 = 1, a_4 = 5, a_5 = 4$. Yes, $\alpha = \rho^3\phi$.

17. (i) $\phi\rho = (25)(34)(12345) = (12)(35)$, $\rho^{-1}\phi = (15432)(25)(34) = (12)(35)$.

Section 2.10, page 103

1. (12)(34), (13)(24), (14)(23). **3.** No.

5. $\{(1), (123), (132)\}, \{(1), (124), (142)\}, \{(1), (134), (143)\}, \{(1), (234), (243)\}$.

7. $\mathbf{A_4}$.

9. (i)

θ	$\{\alpha_1\theta, \alpha_5\theta, \alpha_9\theta\}$
α_1, α_5, or α_9	$\{\alpha_1, \alpha_5, \alpha_9\}$
α_2, α_6, or α_{10}	$\{\alpha_2, \alpha_6, \alpha_{10}\}$
α_3, α_7, or α_{11}	$\{\alpha_3, \alpha_7, \alpha_{11}\}$
α_4, α_8, or α_{12}	$\{\alpha_4, \alpha_8, \alpha_{12}\}$

(ii) 4.

11. (i)

θ	$\{\alpha_1\theta, \alpha_2\theta, \alpha_3\theta, \alpha_4\theta\}$
$\alpha_1, \alpha_2, \alpha_3$, or α_4	$\{\alpha_1, \alpha_2, \alpha_3, \alpha_4\}$
$\alpha_5, \alpha_6, \alpha_7$, or α_8	$\{\alpha_5, \alpha_6, \alpha_7, \alpha_8\}$
$\alpha_9, \alpha_{10}, \alpha_{11}$ or α_{12}	$\{\alpha_9, \alpha_{10}, \alpha_{11}, \alpha_{12}\}$

(ii) 3. (iii) *Hint:* Use Table 2.2.

13. 15; 20; 24. **15.** $\{1, 2, 3, 5\}$.

17.

θ	α_3	α_4	α_5	α_6	α_7	α_8	α_9	α_{10}	α_{11}	α_{12}
$\theta^{-1}\alpha_2\theta$	α_2	α_2	α_4	α_4	α_4	α_4	α_3	α_3	α_3	α_3

19. $\{\alpha_1, \alpha_2, \alpha_3, \alpha_4\}$.

21. (i) Yes.
(ii) No. *Hint:* Square (1), (ab), (abc), $(abcd)$, $(abcde)$, $(abcdef)$, $(ab)(cd)$, $(ab)(cd)(ef)$, $(ab)(cde)$, $(ab)(cdef)$, and $(abc)(def)$.

23. $H_0 = \mathbf{A_4}$; $H_1 = \{\alpha_1, \alpha_2, \alpha_3, \alpha_4\}$; H_2, H_3, H_4, and H_5 are (in any order) the subgroups $\{\alpha_1, \alpha_5, \alpha_9\}, \{\alpha_1, \alpha_6, \alpha_{11}\}, \{\alpha_1, \alpha_7, \alpha_{12}\}$, and $\{\alpha_1, \alpha_8, \alpha_{10}\}$; H_6, H_7, and H_8 are (in any order) the subgroups $\{\alpha_1, \alpha_2\}, \{\alpha_1, \alpha_3\}$, and $\{\alpha_1, \alpha_4\}$; and $H_9 = \{\alpha_1\}$.

Section 2.11, page 108

1. (i) H and $\{(13)(24), (14)(23)\} = \{\alpha_3, \alpha_4\}$.
(ii) $\alpha_5 H = \{\alpha_5, \alpha_6\}$ is different from $H\alpha_5 = \{\alpha_5, \alpha_8\}$.
(iii) K is abelian; hence any subgroup in K is normal in K.

3. (i) Yes. (ii) Yes. (iii) 3.
(iv) N, $\{\alpha_5, \alpha_6, \alpha_7, \alpha_8\}$, and $\{\alpha_9, \alpha_{10}, \alpha_{11}, \alpha_{12}\}$ are the left cosets and also the right cosets of N in $\mathbf{A}_4$.
(v) Yes. $\theta N = N\theta$ for every θ in $\mathbf{A}_4$.

5. (a) For every h of a subgroup H in the center of G and every g in G, $gh = hg$. Hence $gH = Hg$, and H is normal in G.
(b) Yes. If $H = \{\mathbf{e}\}$, then $gH = \{g\} = Hg$ for every g in G.

7. (a) One left (or right) coset consists of the even permutations, and the other consists of the odd permutations.
(b) $\{\bar{0}, \bar{2}, \bar{4}, \bar{6}, \bar{8}, \overline{10}\}$ and $\{\bar{1}, \bar{3}, \bar{5}, \bar{7}, \bar{9}, \overline{11}\}$ are the left cosets, and they also are the right cosets.

9. (i) The only possible orders for elements of G are 1, 2, and 4. If G is not cyclic, each element g has order 1 or 2 and so satisfies $g^2 = \mathbf{e}$.
(ii) This follows from (i) and Problem 9 of Section 2.3.

15. (i) $\{1, 2, 3, 4, 5\}$. (ii) 60.

17. (ii) Yes.

25. 60, 120, 300, 420, 600, 840, 2100, 4200.

31. Let $\rho = (1234)$, $\phi = (24)$, $H = \{(1), \phi\}$, $K = \{(1), \rho^2, \phi, \rho^2\phi\}$, and $G = \mathbf{D}_4$.

Section 2.12, page 115

1. $\alpha_3 H \cdot \alpha_5 H = \{(123), (243), (142), (134)\}$.

3.

	N	P	Q
N	N	P	Q
P	P	Q	N
Q	Q	N	P

5. (i) $\mathbf{Z}$ is abelian; so all of its subgroups are normal.
(ii) m.
(iii) Every integer is in one of these four distinct cosets.
(iv) See the addition table in Example 2 of Section 2.6.
(v) Yes.

7. (i) This follows from Theorem 3(b).
(ii) This follows from $(abc) = (acb)^2$, $(acb) \in \mathbf{A}_4$, and part (i).
(iii) Eight elements are too many for a group of order 6.

11. The order of G/N is p, a prime, and all groups of prime order are cyclic.

13. $N = [(123 \ldots n)]$.

15. (a) $\overline{2m\pi} = m \cdot \overline{2\pi} = m \cdot \bar{0} = \bar{0}$ for all integers m.
(b) For every real number a, there is an integer n such that $n\pi < a \le (n+1)\pi$. If n is even, let $b = a - n\pi$; if n is odd, let $b = a - (n+1)\pi$.
(c) $24 \cdot \overline{\pi/12} = \overline{2\pi} = \bar{0}$ and $m \cdot \overline{\pi/12} \neq \bar{0}$ for $0 < m < 24$.
(d) No positive integral multiple of 1 is a positive integral multiple of 2π.

17. (i) This follows from Lagrange's Theorem.
(ii) If g has order 8, then g^2 has order 4. If g has order 1 or 2, then $g^2 = \mathbf{e}$.
(iii) Such a group is abelian by Problem 11 of Section 2.3. Let a, b, and c be

elements of G such that $a \neq \mathbf{e}$, b is not in $\{\mathbf{e}, a\}$, and c is not in $\{\mathbf{e}, a, b, ab\}$. Then one can see that

$$G = \{\mathbf{e}, a, b, c, ab, ac, bc, abc\}.$$

19. (i) This is true since $a^k \neq \mathbf{e}$ for k in $\{1, 2, 3\}$.
(ii) This follows from part (i).
(iii) Everything but ba^kb^{-1} cancels out in

$$(bab^{-1})^k = (bab^{-1})(bab^{-1}) \cdots (bab^{-1}).$$

(iv) Using parts (ii) and (iii) and the fact that a has order 4, one finds that bab^{-1} has order 4. Then $bab^{-1} \neq a^2$, since a^2 has order 2; and it follows that $ba \neq a^2b$.
(v) Clearly, ba is in bN, which equals Nb by Problem 18(i). Thus, ba is in $\{b, ab, a^2b, a^3b\}$. But $ba \neq b$ by part (i), and $ba \neq a^2b$ by part (iv).

21. This follows from Problems 17 through 20.

Section 2.13, page 121

1. Let $G_0 = \mathbf{D_4}$, $G_1 = [(1234)]$, $G_2 = \{(1), (13)(24)\}$, and $G_3 = \{(1)\}$. For $i = 0$, 1, and 2, the group G_{i+1} has index 2 in G_i and hence is normal in G_i. Since 2 is a prime, the sequence has the needed properties.

3. Let $G_0 = \mathbf{D_6}$, $G_1 = [(123456)]$, $G_2 = \{(1), (14)(25)(36)\}$, and $G_3 = \{(1)\}$. G_1 has index 2 in G_0 and hence is normal in G_0. Also, G_2 is normal in G_1, since G_1 is abelian; and similarly G_3 is normal in G_2. The indices are 2, 3, and 2, all primes.

5. (ii) The identity (1), the twenty-four 5-cycles, and the fifteen products of two disjoint 2-cycles are all the cubes in $\mathbf{A_5}$. (The twenty 3-cycles are not cubes.)
(iii) Theorem 2 of Section 2.12 tells us that a normal subgroup with index 3 would contain the 40 cubes of part (ii). But such a subgroup would only have $60/3 = 20$ elements.

7. (i) Since $(xyz) = (xzy)^2$, (xyz) is in N by Theorem 2 of Section 2.12.
(ii) Since $(abcde) = (adbec)^2$, $(abcde)$ is in N.
(iii) This follows from closure of N and the hint.

9. Such a subgroup would have order $120/5 = 24$ but would also have to contain the 96 fifth powers in $\mathbf{S_5}$. [Every permutation with a standard form of the type (1), (ab), (abc), $(abcd)$, $(ab)(cd)$, $(ab)(cde)$, or $(abc)(de)$ is a fifth power.]

13. Let $G_0 = G$, $G_1 = \{\mathbf{e}, a^5, a^{10}, a^{15}, a^{20}\}$, and $G_2 = \{\mathbf{e}\}$. Since G is cyclic, it is abelian. This helps us see that G_{i+1} is a normal subgroup with prime index 5 in G_i for $i = 0$, 1.

17. Let $G_0 = G$, and let the remainder of the subgroups for G be the subgroups for N (which exist by Problem 16).

Section 2.14, page 127

1.

	order	1	2	3	4
(a)					
(b)	no. of given order	1	9	8	6

3. $\{\beta_1, \beta_8, \beta_9, \beta_{10}, \beta_{17}, \beta_{18}, \ldots, \beta_{24}\}$.

5. (a) {(1), (1234), (13)(24), (1432)}.

(b)

	(1)	(1234)	(13)(24)	(1432)
(1)	(1)	(1234)	(13)(24)	(1432)
(1234)	(1234)	(13)(24)	(1432)	(1)
(13)(24)	(13)(24)	(1432)	(1)	(1234)
(1432)	(1432)	(1)	(1234)	(13)(24)

7. (a) {(1), (123)(456), (132)(465), (14)(26)(35), (15)(24)(36), (16)(25)(34)}.

(b) and (c) The table is obtained from the table for $\mathbf{S_3}$ by replacing (123) with (123)(456), (132) with (132)(465), (23) with (14)(26)(35), (13) with (16)(25)(34), and (12) with (15)(24)(36).

9. It is essentially the dihedral group $\mathbf{D_n}$, as is illustrated by Problems 7 and 8.

Review Problems (Chapter 2), page 130

1. Proof by contradiction: $ab^{-1} = ac^{-1}$ implies that $b^{-1} = c^{-1}$, which in turn implies that $b = c$.

3. $\alpha = (1352)(46)$ and $\alpha^{-1} = (1253)(46)$ have order 4; $\beta = (163254)$ has order 6; and $\beta^{99} = \beta^3 = (12)(34)(56)$ has order 2.

5. (a) 20; the smallest positive integer n with $(a^{36})^n = \mathbf{e}$ is 20.

(b) $\mathbf{e}, a^{16}, a^{32}, a^{48}, a^{64}$.

(c) $\mathbf{e}, a^{10}, a^{20}, a^{30}, a^{40}, a^{50}, a^{60}, a^{70}$.

(d) $a^8, a^{24}, a^{56}, a^{72}$.

7. $\{\mathbf{e}\}, \{\mathbf{e}, a^3\}, \{\mathbf{e}, a^2, a^4\}, G$. Yes; each is finite and closed under the operation of G.

9. (a) 18. (b) 9. (c) 20.

11. Yes. Since (ord H)$\,|\,$12 by Lagrange's Theorem and ord $H \geq 7$, one must have ord $H = 12$ and $H = \mathbf{A_4}$.

19. 420.

Section 3.1, page 139

1. $\theta(n) = n + 1$.

3. (i) One such θ has $\theta(n) = n$ for all n in X.

(ii) No. The element 4 of Y is not the image under θ of any x in X.

5. (i) No. $\theta(-1) = \theta(1)$ but $-1 \neq 1$, for example.

(ii) Yes. Every x in N is an image (of x or $-x$).

(iii) $-5, 5$.

7. (i) No. (ii) No. **9.** $3^2 = 9$.

13. (i) Yes. See Lemma 1 of Section 2.12.

(ii)

α	ε	ρ	ρ^2	ρ^3	ϕ	$\rho\phi$	$\rho^2\phi$	$\rho^3\phi$
$f(\alpha)$	N	B	N	B	C	D	C	D

(iii) Yes. N is the identity of G/N.
(iv) Yes. $f(\alpha)f(\alpha^{-1}) = f(\alpha\alpha^{-1}) = f(\varepsilon) = N$, the identity of G/N.
(v) ε, ρ^2.

15. (i)

n	4	5	6	7
$\bar{r}_n$	$\bar{1}$	$\bar{2}$	$\bar{0}$	$\bar{1}$

(ii) Yes.

17. The mapping γ such that $\gamma(x) = z$ whenever $\alpha(x) = y$ and $\beta(y) = z$.

Section 3.2, page 146

1. (i)

x	$\mathbf{e}$	a	a^2	a^3	a^4
$\theta_1(x)$	$\mathbf{e}'$	b	b^2	b^3	b^4
$\theta_2(x)$	$\mathbf{e}'$	b^2	b^4	b	b^3
$\theta_3(x)$	$\mathbf{e}'$	b^3	b	b^4	b^2
$\theta_4(x)$	$\mathbf{e}'$	b^4	b^3	b^2	b

(ii) An isomorphism θ from $[a]$ to $[b]$ is completely determined once $\theta(a)$ is chosen as one of the four generators of $[b]$.

3. $\theta_2 : \mathbf{e} \mapsto \mathbf{e}', b \mapsto u, c \mapsto w, d \mapsto v.$
$\theta_3 : \mathbf{e} \mapsto \mathbf{e}', b \mapsto v, c \mapsto u, d \mapsto w.$
$\theta_4 : \mathbf{e} \mapsto \mathbf{e}', b \mapsto v, c \mapsto w, d \mapsto u.$
$\theta_5 : \mathbf{e} \mapsto \mathbf{e}', b \mapsto w, c \mapsto u, d \mapsto v.$
$\theta_6 : \mathbf{e} \mapsto \mathbf{e}', b \mapsto w, c \mapsto v, d \mapsto u.$

7. Yes. *Hint:* Use $a'b' = (ab)' = (ba)' = b'a'$.

9.

α	(1)	(123)	(132)	(23)	(13)	(12)
$f(\alpha)$	(1)	(123)	(132)	(23)	(13)	(12)
$g(\alpha)$	(1)	(123)	(132)	(13)	(12)	(23)
$h(\alpha)$	(1)	(123)	(132)	(12)	(23)	(13)

11. Once the images of (123) and (23) are chosen, as in Problems 9 and 10, the images of the other permutations in $\mathbf{S}_3$ are uniquely determined.

13. $\beta = (123)$. **19.** θ_1 and θ_2, with $\theta_1(x) = x$ and $\theta_2(x) = -x$ for all x in $\mathbf{Z}$.

25. 4.

31. (b) L is the conjugate of H by a^{-1}.

Section 3.3, page 154

1. (i)

g	$\mathbf{e}$	a	a^2	a^3	a^4	a^5	a^6	a^7	a^8
$\theta(g)$	(1)	(123)	(132)	(1)	(123)	(132)	(1)	(123)	(132)

(ii) $\{\mathbf{e}, a^3, a^6\}$.
(iii) $\theta^{-1}[(123)] = \{a, a^4, a^7\}$, $\theta^{-1}[(132)] = \{a^2, a^5, a^8\}$.

5. (i) Let $\bar{a} = \bar{b}$. Then b is in $\bar{a} = a + [m]$. Hence $b = a + sm$, with s an integer and

$a \equiv b \pmod{m}$. Conversely, let $a \equiv b \pmod{m}$. Then $b - a = sm$, with s an integer, and $b = a + sm$ is in $\bar{a}$. Now the cosets $\bar{a}$ and $\bar{b}$ have b as a common element, and hence $\bar{a} = \bar{b}$.

(ii) Let $n = qm + r$, with q and r integers and $0 \leq r < m$. Then $n \equiv r \pmod{m}$, and $\bar{n} = \bar{r}$ by part (i).

(iii) *Hint:* See the proof of Theorem 2(ii) in Section 1.3.

(iv) The sum $\bar{a} + \bar{b}$ is the coset $\overline{a + b}$ containing the sum $a + b$ of the element a from $\bar{a}$ and the element b from $\bar{b}$. By part (ii), we have $\overline{a + b} = \bar{r}$, where r is the remainder in the division of $a + b$ by m.

(v) $\bar{2} = \overline{1 + 1} = \bar{1} + \bar{1} = 2 \cdot \bar{1}$ is in $[\bar{1}]$. An inductive argument then shows that each $\bar{n}$ in $\mathbf{Z}/[m]$ is in $[\bar{1}]$.

7. (i) $\theta(ab) = |ab| = |a| \cdot |b| = \theta(a)\theta(b)$ for all a and b in $\mathbf{R}^{\pm}$.
(ii) $\{-1, 1\}$. (iii) $\{-6, 6\}$.

9. (i) $\theta(a)\theta(b) = \mathbf{e}'\mathbf{e}' = \mathbf{e}' = \theta(ab)$ for all a and b in G. (ii) G.

15. (ii) One example has $G = \mathbf{S_3}$ and $s = 2$.

17. (i) $\theta(m + n) = a^{m+n} = a^m a^n = \theta(m)\theta(n)$.
(ii) The set $q\mathbf{Z}$ of integral multiples of q.
(iii) $\{0\}$.

19. (iii) a and b' must have the same order.

23. (c) K consists of all α with $f(\alpha) = 1$.

25. Let a have order 2, $G = [a] = \{\mathbf{e}, a\}$, and $G' = \{\mathbf{e}\}$.

Section 3.4, page 160

1.

θ	(1)	(123)	(132)	(23)	(13)	(12)
$f(\theta)$	(1)	(123)(465)	(132)(456)	(14)(25)(36)	(15)(26)(34)	(16)(24)(35)

3. 1, 2, 3, 4, 5, 6.

5. (i) q_3. (ii) $q_2, q_4, q_5, q_6, q_7, q_8$. (iii) $\{q_1, q_3\}$.
(iv) $\{q_1, q_2, q_3, q_4\}$, $\{q_1, q_5, q_3, q_7\}$, $\{q_1, q_6, q_3, q_8\}$.

9. $\{q_1, q_3\}$.

Section 3.5, page 166

1. $G = \{\mathbf{e}, b\} \cup \{\mathbf{e}, c\} \cup \{\mathbf{e}, d\}$.

3. (a) $\{(1), \rho, \rho^2, \rho^3, \rho^4, \rho^5, \phi, \rho\phi, \rho^2\phi, \rho^3\phi, \rho^4\phi, \rho^5\phi\}$;
(b) $\{(1), \rho^2, \rho^4, \phi, \rho^2\phi, \rho^4\phi\}$; (c) $\{(1), \rho^3, \phi, \rho^3\phi\}$;
(d) $\{(1), \rho^2, \rho^4, \rho\phi, \rho^3\phi, \rho^5\phi\}$; (e) $\{(1), \rho^3, \rho\phi, \rho^4\phi\}$;
(f) $\{(1), \rho^3, \rho^2\phi, \rho^5\phi\}$; (g) same as (a); (h) same as (b).

11. $\{\varepsilon, \rho^2\}$.

31. If a and b are in G, then a is in G_i and b is in G_j for some i and j in $\mathbf{Z}^+$. If $j \leq i$, $G_j \subseteq G_i$ and we define the product of a and b in G to be their product in G_i. If $j > i$, the product in G is as in G_j.

33. $B = \{(1)\}$, C and D are $\{\alpha_5, \alpha_6, \alpha_7, \alpha_8\}$ and $\{\alpha_9, \alpha_{10}, \alpha_{11}, \alpha_{12}\}$ and $E = \{\alpha_2, \alpha_3, \alpha_4\}$.

37. H, $\{(123), (243), (142), (134)\}$, $\{(132), (143), (234), (124)\}$.

43. 18.

Section 3.6, page 170

1. (i) (1, 1), (1, 3), (1, 4), (2, 1), (2, 3), (2, 4).
(ii) (1, 1), (3, 1), (4, 1), (1, 2), (3, 2), (4, 2).
(iii) No.

3. mn.

5. (i) The table is that of a cyclic group $[g]$ of order 6 with $(\mathbf{e}, \mathbf{e}')$ the identity, $(a, b) = g$, $(a^2, \mathbf{e}') = g^2$, $(\mathbf{e}, b) = g^3$, $(a, \mathbf{e}') = g^4$, and $(a^2, b) = g^5$.
(ii) Yes. See answer to part (i).
(iii) No. G is cyclic while $\mathbf{S}_3$ is not. Hence it follows from Lemma 1 of Section 3.2 that G and $\mathbf{S}_3$ are not isomorphic.
(iv)

x	$(\mathbf{e}, \mathbf{e}')$	$(\mathbf{e}, b)$	$(a, \mathbf{e}')$	(a, b)	$(a^2, \mathbf{e}')$	(a^2, b)
$h(x)$	$\mathbf{e}$	$\mathbf{e}$	a	a	a^2	a^2

9. Yes. Projection preserves the operations.

Section 3.7, page 176

7. (i) Yes. (ii) C(a) is the singleton subset $\{a\}$.

9. (a) Yes. (b) No, the relation $\leq$ is not symmetric if X has at least two numbers. No, $\leq$ is not a linear order relation, since one can have $a = b$ and $a \leq b$; these contradict trichotomy.

11. 16.

13. (a) The relation | on $\mathbf{Z}$ is reflexive and transitive but is neither symmetric nor a linear ordering.
(b) P is symmetric but is not reflexive, transitive, or a linear ordering.

15. We replace x by t in sAx and by s in tAx and get sAt and tAs. These and antisymmetry of A give us $s = t$.

17. Yes.

23. (ii) Let $S = \{1, 2\}$ and L be the subset $\{(2, 2)\}$ of $S \times S$.

Section 3.8, page 179

1. (2, 2), (2, 4), (2, 6), (2, 12), (3, 3), (3, 6), (3, 12), (4, 4), (4, 12), (6, 6), (6, 12), (12, 12).

3. There is no least element; 12 is the greatest element.

5. m is the greatest element for $(S, \geq)$.

7. The relation of a poset is antisymmetric.

11. (i)

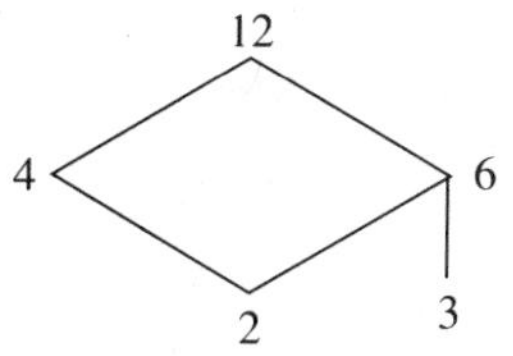

(ii) One does not have segments for the ordered pairs (a, a) implied by reflexiveness nor for the ordered pairs (2, 12) and (3, 12) implied by transitivity.

13. M is reflexive, antisymmetric, and transitive and hence is a partial ordering. M is neither symmetric nor a linear ordering.

15. No. If a and b are distinct elements of S, trichotomy fails for the members $\{a\}$ and $\{b\}$ of P.

Section 3.9, page 182

1. (a) 10010; $\bar{A} = \{s_2, s_3, s_5\}$ has 01101 as its binary numeral.
(b) $B = \{s_2, s_3, s_4\}$; $\bar{B} = \{s_1, s_5\}$.
(c) 00000. (d) $c_i = 1$ unless $a_i = 0 = b_i$. (e) Yes.

3. Yes. **5.** Yes. **7.** Yes, S. **9.** 3^n. **11.** Yes.

13. (a) This is just a different notation for $(P(S), \subseteq)$, where S has n elements.
(b) $00\ldots0$. (c) $11\ldots1$.

15. $(T, \subseteq)$ is reflexive, antisymmetric, and transitive, since $(P(S), \subseteq)$ has these properties and T is a subcollection of $P(S)$.

Section 3.10, page 186

1. (a)

A	$\varnothing$	$\{1\}$	$\{2\}$	$\{1, 2\}$
$\bar{A}$	$\{1, 2\}$	$\{2\}$	$\{1\}$	$\varnothing$

(b)

$\cup$	$\varnothing$	$\{1\}$	$\{2\}$	$\{1, 2\}$
$\varnothing$	$\varnothing$	$\{1\}$	$\{2\}$	$\{1, 2\}$
$\{1\}$	$\{1\}$	$\{1\}$	$\{1, 2\}$	$\{1, 2\}$
$\{2\}$	$\{2\}$	$\{1, 2\}$	$\{2\}$	$\{1, 2\}$
$\{1, 2\}$	$\{1, 2\}$	$\{1, 2\}$	$\{1, 2\}$	$\{1, 2\}$

(c) $\varnothing$.

3. (a)

$\cap$	000	001	010	011	100	101	110	111
000	000	000	000	000	000	000	000	000
001	000	001	000	001	000	001	000	001
010	000	000	010	010	000	000	010	010
011	000	001	010	011	000	001	010	011
100	000	000	000	000	100	100	100	100
101	000	001	000	001	100	101	100	101
110	000	000	010	010	100	100	110	110
111	000	001	010	011	100	101	110	111

(b) 111.

5. (i)

lcm	1	2	3	6
1	1	2	3	6
2	2	2	6	6
3	3	6	3	6
6	6	6	6	6

(ii) 1.
(iii) No; for example, 6 does not have an inverse, since lcm $[6, x] \neq 1$ for all x in X.

7. No.

9. (i)

min	2	3	4
2	2	2	2
3	2	3	3
4	2	3	4

(ii) 4.
(iii) No; $\min(2, s) \neq 4$ for all s in S, and hence 2 has no inverse.
(iv) No. (v) No.

13. Let $t = r$ if $r \leq s$ and let $t = s$ if $r > s$. Also, let $g_i = \min(e_i, f_i)$. Then $\gcd(a, b) = (p_1)^{g_1}(p_2)^{g_2} \cdots (p_t)^{g_t}$.

15. (a) 3^3. (b) 3^9. (c) 3^6. (d) 3^{3^n}.

17. $c_i = \max(a_i, b_i)$ for $1 \leq i \leq n$.

19. (ii) The right identities are 1 and -1.

Section 3.11, page 190

1. (a) $(\mathbf{Z}^+, +)$ is a semigroup. It is neither a monoid nor a group.
(b) $(\mathbf{Z}^+, \cdot)$ is a semigroup and is a monoid but is not a group.

3. No. No. No.

5. (a) $(\mathbf{Z}^+, +)$. (b) No; a monoid is a semigroup with an identity.

7. $\langle\bar{2}\rangle = \{\bar{2}, \bar{4}, \bar{8}, \overline{16}, \overline{32}, \overline{64}, \overline{28}, \overline{56}, \overline{12}, \overline{24}, \overline{48}, \overline{96}, \overline{92}, \overline{84}, \overline{68}, \overline{36}, \overline{72}, \overline{44}, \overline{88}, \overline{76}, \overline{52}\}$.

9. Yes; gcd is an associative binary operation on $\mathbf{Z}^+$. There is no identity.

11. (i) Yes. (ii) Yes. (iii) Yes. The identity is 1.

13. (i) Yes. (ii) Yes. (iii) Yes. There is no identity.

15. (i) Yes. (ii) No; the semigroup $(\mathbf{Z}^+, \min)$ has no identity.
(iii) No. The same reason as for part (ii).

17. $(\mathbf{Z}, \max)$ is a semigroup but has no identity, since there is no e in $\mathbf{Z}$ with $\max(n, e) = n$ for all n in $\mathbf{Z}$.

19. The largest integer in X.

Section 3.12, page 196

1. (i) O is the identity under $\vee$.
(ii) I is the identity under $\wedge$.

(iii) These follow from Theorem 3(a).
(iv) $O' = I$ by Theorem 1, and then $I' = O$ by duality.

3.

x	1	3	5	15
x'	15	5	3	1

lcm	1	3	5	15
1	1	3	5	15
3	3	3	15	15
5	5	15	5	15
15	15	15	15	15

gcd	1	3	5	15
1	1	1	1	1
3	1	3	1	3
5	1	1	5	5
15	1	3	5	15

5. $f(5) = \{b\}, f(1) = \varnothing, f(15) = S = \{a, b\}$.

7.

y	1	2	3	5	6	10	15	30
y'	30	15	10	6	5	3	2	1

lcm	1	2	3	5	6	10	15	30
1	1	2	3	5	6	10	15	30
2	2	2	6	10	6	10	30	30
3	3	6	3	15	6	30	15	30
5	5	10	15	5	30	10	15	30
6	6	6	6	30	6	30	30	30
10	10	10	30	10	30	10	30	30
15	15	30	15	15	30	30	15	30
30	30	30	30	30	30	30	30	30

23. This follows from Problems 15, 21, and 22.

27.

x	000	011	101	110	111
$f(x)$	1	6	10	15	30

29. Let $d_1 d_2 \ldots d_n$ be the binary numeral for a subset T of S, that is, a member of $P(S)$. Then $f(T) = (c_1, c_2, \ldots, c_n)$, where $c_i = O$ when $d_i = 0$ and $c_i = I$ when $d_i = 1$.

33. The least element is the identity O under $\vee$, and the greatest element is the identity I under $\wedge$.

35. $A \leq B$ if and only if A is a subset of B.

Section 3.13, page 204

1. Yes; $x \vee (y \wedge y') = x$.

3. No; $\alpha(0, 1, 1) = 0$ and $\beta(0, 1, 1) = 1$, for example.

5.

(x, y)	$x \Rightarrow y$	y'	x'	$y' \Rightarrow x'$
(0, 0)	1	1	1	1
(0, 1)	1	0	1	1
(1, 0)	0	1	0	0
(1, 1)	1	0	0	1

7. (i) $\beta(0, 0) = 1 \vee 0 = 1$. (ii) $\alpha(0, 0) = 1$. (iii) True.

9. Yes; $\alpha(x, y) = \beta(x, y)$ for all x and y in $\{0, 1\}$.

11. $x \Leftrightarrow y'$ (or $x' \Leftrightarrow y$, among other possibilities).

13. $\alpha(0, 1, 1) = 0$; the check is not valid.

15. No; neither (i) nor (ii) of Example 4 is satisfied.

17. Let R, S, and T denote $p \mid (ab)$, $p \mid a$, and $p \mid b$, respectively.

19. (a) Contingency; true for $(x, y) = (0, 0)$ and false for $(x, y) = (1, 0)$.
(b) Tautology; true for all x and y.
(c) Contradiction; false for all x and y.
(d) Tautology; always true.
(e) Contingency; true for $(x, y, z, w) = (1, 1, 1, 1)$, false for $(x, y, z, w) = (1, 0, 0, 0)$.

Section 3.14, page 208

1.

x	e	a	a^2	a^3	a^4	a^5	a^6	a^7	a^8	a^9	a^{10}	a^{11}
$\gamma(x)$	e	c	e	c	e	c	e	c	e	c	e	c

α, β, and γ are all homomorphisms.

13. We define β as follows. If y is in the image set of the injection α, we let $\beta(y)$ be the unique x in X with $\alpha(x) = y$. For the other elements y in Y, we let $\beta(y) = u$. Clearly $\alpha\beta$ is the identity mapping $X \to X$, $x \mapsto x$. Then β is a surjection, by Problem 12.

Review Problems (Chapter 3), page 212

1. (i)

x	e	a	a^2	a^3
$\theta_1(x)$	1	i	-1	$-$i
$\theta_2(x)$	1	$-$i	-1	i

(ii) The orders of elements must be preserved under an isomorphism.

3. (a) Using the fact that $f(\theta) = \rho^2$ for θ in $\{\rho, \rho^3\}$ and $f(\theta) = \varepsilon$ otherwise, one sees easily that f is neither an injection nor a surjection. It is not a homomorphism, since (for example) $f(\rho\phi) \neq f(\rho)f(\phi)$.
(b) g is both injective and surjective (that is, bijective) but is not a homomorphism, because $g(\rho\phi) \neq g(\rho)g(\phi)$.

9.

(x, y)	(e, e)	(e, b)	(e, c)	(b, e)	(b, b)	(b, c)	(c, e)	(c, b)	(c, c)
$(b, c)(x, y)$	(b, c)	(b, e)	(b, b)	(c, c)	(c, e)	(c, b)	(e, c)	(e, e)	(e, b)

11. 12 antisymmetric relations, of which 3 are reflexive.

13. D is reflexive, antisymmetric, and transitive and hence is a partial ordering. It is not symmetric.

15. (a) $2^4 = 16$. (b) 4^{16}. (c) 16^4, of which $16 \cdot 15 \cdot 14 \cdot 13$ are injective.

17. (a) 4^4. (b) 4^{16}. (c) 4^{10}.

19. (a) It is a monoid with $\bar{1}$ as identity. It is not a group, since $\bar{0}$ has no inverse.
(b) 102. (c) No; no; it has no identity.

21. $(P(S), \cup, \cap, ^-, \varnothing, S)$ with S having 5 elements.

25. (a) Tautology; $\alpha(x, y) = 1$ for all x and y.
(b) Tautology; $\beta(x, y, z) = 1$ for all x, y, and z.

Section 4.1, page 219

1. (i)

+	0	b
0	0	b
b	b	0

(ii)

·	0	b
0	0	0
b	0	b

(iii)

·	0	b
0	0	0
b	0	0

(iv) In part (ii), b is the unity. There is no unity in part (iii).

3. Yes. It meets the conditions in the definition of a ring.

11. (i)

+	$\bar{0}$	$\bar{2}$	$\bar{4}$	$\bar{6}$
$\bar{0}$	$\bar{0}$	$\bar{2}$	$\bar{4}$	$\bar{6}$
$\bar{2}$	$\bar{2}$	$\bar{4}$	$\bar{6}$	$\bar{0}$
$\bar{4}$	$\bar{4}$	$\bar{6}$	$\bar{0}$	$\bar{2}$
$\bar{6}$	$\bar{6}$	$\bar{0}$	$\bar{2}$	$\bar{4}$

·	$\bar{0}$	$\bar{2}$	$\bar{4}$	$\bar{6}$
$\bar{0}$	$\bar{0}$	$\bar{0}$	$\bar{0}$	$\bar{0}$
$\bar{2}$	$\bar{0}$	$\bar{4}$	$\bar{0}$	$\bar{4}$
$\bar{4}$	$\bar{0}$	$\bar{0}$	$\bar{0}$	$\bar{0}$
$\bar{6}$	$\bar{0}$	$\bar{4}$	$\bar{0}$	$\bar{4}$

(ii) S is a subring but does not have a unity.
(iii) Yes. Each of $\bar{2}$ and $\bar{6}$ is a generator.
(iv) Yes. (v) Yes.

13. Only the subsets (c) and (d).

19. (ii) Yes.

23. Since $\{1, c, d\}$ is a multiplicative group, Example 2 of Section 2.3 helps us see that the table is as follows:

	0	1	c	d
0	0	0	0	0
1	0	1	c	d
c	0	c	d	1
d	0	d	1	c

25. *Hint:* Use distributivity and the fact that the additive group of R is cyclic.

Section 4.2, page 232

3. θ does not preserve multiplications, since (for example) $\theta(1 \cdot 1) = 2$ and $\theta(1)\theta(1) = 2 \cdot 2 = 4 \neq 2$.

5. (a) Since $\bar{0} = \bar{4}$ in $\mathbf{Z}_4$, one would have $\bar{0} = \theta(\bar{0}) = \theta(\bar{4}) = \bar{4} =$ in $\mathbf{Z}_3$ if such a mapping θ existed. As $\bar{0} \neq \bar{4}$ in $\mathbf{Z}_3$, there is no such mapping.
(b) If $\bar{a} = b$ in $\mathbf{Z}_{15}$, then $15 | (a - b)$ and hence $3 | (a - b)$, which implies that $\bar{a} = b$ in $\mathbf{Z}_3$. This means that a mapping θ from $\mathbf{Z}_{15}$ to $\mathbf{Z}_3$ is well defined by $\theta(\bar{n}) = \bar{n}$ for all n in $\mathbf{Z}$.
(c) The kernel is $(\bar{3}) = \{\bar{0}, \bar{3}, \bar{6}, \bar{9}, \overline{12}\}$.

9. $m = 0$.

11. (i) 0. (ii) Yes, $\bar{1}$. (iii) Yes, additively $\mathbf{Z_m} = [\bar{1}]$. (iv) No.

13. (i) The integral multiples of 14.
(ii) The integers of the form $14n + 1$, with n in $\mathbf{Z}$.

15. (a)

x	$\bar{0}$	$\bar{1}$	$\bar{2}$	$\bar{3}$	$\bar{4}$	$\bar{5}$
$-x$	$\bar{0}$	$\bar{5}$	$\bar{4}$	$\bar{3}$	$\bar{2}$	$\bar{1}$

(b)

x	$\bar{0}$	$\bar{1}$	$\bar{2}$	$\bar{3}$	$\bar{4}$	$\bar{5}$
order of x	1	6	3	2	3	6

(c) $(\bar{0}) = \{\bar{0}\}, (\bar{1}) = \mathbf{Z_6} = (\bar{5}), (\bar{2}) = \{\bar{0}, \bar{2}, \bar{4}\} = (\bar{4}), (\bar{3}) = \{\bar{0}, \bar{3}\}$.
(d) In $\mathbf{Z_6}$, $\bar{1}$ is not in $(\bar{2})$, and hence $\bar{2}$ is not an invertible.
(e) In $\mathbf{Z_6}$, the only examples of $\bar{a} \cdot \bar{b} = \bar{1}$ are $\bar{1} \cdot \bar{1} = \bar{1} = \bar{5} \cdot \bar{5}$. Hence the only invertibles in $\mathbf{Z_6}$ are $\bar{1}$ and $\bar{5}$ (and each is its own reciprocal).
(f)

$\bar{a}$	$\bar{2}$	$\bar{3}$	$\bar{3}$	$\bar{4}$
$\bar{b}$	$\bar{3}$	$\bar{2}$	$\bar{4}$	$\bar{3}$

(g) $\{\bar{0}, \bar{1}, \bar{3}, \bar{4}\}$.
(h)

x	$\bar{0}$	$\bar{1}$	$\bar{2}$	$\bar{3}$	$\bar{4}$	$\bar{5}$
$y = z$	$\bar{0}$	$\bar{0}$	$\bar{2}$	$\bar{0}$	$\bar{0}$	$\bar{2}$

(i) $\bar{0}, \bar{1}, \bar{3}, \bar{4}$.

17. (ii) 7, 13, 19, 31, 37, 43, 61, 67, 73, 79, 97.

19. (i)

	$\bar{0}$	$\bar{1}$	$\bar{2}$	$\bar{3}$	$\bar{4}$
$\bar{0}$	$\bar{0}$	$\bar{0}$	$\bar{0}$	$\bar{0}$	$\bar{0}$
$\bar{1}$	$\bar{0}$	$\bar{1}$	$\bar{2}$	$\bar{3}$	$\bar{4}$
$\bar{2}$	$\bar{0}$	$\bar{2}$	$\bar{4}$	$\bar{1}$	$\bar{3}$
$\bar{3}$	$\bar{0}$	$\bar{3}$	$\bar{1}$	$\bar{4}$	$\bar{2}$
$\bar{4}$	$\bar{0}$	$\bar{4}$	$\bar{3}$	$\bar{2}$	$\bar{1}$

(ii)

x	$\bar{1}$	$\bar{2}$	$\bar{3}$	$\bar{4}$
x^{-1}	$\bar{1}$	$\bar{3}$	$\bar{2}$	$\bar{4}$

(iii) $\mathbf{V_5}$ has order 4.
(iv) This follows from (iii) and Lagrange's Theorem.
(v) and (vi) These follow from part (iv).
(vii) If $\bar{b} = \bar{0}$, this is clear; and if $\bar{b} \neq \bar{0}$, then it follows from part (v).
(viii) This is a restatement of part (vii).
(ix) $(\bar{0}) = \{\bar{0}\}, (\bar{1}) = (\bar{2}) = (\bar{3}) = (\bar{4}) = \mathbf{Z_5}$.

21. (a) $(\bar{2}) = \{\bar{0}, \bar{2}, \bar{4}, \bar{6}\}$. (b) Same as part (a).
(c) $(\bar{3}) = \{\bar{0}, \bar{1}, \bar{2}, \bar{3}, \bar{4}, \bar{5}, \bar{6}, \bar{7}\}$. (d) Same as part (c).

23. (ii) 6.

25. (a) $\bar{0}, \bar{1}, \bar{5}$, and $\bar{6}$ satisfy $x^2 = x$; $\bar{5}$ is the unity of $\{\bar{0}, \bar{5}\}$.
(c) $(\bar{0}), (\bar{1}), (\bar{6}), (\overline{10}), (\overline{15}), (\overline{16}), (\overline{25})$.

27. (i)

$\bar{a}$	$\bar{0}$	$\bar{1}$	$\bar{2}$	$\bar{3}$	$\bar{4}$	$\bar{5}$	$\bar{6}$	$\bar{7}$	$\bar{8}$	$\bar{9}$	$\overline{10}$	$\overline{11}$
$\overline{4a}$	$\bar{0}$	$\bar{4}$	$\bar{2}$	$\bar{0}$	$\bar{4}$	$\bar{2}$	$\bar{0}$	$\bar{4}$	$\bar{2}$	$\bar{0}$	$\bar{4}$	$\bar{2}$

29. This follows from Problem 38 of Section 2.7.

31. $\{\bar{0}\}, \{\bar{0}, \overline{12}\}, \{\bar{0}, \bar{8}, \overline{16}\}, \{\bar{0}, \bar{6}, \overline{12}, \overline{18}\}, \{\bar{0}, \bar{4}, \bar{8}, \overline{12}, \overline{16}, \overline{20}\}, \{\bar{0}, \bar{3}, \bar{6}, \bar{9}, \overline{12}, \overline{15}, \overline{18}, \overline{21}\}$. $\{\bar{0}, \bar{2}, \bar{4}, \bar{6}, \bar{8}, \overline{10}, \overline{12}, \overline{14}, \overline{16}, \overline{18}, \overline{20}, \overline{22}\}$, and $\mathbf{Z}_{24} = \{\bar{0}, \bar{1}, \bar{2}, \bar{3}, \ldots, \overline{23}\}$.

33. (ii) The mapping θ with $\theta(x) = \bar{0}$ for all x in $\mathbf{Z}_{30}$.

45. Using Theorem 4 of Section 4.1, one shows that S is a subring in G. But S is not an ideal in G, since $(1 + 2\mathbf{i})(1 + \mathbf{i}) = -1 + 3\mathbf{i}$, for example, shows that S is not closed under left and right multiplication.

Section 4.3, page 241

1. (i)

x	$\bar{1}$	$\bar{5}$	$\bar{7}$	$\overline{11}$	$\overline{13}$	$\overline{17}$
order of x	1	6	3	6	3	2

(ii) Yes. $\bar{5}$ and $\overline{11}$ are generators.

3. (iii) $\bar{1} \cdot \bar{2} \cdot \bar{3} \cdot \bar{4} \cdot \bar{5} \cdot \bar{6} = \overline{-1}$ in $\mathbf{V}_7$. (iv) $7 \mid (6! + 1)$.

11.

m	1	2	3	4	5	6	7	8	9	10
$\phi(m)$	1	1	2	2	4	2	6	4	6	4

13. $\phi(3^n) = 2 \cdot 3^{n-1}$. **15.** $x = 19$.

17. (ii) For all integers a and h, the integers a and 1 are relatively prime and $ah \equiv 1 \pmod{1}$.

19. $\bar{1}, \bar{5}, \bar{7}, \overline{11}$.

21. (i) One example is $a = 3$, $b = 2$, $c = 4$, and $m = 6$.

25. (i) 4. (ii) 3. (iii) 1. (iv) 2. (v) 4. (vi) 2.

27. $x = 9$. Since $94 \equiv 3 \pmod{13}$, y is also 9.

35.

m	11	12	13	14	15	16	17	18	19	20
$\phi(m)$	10	4	12	6	8	8	16	6	18	8

41. 0, 1, 4.

49. *Hint:* See Examples 4 and 5 of this section.

55. (i) .00990099 ..., in which the block 0099 repeats endlessly;
(ii) .01980198 ...; (iii) .02970297 ...;
(iv) .$cdefcdef$ ..., where $10a + b - 1 = 10c + d$ with c and d digits, $e = 9 - c$, and $f = 9 - d$. (If $a = 0 = b$, the answer is .0000 ....)

57. $1/7 = .142857142857\ldots$, with the block 142857 repeating endlessly.

Section 4.4, page 250

1. (v) Neither c nor d is 0, and neither is a 0-divisor.
(vi) $cd = d = 1 \cdot d$ would give the contradiction $c = 1$, since d is not a 0-divisor. Similarly $cd = c$ leads to the contradiction $d = 1$.

(vii) Since cd is in D but is not in $\{0, c, d\}$, we have $cd = 1$.
(viii) $\{1, c, d\}$ is the multiplicative group of invertibles in D. For the multiplication table, see the answer to Problem 23 of Section 4.1. The addition table is

+	0	1	c	d
0	0	1	c	d
1	1	0	d	c
c	c	d	0	1
d	d	c	1	0

7. (iii) p.

11. (i) *Hint:* Use the Corollary to Theorem 5 of Section 4.1 with $a = b = 1$.

19. A subring S in D is a subdomain if and only if the unity of D is in S.

21. *Hint:* Every abelian group of order 6 is cyclic and so has elements of orders 1, 2, 3, and 6.

27. (a) 100. (b) 1.

33. No; $(0, 1)(1, 0) = (0, 0)$ shows that $D \times D'$ has 0-divisors.

Section 4.5, page 255

1. In $\mathbf{Z}_5$, $\bar{3}^{-1} = \bar{2}$ and $x = \bar{3}^{-1}\bar{2} = \bar{2} \cdot \bar{2} = \bar{4}$.

3.

x	$\bar{0}$	$\bar{1}$	$\bar{2}$	$\bar{3}$	$\bar{4}$
x^2	$\bar{0}$	$\bar{1}$	$\bar{4}$	$\bar{4}$	$\bar{1}$

(a) $\bar{0}$; (b) $\bar{1}, \bar{4}$; (c) no solutions; (d) no solutions; (e) $\bar{2}, \bar{3}$.

5. The solution x must be $a^{-1}(-b)$.

11. (iii)

+	0	1	c
0	0	1	c
1	1	c	0
c	c	0	1

·	0	1	c
0	0	0	0
1	0	1	c
c	0	c	1

(v) $(0) = \{0\}$, $(1) = F = \{0, 1, c\} = (c)$.

17. $x = \overline{14}$, $y = \bar{2}$.

21. *Hint:* See Problem 24 of Section 4.2.

23. No, $\theta(x)$ may be 0 for all x in F. [Also, see Problem 28(c) of Section 4.2.]

Section 4.6, page 261

25. 5/2 and 17/2.

31. (i) $\{x, y, z\} = \{2, 3, 5\}$. (ii) $\{x, y, z\} = \{2, 3, 5\}$.

35.

x	-2	-1	2
y	± 3	± 2	± 7

Section 4.7, page 269

1. The ring $M(\mathbf{Z}_2)$ (of 2×2 matrices with entries in $\mathbf{Z}_2$).

3. (i) $\begin{pmatrix} 0 & 0 \\ 0 & 0 \end{pmatrix}$. (ii) Yes; see part (i).

5. Each entry is 0.

7. Let $a_{ij} = 1$ for each i and j.

9. (i) One choice is $A = \begin{pmatrix} 1 & 0 \\ 1 & 0 \end{pmatrix}$, $B = \begin{pmatrix} 0 & 0 \\ 0 & 0 \end{pmatrix}$, $C = \begin{pmatrix} 0 & 0 \\ 1 & 1 \end{pmatrix}$.

(ii) No; A can be the zero matrix.

11. $(1 \quad 1 \quad 0 \quad 0)$.

13. *Hint:* Use displays (7) and (8) to convert to form (6), and then use matrix multiplication.

Review Problems (Chapter 4), page 277

3. $\bar{1}, \bar{5}, \overline{11}, \overline{13}, \overline{17}, \overline{19}, \overline{23}, \overline{25}, \overline{29}, \overline{31}, \overline{37}$, and $\overline{41}$.

5. 101, 107, 113, 131, 137, 149, 167, 173, 179, 191, and 197.

7. (i) $\bar{0}, \bar{1}, \bar{5}$, and $\overline{16}$.
(ii) $\bar{5}$ is the unity of $(\bar{5}) = \{\bar{0}, \bar{5}, \overline{10}, \overline{15}\}$; also, $\overline{16}$ is the unity of $(\bar{4}) = \{\bar{0}, \bar{4}, \bar{8}, \overline{12}, \overline{16}\}$.

11. $x = 40$.

13. (i)

v	$\bar{1}$	$\bar{7}$	$\overline{11}$	$\overline{13}$	$\overline{17}$	$\overline{19}$	$\overline{23}$	$\overline{29}$
v^{-1}	$\bar{1}$	$\overline{13}$	$\overline{11}$	$\bar{7}$	$\overline{23}$	$\overline{19}$	$\overline{17}$	$\overline{29}$

17. (i) 3. (ii) Since $943 \equiv 3 \pmod{10}$, the answer is the same as in (i).

31. (i)

	$\bar{1}$	$\bar{3}$	$\bar{5}$	$\bar{7}$
$\bar{1}$	$\bar{1}$	$\bar{3}$	$\bar{5}$	$\bar{7}$
$\bar{3}$	$\bar{3}$	$\bar{1}$	$\bar{7}$	$\bar{5}$
$\bar{5}$	$\bar{5}$	$\bar{7}$	$\bar{1}$	$\bar{3}$
$\bar{7}$	$\bar{7}$	$\bar{5}$	$\bar{3}$	$\bar{1}$

35. (ii) $\{\mathbf{V}_5, \mathbf{V}_{10}\}$ and $\{\mathbf{V}_8, \mathbf{V}_{12}\}$.

41. (i)

x	$\bar{1}$	$\bar{3}$	$\bar{5}$	$\bar{9}$	$\overline{11}$	$\overline{13}$	$\overline{15}$	$\overline{17}$	$\overline{19}$	$\overline{23}$	$\overline{25}$	$\overline{27}$
x^2	$\bar{1}$	$\bar{9}$	$\overline{25}$	$\overline{25}$	$\bar{9}$	$\bar{1}$	$\bar{1}$	$\bar{9}$	$\overline{25}$	$\overline{25}$	$\bar{9}$	$\bar{1}$

(ii) $K = \{\bar{1}, \overline{13}, \overline{15}, \overline{27}\}$.
(iii) Let $L = \{\bar{3}, \overline{11}, \overline{17}, \overline{25}\}$ and $M = \{\bar{5}, \bar{9}, \overline{19}, \overline{23}\}$. Then the natural map is given by

x	$\bar{1}$	$\bar{3}$	$\bar{5}$	$\bar{9}$	$\overline{11}$	$\overline{13}$	$\overline{15}$	$\overline{17}$	$\overline{19}$	$\overline{23}$	$\overline{25}$	$\overline{27}$
$\alpha(x)$	K	L	M	M	L	K	K	L	M	M	L	K

(iv)

	K	L	M
K	K	L	M
L	L	M	K
M	M	K	L

43.

x	$\bar{1}$	$\bar{3}$	$\bar{5}$	$\bar{9}$	$\overline{11}$	$\overline{13}$
$\theta_1(x)$	$\bar{1}$	$\bar{5}$	$\overline{11}$	$\bar{7}$	$\overline{13}$	$\overline{17}$
$\theta_2(x)$	$\bar{1}$	$\overline{11}$	$\bar{5}$	$\overline{13}$	$\bar{7}$	$\overline{17}$

45. (b) Yes. $\mathbf{V}_{12}$ is isomorphic to $\mathbf{V}_8$.
(c) No. $\mathbf{V}_5$ and $\mathbf{V}_{10}$ are cyclic, while a direct product of two groups of order 2 is not cyclic.

55. $p = 2$. **59.** (a) 138. (b) 1.

63. Four. The distinct ideals are $(\bar{0})$, $(\bar{1})$, $(\bar{3})$, and $(\overline{13})$.

65. Two. [*Hint:* Problems 28(c) and 26(i) of Section 4.2 may facilitate the proof.]

67. One such homomorphism is $\alpha\colon \mathbf{Z}_{54} \to \mathbf{Z}_{54}$; $\bar{x} \mapsto \overline{18x}$. Another is $\beta\colon \mathbf{Z}_{54} \to \mathbf{Z}_3 = \{\bar{0}, \bar{1}, \bar{2}\}$; $\bar{x} \mapsto \bar{x}$.

69. They are all the same subring in U.

Section 5.1, page 286

3. $a_1 = -4$, $a_0 = 13$. **11.** S is the set of all $m + n\sqrt{2}$ with m and n in $\mathbf{Z}$.

13. (a) $(3 - \sqrt{2})^{-1} = (3/7) + (1/7)\sqrt{2}$.

Section 5.2, page 294

1. (ii) $x^3, x^3 + x^2, x^3 + x, x^3 + \bar{1}, x^3 + x^2 + x, x^3 + x^2 + x + \bar{1}$.
(iii) $x^3 + x^2 + \bar{1}, x^3 + x + \bar{1}$.

3. (i) 100, 102, 110, 111, 120, 121. (ii) 101, 112, 122.

5. m^d.

11. No. If $n > \deg \alpha$ and $n > \deg \beta$, the coefficient of x^n is 0 in α and in β and hence in their sum.

15. $x^2 = x \cdot x = (x + \bar{3})(x + \bar{6})$. **29.** (ii) x. **35.** (i) Yes. (ii) No. (iii) Yes.

Section 5.3, page 302

1. (i) $\bar{0}, \bar{2}, \bar{4}, \bar{6}, \bar{8}, \overline{10}$; (ii) $\bar{1}, \bar{2}, \bar{5}, \bar{7}, \overline{10}, \overline{11}$; (iii) $\bar{2}, \overline{10}$; (iv) $\bar{2}x^3 + x, x^3 + \bar{2}x$.

3. (i) 4. (ii) 8. (iii) 16. (iv) 32.

5. $\gamma_1 = x$, $\alpha_2 = \bar{2}x^3 + x^2 + x + \bar{2}$, $\gamma_2 = \bar{2}x + \bar{1}$, $\alpha_3 = \bar{2}x^2 + \bar{1}$, and $\gamma_3 = x + \bar{2}$.

7. $\beta = x^{n-1} + ax^{n-2} + \cdots + a^{n-2}x + a^{n-1}$.

15. The principal ideal generated by $2x$.

17. Yes. A is symmetric, reflexive, and transitive. **27.** $(81 - 9)/2 = 36$.

Section 5.4, page 310

1. 69. **9.** (ii) Four.

17. (a) 2. (b) 1. (c) 2. (d) 9. (e) 10. (f) 2.

25. (a) $r_1 + r_2 + \cdots + r_d = -a_{d-1}$.
(b) $r_1 r_2 \cdots r_d = (-1)^d a_0$.

29. $h_5 = (5ab^3c - 5a^2bc^2 - b^5)/a^5$.

31. (i) $\rho = 2x$.
(ii) $\alpha(\mathbf{i}) = 2\mathbf{i}$.
(iii) $\alpha(\mathbf{i}) = \rho(\mathbf{i})$, since $\alpha(\mathbf{i}) = \gamma(\mathbf{i})\beta(\mathbf{i}) + \rho(\mathbf{i})$ and $\beta(\mathbf{i}) = 0$.

35. (ii) $c = -1/3$.

39. Reduce α to a second-degree polynomial γ by replacing $x^3, x^6, x^9, \ldots$ by 1; replacing $x^4, x^7, \ldots$ by x; and replacing $x^5, x^8, \ldots$ by x^2. Then it is easy to divide γ by β, and the remainder is ρ.

Section 5.5, page 316

1. The integer root is -1; the others are $-1 \pm \sqrt{3}$.

3. $1, -3, \mathbf{i}\sqrt{2}, -\mathbf{i}\sqrt{2}$.

5. (a) $5/3, (-3 \pm \sqrt{17})/2$. (b) $-1, 2/3, 5/4$.
(c) 1 is a root with multiplicity 3, and -2 is a simple root.
(d) $0, 2/3, -1/3, (1 \pm \mathbf{i}\sqrt{3})/6$. (e) $\pm\mathbf{i}, \pm\mathbf{i}\sqrt{2}$.

7. $\pm 1, \pm 3, \pm 7, \pm 21$.

9. Yes. In lowest terms, $35/21 = 5/3$; and this meets the conditions of Theorem 2.

13. Either $s = 0$ or both $s \mid a_1$ and $t \mid a_d$. **17.** (i) $\alpha = x^4 - 10x^2 + 1$.

Section 5.6, page 325

1. $3(x-1)(x-2)(x-3)(x-5)$. **3.** $\bar{2}x^3 + x^2 + x + \bar{1}$.

7. $\alpha = 1 - x + 10x(x-1) + 2x(x-1)(x-4) = 2x^3 - 3x + 1$.

9. (a) $(n+1)n(n-1)(n-2)/4 = (n^4 - 2n^3 - n^2 + 2n)/4$.
(b) $(n+1)n(n-1)(n-2)(n-3)/5 = (n^5 - 5n^4 + 5n^3 + 5n^2 - 6n)/5$.

11. (i) $A_1 = 2, A_2 = 4, A_3 = 7, A_4 = 11$.
(iii) $(n^2 + n + 2)/2$.

25. $\beta = 1 + 6B_2 + 6B_3$.

27. *Hint:* See Problem 16 of Section 5.3.

Section 5.7, page 335

1. (i) $\gamma_1 = \bar{1}$ and $\alpha_2 = x^4 + x$.
(ii) $\gamma_2 = x$ and $\alpha_3 = x^2 + x + \bar{1}$. (iii) $\gamma_3 = x^2 + x$.
(iv) $\delta = \alpha_3 = x^2 + x + \bar{1}$. (v) $\xi = x$ and $\eta = x + \bar{1}$.

3. (i) 1. (ii) 1, 4, 7. (iii) 2, 3, 5.
(iv) Let $\alpha = n^2 + n + 1$ and $\beta = n - 1$. Since $\alpha - (n + 2)\beta = 3$, $\gcd(\alpha, \beta)$ must be a positive integral divisor of 3 by Theorem 3 of Section 1.4.

5. (iii) 1. (iv) 1 and 2.

13. The element $\bar{\alpha} = \alpha + I$ is a 0-divisor, since $\bar{\alpha} \neq 0$ and $\bar{\alpha} \cdot \bar{\alpha} = 0$. A ring with 0-divisors is not an integral domain and hence is not a field.

15. (ii) Every polynomial in K has s as a zero.

17. *Hint:* See Section 4.8.

Section 5.8, page 339

5. (a) $x = (\bar{2}x + \bar{3})(\bar{3}x + \bar{2})$.
(b) $x + \bar{2} = (\bar{2}x + \bar{1})(\bar{3}x + \bar{2})$.
(c) $x + \bar{3} = (\bar{2}x + \bar{3})(\bar{3}x + \bar{5})$.

Section 5.9, page 342

1. (a) $[(17/5)/(x - 2)] + [(8/5)/(x + 3)]$.
(b) $[2/(x + 1)] - [3/(x + 1)^2] - [4/(x^2 + 4)]$.

3. $x + [\bar{4}/(x - \bar{1})] + [\bar{4}/(x + \bar{1})]$.

7. $[(x - r_1)(x - r_2) \cdots (x - r_n)]^{-1} = a_1(x - r_1)^{-1} + \cdots + a_n(x - r_n)^{-1}$, where
$$a_k = [(r_k - r_1)(r_k - r_2)(r_k - r_3) \cdots (r_k - r_{k-1})(r_k - r_{k+1}) \cdots (r_k - r_n)]^{-1}.$$

11. Let a ppd-term in $\mathbf{Q}$ be of the form c/p^k with p a positive prime in $\mathbf{Z}$, k a positive integer, and c an integer satisfying $|c| < p$. Then every rational number is the sum of an integer and a finite number of ppd-terms in $\mathbf{Q}$.

Section 5.10, page 348

1. (iii) $3^3 = 27$.

3. Since $25 = 5^2$, we seek a prime polynomial π of degree 2 in $\mathbf{Z}_5[x]$ and find that $x^2 + \bar{2}$ is such a π. Then Theorem 1 tells us that
$$\mathbf{Z}_5[x]/(x^2 + \bar{2})$$
is a field with 25 elements.

11. $x^3 - 6x - 6$.

Section 5.11, page 354

3. $3 + (4 + \sqrt{7})\mathbf{i}$, $-3 + (-4 + \sqrt{7})\mathbf{i}$.

9. (a) $D = -23$; one root is real and the others are complex conjugates.
(b) $D = 0$; there is a root with multiplicity greater than 1.
(c) $D = 392$; the roots are real and distinct.

11. $-(r+s)$, $-(wr+w^2s)$, and $-(w^2r+ws)$ where $r^3=(-1+\sqrt{5})/2$, $s=-1/r$, and $w=(-1+\mathbf{i}\sqrt{3})/2$.

15. (a) $-1\pm\mathbf{i}\sqrt{2}$, $3\pm\sqrt{6}$.
(b) $(5\pm\sqrt{17})/2$, $(5\pm\sqrt{29})/2$.
(c) The only root is -1, which has multiplicity 4.

23. If r is a root of $ax^2+(b-1)x+c=0$, then $\alpha(r)=r$ and $\beta(r)=\alpha[\alpha(r)]-r=\alpha(r)-r=r-r=0$. Dividing $\beta(x)$ by $ax^2+(b-1)x+c$, one gets the quotient $\gamma(x)=a^2x^2+(ab+a)x+ac+b+1$.

Section 5.12, page 359

3. (i) Two.
(ii) One. [There is only one cube root of 2 (the real number $\sqrt[3]{2}$) in $\mathbf{Q}(\sqrt[3]{2})$.]

5. (i) 6. (ii) 2.

Review Problems (Chapter 5), page 359

5. x^5-x. **9.** (i) $\rho=3x-9$. (ii) $\rho=\bar{3}x+\bar{1}$.

11. (i) Eight. (ii) Two. **13.** $x^{14}-\bar{2}$.

15. The monic generator is $x^{17}-x$.

17. The rational root is 5/2; the other roots are $\pm 3\mathbf{i}$.

19. $\bar{2}x^3+x^2+\bar{2}$. **21.** $\alpha=5(x-2)^2+(x-2)^3$. The multiplicity of the zero 2 is 2.

25. (ii) Yes. (iii) $x-r$. **27.** 406.

29. No; 99 is not a power of a prime in $\mathbf{Z}$.

33. Yes. (For the proof, show that θ sends 1 to 1 and $1+1$ to $1+1$.)

35. Let $r^3=(-7+\mathbf{i}\sqrt{1323})/2$, $s=7/r$, and $w=(-1+\mathbf{i}\sqrt{3})/2$. Also, let u take on the values $-(r+s)$, $-(wr+w^2s)$, and $-(w^2r+ws)$. Then the zeros are given by $x=(u-1)/3$.

Section 6.2, page 366

9. Let $a=\sqrt{7}$, $b=\sqrt{3+a}$, and $c=\sqrt{3-a}$. Then the four roots, $\pm b$ and $\pm c$, are in $\mathbf{Q}(a,b,c)$.

Section 7.1, page 379

1. $8x\equiv 37 \pmod{141}$. **3.** 209 and 419.

5. 59 and 119. **7.** 548. **11.** 135.

13. (i) $a=21, r=77$. (ii) $b=42, s=154$. (iii) $c=24374, t=30030$.

Section 7.2, page 382

13. $\binom{k+n-1}{n} = (k+n-1)!/n!(k-1)!$

17. $A_n = (n^2 + 11n + 12)/2.$ **31.** *Hint:* See Problem 13.

Section 7.3, page 391

1. (a) 16; (b) 1680; (c) 16. **3.** (a) 5, 8, 10, 12; (b) 7, 9, 14, 18.

13. (iii) 65, 104, 105, 112, 130, 140, 144, 156, 168, 180, 210.

25. (ii) 1, 4, 9, 16.

27. $n = q^{p-1}$, with q a positive prime.

29. $2^4 \cdot 3 = 48.$

33.

n	1	2	3	4	5	6	7	8
$p(n)$	1	2	3	5	7	11	15	22

35. $n = 40.$ **39.** (a) Yes. (b) Yes.

Section 7.4, page 397

1. (a) 1, 2, 4, 8, 9, 13, 15, 16.
(b) 1, 4, 5, 6, 7, 9, 11, 16, 17.

Review Problems (Chapter 7), page 398

1. 20, 60. **3.** 1, 29, 41, 71.

11.

a	1	2	2	4	4	6	6	8	10	12	16	16	18	18	20
p	2	2	3	2	5	3	7	2	11	13	2	17	3	19	5
n	1	2	1	3	1	2	1	4	1	1	5	1	3	1	2

15. Only $n = 10.$ **19.** $2^6 \cdot 3 = 192.$

21. The squares of positive primes in **Z**. **23.** 30.

Section 8.1, page 406

1. (a) 11101; (b) 10100. **3.** (a) "error"; (b) 0101.

5. No. Although 1000100 is wrong, it is a codeword.

7. (a) 000000000; (b) 110110110; (c) 100100100.

9. (a) 1011; (b) 0000; (c) 1010.

11. (a) Yes, $D(100100) = 00.$ (b) No, $D(000101) = 01 \neq 00.$

13. (a) 0000000; (b) 1000111; (c) 0100110; (d) 0001011.

15. (a) 1110100; (b) 1101010; (c) 1011001.

17. (a) Codeword, since $E(1110) = 1110100$.
(b) Not a codeword, since $E(1110) \neq 1110011$.

Section 8.2, page 412

1. (a) $(1\ \ 0\ \ 0\ \ 0\ \ 1\ \ 1\ \ 1)$; (b) $(1\ \ 0\ \ 1\ \ 0\ \ 0\ \ 1\ \ 0)$;
(c) $(1\ \ 1\ \ 0\ \ 1\ \ 0\ \ 1\ \ 0)$; (d) $(0\ \ 0\ \ 0\ \ 0\ \ 0\ \ 0\ \ 0)$.

3. $2^4 = 16$.

5. (a) $\gamma K = (0\ \ 1\ \ 0)$, $D(\gamma) = (0\ \ 0\ \ 1\ \ 0)$;
(b) $\gamma K = (0\ \ 0\ \ 0)$, $D(\gamma) = (1\ \ 1\ \ 1\ \ 1)$;
(c) $\gamma K = (1\ \ 1\ \ 0)$, $D(\gamma) = (0\ \ 0\ \ 1\ \ 0)$.

7. (a) $\beta = (0\ \ 0\ \ 0\ \ 1\ \ 0\ \ 1\ \ 1)$.
(b) $\gamma K = (1\ \ 0\ \ 0)$, $D(\gamma) = (0\ \ 0\ \ 1\ \ 1) \neq \alpha$.

9. No, see Problem 7 and its answer.

Section 8.3, page 415

1. (a) $\bar{8}$; (b) $\bar{4}$; (c) $\overline{13}$. **3.** $d = 37$.

5. 7, 11, 13, 17, 19, 23, 29, 31, 37, 41, 43, 47, 49, 53, 59, 61, 67, 71, 73.

7. 9.

11. Yes; one could use $\bar{2}, \bar{3}, \ldots, \overline{677}$ to represent the 676 $(=26^2)$ ordered pairs of letters.

Review Problems (Chapter 8), page 415

1. 01100, 01101, 01110, 01000, 00100, 11100.

3. (a) 010101; (b) 11; (c) 10.

5. (a) $(0\ \ 1\ \ 0\ \ 0\ \ 1\ \ 1\ \ 0)$;
(b) $(1\ \ 0\ \ 0\ \ 1\ \ 1\ \ 0\ \ 0)$;
(c) $(1\ \ 0\ \ 1\ \ 1\ \ 0\ \ 0\ \ 1)$.

7. (a) $\gamma K = (1\ \ 1\ \ 0)$, $D(\gamma) = (0\ \ 1\ \ 1\ \ 0)$.
(b) $\gamma K = (0\ \ 0\ \ 0)$, $D(\gamma) = (0\ \ 1\ \ 1\ \ 0)$.
(c) $\gamma K = (0\ \ 0\ \ 1)$, $D(\gamma) = (0\ \ 1\ \ 1\ \ 0)$.

9. (i) 1, 5, 7, 11, 13, 17, 19, 23. (ii) $d = e$.

INDEX

Summary of Axioms for Groups, Rings, and Fields

Properties Under Addition	Properties Under Multiplication
1. Closure	6. Closure
2. Associativity	7. Associativity
3. Existence of an identity	8. Existence of an identity
4. Existence of inverses	9. Existence of inverses (for nonzero elements)
5. Commutativity	9′. Cancellation law (or no 0-divisors)
	10. Commutativity

Properties Under Addition and Multiplication
11. Distributivity
12. $0 \neq 1$

Algebraic Structure	Required Properties
Additive group	1–5
Multiplicative group	6–9
Ring	1–7, 11
Ring with unity	1–8, 11
Commutative ring	1–7, 10, 11
Integral domain	1–8, 9′, 10–12
Division ring	1–9, 11, 12
Skew-field	1–9, 11, 12, and negation of 10
Field	1–12